Spacecraft Design, Development, and Operations

Short Course Lecture Slides

Don Edberg, Ph.D.
Professor of Aerospace Engineering
California State Polytechnic University, Pomona

Adjunct Lecturer,
Astronautical Engineering,
University of Southern California

October 2022

Rev. 02: SC04A missing slides added

About this book:

This book is a collection of the slides presented in a short course on the design, development, and operation of spacecraft. The course is offered by the American Institute of Aeronautics and Astronautics (AIAA), UCLA, and Taksha Institute as well as given by its author, Don Edberg, as a consultant. The materials are also taught in a two-semester course "Introduction to Space Vehicle Design" given by the author at California State Polytechnic University, Pomona. The short course has been given 23 times to over 400 registrants, and the college course has been given for 17 years to over 350 college seniors.

The informational materials in these notes have been gathered from the author's experience as well as open sources such as engineering textbooks, technical papers, manufacturer websites, and other sources to tell a complete and unified story on how spacecraft are designed, built, and operated. The materials are all in the public domain but are rarely available together in a cohesive collection such as this.

This book contains information that includes the beginning of the design process all the way to end-of-flight disposal or decommissioning. There are slide presentations for 26 oral lectures on all aspects of design and operation of spacecraft, each providing a combination of theoretical concepts along with practical, easy-to-understand diagrams and photos of illustrative hardware and concepts. A multitude of practical step-by-step numerical examples is used to demonstrate important ideas and practices.

The goals of this short course are:

- to present the history of spacecraft as a basis of understanding current design and operational practices
- to provide and explain basic technological vocabulary for spacecraft
- to provide knowledge of current spacecraft technology
- to provide the basic principles of spacecraft design and engineering
- to demonstrate all the various factors and technical disciplines that can influence a spacecraft design, and how disciplines can interact with each other.

By studying these notes, the reader will learn important factors of a space vehicle systems design, be able to read and follow most technical papers and presentations on the subject and know where to look for further expertise.

Please note: some presented materials have been simplified to an introductory format, and their reality is more complex. If you are an expert practitioner, please understand liberties have been taken for brevity.

Table of Contents

About the course	iv
Spacecraft (SC) Class Introduction	1
1SC: Introduction to SC Missions, Operations, Launch sites	3
2SC*: History; current and future spacecraft	11
3SC: Ground and Launch Environment	31
ORB0: Introduction to Orbits (ORB)	55
ORB1A: The Space Environment	67
ORB1B: Effects of the Environment on Spacecraft	81
ORB2: Orbital Mechanics	97
ORB3*: Orbital Rendezvous	111
ORB4*: Low-Thrust Orbital Maneuvers & Atmospheric Entry	121
SC01: Payload and Systems Engineering	133
SC02: Systems Design	149
SC03: Propulsion	163
SC04A: Structures	181
SC04B: Mechanisms	195
SC05A: Structural Analysis	205
SC05B*: Mass properties	219
SC06: Spacecraft Attitude Sensing	229
SC07: Dynamics and Control Systems	239
SC08A: Thermal Control	255
SC08B*: Environmental Control & Life Support Systems (ECLSS)	271
SC09: Power Systems	279
SC10: Telecommunications	299
SC11: Command and Data Systems	313
SC12: Testing	325
SC13: Failures and Lessons Learned	337
SC14*: Cost Estimation	349

*Depending on attendees' interests, these sections may be omitted without serious loss of information

About the Course…

The course lectures are laid out as shown in the figure below. Although a major portion of the class consists of following the spacecraft design process shown in the large box to the lower left, we surround that process with useful background information that is important to the proper use of the design process.

After an introduction, we look at space projects, launching locations, and the space economy. Next, we look at the vast history of spacecraft from the dawn of spaceflight in 1957 until the current time and look at current spacecraft and future planned projects. Then, before the actual design process begins, we provide information on the launch and on-orbit environments, important aspects of orbits and maneuvering, and the operation on payloads to be carried by a spacecraft.

Next, we provide a series of lectures on the design process itself, beginning with the mission requirements and design constraints. Then we discuss how a spacecraft will operate, and how operation will dictate some of the design parameters that need to be met, such as propulsion system requirements. Next, we look at payload selection, and use the payload to estimate initial mass and power requirements so we can select a launch vehicle and begin laying out the internals of the spacecraft. We then check the configuration for proper operation and determine its stability and control characteristics. Then, we investigate the thermal control system, the power system, telecommunications, and the command and data system.

After passing through the design process, we look at the battery of testing that the spacecraft must pass, followed by a discussion of some failures and lessons to be learned, ending with an introduction to cost estimating.

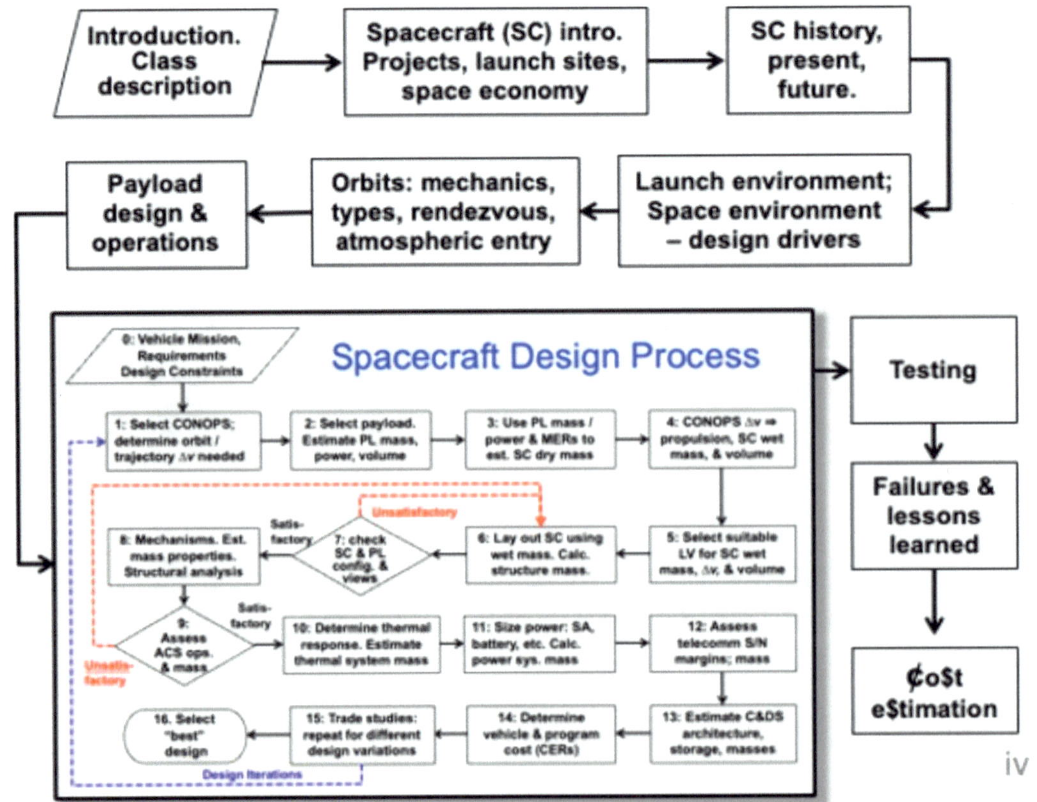

Introduction. Class description
Spacecraft (SC) intro. Projects, launch sites, space economy
SC history, present, future.
Launch environment; Space environment – design drivers
Orbits: mechanics, types, rendezvous, atmospheric entry
Payload design & operations
Spacecraft Design Process
0: Vehicle Mission, Requirements Design Constraints
1: Select CONOPS; determine orbit / trajectory Δv needed
2: Select payload. Estimate PL mass, power, volume
3: Use PL mass / power & MERs to est. SC dry mass
4: CONOPS Δv ⇒ propulsion, SC wet mass, & volume
5: Select suitable LV for SC wet mass, Δv, & volume
6: Lay out SC using wet mass. Calc. structure mass.
Unsatisfactory
7: check SC & PL config. & views
Satis-factory
8: Mechanisms. Est. mass properties. Structural analysis
9: Assess ACS ops. & mass
Satis-factory
Unsatis-factory
10: Determine thermal response. Estimate thermal system mass
11: Size power: SA, battery, etc. Calc. power sys. mass
12: Assess telecomm S/N margins; mass
13: Estimate C&DS architecture, storage, masses
14: Determine vehicle & program cost (CERs)
15: Trade studies: repeat for different design variations
16. Select "best" design
Design Iterations
Testing
Failures & lessons learned
¢o$t e$timation

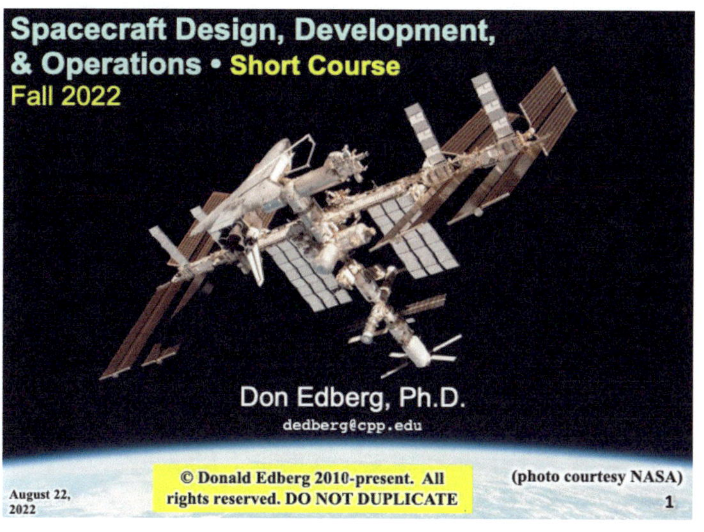

Short Course Contents

0: Title, contents, instructor info, intro to course
1: Intro to spacecraft (SC), industry economics, launch sites
2 SC: SC history, current, future, LVs
3 SC: Ground & launch environment
4 ORB0: Introduction to orbits
5 ORB1A: Space environment
6 ORB1B: Space's effects on SC
7 ORB2: Orbital mechanics
8 ORB3: Rendezvous & docking
9 ORB4: Low-thrust orbits; atmospheric entry
10 SC01: Systems engineering
11 SC02: Systems design
12 SC03: Propulsion
13 SC04A: Structures
14 SC04B: Mechanisms
15 SC05A: Structural analysis
16 SC05B: Mass properties
17 SC06: SC types & sensing
18 SC07: SC dynamics & control
19 SC08A: Thermal control
20 SC08B: Environmental Control & Life Support Systems (ECLSS; for crewed SC)
21 SC09: Power systems
22 SC10: Telecommunication system
23 SC11: Command & Data systems
24 SC12: SC & component testing
25 SC13: SC failures, lessons learned
26 SC14: SC cost estimation
27 SC15: Conclusions

About the Course …

- The slides in this book are copyrighted © lectures of a short course by Don Edberg on *spacecraft design, development, & operation*
- They are also used for a year-long design course at California State Polytechnic University, Pomona
- **The course materials may not be reproduced, duplicated, or transmitted without author's permission**
- Some material (history, orbits, testing, failures, etc.) included for background and are not parts of the formal design process

Who Is Don Edberg?

- **Professor of Aerospace Engineering**, Cal State Polytechnic Univ., Pomona
- **Faculty Fellow**, NASA MSFC, 2016
- **Part-Time Lecturer**, USC & UCLA
- **Visiting Research Engineer**, Air Force Research Lab, 2011
- **Boeing Technical Fellow**, 1989-2009 (& McDonnell Douglas)
- **Visiting Associate Professor**, UCSD, UCI
- **Work**: AeroVironment, JPL, Convair.
- Hold 10 U.S. patents.
- **Rocket & space nut**: since early memory!

Marines Test AeroVironment's Pointer Man-Portable RPV
AvWeek Aug '88
Don & Vostok 1
Don Chief Engr. AV Pointer UAV
Don & C-17
Don-Chief Engr. STABLE
Don & X-33 model
Don @ 1981 FAI World Champs
Don @ 12
STABLE Installed in Spacelab Rack
Don @ STS-135

Spacecraft Design, Development, & Operations

• • •

The Short Course

Course Objectives

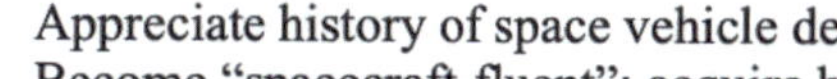

1. Appreciate history of space vehicle design
2. Become "spacecraft-fluent": acquire basic vocabulary & knowledge of current technology
3. Understand the basic principles of spacecraft design & systems engineering
4. Appreciate all the various factors which influence a particular design, & how they interact with other disciplines

By the end of class, you should know important factors of a vehicle systems design, be able to read and follow most papers and presentations, and know where to look for further expertise.

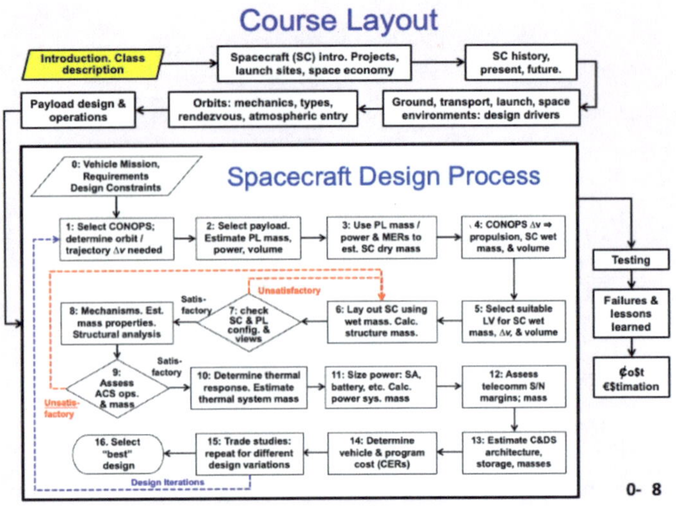
Course Layout
Introduction. Class description
Spacecraft (SC) intro. Projects, launch sites, space economy
SC history, present, future.
Ground, transport, launch, space environments: design drivers
Orbits: mechanics, types, rendezvous, atmospheric entry
Payload design & operations
Spacecraft Design Process
0: Vehicle Mission, Requirements Design Constraints
1: Select CONOPS; determine orbit / trajectory Δv needed
2: Select payload. Estimate PL mass, power, volume
3: Use PL mass / power & MERs to est. SC dry mass
4: CONOPS Δv ⇒ propulsion, SC wet mass, & volume
5: Select suitable LV for SC wet mass, Δv, & volume
6: Lay out SC using wet mass. Calc. structure mass.
7: check SC & PL config. & views
Unsatisfactory
Satis-factory
8: Mechanisms. Est. mass properties. Structural analysis
9: Assess ACS ops. & mass
Satis-factory
Unsatis-factory
10: Determine thermal response. Estimate thermal system mass
11: Size power: SA, battery, etc. Calc. power sys. mass
12: Assess telecomm S/N margins; mass
13: Estimate C&DS architecture, storage, masses
14: Determine vehicle & program cost (CERs)
15: Trade studies: repeat for different design variations
16. Select "best" design
Design Iterations
Testing
Failures & lessons learned
¢o$t €$timation
0- 8

This Is An Introduction – *Not* A Complete Design Course!

- It takes an *expert* to do the design of any feature of any SC, experts have *years* of learning and practice
- All subsystems must work together
- The devil is in the details!
- The course will give you enough knowledge:
  - To ask intelligent questions
  - To understand why things are done the way they are
  - To facilitate deeper learning
- Note: some presented materials have been simplified to an introductory format, and their reality is more complex. If you are an expert practitioner, please understand liberties have been taken for brevity.

Legal Notice ©

- All materials & entire course copyright © Don Edberg
- No reproduction in any form, including photographically, electronically, or by Internet, without written permission from the author
- This course contains many images. Some images are reproduced from references and are credited. Other materials taken from the Internet under "fair use" doctrine for educational purposes. We have made a good-faith effort to obey copyright laws. If you notice that we have erred on a copyright, or if you are the owner of a copyright and have not yet received our request for permission, please contact us at dedberg@cpp.edu

Slides With Green Background Provide Topics of Presentation

1. Topics previously covered are gray
2. The topic to be covered in the following slides is blue
3. And the next subject is gray
4. And so is the one after that

Slides With Orange or Yellow or Violet Background Provide Additional Material or Corrections to Previous Notes

1. Updated material
2. Supplemental materials
3. Possible corrections

Blue backgrounds indicate videos or movies

Let's Get Started!

Spacecraft Design, Development, & Operations

Introduction, Operation, and the Design Process

Lecture 1

Introduction: Topics

1. Need for spacecraft (SC) or satellite
2. SC parts & purposes
3. SC integration, launch, & delivery
4. SC project elements
5. Launch sites
6. World space industry

Course Layout

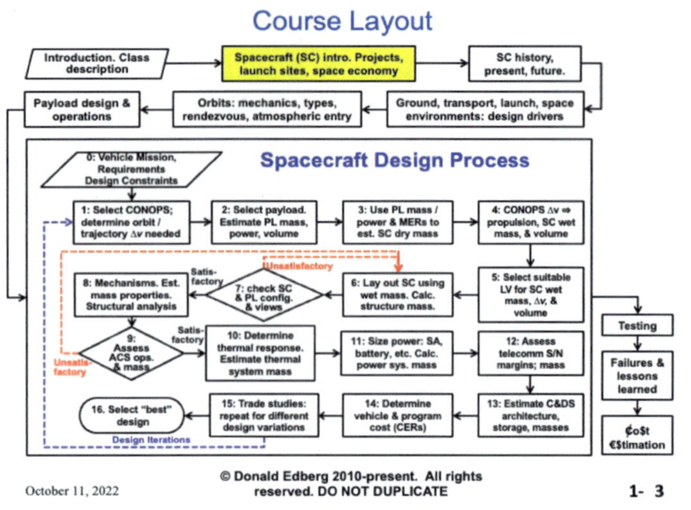

Introduction. Class description
Spacecraft (SC) intro. Projects, launch sites, space economy
SC history, present, future.
Payload design & operations
Orbits: mechanics, types, rendezvous, atmospheric entry
Ground, transport, launch, space environments: design drivers
Spacecraft Design Process
0: Vehicle Mission, Requirements Design Constraints
1: Select CONOPS; determine orbit / trajectory Δv needed
2: Select payload. Estimate PL mass, power, volume
3: Use PL mass / power & MERs to est. SC dry mass
4: CONOPS Δv ⇒ propulsion, SC wet mass, & volume
5: Select suitable LV for SC wet mass, Δv, & volume
6: Lay out SC using wet mass. Calc. structure mass.
7: check SC & PL config. & views
8: Mechanisms. Est. mass properties. Structural analysis
9: Assess ACS ops. & mass
10: Determine thermal response. Estimate thermal system mass
11: Size power: SA, battery, etc. Calc. power sys. mass
12: Assess telecomm S/N margins; mass
13: Estimate C&DS architecture, storage, masses
14: Determine vehicle & program cost (CERs)
15: Trade studies: repeat for different design variations
16. Select "best" design
Satis-factory
Unsatisfactory
Design Iterations
Testing
Failures & lessons learned
¢o$t €$timation

Introduction: Topics

1. Need for spacecraft (SC) or satellite
2. SC parts & purposes
3. SC integration, launch, & delivery
4. SC project elements
5. Launch sites
6. World space industry

Missions or Purposes of Spacecraft (SC) or Satellites

- Communication
- Entertainment (DirecTV, XM radio, etc.)
- Navigation (GPS)
- Remote sensing: weather, Earth observation
- Military (surveillance, intelligence, nuclear monitoring/early warning, anti-satellite)
- Exploration (planetary probes, orbiters, landers)
- Crew & cargo transportation (ISS, exploration)
- Tourism

Spacecraft Size Classifications

Size / Class	Mass (kg)
Pico	0.1 – 1.0
Nano	1.1 – 10 ("CubeSats")
Micro	11 – 200
Mini	201 – 600
Small	601 – 1,200
Medium	1,201 – 2,500
Intermediate	2,501 – 4,200
Large	4,201 – 5,400
Heavy	5,401 – 7,000
Extra Heavy	> 7,000

Courtesy BryceTech

Introduction: Topics

1. Need for spacecraft (SC) or satellite
2. SC parts & purposes
3. SC integration, launch, & delivery
4. SC project elements
5. Launch sites
6. World space industry

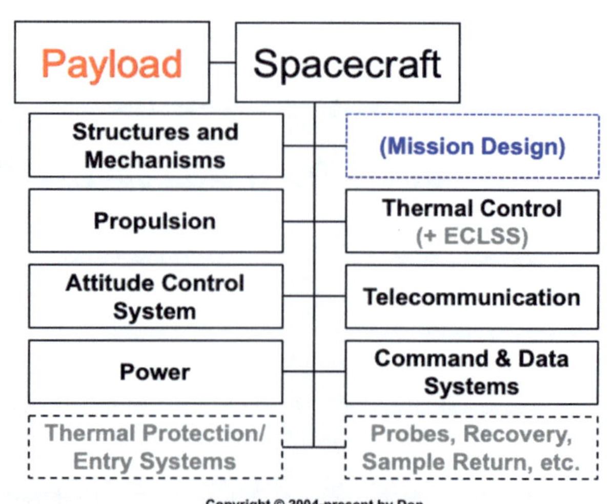

Payload
Spacecraft
Structures and Mechanisms
(Mission Design)
Propulsion
Thermal Control (+ ECLSS)
Attitude Control System
Telecommunication
Power
Command & Data Systems
Thermal Protection/ Entry Systems
Probes, Recovery, Sample Return, etc.

Spacecraft Design Process

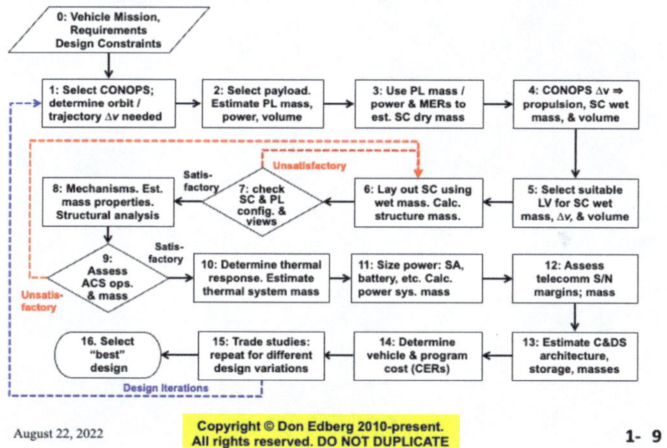
0: Vehicle Mission, Requirements Design Constraints
1: Select CONOPS; determine orbit / trajectory Δv needed
2: Select payload. Estimate PL mass, power, volume
3: Use PL mass / power & MERs to est. SC dry mass
4: CONOPS Δv ⇒ propulsion, SC wet mass, & volume
5: Select suitable LV for SC wet mass, Δv, & volume
6: Lay out SC using wet mass. Calc. structure mass.
7: check SC & PL config. & views
Unsatisfactory
Satis-factory
8: Mechanisms. Est. mass properties. Structural analysis
9: Assess ACS ops. & mass
Unsatis-factory
Satis-factory
10: Determine thermal response. Estimate thermal system mass
11: Size power: SA, battery, etc. Calc. power sys. mass
12: Assess telecomm S/N margins; mass
13: Estimate C&DS architecture, storage, masses
14: Determine vehicle & program cost (CERs)
15: Trade studies: repeat for different design variations
16. Select "best" design
Design Iterations

August 22, 2022 1- 9

Bus Example: *Pioneer*

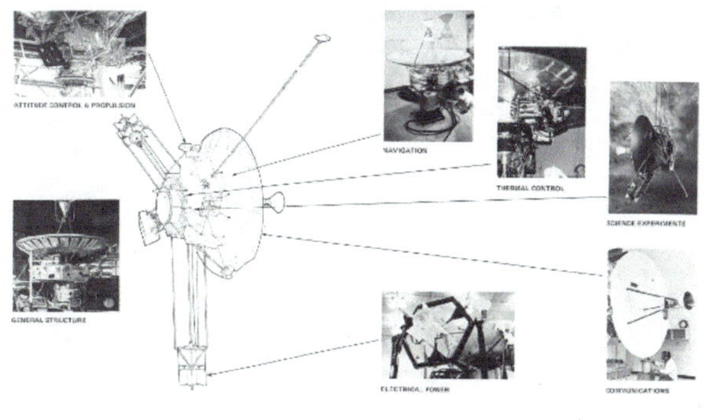

August 22, 2022 Courtesy JPL 1-11

Introduction: Topics

1. Need for spacecraft (SC) or satellite
2. SC parts & purposes
3. SC integration, launch, & delivery
4. SC project elements
5. Launch sites
6. World space industry

August 22, 2022 1-12

Spacecraft Integrated with LV

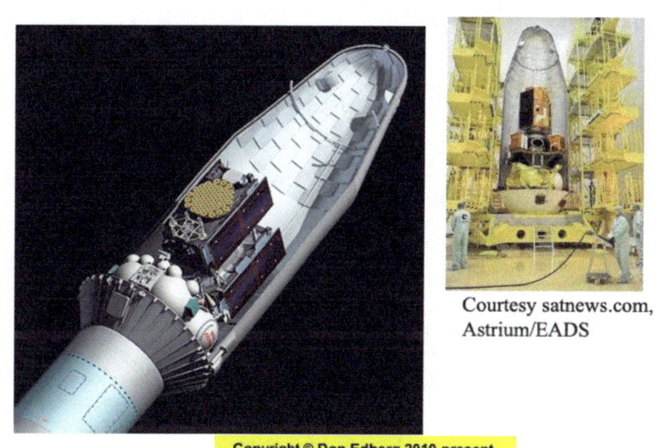

Courtesy satnews.com, Astrium/EADS

August 22, 2022 1-13

Launch & Ascent Profile: Ground Launch

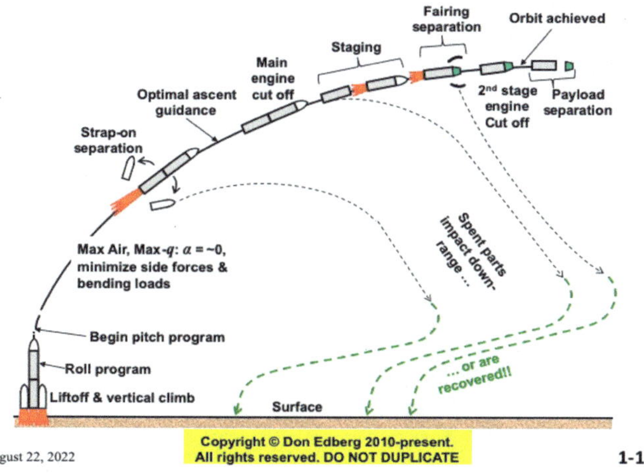
Fairing separation
Orbit achieved
Staging
Main engine cut off
Optimal ascent guidance
2nd stage engine Cut off
Payload separation
Strap-on separation
Spent parts impact down-range ...
Max Air, Max-q: α = ~0, minimize side forces & bending loads
Begin pitch program
Roll program
Liftoff & vertical climb
Surface
... or are recovered!!

August 22, 2022 1-14

Launch & Ascent Profile: Air Launch

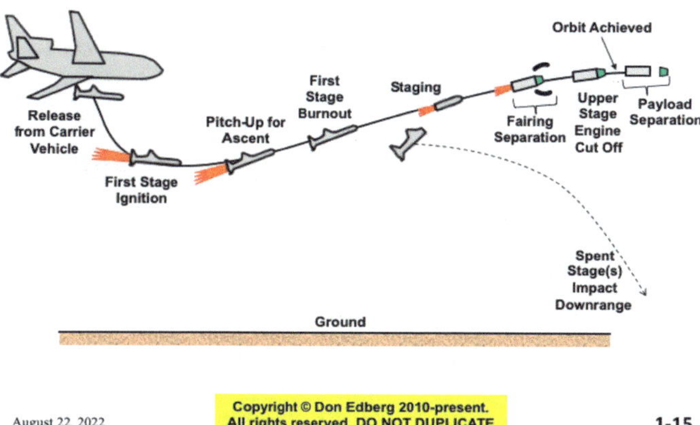
Orbit Achieved
Release from Carrier Vehicle
First Stage Burnout
Staging
Pitch-Up for Ascent
Fairing Separation
Upper Stage Engine Cut Off
Payload Separation
First Stage Ignition
Spent Stage(s) Impact Downrange
Ground

August 22, 2022 1-15

Launch Movie: *MER/Delta II*
MER-mars-HD 6m32s.mp4
August 22, 2022
Copyright © Don Edberg 2010-present. All rights reserved. DO NOT DUPLICATE
1-16

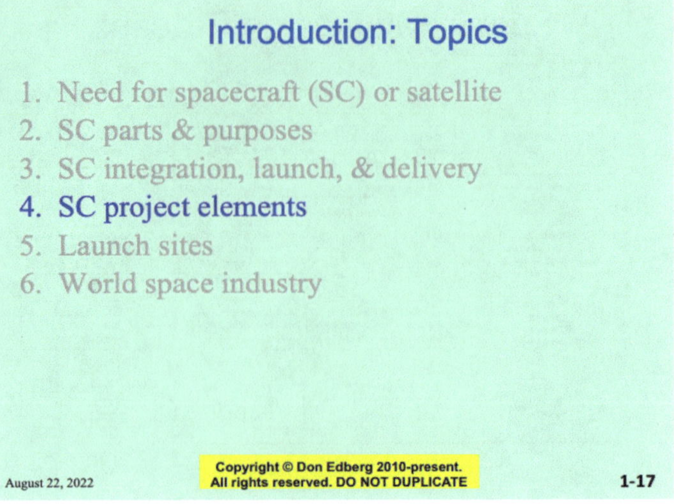

Spacecraft Project Elements

Spacecraft

PAYLOAD

Mission Operations

Launch System/Trajectory

Space Mission

Tracking & Data (Communication Network)

Courtesy NASA

August 22, 2022

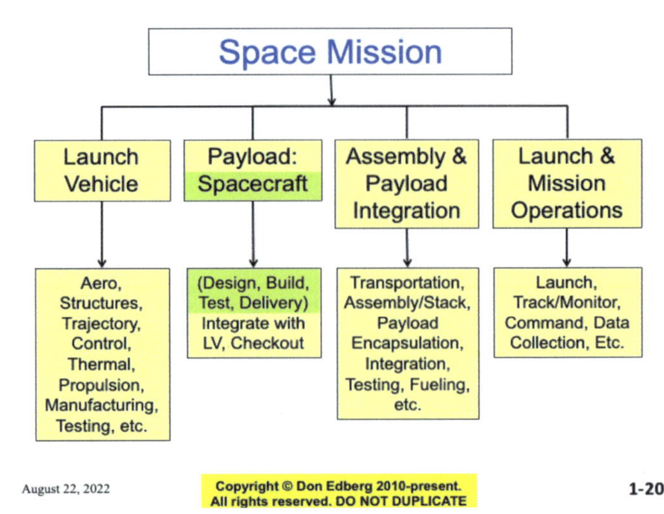

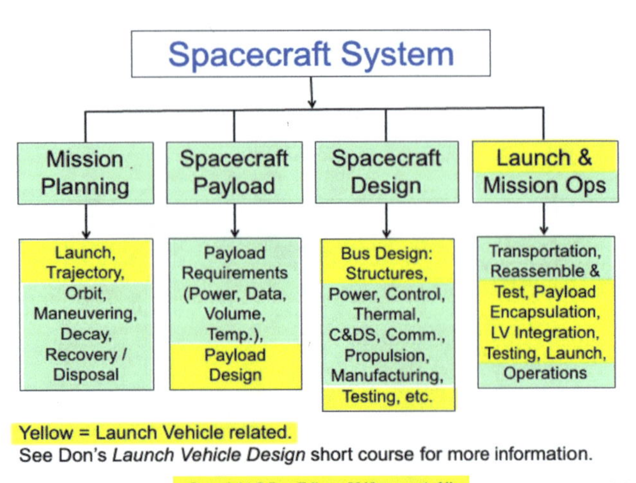

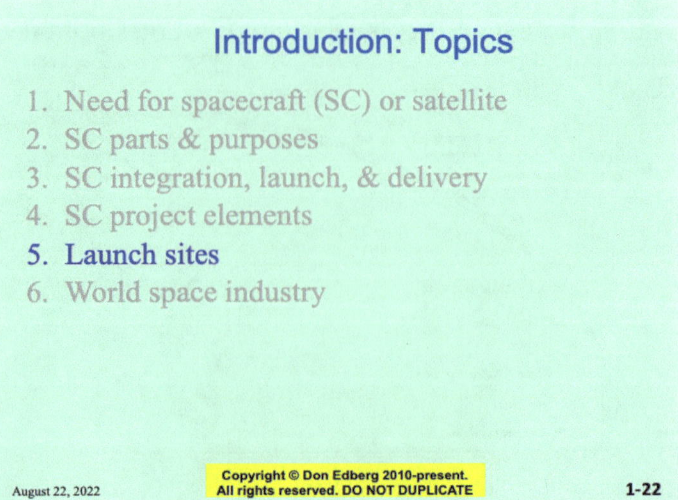

World Launch Sites (courtesy FAA)

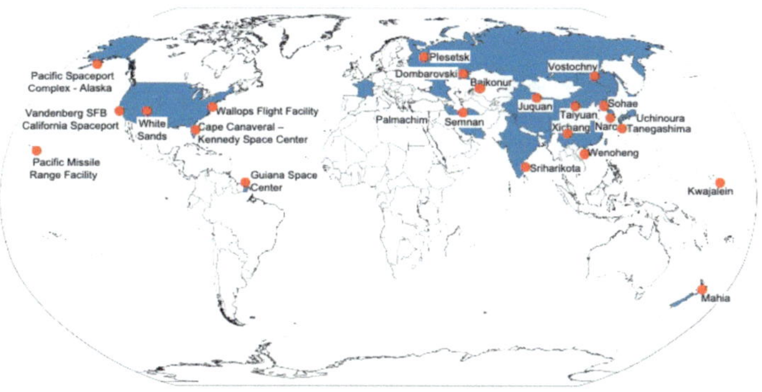

Source: FAA *2018 Annual Compendium of Commercial Space Transportation*

Major CONUS Launch Sites

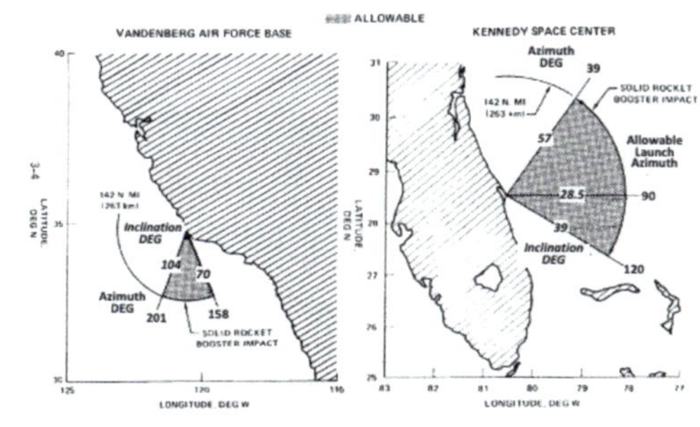

- WTR = Western Test Range (Vandenberg Space Force Base)
- ETR = Eastern Test Range (Cape Canaveral, KSC)
- Launch azimuth limited by range safety considerations, i.e. flights over populated area

WTR coordinates: 34.4° N, 120.35° W
ETR coordinates: 28.5° N, 81° W

Cape Canaveral / ETR / KSC Trajectory Info

Orbits:

- Low Earth orbit (LEO)
- Geostationary Transfer Orbit (GTO)
- Geostationary Orbit (GEO)
- *Atlas V*, *Delta IV*, *Falcon* launches

Source: Pisacane

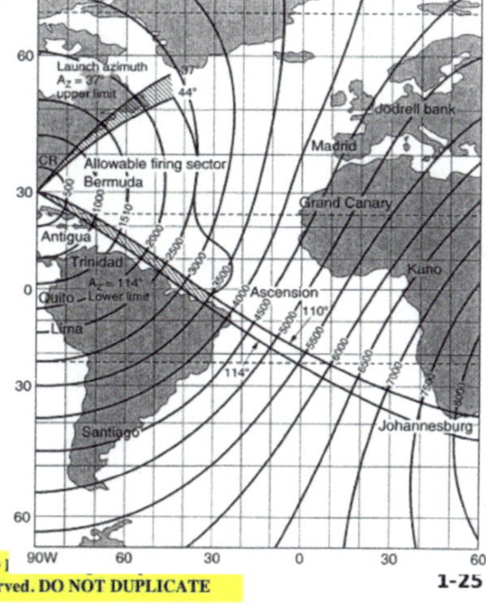

Kodiak Island, AK

- Kodiak Launch Complex (KLC) located 30 miles off the coast of Alaska (a.k.a. Pacific Spaceport Complex Alaska, PSCA)
- Wide-open launch corridor, unobstructed down-range flight path
- Ideal for PEO, SSO
- *Athena & Minotaur* rockets, small sats

Courtesy SpaceToday.org

Wallops Flight Facility (WFF), VA

- *Antares*, *Minotaur*
- Suborbital sounding rockets
- 3 min tour: www.youtube.com/watch?v=_NMJ_jE-slo

Other U.S. Gov't. Launch Sites

- Kwajalein, Pacific: *Falcon 1*, interceptors
- White Sands Missile Range (WSMR), NM: sounding rockets
- Barking Sands, HI: interceptors. Orbital? One attempt, *Super Strypi*, failed.

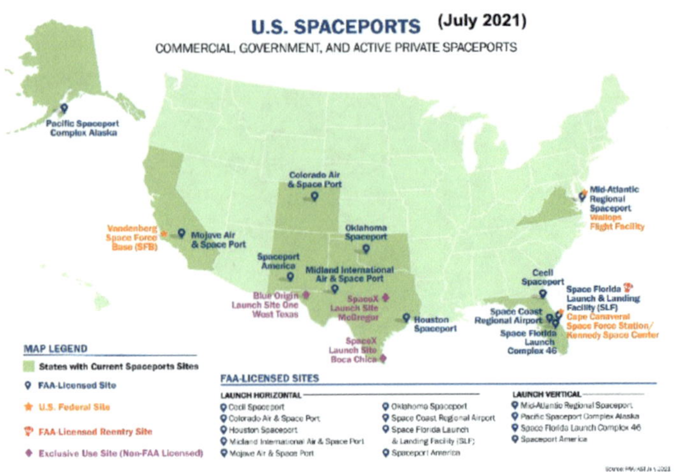

ESA: Kourou, French Guiana

- "Centre Spatial Guyanais" = CSG used by European Space Agency (ESA)
- Latitude 5.2° N Longitude 52.8° W
- Vehicles: *Ariane 5*, *Soyuz*, *Vega*
- Near-equatorial site favors LEO, GTO, GEO, SSO

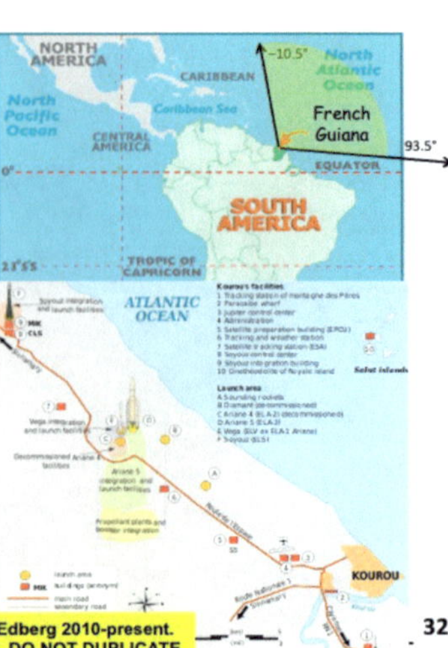

Courtesy: worldatlas, wikipedia

Russia Launch Sites

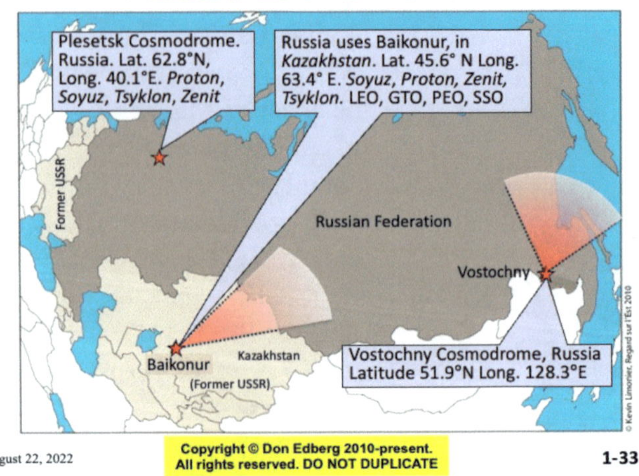

China Launch Sites

- Xichang: Lat 28.25° N, Long 102.0° E
- Taiyuan Space Launch Center: Lat 37.5° N Long 112.6° E. *Long March.*

Japan Launch Sites

- Tanegashima Space Center, Lat. 30.4° N, Long. 131.0° E. *H-IIA* and *H–IIB* rockets
- Uchinoura Space Center, Lat. 31.2° N, Long. 131.1° E. *Epsilon* rockets

India Launch Site

- India: Sriharikota: Lat 13.9° N, Long 80.4° E. *SLV*, *PLSV*, *GSLV* launches. Courtesy en.people.cn

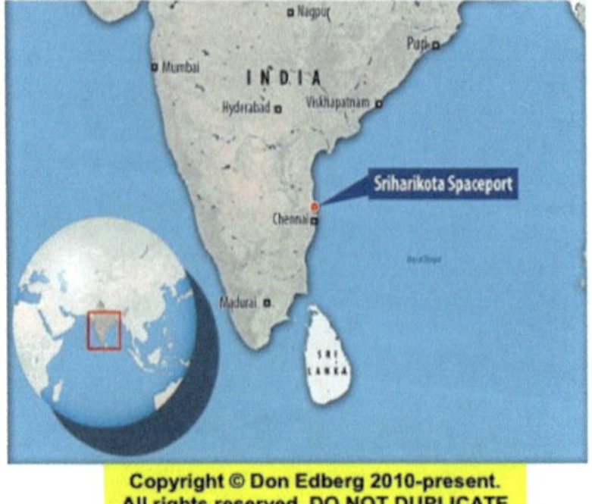

New Zealand Launch Site

- Mahia Peninsula, NZ: Lat 39°90' S, Long 177°54' E. Rocket Lab *Electron* launches.

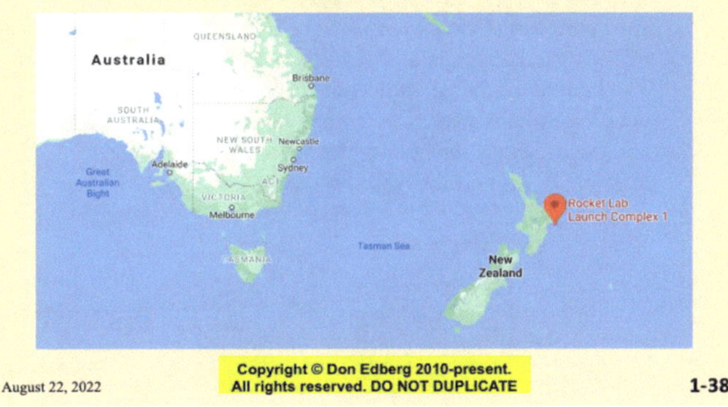

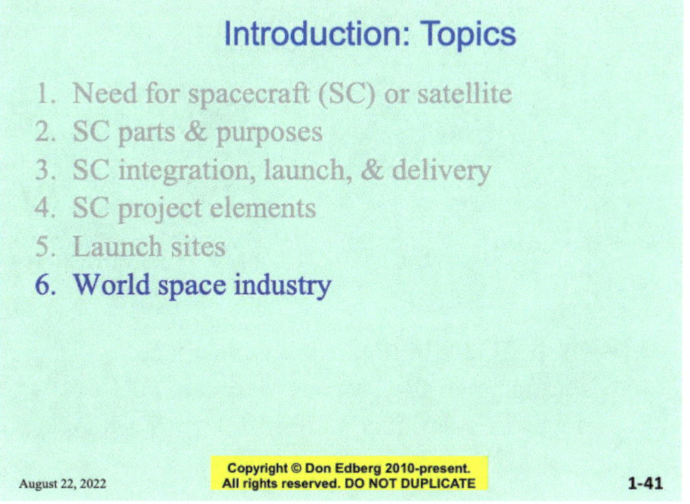

Introduction: Topics

1. Need for spacecraft (SC) or satellite
2. SC parts & purposes
3. SC integration, launch, & delivery
4. SC project elements
5. Launch sites
6. World space industry

The World Space Industry: 2019

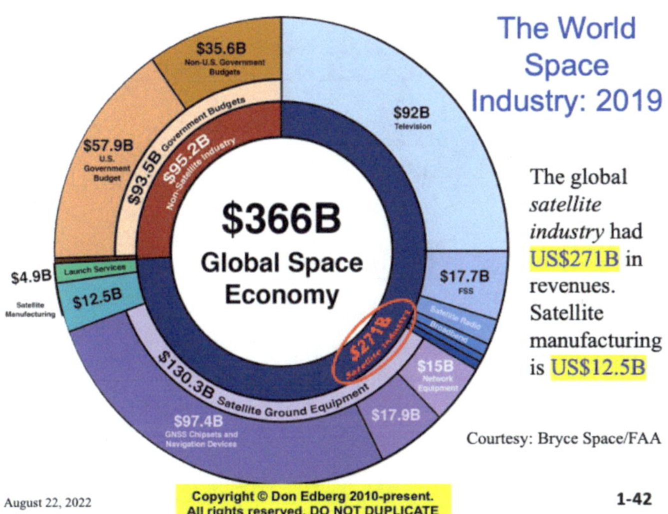

The global *satellite industry* had US$271B in revenues. Satellite manufacturing is US$12.5B

Courtesy: Bryce Space/FAA

Private Investment in Space Services Companies

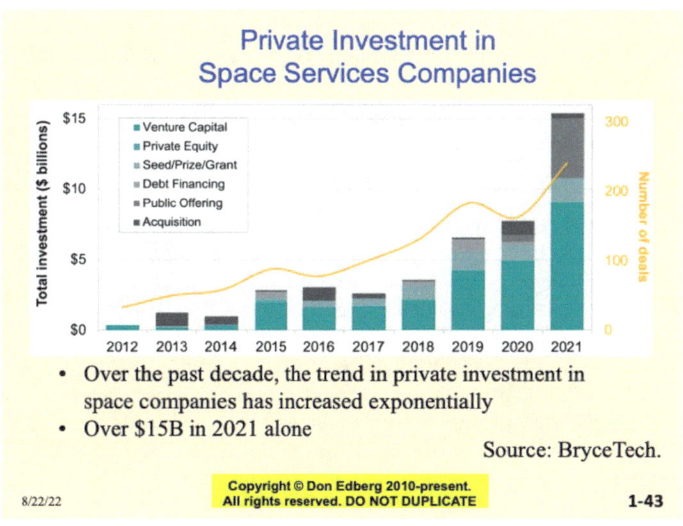

- Over the past decade, the trend in private investment in space companies has increased exponentially
- Over $15B in 2021 alone

Source: BryceTech.

Private Investment in Space Services Companies

- Rise in popularity of special purpose acquisition companies (SPACs)
- Emergence of newer and more junior space companies within public markets

Has driven:

- Growth in number of space companies founded & funded

AND

- Increased number of mergers & acquisitions driven by space-focused holding companies and legacy industry incumbents

Space Startup Investing

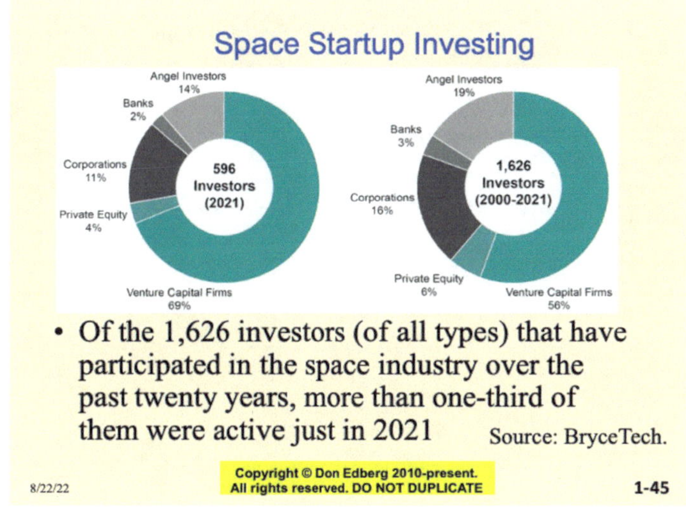

- Of the 1,626 investors (of all types) that have participated in the space industry over the past twenty years, more than one-third of them were active just in 2021

Source: BryceTech.

Space Startup Investing

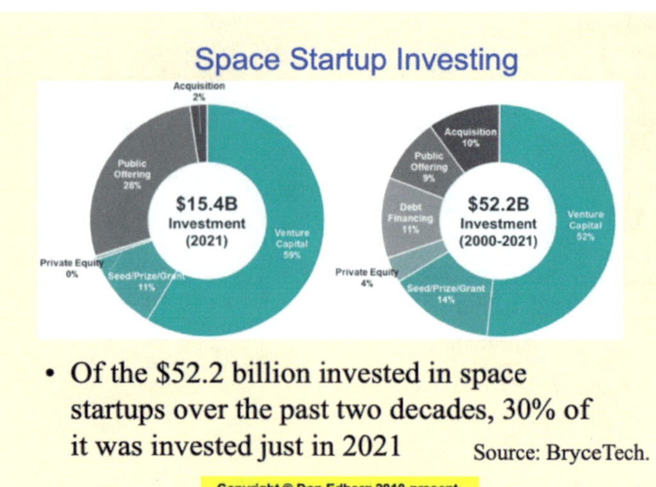

- Of the $52.2 billion invested in space startups over the past two decades, 30% of it was invested just in 2021

Source: BryceTech.

Results Should Not Be Surprising…

- Successes of companies like SpaceX make "space cool"
- Inspired numerous engineers, businesspeople, entrepreneurs, and visionaries to re-examine the problems of LVs, satellites, space utilization in new fashions
- The amount of data & connectivity required by most people – weather, satellite radio, internet, television, navigation, earth science, exploration, etc. – relies on efficient & economically feasible access to space
- Technologies & applications are helping to form a new Golden Age of aerospace

Why Spacecraft: Take-aways

1. Purpose: need spacecraft for all types of missions: communication, entertainment, navigation, weather, military/surveillance, exploration (including crews)
2. Project elements: SC supports payload with subsystems; operational support critical
3. Design process: SC design must support payload while integrating with and surviving environment of LV, and space
4. Worldwide launch sites: often on east coasts, but located by national interests and desires
5. World's space industry: spacecraft provide $271B of $366B space industry in 2019

Legal Stuff Concerning Flying …

https://www.law.cornell.edu/cfr/text/14/chapter-III

- Cornell Law's *Legal Information Institute* offers searchable online source for Federal Aviation Regulations (FARs); FAA publishes PDFs
- *Note: I am not a lawyer & I do not play one on TV. My current gig is pretending to be a professor in SoCal*

Legal Information Institute [LII]

14 CFR Chapter III - COMMERCIAL SPACE TRANSPORTATION, FEDERAL AVIATION ADMINISTRATION, DEPARTMENT OF TRANSPORTATION

Legal Stuff Concerning Flying

FAA regulations for space trans. are under Volume 4, Chapter III for Commercial Space Transportation:
Code of Federal Regulations (CFR)
Title 14: Aeronautics and Space
Commercial Space Section Link: https://www.ecfr.gov/current/title-14/chapter-III

Title	Volume	Chapter	Browse Parts	Regulatory Entity
Title 14 Aeronautics and Space	1	I	1-59	FEDERAL AVIATION ADMINISTRATION, DEPARTMENT OF TRANSPORTATION
	2		60-109	
	3		110-199	
	4	II	200-399	OFFICE OF THE SECRETARY, DEPARTMENT OF TRANSPORTATION (AVIATION PROCEEDINGS)
		III	400-1199	COMMERCIAL SPACE TRANSPORTATION, FEDERAL AVIATION ADMINISTRATION, DEPARTMENT OF TRANSPORTATION
	5	V	1200-1299	NATIONAL AERONAUTICS AND SPACE ADMINISTRATION
		VI	1300-1399	AIR TRANSPORTATION SYSTEM STABILIZATION

Last-Minute Updates

Some changes in the industry… may affect searches for information and products

- 2021 May: *Vandenberg Air Force Base* (VAFB) renamed *Vandenberg Space Force Base* (VSFB)
- ~~2020 Dec.: Lockheed Martin is acquiring Aerojet Rocketdyne~~
- 2018: Northrop Grumman acquired Orbital ATK, which included Orbital Sciences & Thiokol

Spacecraft: History & Future

Topics

- The Cold War
- The Moon Race
- Space stations, shuttles, & cooperation
- Planetary probes
- Earth orbiters: observation, spying, research
- Capsules & transfer vehicles
- Recent spacecraft developments
- Launch vehicles

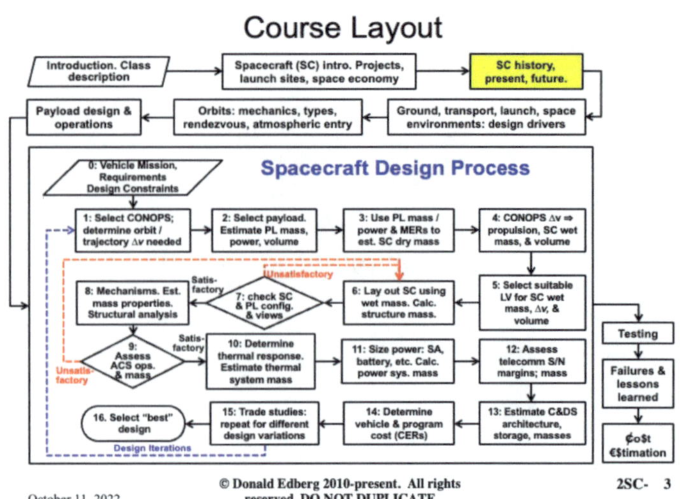
Course Layout
Introduction. Class description
Spacecraft (SC) intro. Projects, launch sites, space economy
SC history, present, future.
Ground, transport, launch, space environments: design drivers
Orbits: mechanics, types, rendezvous, atmospheric entry
Payload design & operations
Spacecraft Design Process
0: Vehicle Mission, Requirements Design Constraints
1: Select CONOPS; determine orbit / trajectory Δv needed
2: Select payload. Estimate PL mass, power, volume
3: Use PL mass / power & MERs to est. SC dry mass
4: CONOPS Δv ⇒ propulsion, SC wet mass, & volume
5: Select suitable LV for SC wet mass, Δv, & volume
6: Lay out SC using wet mass. Calc. structure mass.
7: check SC & PL config. & views
8: Mechanisms. Est. mass properties. Structural analysis
9: Assess ACS ops. & mass
10: Determine thermal response. Estimate thermal system mass
11: Size power: SA, battery, etc. Calc. power sys. mass
12: Assess telecomm S/N margins; mass
13: Estimate C&DS architecture, storage, masses
14: Determine vehicle & program cost (CERs)
15: Trade studies: repeat for different design variations
16. Select "best" design
Satis-factory
Unsatisfactory
Unsatis-factory
Design Iterations
Testing
Failures & lessons learned
¢o$t €$timation

October 11, 2022
© Donald Edberg 2010-present. All rights reserved. DO NOT DUPLICATE
2SC- 3

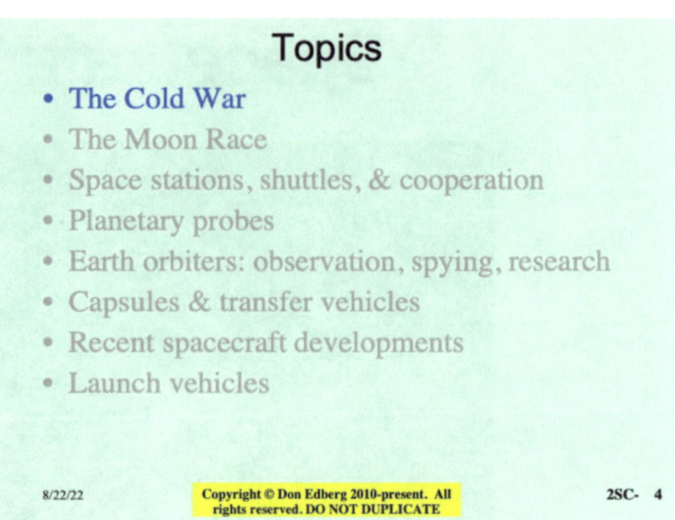
Topics
The Cold War
The Moon Race
Space stations, shuttles, & cooperation
Planetary probes
Earth orbiters: observation, spying, research
Capsules & transfer vehicles
Recent spacecraft developments
Launch vehicles
8/22/22
Copyright © Don Edberg 2010-present. All rights reserved. DO NOT DUPLICATE
2SC- 4

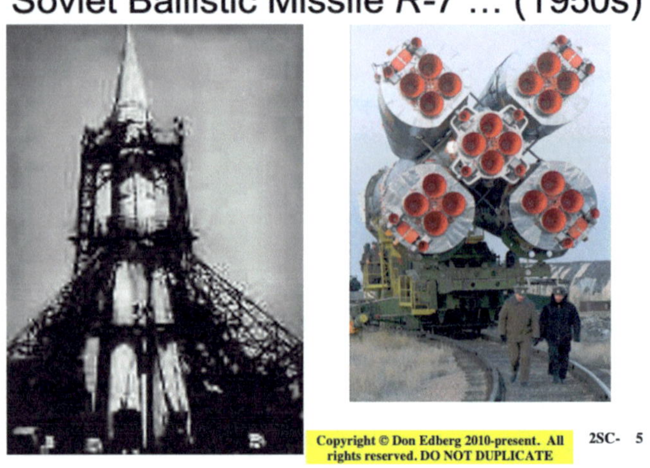
Soviet Ballistic Missile R-7 … (1950s)
Copyright © Don Edberg 2010-present. All rights reserved. DO NOT DUPLICATE
2SC- 5

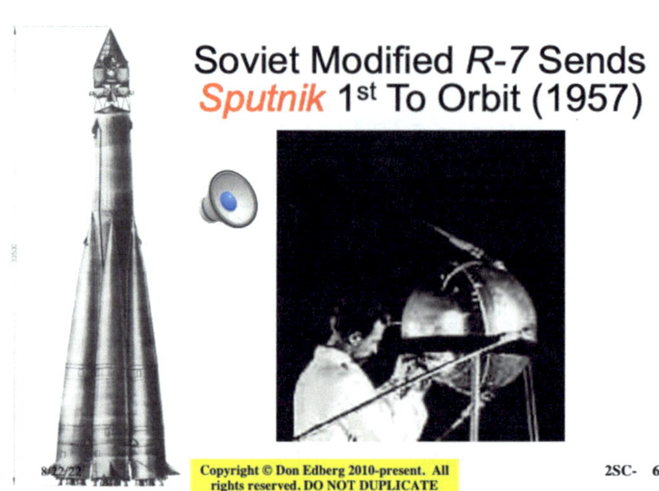
Soviet Modified R-7 Sends Sputnik 1st To Orbit (1957)
8/22/22
Copyright © Don Edberg 2010-present. All rights reserved. DO NOT DUPLICATE
2SC- 6

Vanguard Failure — USA (1958) Eventually Successful

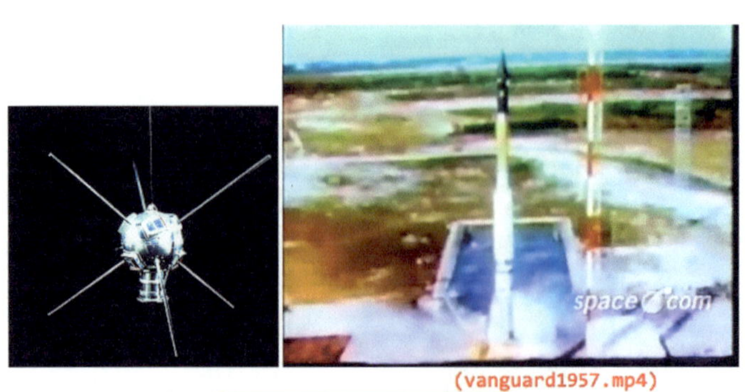
space.com

(vanguard1957.mp4)

Jupiter-C (Modified *Redstone*) Launched *Explorer I* (1958), Van Allen Belts Discovered

Courtesy NASA

SCIENTISTS WILLIAM PICKERING, JAMES VAN ALLEN AND WERNHER von BRAUN (LEFT TO RIGHT) DISPLAY THE UNITED STATES' FIRST SATELLITE, EXPLORER 1

Soviets Orbit First Human, Gagarin in *Vostok* (1961)

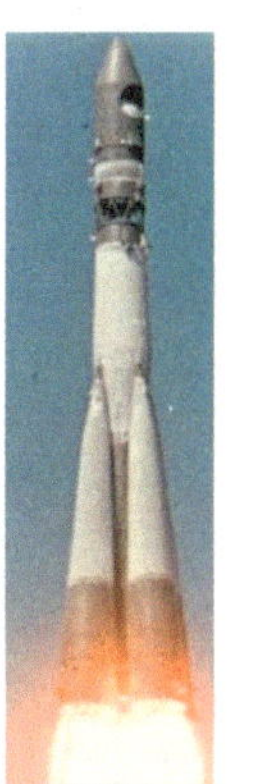

Gagarin, *Vostok*; *Voshkhod*

8/22/22

2SC- 10

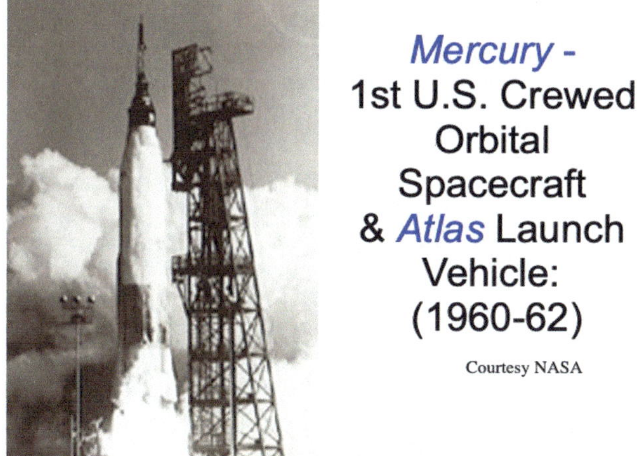

Mercury - 1st U.S. Crewed Orbital Spacecraft & *Atlas* Launch Vehicle: (1960-62)

Courtesy NASA

8/22/22

2SC- 11

Titan II / *Gemini* Launch Vehicle (3,600 kg to LEO, 1966-67)

Courtesy NASA

8/22/22

2SC- 12

Soviet Multi-Person Spacecraft — (1970s-Present)

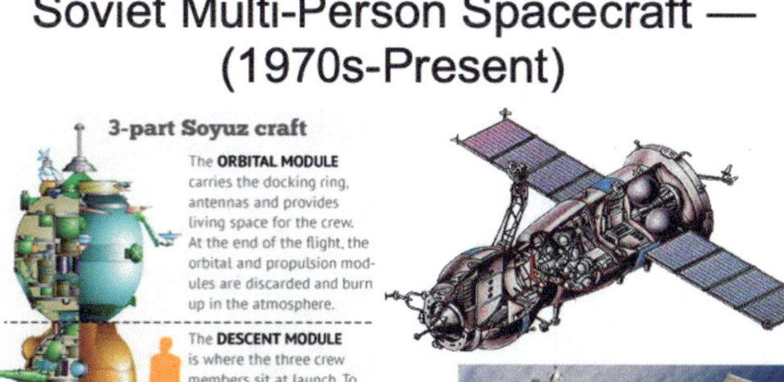
3-part Soyuz craft
The ORBITAL MODULE carries the docking ring, antennas and provides living space for the crew. At the end of the flight, the orbital and propulsion modules are discarded and burn up in the atmosphere.
The DESCENT MODULE is where the three crew members sit at launch. To save weight, the module is only as big as is needed to return the crew to Earth.
The PROPULSION MODULE carries oxygen and fuel, rocket engines, solar power "wings" and also radiator panels for expelling excess heat into space.
Drawing: RSC Energia

PHOTO: SOYUZ TMA-7 IN ORBIT, APRIL 8, 2006 (CREDIT: NASA)

2SC- 13

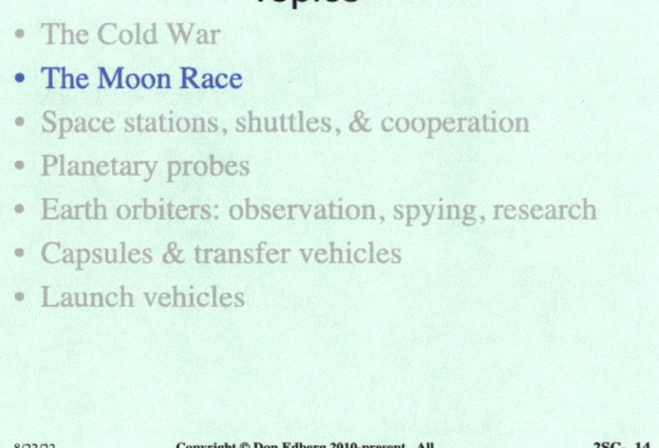

Topics

- The Cold War
- The Moon Race
- Space stations, shuttles, & cooperation
- Planetary probes
- Earth orbiters: observation, spying, research
- Capsules & transfer vehicles
- Launch vehicles

8/22/22

2SC- 14

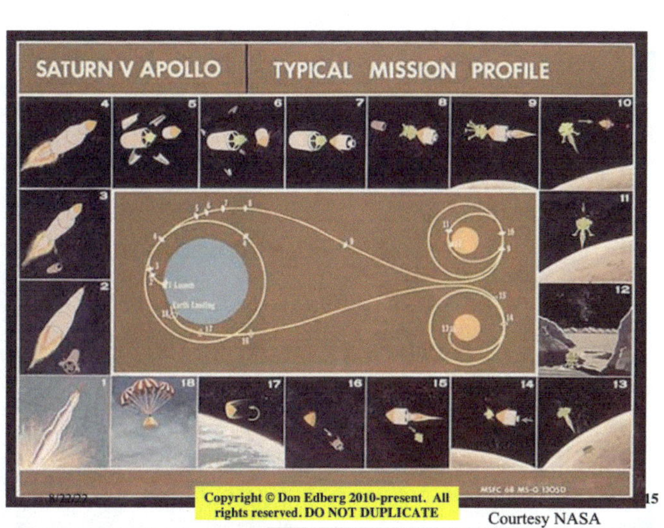
SATURN V APOLLO
TYPICAL MISSION PROFILE

Courtesy NASA

U.S. Crewed Spacecraft before Shuttle

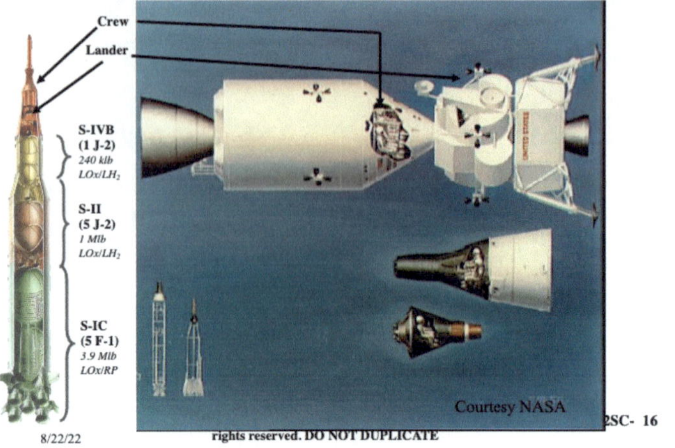

8/22/22 2SC- 16

U.S. Lunar Module for Moon Landing (1969-72)

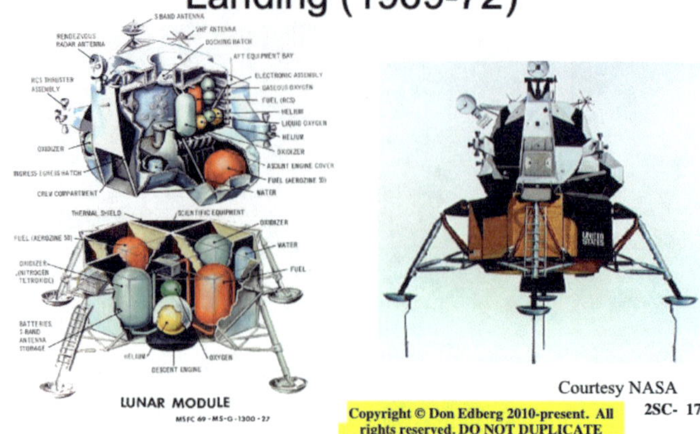

Courtesy NASA

 2SC- 17

Walk on Moon, Visit Old Friend

8/22/22 2SC- 19

Courtesy NASA

Lunar Rover Drives on the Moon

8/22/22 Courtesy NASA 2SC- 20

N-1: Soviet Moon Rocket (1968)

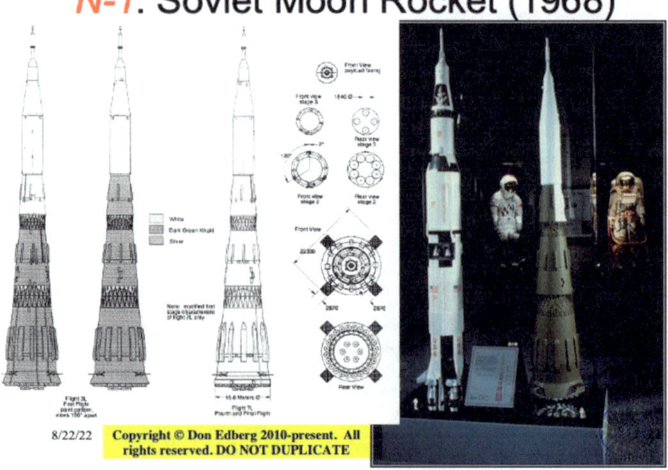

8/22/22

Soviet Lunar Lander

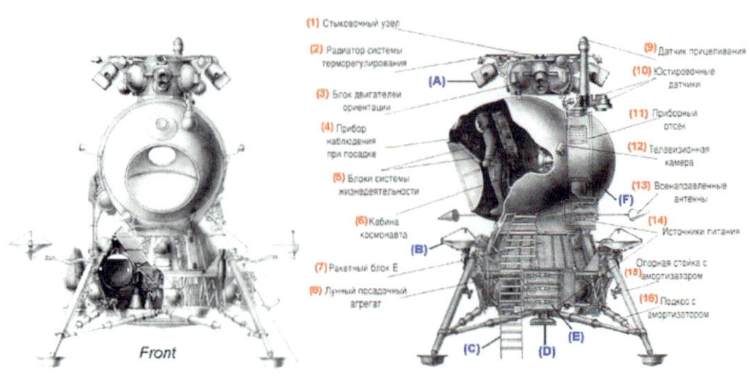

Source: http://www.myspacemuseum.com/SOVPHOT2.HTM

8/22/22 2SC- 22

Topics

- The Cold War
- The Moon Race
- **Space stations, shuttles, & cooperation**
- Planetary probes
- Earth orbiters: observation, spying, research
- Capsules & transfer vehicles
- Recent spacecraft developments
- Launch vehicles

8/22/22 2SC- 23

Salyut Space Station 1971-82

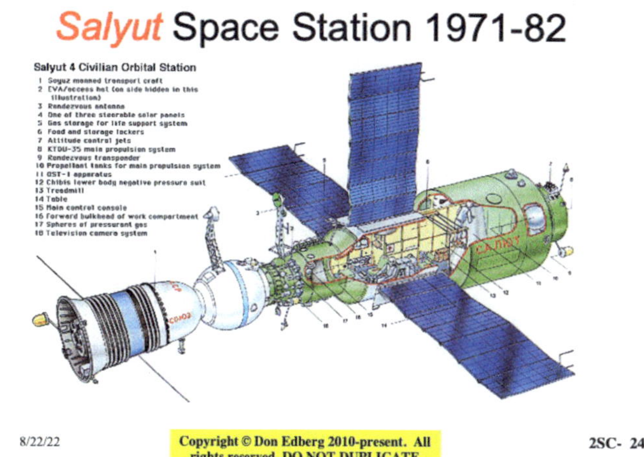

8/22/22 2SC- 24

Apollo-Soyuz Test Project (1974)

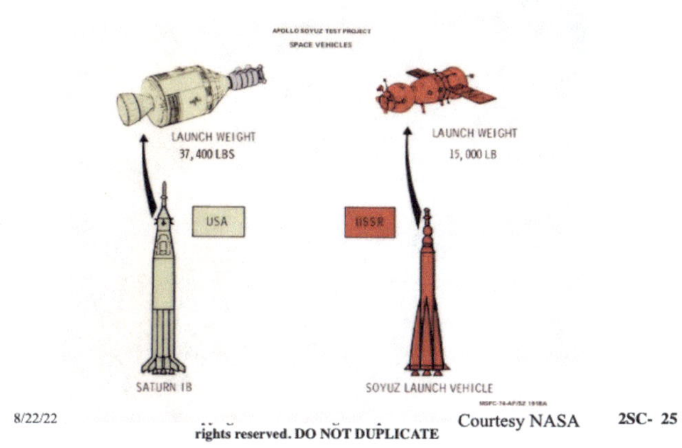

Courtesy NASA

8/22/22 2SC- 25

Apollo-Soyuz Test Project (2)

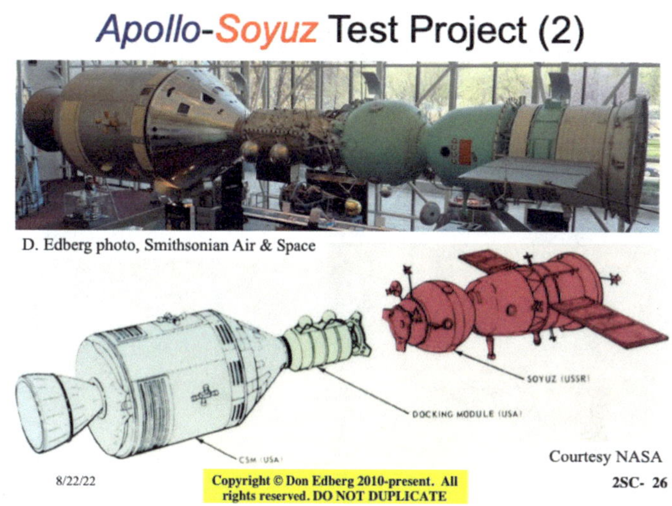

D. Edberg photo, Smithsonian Air & Space

Courtesy NASA

8/22/22 2SC- 26

Skylab Space Station (1974)

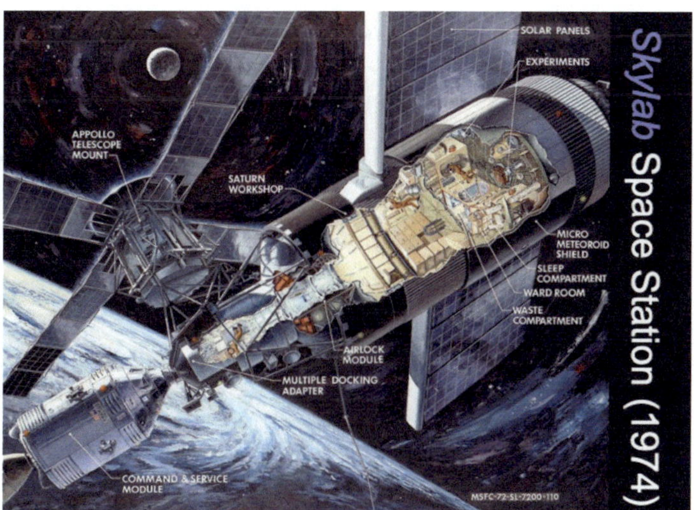

Soviet *Mir* Space Station (1986-2001)

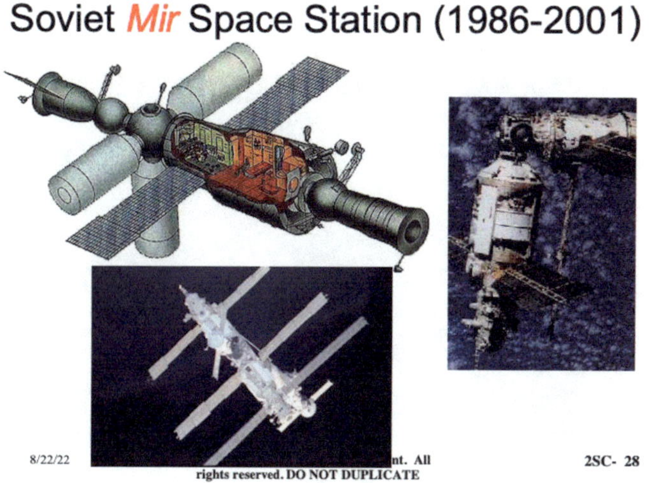

8/22/22 2SC- 28

Space Shuttle — 1st "Reusable" SC (1981 - 2011)

Courtesy NASA 29

Shuttle Docks With *Mir*

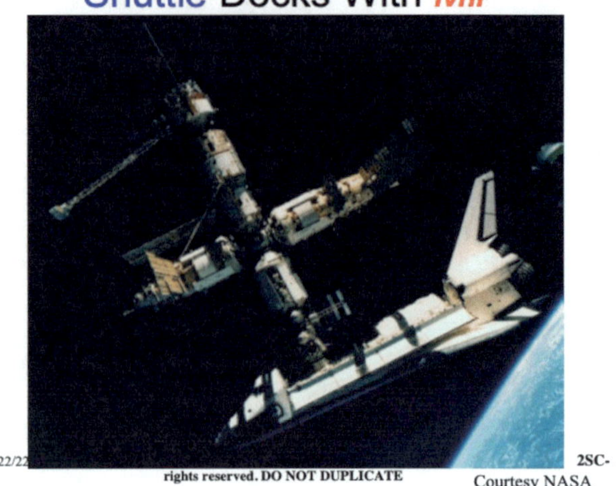

8/22/22 2SC- 30

Courtesy NASA

Soviet *Buran & Energia* (1980s)

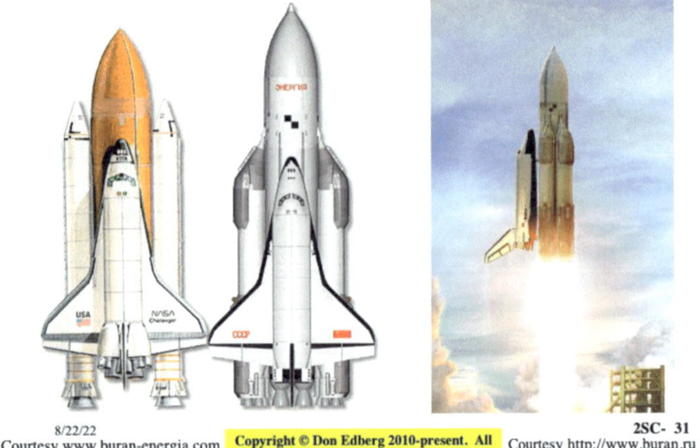

8/22/22 2SC- 31

Courtesy www.buran-energia.com

Courtesy http://www.buran.ru

International Space Station — ISS: 1998 – present

S119E010027

8/22/22

Courtesy NASA 2SC- 32

Tianggong-1 (China)

8/22/22

Courtesy Quora 2SC- 33

NGC Awarded $187M to design Habitation & Logistics Outpost for Lunar Orbit Gateway (2020)

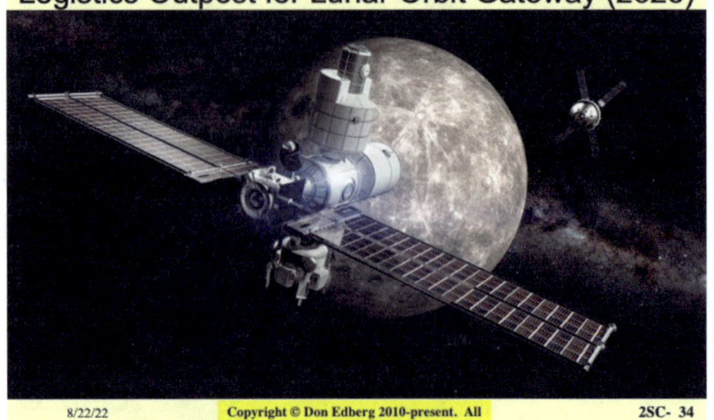

8/22/22

Courtesy: NGC

2SC- 34

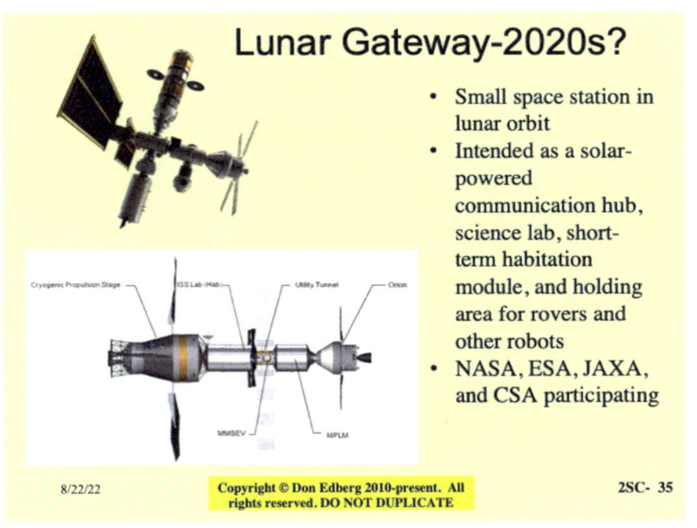
Lunar Gateway-2020s?
• Small space station in lunar orbit
• Intended as a solar-powered communication hub, science lab, short-term habitation module, and holding area for rovers and other robots
• NASA, ESA, JAXA, and CSA participating
Cryogenic Propulsion Stage
ISS Lab (Hab)
Utility Tunnel
Orion
MMSEV
MPLM
8/22/22
Copyright © Don Edberg 2010-present. All rights reserved. DO NOT DUPLICATE
2SC- 35

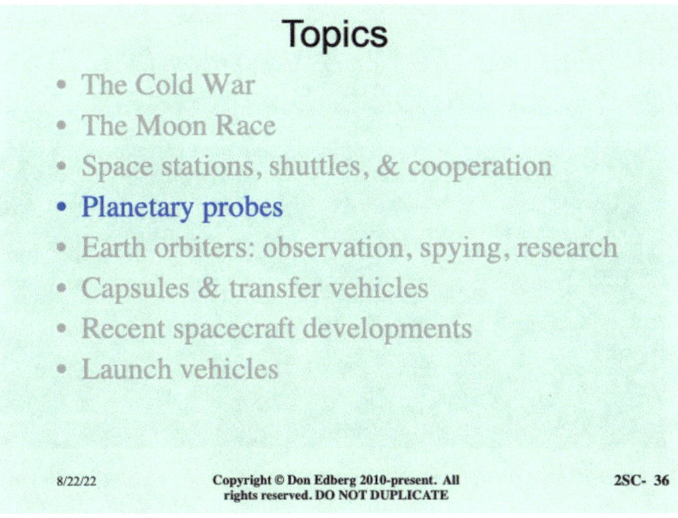
Topics
• The Cold War
• The Moon Race
• Space stations, shuttles, & cooperation
• Planetary probes
• Earth orbiters: observation, spying, research
• Capsules & transfer vehicles
• Recent spacecraft developments
• Launch vehicles
8/22/22
Copyright © Don Edberg 2010-present. All rights reserved. DO NOT DUPLICATE
2SC- 36

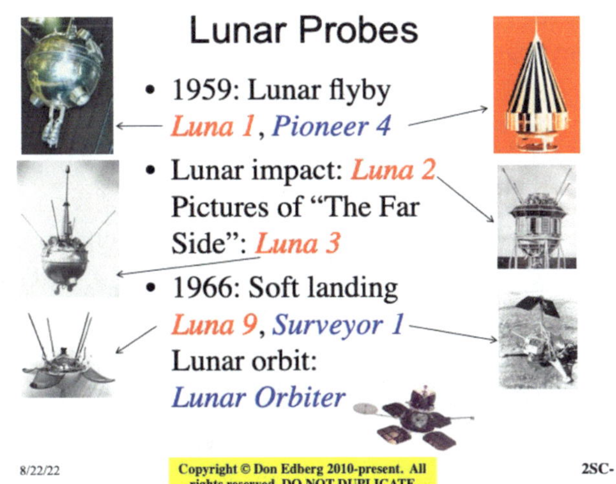
Lunar Probes
• 1959: Lunar flyby
Luna 1, Pioneer 4
• Lunar impact: Luna 2
Pictures of "The Far Side": Luna 3
• 1966: Soft landing
Luna 9, Surveyor 1
Lunar orbit:
Lunar Orbiter
8/22/22
Copyright © Don Edberg 2010-present. All rights reserved. DO NOT DUPLICATE
2SC- 37

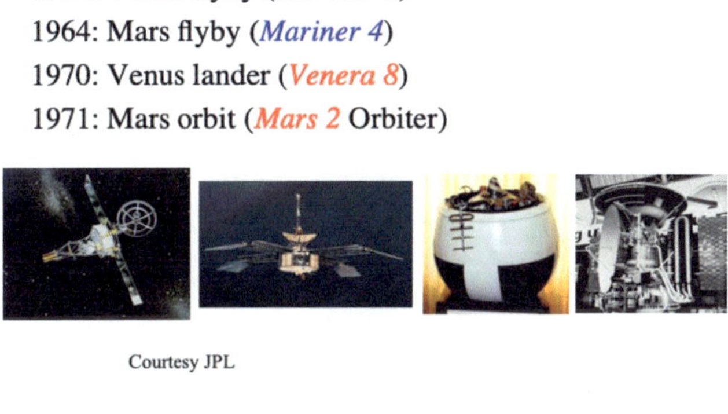
Inner Planet Probes
1962: Venus flyby (Mariner 2)
1964: Mars flyby (Mariner 4)
1970: Venus lander (Venera 8)
1971: Mars orbit (Mars 2 Orbiter)
Courtesy JPL
8/22/22
Copyright © Don Edberg 2010-present. All rights reserved. DO NOT DUPLICATE
2SC- 38

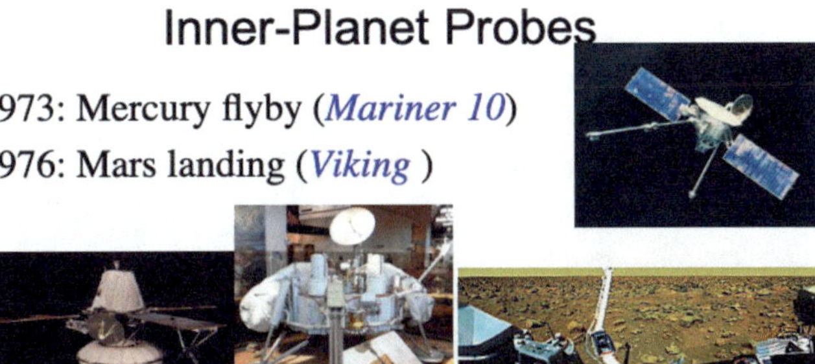
Inner-Planet Probes
1973: Mercury flyby (Mariner 10)
1976: Mars landing (Viking)
1978: Venus orbit (Pioneer Venus 1)
Courtesy NASA
8/22/22
Copyright © Don Edberg 2010-present. All rights reserved. DO NOT DUPLICATE
2SC- 39

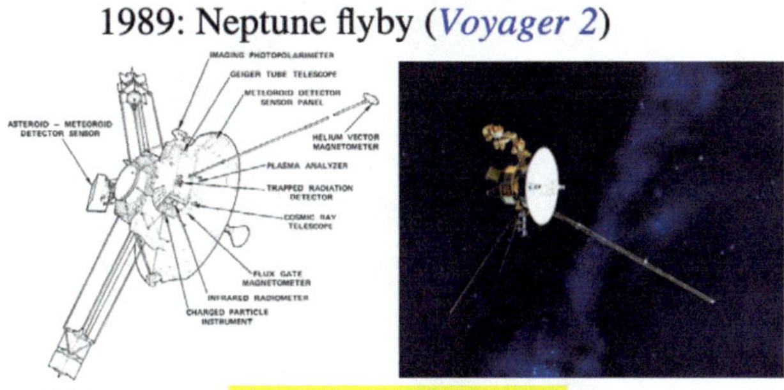
Outer-Planet Probes
1972: Jupiter flyby (Pioneer 10)
1973: Saturn flyby (Pioneer 11)
1986: Uranus flyby (Voyager 2, '77)
1989: Neptune flyby (Voyager 2)
IMAGING PHOTOPOLARIMETER
GEIGER TUBE TELESCOPE
METEOROID DETECTOR SENSOR PANEL
ASTEROID – METEOROID DETECTOR SENSOR
HELIUM VECTOR MAGNETOMETER
PLASMA ANALYZER
TRAPPED RADIATION DETECTOR
COSMIC RAY TELESCOPE
FLUX GATE MAGNETOMETER
INFRARED RADIOMETER
CHARGED PARTICLE INSTRUMENT
8/22/22
Copyright © Don Edberg 2010-present. All rights reserved. DO NOT DUPLICATE
2SC- 40

Pioneer (1972) & *Voyager* (1977) Explore Outer Solar System

8/22/22 2SC- 41

Courtesy NASA

Magellan Explores Venus (1989)

8/22/22 Courtesy NASA 2SC- 42

Galileo Spacecraft (1989-2003)

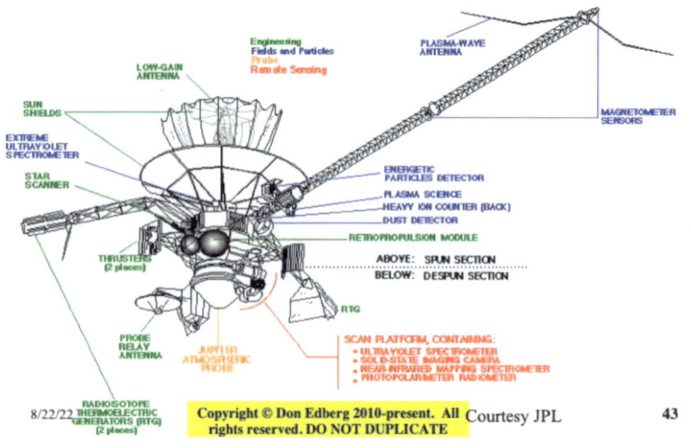

8/22/22 Courtesy JPL 43

Galileo at Jupiter Planned vs. Actual (Jammed Antenna)!

8/22/22 Courtesy JPL 2SC- 44

Ulysses: Solar Polar Mission (1990)

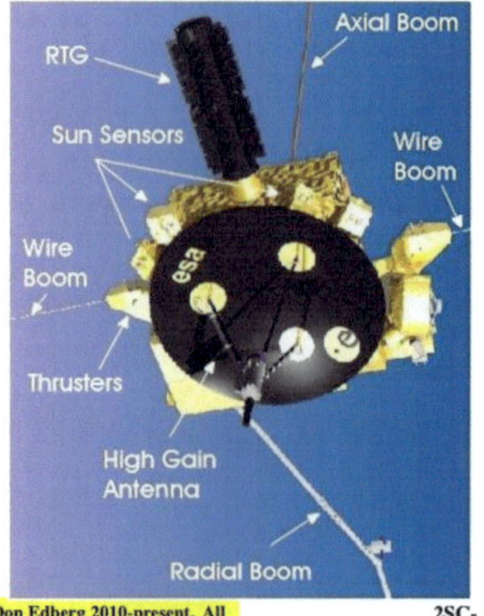

Courtesy NASA/ESA

8/22/22 2SC- 45

Cassini Spacecraft & *Titan* LV (1997-2017)

8/22/22 2SC- 46

Courtesy JPL Courtesy Lockmart

Mars Global Surveyor (1996)

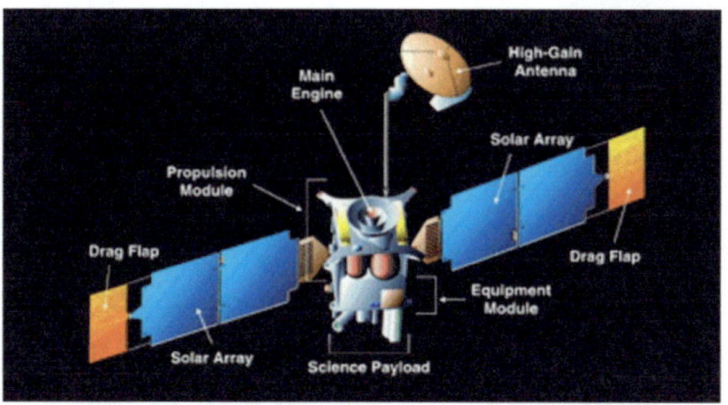

8/22/22 Courtesy JPL 2SC- 47

Deep Space 1 — Ion Propelled (1998)

8/22/22 Courtesy NASA 2SC- 48

Comet & Asteroid Rendezvous

1999: Comet sample return (*Stardust*)

2005: Asteroid landing (*Muses-C/Hayabusa*)

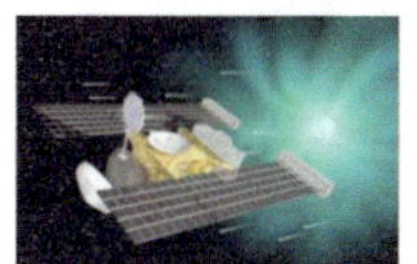

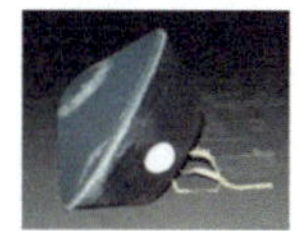

8/22/22 2SC- 49

Mars Climate Orbiter RIP 1999

- Lost at Mars (units error!)

8/22/22 Courtesy JPL 2SC- 50

Chinese 1st Crewed Flight 11/2003

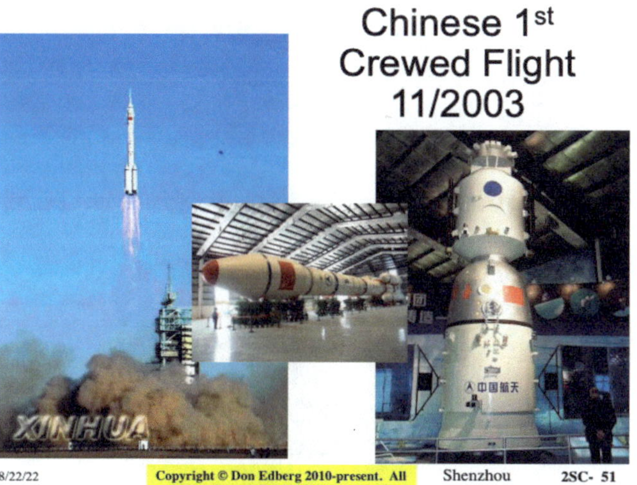

Shenzhou Spacecraft

8/22/22 2SC- 51

Mars Exploration Rover & *Delta II* (2004)

Mars Pathfinder, 2000. Mars Exploration Rovers, 2004

8/22/22 Courtesy JPL 2SC- 52

Messenger: MErcury Surface, Space ENvironment, Geochemistry, & Ranging — to Mercury, 2004 (Orbit: 2011; RIP 4/2015)

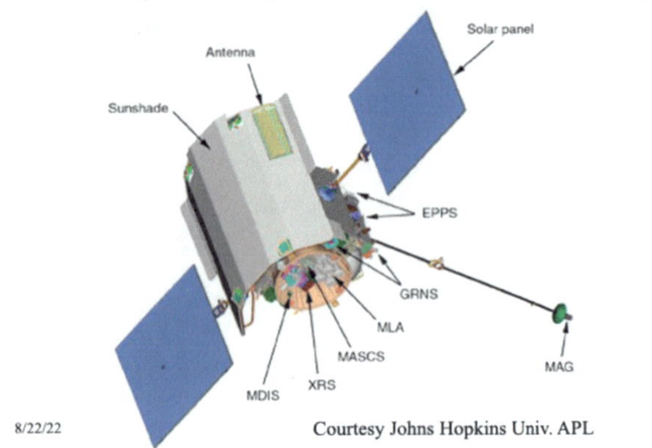

8/22/22 Courtesy Johns Hopkins Univ. APL 2SC- 53

New Horizons — Launched in 2006 Pluto arrival 2015!

8/22/22 Courtesy NASA 2SC- 54

DAWN, Ion-powered (2007) to Vesta & Ceres Asteroids

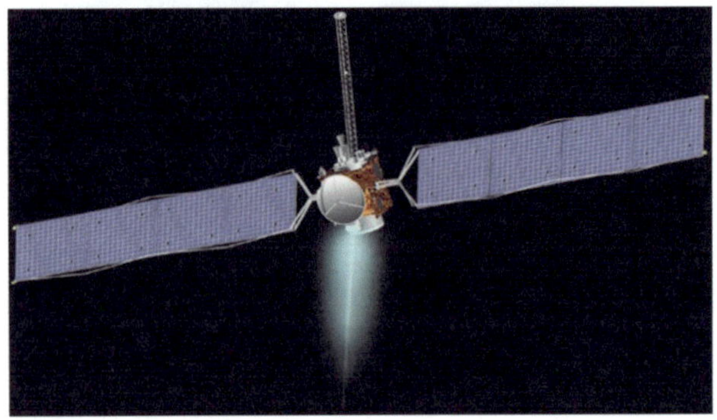

8/22/22 Courtesy NASA 2SC- 55

Juno to Jupiter – Launch Aug. 2011, Arrived 2016

Courtesy JPL

8/22/22 2SC- 56

China *Jade Rabbit* Moon Rover (2014)

Credit: CNSA / CCTV

8/22/22 2SC- 58

Rosetta Orbits & *Philae* Probe Lands On Comet Churyumov/ Gerasimenko (ESA, 2014)

Courtesy ESA, C. Carreau

8/22/22 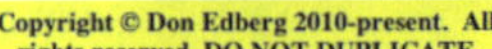 2SC- 59

Mars *INSIGHT* (Interior exploratioN using Seismic Investigation, Geodesy, and Heat Transport) 2018

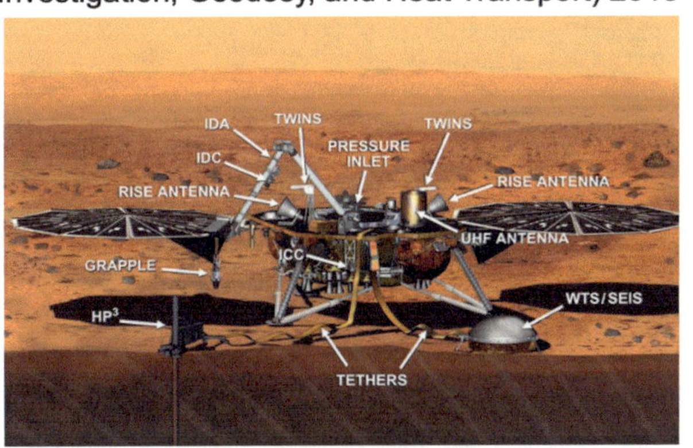

8/22/22 2SC- 60

Parker Solar Probe (2018)

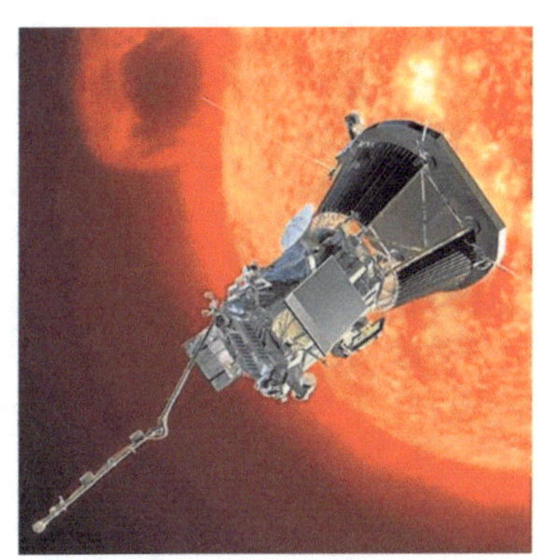

- Perihelion 3.83 million mi; peak temp 1400 C
- 25 × solar energy while orbiting Earth
- Max speed: 125 mi/sec (25 × Earth orbit speed)
- Formerly named *Solar Probe Plus*

Courtesy NASA

8/22/22 2SC- 61

Transiting Exoplanet Survey Satellite – TESS (2018)

Courtesy NASA

8/22/22 2SC- 62

Perseverance & Ingenuity (2021)

NASA/JPL-Caltech

8/22/22 2SC- 63

China Tianwen-1 Rover Mars Landing (2021 May)

8/22/22 2SC- 64

Topics

- The Cold War
- The Moon Race
- Space stations, shuttles, & cooperation
- Planetary probes
- **Earth orbiters: observation, spying, research**
- Capsules & transfer vehicles
- Recent spacecraft developments
- Launch vehicles

8/22/22 2SC- 67

Earth Orbit Satellites

Courtesy Lockmart, Boeing, D. Edberg photo, Smithsonian Udvar-Hazy

Clockwise: GPS; Boeing 601, TDRSS

8/22/22 2SC- 68

KH-9 *Hexagon* Spy satellite (1971-89. 0.6 - 1 m resolution, 20 flights)

Courtesy space.com

8/22/22 2SC- 69

Hubble Space Telescope (1990-present)

 Courtesy NASA 2SC- 70

Spitzer (formerly SIRTF, Space Infrared Telescope Facility, 2003)

8/22/22 Courtesy NASA 2SC- 71

Orbital Express: In-Flight Propellant Reloading 2007

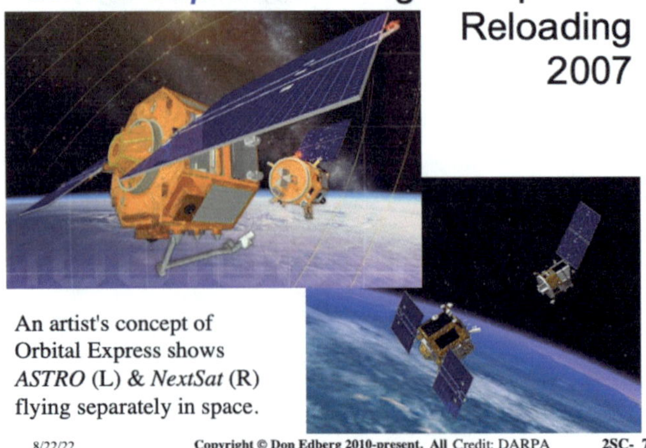

An artist's concept of Orbital Express shows *ASTRO* (L) & *NextSat* (R) flying separately in space.

8/22/22 Credit: DARPA 2SC- 72

X-37B (2010)

X-37B Orbital Test Vehicle

The X-37B is an unmanned space test vehicle for the United States Air Force, based on NASA's original X-37 design.

X-37B

PAYLOAD FAIRING

CENTAUR

INTERSTAGE ADAPTERS

ATLAS V BOOSTER

RD-180 ENGINES

Atlas V Booster

Main engine

Hydrogen peroxide tank

JP-8 kerosene-based jet fuel tank

Maneuvering thrusters

Experiment bay

Avionics equipment

Maneuvering thrusters

6-ft. human to scale

Height:	9 ft 6 in (2.9 m)
Length:	29 ft 3 in (8.9 m)
Wingspan:	14 ft 11 in (4.5 m)
Launch weight:	11,000 lb (4,990 kg)

X-37B

Space Shuttle Orbiter

SOURCE: NASA, United Launch Alliance

Graphic by Karl Tate

2SC- 73

Ikaros Solar Sail, 2010 (Japan Aerospace Exploration Agency JAXA)

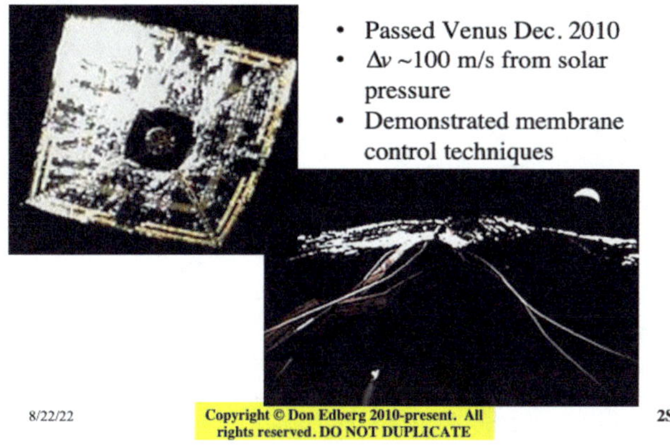

- Passed Venus Dec. 2010
- Δv ~100 m/s from solar pressure
- Demonstrated membrane control techniques

8/22/22 2SC- 74

Northrop Grumman *Mission Extension Vehicle*-1 (*MEV*-1) Space Tow Truck Docks With INTELSAT-901(2020)

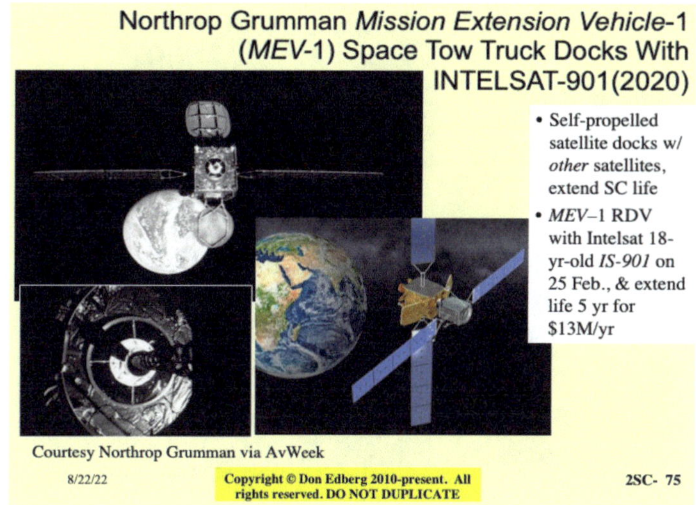

- Self-propelled satellite docks w/ *other* satellites, extend SC life
- *MEV*–1 RDV with Intelsat 18-yr-old *IS-901* on 25 Feb., & extend life 5 yr for $13M/yr

Courtesy Northrop Grumman via AvWeek

8/22/22 2SC- 75

James Webb Space Telescope (*JWST*) to Earth-Moon *L2* Region, 2021 Launch

Courtesy of NASA

8/22/22 2SC- 76

Topics

- The Cold War
- The Moon Race
- Space stations, shuttles, & cooperation
- Planetary probes
- Earth orbiters: observation, spying, research
- **Capsules & transfer vehicles**
- Recent spacecraft developments
- Launch vehicles

8/22/22 2SC- 77

Apollo CSM & LM: Trans. to Moon & Skylab

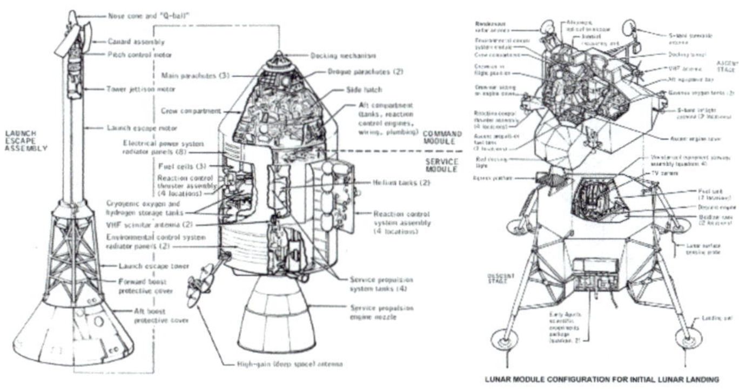

8/22/22 2SC- 78

Private Rocket: Scaled Composites *Spaceship 1* (2004, X Prize 2005)

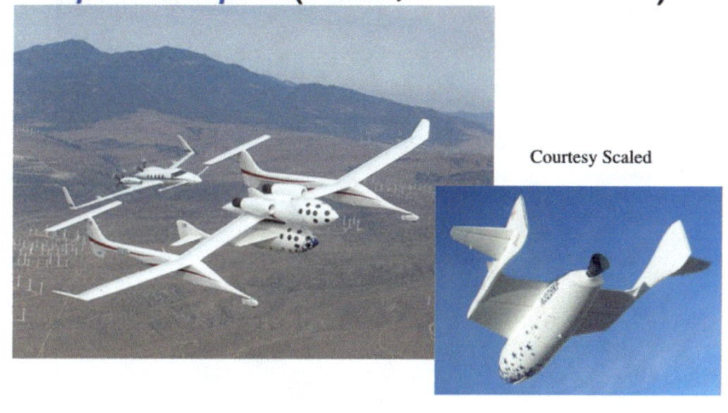

Courtesy Scaled

8/22/22 2SC- 79

Private Rocket/SC: Virgin Galactic *Spaceship 2* = SS2 (2009)

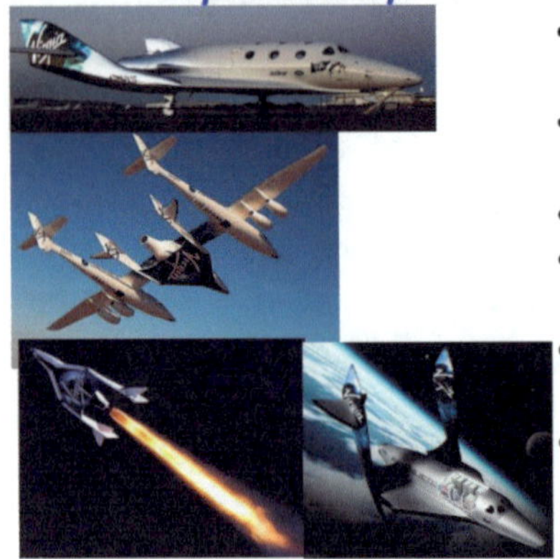

- Drop vehicle: *WK2 = WhiteKnight 2*
- SS2: 6 pax, 2 crew, suborbital to 100 km
- Hybrid engine power
- Unique variable-geometry recovery
- Redesigned after fatal vehicle breakup 2014
- Branson & 3 pax fly July 2021

Courtesy Virgin

Dragon Cargo Spacecraft, SpaceX, 2010+

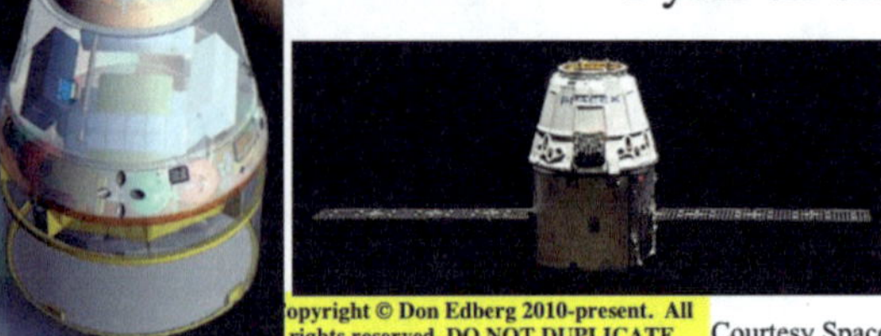

- Flies to ISS uncrewed
- 6T, 25 m^3 volume
- 3T, 11 m^3 return
- 2 year on-orbit ok

Courtesy SpaceX

ESA's Automated Transfer Vehicle *ATV-2*, aka *Johannes Kepler* (2011)

Courtesy ESA

ESA's Automated Transfer Vehicle ATV

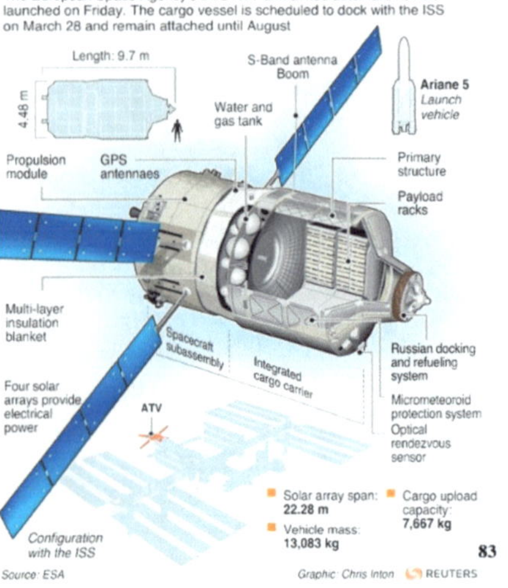

Courtesy ESA, Reuters

Orion Crew Exploration Vehicle

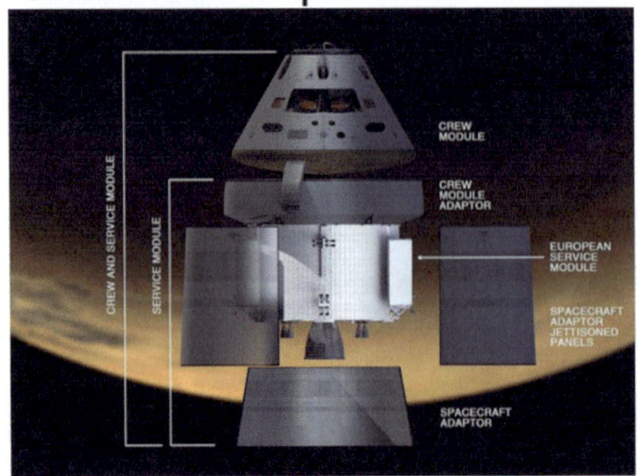

Courtesy NASA

Orion Crew Exploration Vehicle (CEV) (2011), 1st flight 2014

Courtesy NASA; ULA

Apollo vs. Orion

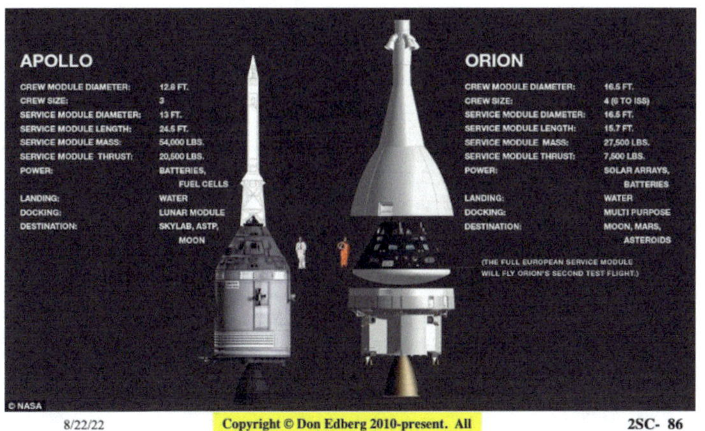

8/22/22 2SC- 86

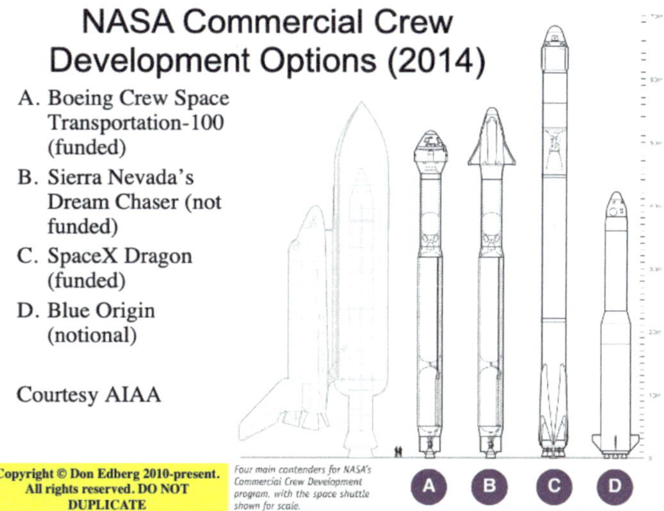

Blue Origin's *New Shepherd* (2015)

- Fully reusable *BE-3* flies 6-pax capsule, 50 lb payload boxes to suborbital trajectory: Altitude 100.5 km, speed $M = 3.7$
- LH_2/LOx engine throttles 110 – 20 klb_f thrust
- Automatically guided by forward & aft steering fins
- Propulsion module: Base-first descent stabilized by upper aerodynamic "ring fin", rocket-powered landing
- Capsule: parachute landing
- 1st crewed flight July 2021 to ~100 km

Video: Blue Origin Landing '15.mp4

Courtesy Blue Origin

8/22/22 2SC- 88

SpaceX Human-Rated Falcon 9 Launches Dragon to ISS (May 2020)

- Two U.S. astronauts launch on F9 in crew Dragon capsule from U.S. soil, dock with ISS
- 1st step of F9 recovered
- Reduced need for Russian Soyuz seats!

8/22/22 NASA photos 2SC- 89

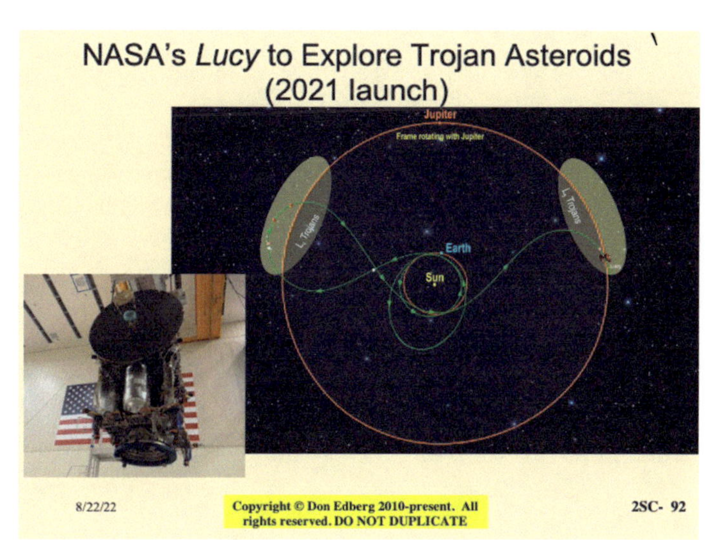

SpaceX *Starship* Lunar Lander (NASA awarded 2021)

Jeff Bezos attempts to buy Blue Origin's way back into "return to moon" by providing NASA $2B!

Tiangong Space Station: China, 2022?

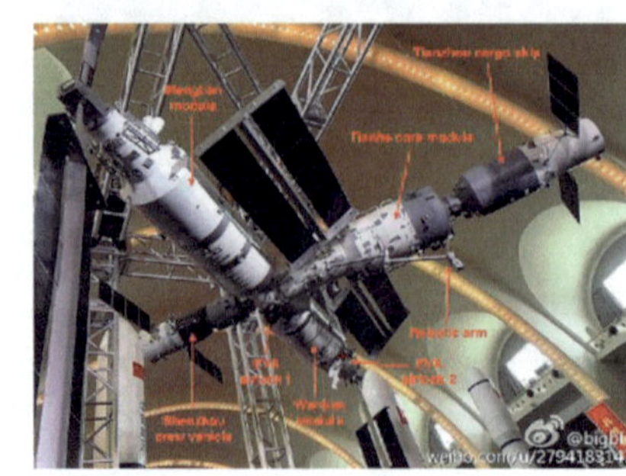

Courtesy astronautika.lt

- Core module *Tiahne* ("Harmony of the Heavens")
- 14-m experiment modules *Wentian* ("Quest for the Heavens") & *Mengtian* ("Dreaming of the Heavens") on either side
- 10-m robotic arm

NEA Scout (on 2022 SLS Launch)

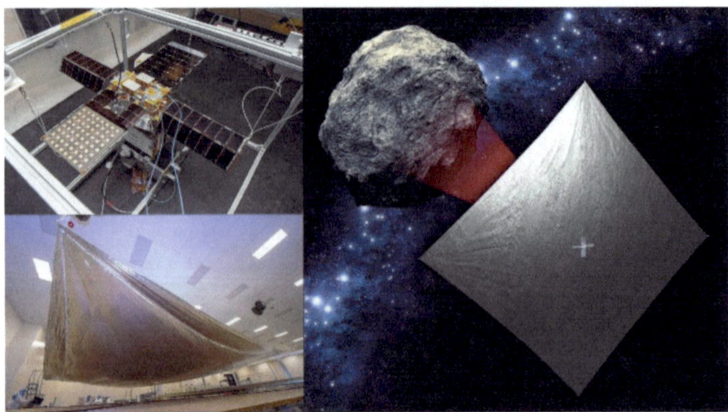

Courtesy JPL

Europa Clipper (2024 Launch F9H)

https://europa.nasa.gov/spacecraft/meet-europa-clipper/

Topics

- The Cold War
- The Moon Race
- Space stations, shuttles, & cooperation
- Planetary probes
- Earth orbiters: observation, spying, research
- Capsules & transfer vehicles
- Recent spacecraft developments
- **Launch vehicles**

Delta IV (Boeing, 2002 - present)

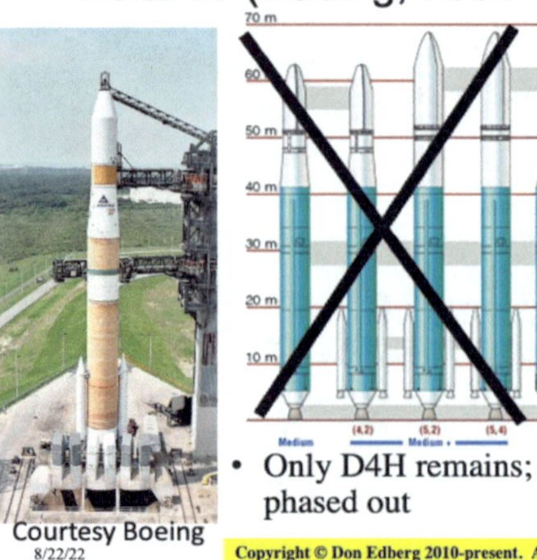

- Only D4H remains; others phased out

Courtesy Boeing

Atlas V (USA, 2002 - present)

- Capable of lifting varying payloads
- Zero to five strap-ons
- Russian engines on first stage creating controversy!

Courtesy Lockheed Martin

Soyuz (Russia, 1957-present)

Ariane 5 (European Consortium, 1996)

- Manufacturer: Arianespace
- Launches from French Guiana

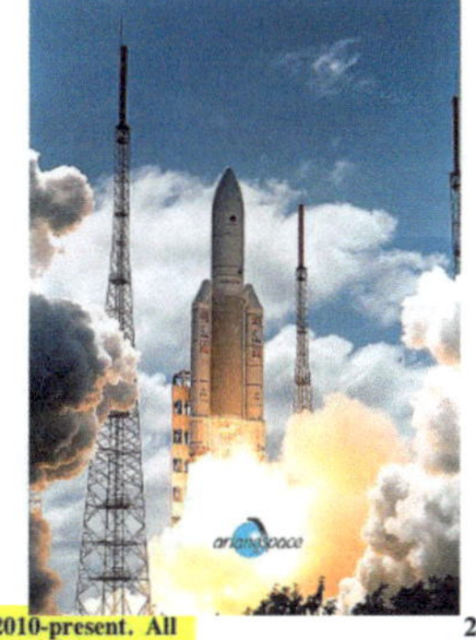

Long March Series (China)

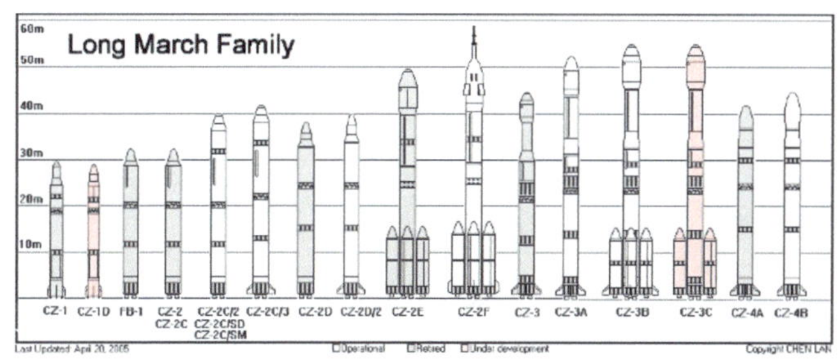

Courtesy Chen Lan

Pegasus (USA, 1989-present)

- Manufacturer: Orbital Sciences Corp
- Air launch
- 1000 lb_m to 100 nm Low Earth Orbit (LEO)

pegasus.mp4 Courtesy OSC

Proton (ILS/Russia, 1965-present)

Courtesy ILS

Minotaur Space Launch Vehicle (2000-present)

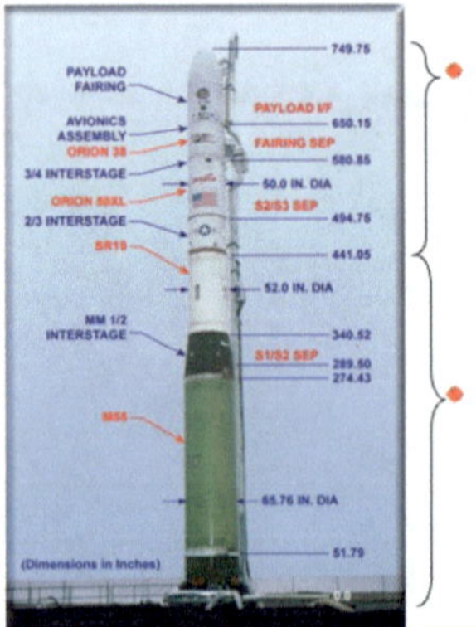

- Upper Stack Assembly (USA)
  - 50 In. Pegasus Payload Fairing
  - OSP-Standard Avionics
    - Inertial Guidance
    - Modular Avionics Components
  - Pegasus Avionics Structure and RCS
  - Orion-38 Insertion Stage 4
  - Orion-50XL Stage 3
  - Interstage
- Lower Stack Assembly (LSA)
  - Minuteman II Boosters: Stages 1 & 2
    - Solid Rocket Motors
    - Unmodified, GFE Systems

Courtesy Scott Schoneman UAz

8/22/22 2SC-107

Falcon 9 (2010)

http://spacex.com/multimedia/videos.php?id=36

D. Edberg photo (below)

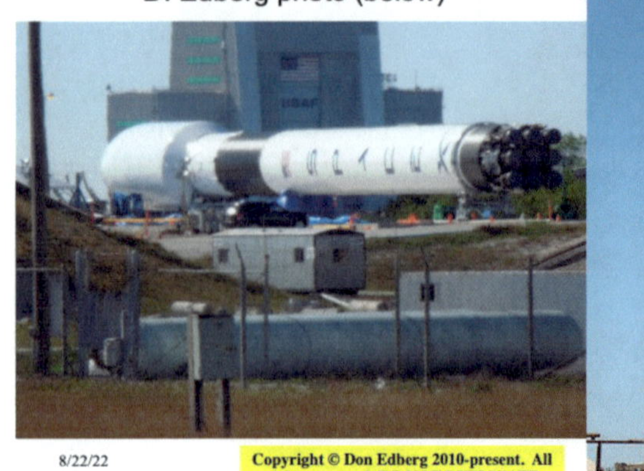

8/22/22 108

Arianespace *Vega* & *Vega-C* (2012 & 2018)

- 1st three steps solid, last step liquid propellant
  - Vega-C first step is SRM for Ariane 6
- 1.5T & 2.2T to 700 km polar orbit
- Launch site: Kourou

Courtesy Arianespace

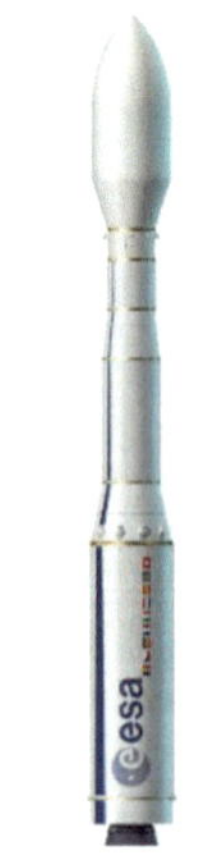

8/22/22 2SC-109

Falcon 9 Heavy (aka *Falcon 9H*, F9H)

- Announced April 2011
- Two F9 boosters strapped to F9 core
- 53 T to LEO
- First flight ~~2015 2016 2017~~ 2018
- Originally 'crossfed,' now dropped

8/22/22 Courtesy SpaceX Photos 2SC-111

Electron: Rocket Lab (2018)

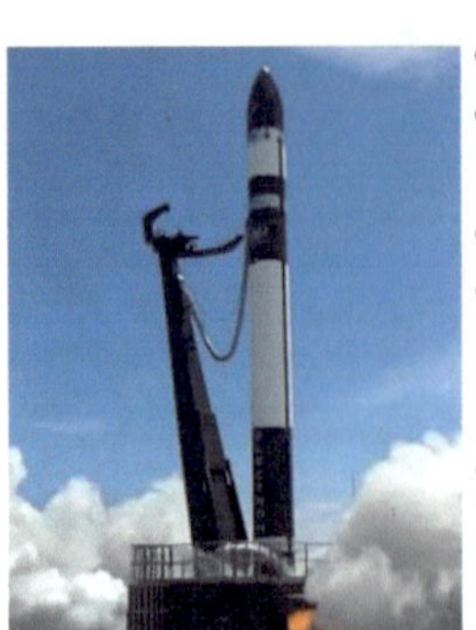

- 150 kg to 500 km orbit, m_0 13 T
- 17 m tall, 1.3 m dia. carbon all composite structure
- 2 step, $5M launch costs
- Electric pump-fed LOx/RP-1 *Rutherford* engines, 4900 lb$_f$ each, 9 + 1 on 1st + 2nd steps
- Launch: Mahia Peninsula, NZ; Wallops Flt. Facility, VA

Courtesy: website, spacelaunchreport

8/22/22 2SC-112

Virgin Orbit / VOX Space *LauncherOne* 747 Air-Launched System (2020)

- All composite
- Drop from B747
- 73.5 klb$_f$ LOx/RP NewtonThree 1st stage; 5 klb$_f$ NewtonFour LOx/RP 2nd stage
- 200 kg to LEO, $10M
- Awarded 39 of 60 launches for OneWeb

8/22/22 Courtesy Virgin Orbit, voxspace.com 2SC-113

Astra: *LV0007* Reaches Orbit 11/2021

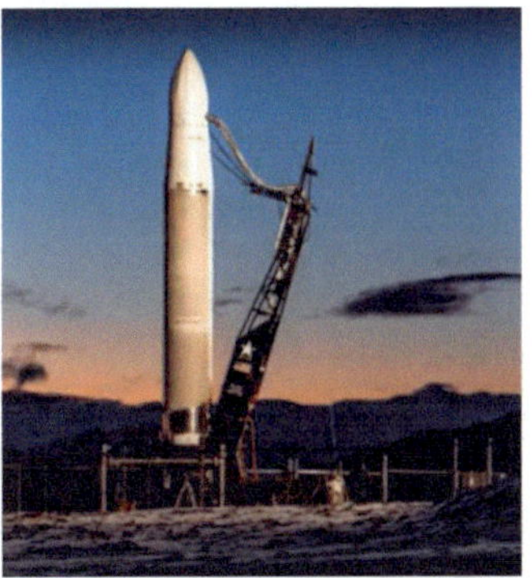
Courtesy: Scott Manley

- 2-step kerolox
- Kodiak, AK launch site
- Achieved 310 mile orbit
- Offers dedicated launch services for 50 – 150 kg payloads to 500 km SSO
- (LV0006 had "interesting" launch failure 08/2021! See "Failures" Lecture 17 for more info)

8/22/22 2SC-114

Topics

- The Cold War
- The Moon Race
- Space stations, shuttles, & cooperation
- Planetary probes
- Earth orbiters: observation, spying, research
- Capsules & transfer vehicles
- Recent spacecraft developments
- Launch vehicles

8/22/22 2SC-115

The Future?

SpaceX

8/22/22 2SC-116

NASA *Space Launch System, SLS*
(announced 09/2011, 1st flight ~~2017~~ 202x)

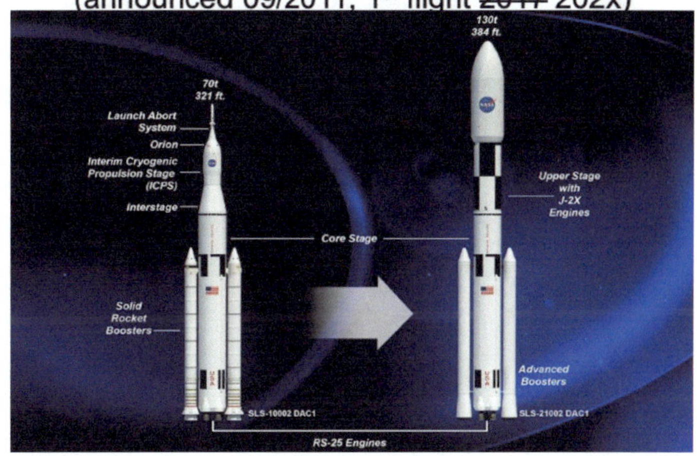

SpaceX *Starship-Superheavy* (2021)

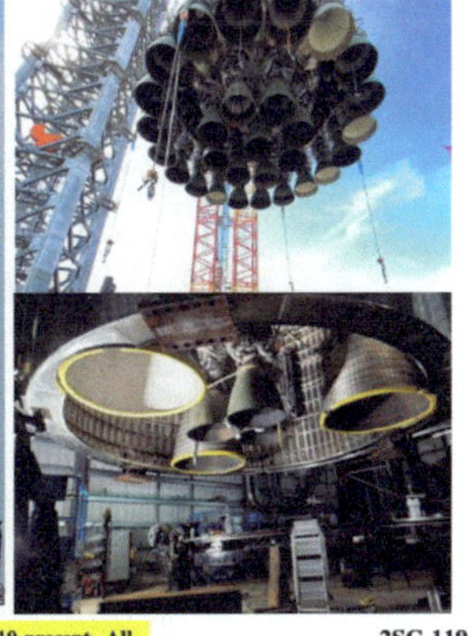
Courtesy twitter

8/22/22 2SC-119

ULA Announces *Vulcan* (2015)

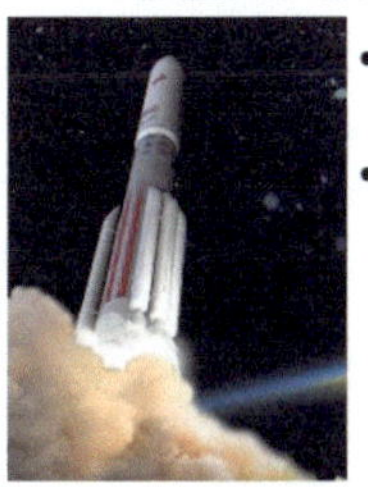

- Single booster stage with SRBs, high-energy Centaur 2nd stage, 4- or 5-meter-dia. fairing
- U.S.-made *BE-4* rocket engine uses low-cost liquid natural gas (Methane, LCH_4) fuel

- Initial launch capable ~~2019~~ ~~2021~~ 2022?
- "More affordable" launches
- "Sensible, Modular, Autonomous Return Technology" (SMART) reuses booster engines via ocean splashdown ~~mid-air capture~~

8/22/22 2SC-120

Ariane 62 & *64* (1st flight 2022?)

- *Ariane 62* has two P120 SRMs; *A64* has four for double launch up to 4.5 – 5 T
- LOx/LH_2 lower step based on *Ariane 5 ECA* and *ME*, *Vulcain* engine
- LOx/LH_2 upper step based on the *A5ME* upper step propelled by a *Vinci* engine
- Courtesy ESA

121

8/22/22

DreamChaser, 201x (Sierra Nevada)

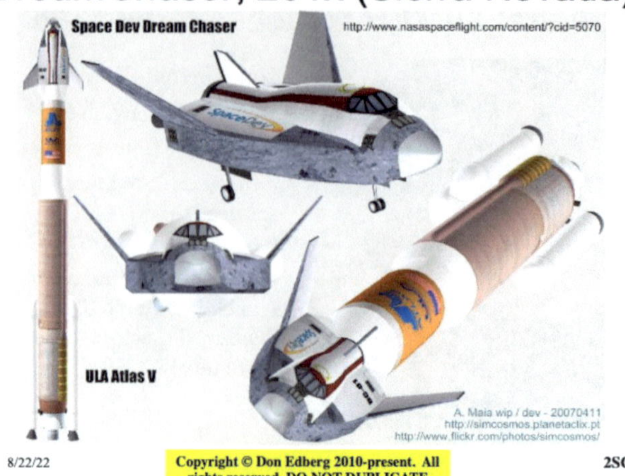

8/22/22

2SC-122

The End … For Now!

8/22/22

2SC-124

Pre-Launch & Launch Environment

8/22/22

Courtesy: KTLA news, S. Edberg, NASA

Topics

- Transportation & launch loads cases
  - Mechanical transportation loads: air, ship, rail; typical load factors
- Anatomy of a launch vehicle (LV)
- Launch trajectory phases
- Flight environment (mechanical & acoustic)
- Load factors
- Flight events: mechanical & acoustic loads: ignition, liftoff, max-q, staging/separation, pogo, resonant burn oscillation
- Separation systems: pyro & non-pyro
- Spacecraft flight loads environment: quasi-static, vibration, shock, acoustic, thermal
  - Payload isolation

8/22/22

Course Layout

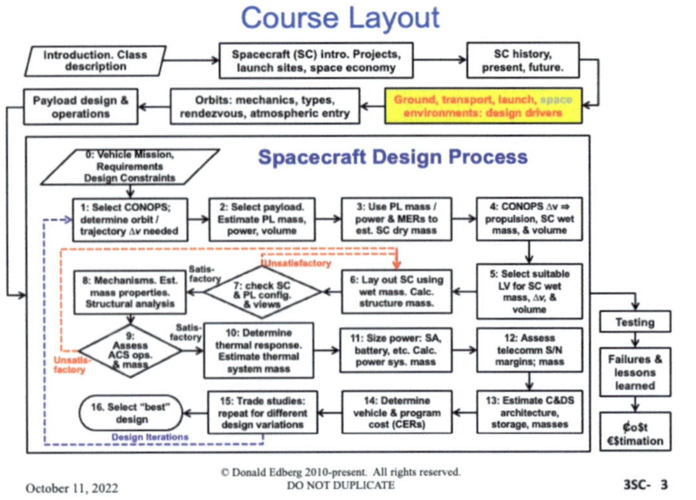
Introduction. Class description
Spacecraft (SC) intro. Projects, launch sites, space economy
SC history, present, future.
Payload design & operations
Orbits: mechanics, types, rendezvous, atmospheric entry
Ground, transport, launch, space environments: design drivers
Spacecraft Design Process
0: Vehicle Mission, Requirements Design Constraints
1: Select CONOPS; determine orbit / trajectory Δv needed
2: Select payload. Estimate PL mass, power, volume
3: Use PL mass / power & MERs to est. SC dry mass
4: CONOPS Δv ⇒ propulsion, SC wet mass, & volume
5: Select suitable LV for SC wet mass, Δv, & volume
6: Lay out SC using wet mass. Calc. structure mass.
7: check SC & PL config. & views
8: Mechanisms. Est. mass properties. Structural analysis
9: Assess ACS ops. & mass
10: Determine thermal response. Estimate thermal system mass
11: Size power: SA, battery, etc. Calc. power sys. mass
12: Assess telecomm S/N margins; mass
13: Estimate C&DS architecture, storage, masses
14: Determine vehicle & program cost (CERs)
15: Trade studies: repeat for different design variations
16. Select "best" design
Satis-factory
Unsatisfactory
Unsatis-factory
Design Iterations
Testing
Failures & lessons learned
¢o$t €$timation

Topics

- Transportation & launch loads cases
  - Mechanical transportation loads: air, ship, rail; typical load factors
- Anatomy of a launch vehicle (LV)
- Launch trajectory phases
- Flight environment (mechanical & acoustic)
- Load factors
- Flight events: mechanical & acoustic loads: ignition, liftoff, max-q, staging/separation, pogo, resonant burn oscillation
- Separation systems: pyro & non-pyro
- Spacecraft flight loads environment: quasi-static, vibration, shock, acoustic, thermal
  - Payload isolation

Mechanical Loads Cases

Loads need to be analyzed for many cases:

Load Description	Vehicle Condition
Truck, train, air, sea; handling & stacking	Transportation, Assembly
On-launch pad thermal conditions	air-conditioning
Ignition Release jerk (vehicle acceleration axial loads)	Mechanical (axial acceleration) & acoustic
Max-q (due to loads from dynamic pressure) Max-$q\alpha$ (angle of attack & lateral loads, usually combined lateral & axial loads)	Axial + lateral acceleration & acoustic
Pogo and/or slosh or "Tail Wags Dog" (liquid) or resonant burn (solid)	Axial + lateral acceleration
Shutdown/MECO (deceleration axial loads, lateral loads if multi-engine); upper-stage startups and shutdowns	Axial + lateral acceleration
Separation / staging loads; payload fairing separation, payload separation	Pyro or separation shock

Loads Cases

Loads need to be analyzed for many cases:

Load Description	Condition
Transportation (truck, train, air, handling)	Transportation to integrator & launch site
Liftoff: Ignition (mechanical & acoustic) Liftoff: release jerk (vehicle acceleration axial loads) Max-q & Max-$q\alpha$ (angle of attack & lateral loads from crosswind shear, usually combined lateral & axial loads) Vibration: Pogo (liquid) or resonant burn (solid) Shutdown (vehicle acceleration axial loads, potential lateral loads if multi-engine) Separation & staging (pyro shock loads) Second-stage burns & shutdowns Payload fairing separation (pyro shock loads) Payload separation (pyro shock loads) Acoustic, thermal effects during all	Liftoff & Flight Loads Separation: Pyrotechnic shock Spin Stabilization: Upper stage Heat Shield Separation: Mechanical shock, Pyrotechnic shock Staging: Mechanical shock, Vibration, Sustained acceleration Max Winds: S/C/Heat shield relative deflection, Mechanical vibration Liftoff: Acoustic, Ignition transients

Handling, Packing, Unpacking

10

CALIPSO Unpacking at Vandenberg Air Force Base

Courtesy NASA

Truck Transport

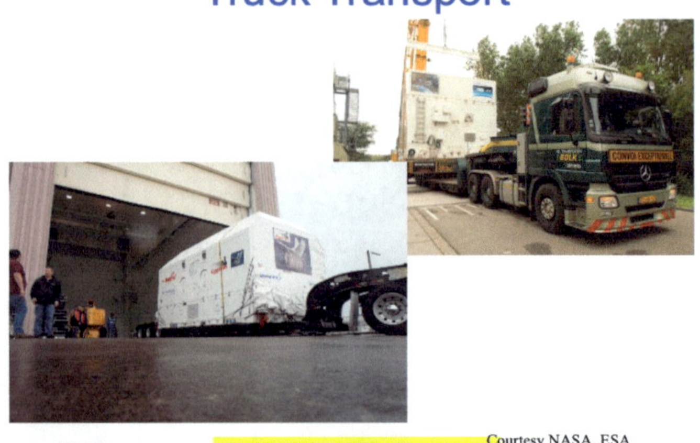

Courtesy NASA, ESA

Air Transport

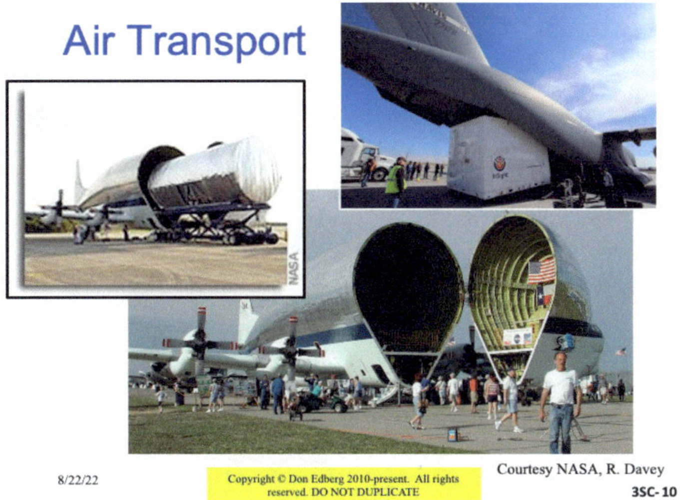

Courtesy NASA, R. Davey

Sea Transport (courtesy NASA)

Train / Railroad Loads

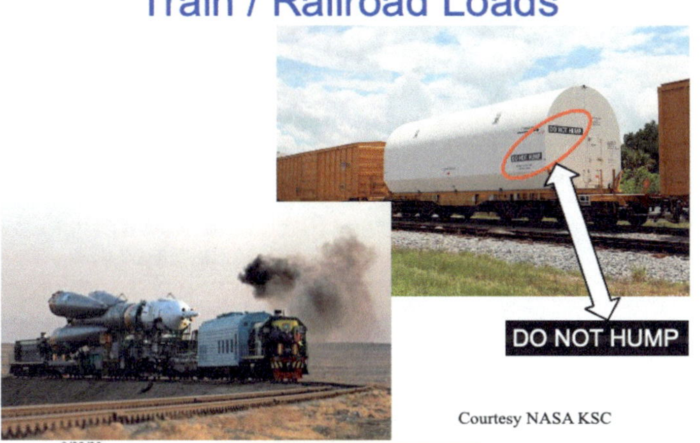

Courtesy NASA KSC

Transportation & Handling Loads

Spacecraft must be transported to launch site somehow
Truck, rail, ship, or air
All have associated loadings

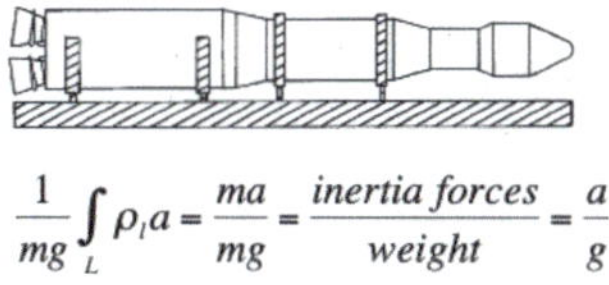

$$\frac{1}{mg}\int_L \rho_l a = \frac{ma}{mg} = \frac{inertia\ forces}{weight} = \frac{a}{g}$$

Courtesy NASA KSC

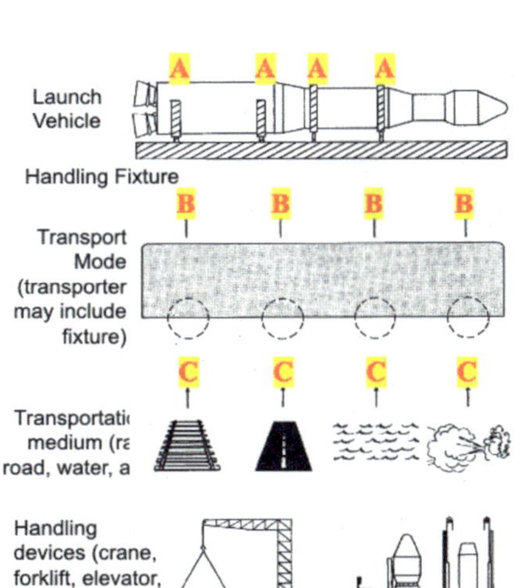

Transportation & Handling Loads

"A" = loads input *to* a LV *from* vehicle's handling fixture

"B" shows loads being applied *from* transport vehicle *to* handling fixture

"C" indicates different forcing functions provided by different transportation modes: rail, road, water, air

"D" indicates inputs from handling devices, such as cranes, forklifts, elevators, etc.

Courtesy NASA

Typical Transportation Loads

Operation		Applied Accel.(x-, y-, z-axes, g)			Remarks
Clean Room Handling	Dolly	±1.0	±0.75	-1 ± 0.5	Any orientation
	Movement	± 0.2	± 0.2	-1 ± 0.5	
	Vertical hoist	± 0.2	± 0.2	-1 ± 0.5	
	LV mate/de-mate	± 0.5	± 0.5	-2 / +0	
Container	Hoisting	± 0.5	± 0.5	-1 ± 0.5	SC horizontal
Road	Quasi-static	± 2	± 2	-3 / +1	40 km/h top speed
Air	Take-off	- 1.5	± 0.1	-2.5 / +1.5	
	Vertical gusts	0	± 1.5	-2.5	
	Lateral gusts	0	± 1.5	-1.0	
	Landing	± 1.5	± 1.5	-2.5	
Barge/ship	Slamming	0.0	0.0	-1.8 / +0.2	
	Waves	± 0.3	± 0.5	-1.6 / +0.4	
Any transportation	Continuous vibration	± 0.1	± 0.1	± 0.1	Below 10 Hz, not including gravity
Transport	Shock load	± 2	± 2	± 3	

Galileo (GLL) Spacecraft (1989-2003)

HIGH-GAIN ANTENNA
LOW-GAIN ANTENNA
SUN SHIELDS
EXTREME ULTRAVIOLET SPECTROMETER
STAR SCANNER
THRUSTERS (2 places)
PROBE RELAY ANTENNA
JUPITER ATMOSPHERIC PROBE
RADIOISOTOPE THERMOELECTRIC GENERATORS (RTG) (2 places)
Engineering
Fields and Particles
Probe
Remote Sensing
PLASMA-WAVE ANTENNA
MAGNETOMETER SENSORS
ENERGETIC PARTICLES DETECTOR
PLASMA SCIENCE
HEAVY ION COUNTER (BACK)
DUST DETECTOR
RETROPROPULSION MODULE
ABOVE: SPUN SECTION
BELOW: DESPUN SECTION
RTG
SCAN PLATFORM, CONTAINING:
- ULTRAVIOLET SPECTROMETER
- SOLID-STATE IMAGING CAMERA
- NEAR-INFRARED MAPPING SPECTROMETER
- PHOTOPOLARIMETER RADIOMETER

Courtesy JPL

Galileo at Jupiter Planned vs. Actual (Jammed Antenna)!

Courtesy JPL

Galileo Antenna Failure Blamed on Transportation Loads

- GLL made *three* trips from California to KSC before launch
- Transportation loads believed to have caused GLL HGA (high gain antenna) to jam from removal of lubrication due to vibration during transport
- Solution: no more deployable antennae on deep-space missions!

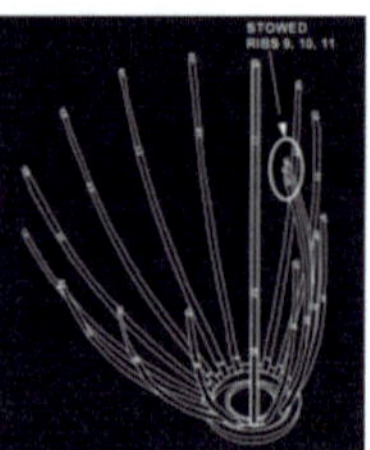

Galileo jammed antenna figure courtesy JPL

Topics

- Transportation & launch loads cases
  - Mechanical transportation loads: air, ship, rail; typical load factors
- Anatomy of a launch vehicle (LV)
- Launch trajectory phases
- Flight environment (mechanical & acoustic)
- Load factors
- Flight events: mechanical & acoustic loads: ignition, liftoff, max-q, staging/separation, pogo, resonant burn oscillation
- Separation systems: pyro & non-pyro
- Spacecraft flight loads environment: quasi-static, vibration, shock, acoustic, thermal
  - Payload isolation

Delta II USA (1962-present)

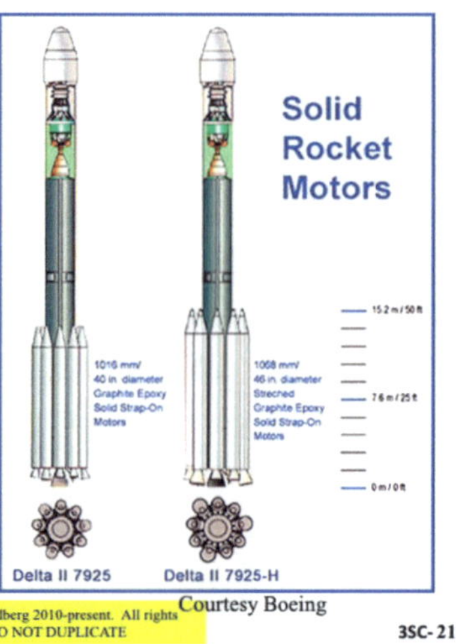

Courtesy Boeing

Delta II "Exploded" View

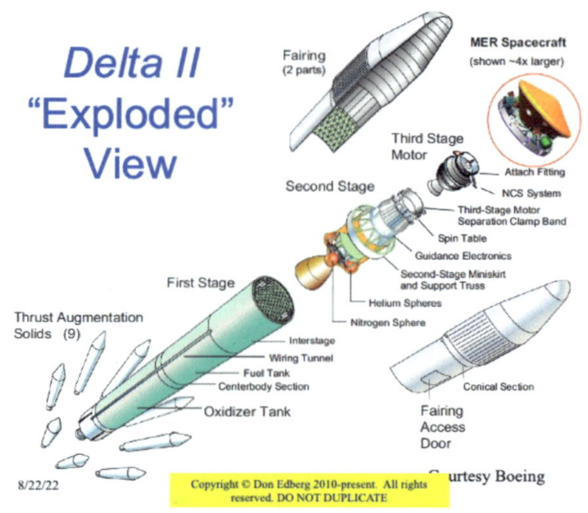

Courtesy Boeing

Delta II First Stage

Interstage
Wiring Tunnel
Fuel Tank
Centerbody Section
Oxidizer Tank
Solid Rocket Motors (9)

Courtesy Boeing

8/22/22

3SC- 23

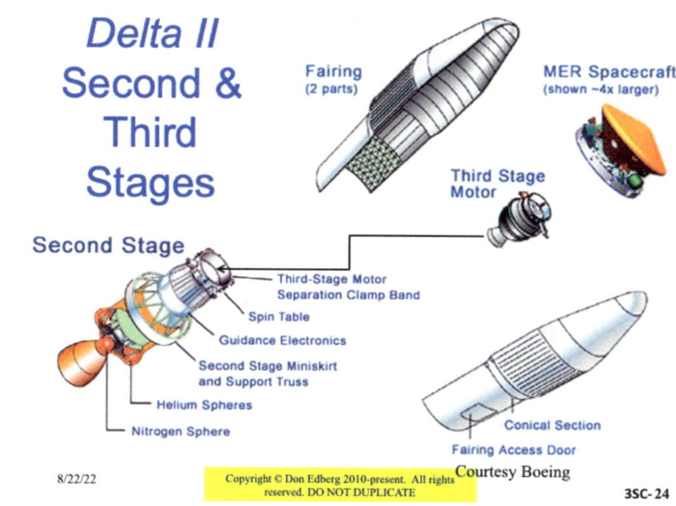

Delta II
Second &
Third
Stages
Fairing
(2 parts)
MER Spacecraft
(shown ~4x larger)
Third Stage
Motor
Second Stage
Third-Stage Motor
Separation Clamp Band
Spin Table
Guidance Electronics
Second Stage Miniskirt
and Support Truss
Helium Spheres
Nitrogen Sphere
Conical Section
Fairing Access Door
Courtesy Boeing
8/22/22
Copyright © Don Edberg 2010-present. All rights reserved. DO NOT DUPLICATE
3SC- 24

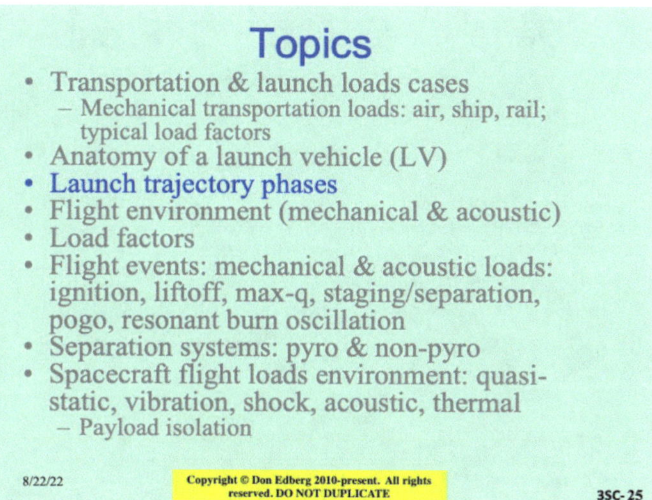

Topics
Transportation & launch loads cases
– Mechanical transportation loads: air, ship, rail; typical load factors
Anatomy of a launch vehicle (LV)
Launch trajectory phases
Flight environment (mechanical & acoustic)
Load factors
Flight events: mechanical & acoustic loads: ignition, liftoff, max-q, staging/separation, pogo, resonant burn oscillation
Separation systems: pyro & non-pyro
Spacecraft flight loads environment: quasi-static, vibration, shock, acoustic, thermal
– Payload isolation
8/22/22
Copyright © Don Edberg 2010-present. All rights reserved. DO NOT DUPLICATE
3SC- 25

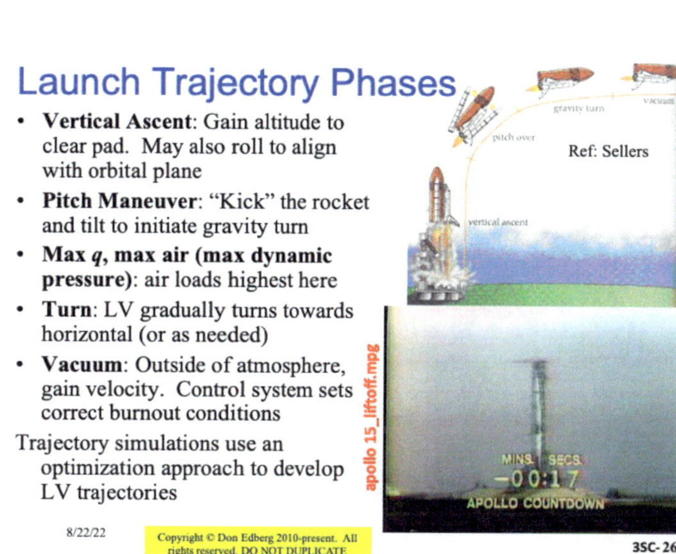

Launch Trajectory Phases
Vertical Ascent: Gain altitude to clear pad. May also roll to align with orbital plane
Pitch Maneuver: "Kick" the rocket and tilt to initiate gravity turn
Max q, max air (max dynamic pressure): air loads highest here
Turn: LV gradually turns towards horizontal (or as needed)
Vacuum: Outside of atmosphere, gain velocity. Control system sets correct burnout conditions
Trajectory simulations use an optimization approach to develop LV trajectories
Ref: Sellers
apollo 15_liftoff.mpg
MINS SECS
APOLLO COUNTDOWN
8/22/22
Copyright © Don Edberg 2010-present. All rights reserved. DO NOT DUPLICATE
3SC- 26

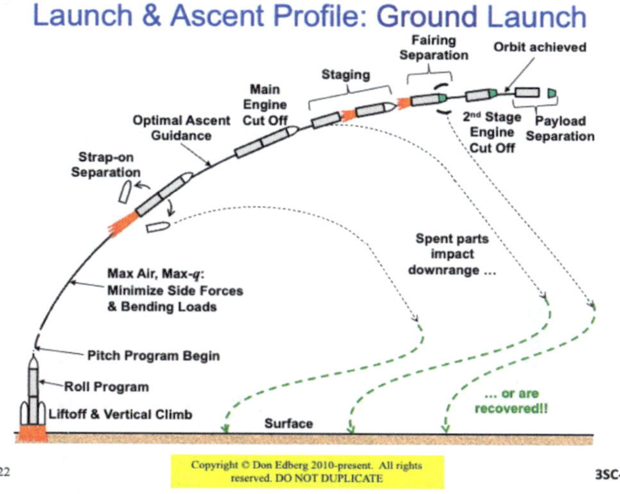

Launch & Ascent Profile: Ground Launch
Fairing Separation
Orbit achieved
Staging
Main Engine Cut Off
Optimal Ascent Guidance
2nd Stage Engine Cut Off
Payload Separation
Strap-on Separation
Spent parts impact downrange ...
Max Air, Max-q: Minimize Side Forces & Bending Loads
Pitch Program Begin
Roll Program
Liftoff & Vertical Climb
Surface
... or are recovered!!

8/22/22

3SC- 27

Launch & Ascent Profile: Air Launch

Orbit Achieved
Release from Carrier Vehicle
First Stage Ignition
Pitch-Up for Ascent
First Stage Burnout
Staging
Fairing Separation
Upper Stage Engine Cut Off
Payload Separation
Spent Stage(s) Impact Downrange
Ground

8/22/22

3SC- 28

Typical Low Earth Orbit Speeds (approx.)

$v_{orbit\ LEO}$	~ 7,900 m/s	~ 17,700 MPH
$\Delta v_{gravity\ loss}$	~1,200 – 1,500 m/s	~ 2,680 – 3,360 MPH
$\Delta v_{drag\ loss}$	180 – 360 m/s	400 – 800 MPH
$\Delta v_{steering\ loss}$	100 – 300 m/s	220 – 660 MPH
Total Δv *required*	~ 9.4 – 9.5 km/s	~ 21,000 – 22,500 MPH
Earth rotation	(350 – 410 m/s)	(780 – 920 MPH)

8/22/22 3SC-29

LV Determines Payload

- Maximum payload (spacecraft mass) determined by velocity/orbit requirements
- Interplanetary missions also have energy/time-of-flight/arrival speed constraints

Determine LV capabilities from:

1. LV Payload Planner's/User's Guides
2. Web sites such as NASA Launch services: http://elvperf.ksc.nasa.gov/elvMap/elvMap.ui.PerfGraph0?ReqType=Graph&OrbitType=C3&Contract=2&Vehicles=4

8/22/22 3SC-30

NASA LV Performance Curves

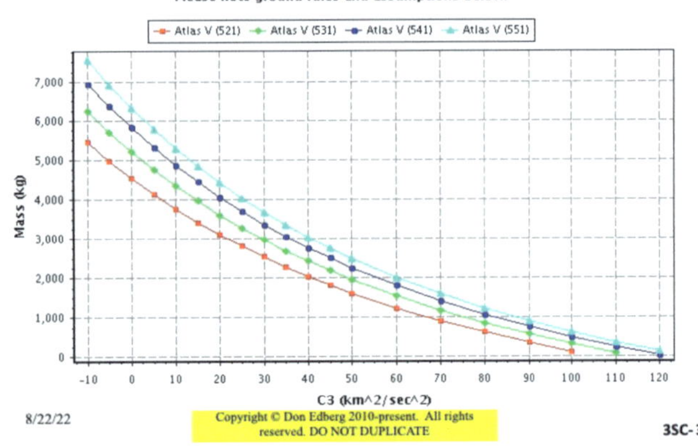

8/22/22 3SC-31

Delta II-MER Launch Profile

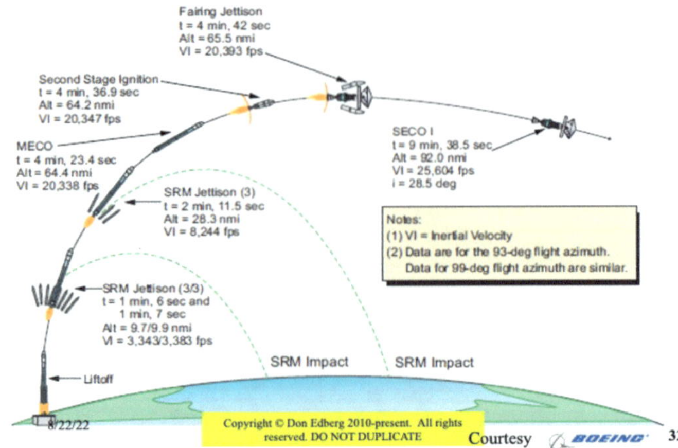

8/22/22 Courtesy BOEING 32

Launch Movie: *MER/Delta II*

MER-mars-HD 6m32s.mp4

8/22/22 3SC-33

Delta II Launch Movie

Delta-II_mars-odyssey-launch.mp4

courtesy Space.com

8/22/22 3SC-34

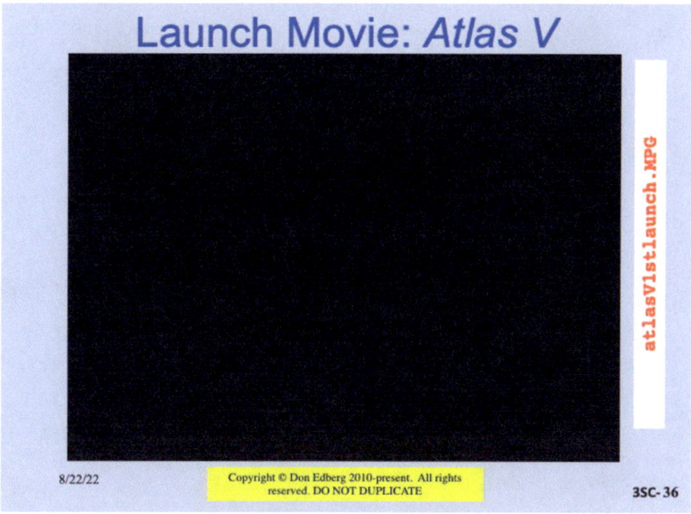

LEASAT Mission Profile

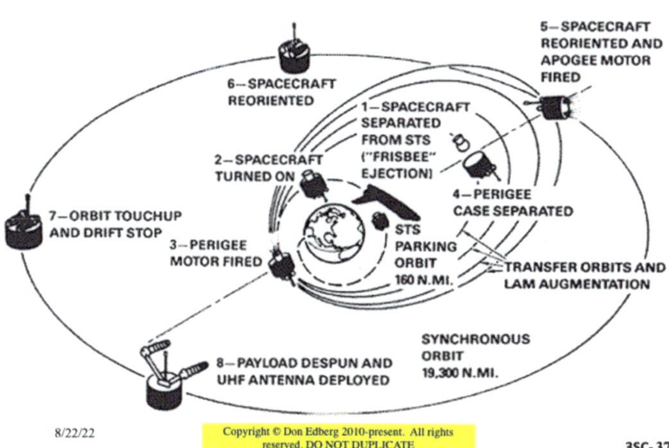

8/22/22 3SC-37

Topics

- Transportation & launch loads cases
  - Mechanical transportation loads: air, ship, rail; typical load factors
- Anatomy of a launch vehicle (LV)
- Launch trajectory phases
- Flight environment (mechanical & acoustic)
- Load factors
- Flight events: mechanical & acoustic loads: ignition, liftoff, max-q, staging/separation, pogo, resonant burn oscillation
- Separation systems: pyro & non-pyro
- Spacecraft flight loads environment: quasi-static, vibration, shock, acoustic, thermal
  - Payload isolation

8/22/22 3SC-38

Loads Imposed on the Spacecraft

- Quasi-static loads from acceleration
- Transient or vibratory loads due to LV maneuvering, winds & aero loads, separations
- Acoustic loads
- Thermal loads

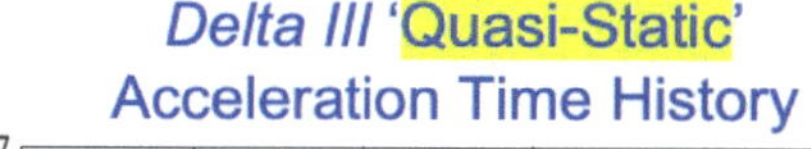

Even though the launch only lasts a few minutes and missions last for years, launch loads can be the design drivers for many spacecraft components.

8/22/22 3SC-39

Delta III Acceleration Time History

(with fictitious vibration – shaded areas near lines)

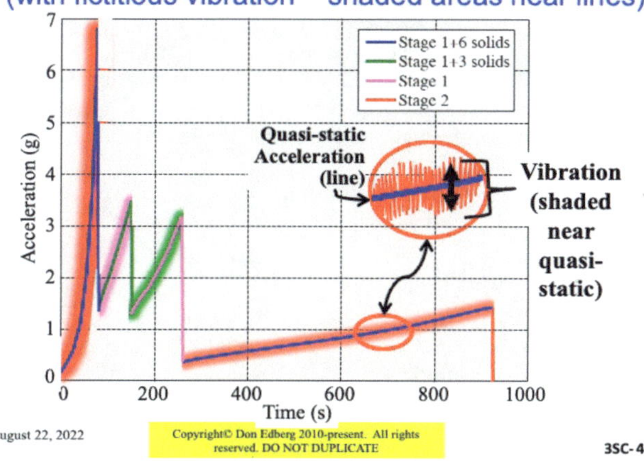

August 22, 2022 3SC-40

Delta III 'Quasi-Static' Acceleration Time History

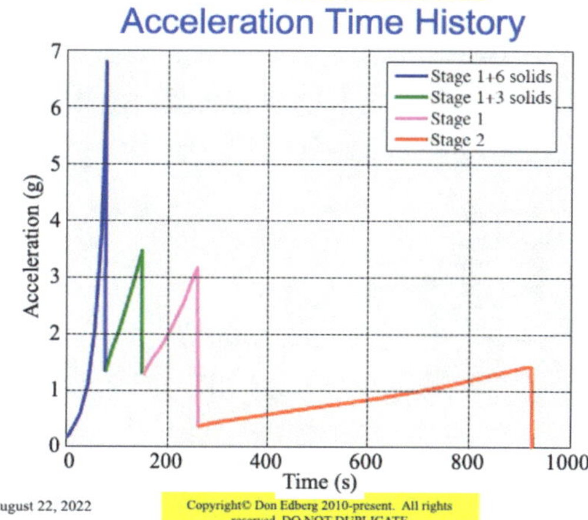

August 22, 2022 3SC-41

Delta II Flight Profile & 'AC' Excitation

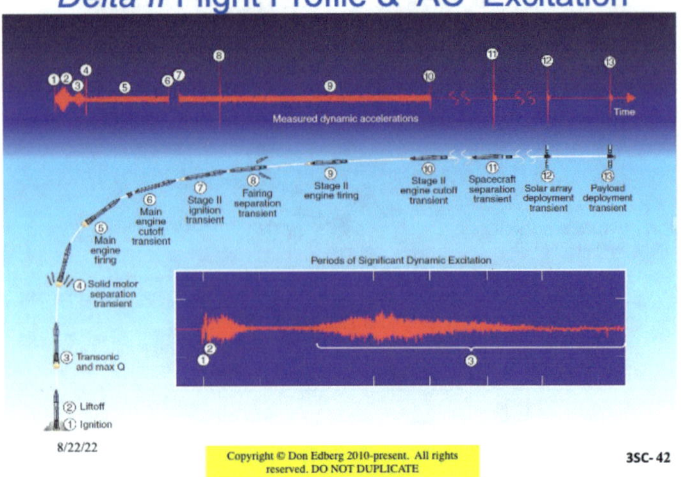

8/22/22

3SC-42

Topics

- Transportation & launch loads cases
  - Mechanical transportation loads: air, ship, rail; typical load factors
- Anatomy of a launch vehicle (LV)
- Launch trajectory phases
- Flight environment (mechanical & acoustic)
- **Load factors**
- Flight events: mechanical & acoustic loads: ignition, liftoff, max-q, staging/separation, pogo, resonant burn oscillation
- Separation systems: pyro & non-pyro
- Spacecraft flight loads environment: quasi-static, vibration, shock, acoustic, thermal
  - Payload isolation

8/22/22

3SC-44

Load Factor *n*

x

Inertial force $-F_x$

Weight $W = mg_0$

Thrust, T

- The *load factor n* represents "inertial force" in units of g_0
- Here g_0 = 1 Earth standard gravity $g_0 = 9.80665$ m/s² = 32.174 ft/s²
- *No-gravity* acceleration = T/m
- For a given acceleration, load factor is opposite sign

$$n_x = -\frac{T/m}{g_0}$$

n_x = no-grav. accel. in g_0 units ("gs" or "gees")

8/22/22

3SC-45

Rigid-Body Loads Calculation Basis: $F = ma$ or $F - ma = 0$

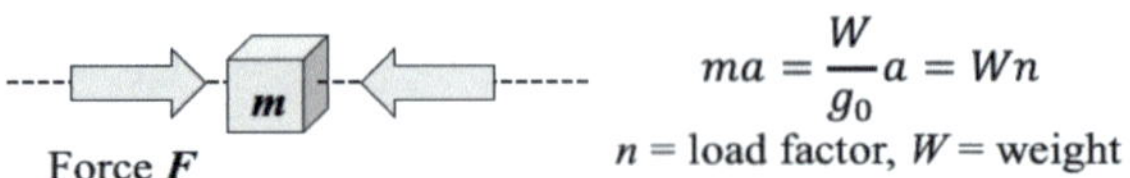

$$ma = \frac{W}{g_0}a = Wn$$

n = load factor, W = weight

Internal masses (i.e. filled propellant tanks, boxes, payload, etc.)

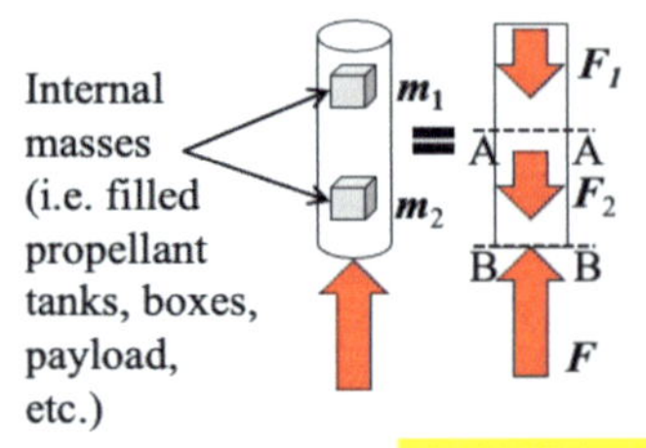

Sum of forces = zero (+ upwards):
$(+)F_B - F_1 - F_2 = 0$

Load above A-A = $F_1 = ngm_1$
Load above B-B is $F_B = F_1 + F_2 = ng_0(m_1 + m_2)$
(total of loads above B-B)

8/22/22

3SC-46

Lateral & Axial Load Definitions

Primary loads are:
AXIAL (THRUST direction) and
LATERAL (SIDE-to-SIDE direction)

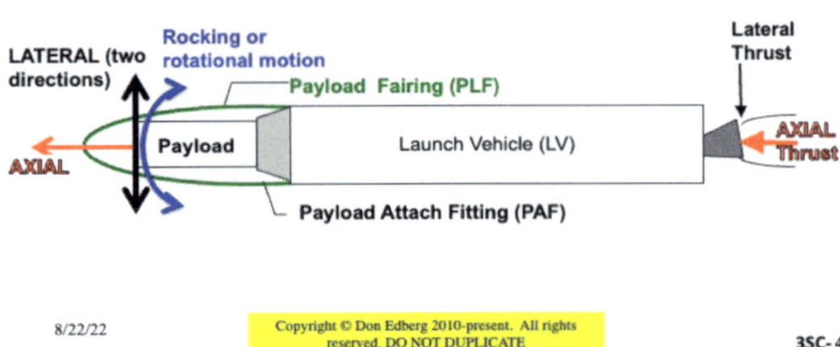

8/22/22

3SC-47

Sources of Flight Loads

- Thrust forces - acceleration
- Air loads (dynamic pressure $q = ½\rho v^2$)
- Control forces & torques
- Winds (produce angle of attack α, bending moment)
- Vibrations, slosh, flex motion
- Thermal
- Acoustic

Design drivers usually:

1. Max-q
2. Max-$q\alpha$
3. Max acceleration
4. MECO (Main Engine Cutoff)

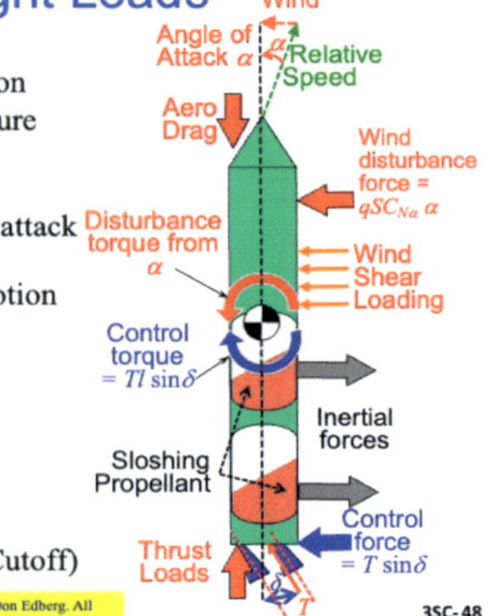

8/22/22

3SC-48

Acoustic Environment

- Rockets produce *huge* noise:
  - Engine ignition pressure pulse
  - Noise can reflect off of ground near liftoff
  - "Boundary layer" and buffeting excites panels during climb

Worst acoustic load periods:
- Ignition & liftoff
- Max-*q*

August 22, 2022

3SC- 49

Measuring Acoustic Loads

- Sound is an oscillating pressure in air
- Acoustic loads measured in logarithmic scale with units of dB or decibels
- Sound pressure level, SPL, defined *logarithmically*: SPL (dB) = $20 \log_{10} (P/P_{ref})$ decibels
- $P_{ref} = 2.9 \times 10^{-9}$ psi $= 2 \times 10^{-5}$ Pa = human hearing threshold
- NOTE: 6 dB = a factor of 2; +6 dB = doubling, –6 dB = halving

Know which noises can cause damage. Wear hearing protection when you are involved in a loud activity.

- **85 dB(A)**
  Regular and prolonged exposures to noise at or above 85 dB(A) (averaged over 8 hours per day) are considered hazardous.
- **100 dB(A)**
  Regular and prolonged unprotected exposure of more than 15 minute per day risks permanent hearing loss.
- **110 dB(A)**
  Regular and prolonged unprotected exposure of more than 1.5 minutes per day risks permanent hearing loss.

Examples of noise levels

- **194 dB** Loudest possible tone
- **180 dB** Rocket launch
- **165 dB** 12-gauge shotgun
- **140 dB** Jet engine at takeoff
- **120 dB** Ambulance siren
- **119 dB** Pneumatic percussion drill
- **114 dB** Hammer drill
- **108 dB** Chain saw
- **108 dB** Continuous miner
- **105 dB** Bulldozer, spray painter
- **103 dB** Impact wrench
- **98 dB** Hand drill
- **96 dB** Tractor
- **93 dB** Belt sander
- **90 dB** Hair dryer/power lawn mower
- **80 dB** Ringing telephone
- **60 dB** Normal conversation

August 22, 2022

Courtesy NIOSH

3SC-50

Topics

- Transportation & launch loads cases
  - Mechanical transportation loads: air, ship, rail; typical load factors
- Anatomy of a launch vehicle (LV)
- Launch trajectory phases
- Flight environment (mechanical & acoustic)
- Load factors
- **Flight events: mechanical & acoustic loads: ignition, liftoff, max-q, staging/separation, pogo, resonant burn oscillation**
- Separation systems: pyro & non-pyro
- Spacecraft flight loads environment: quasi-static, vibration, shock, acoustic, thermal
  - Payload isolation

8/22/22

3SC- 51

Delta II Flight Profile & 'AC' Excitation

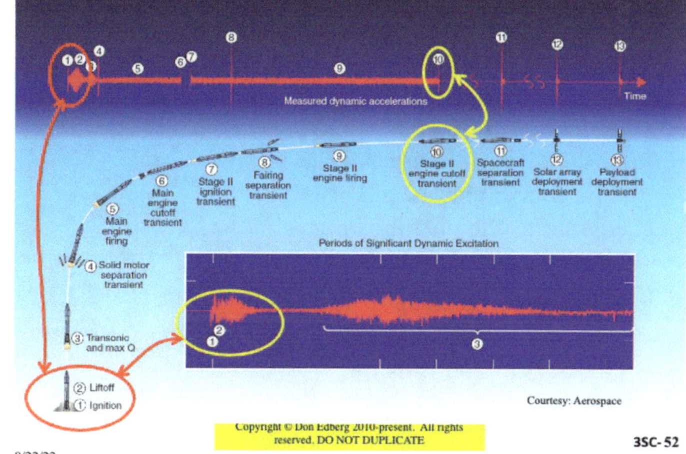

8/22/22

3SC- 52

Engine Startups & Cutoffs

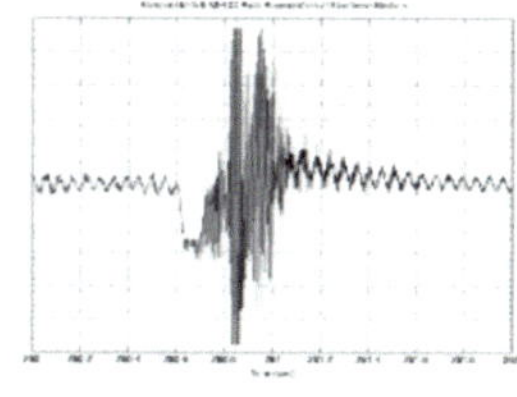

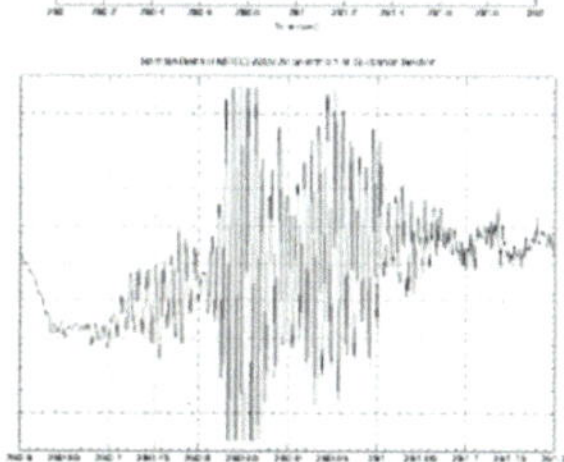

DII Cutoff Transient Courtesy NASA/LeRC

- Here acceleration suddenly changes from positive to ~ zero or vice versa
- Structures suddenly loaded or unloaded
- Multiple engines may not stop simultaneously, providing lateral loads

8/22/22

3SC- 53

Liftoff Loads

- Want smooth transition from zero velocity to begin vertical motion
- "Jerk" refers to time rate of change of acceleration [jerk = d/dt(acceleration) = $d/dt(dv/dt) = d^3z/dt^3$]
- Want jerk = 0 if possible
- LVs use mechanisms to smooth liftoff motion (e.g. slow release)

8/22/22

3SC- 54

Shutdown Loads

5
0
−5
−10
Acceleration (g s)
Transient Event
0 0.1 0.2 0.3 0.4 0.5 0.6 0.7 0.8 0.9 1
Time (Seconds)

Courtesy: CSA Engr.

8/22/22

Startup & Shutdown Loads Video (Loads shown AKA "rail humping")

Clip from movie "Dumbo" from Disney (dumbo.mp4)

8/22/22

Rocket Engine Ignition Overpressure

- Ignition overpressure (IOP) is a significant transient low-frequency pressure event caused by the rapid pressure rise rate of solid rocket motors and/or liquid engines as they begin firing at ignition
- Estimate or scale-up from model tests until actuals are obtained (5% SLS model shown)

Courtesy NASA

August 22, 2022

SSME IOP Environment Update

Presenter: M. Dunham
Date: Oct 27, 2009
Page: 76

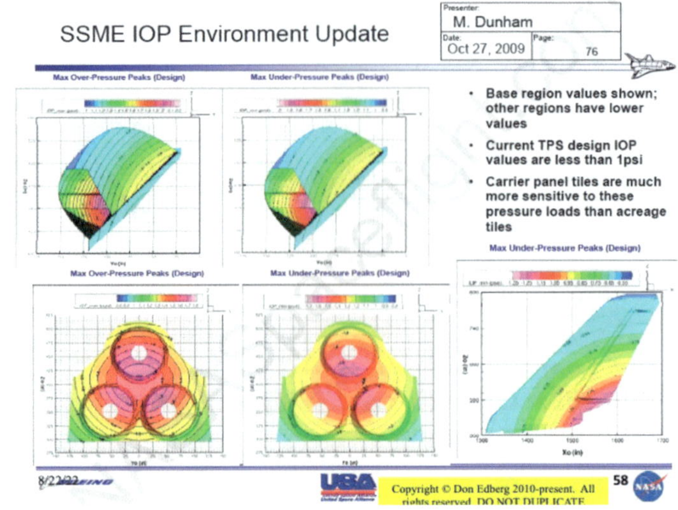

8/22/22

Ground Acoustic Environment

Liftoff

- Overpressure during ignition
- During liftoff, sound reflects from pad & ground
- Water spray suppresses noise intensity
- Ducting channels sound away from LV

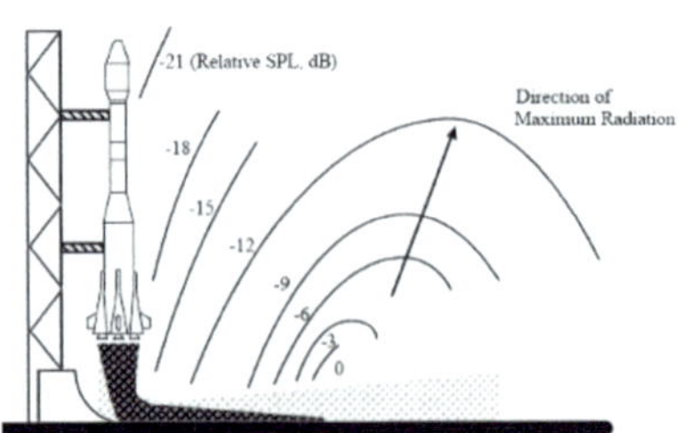

August 22, 2022

Delta II Transonic & Max-q

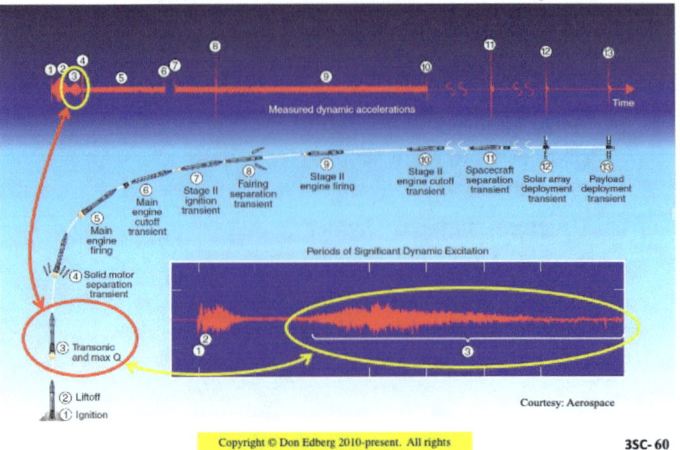

Courtesy: Aerospace

8/22/22

Atmospheric (Max-q) Aero Loads

- Dynamic pressure q is one-half the fluid density times velocity squared: $q = \frac{1}{2}\rho v^2$
- During launch, velocity increases but air density decreases with time/altitude
- "max q" produces largest pressure & aero forces
- "*Buffet*" loading from unsteady aerodynamics

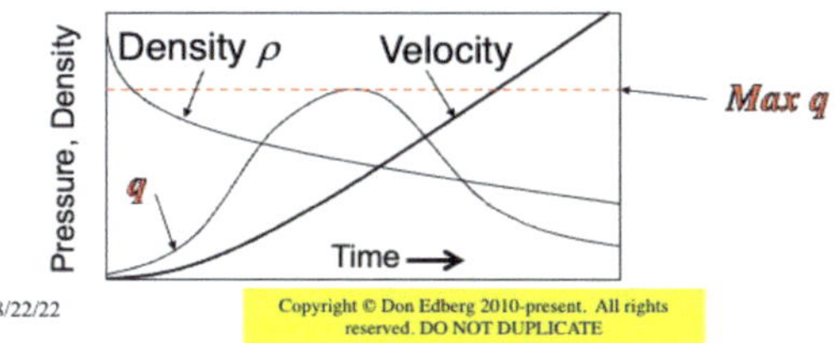

8/22/22

3SC-61

Wind Shear Loads

- LV flies through horizontal gust producing angle of attack α, generating significant lateral loads, engine gimbals to trim
- Called "max air," "max $q\alpha$," gust, "STEL"
  - Gust = vehicle loading due to transient wind
  - STEL = "STatic-ELastic" loading due to high altitude winds resulting in vehicle angle of attack generating side force/acceleration
- Structure loads calculated during coupled-loads analysis

8/22/22

3SC-62

Forces on a Launch Vehicle

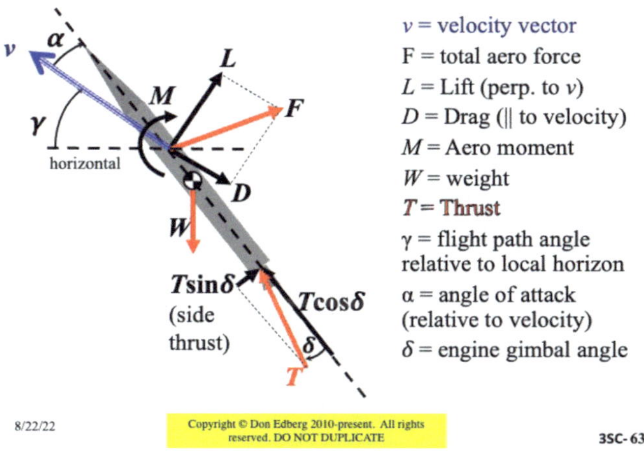

v = velocity vector
F = total aero force
L = Lift (perp. to v)
D = Drag (|| to velocity)
M = Aero moment
W = weight
T = Thrust
γ = flight path angle relative to local horizon
α = angle of attack (relative to velocity)
δ = engine gimbal angle

8/22/22

3SC-63

Vehicle in Crosswind: How to Trim

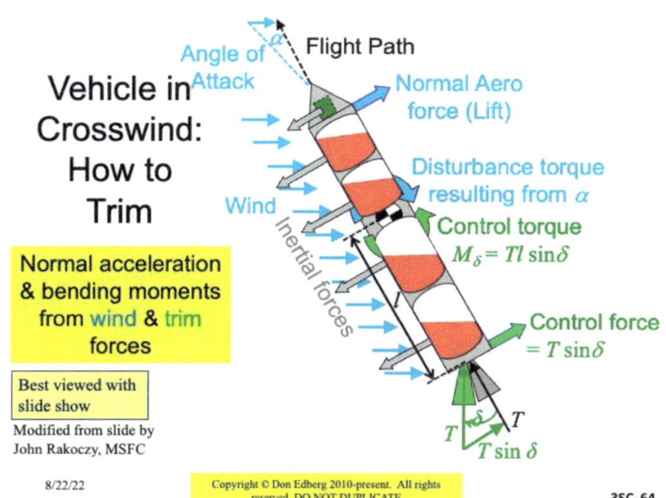

Normal acceleration & bending moments from wind & trim forces

Best viewed with slide show

Modified from slide by John Rakoczy, MSFC

8/22/22

3SC-64

TVC System Behavior: AoA, Flt Path & TVC Angle Time Response for LV like *Ares 1* to Triangular Wind Gust at Max-q

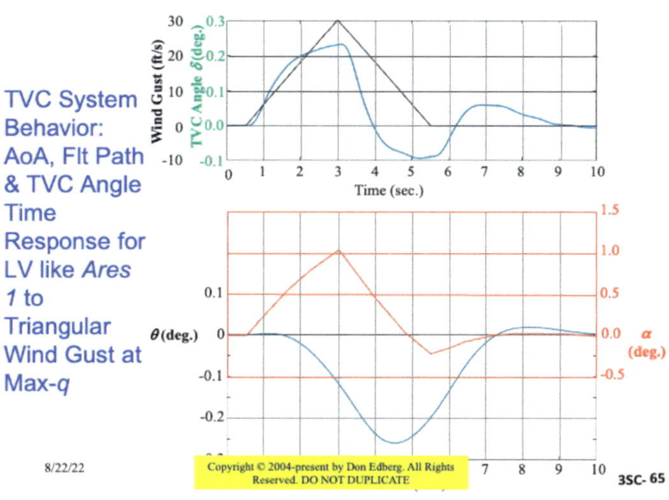

8/22/22

3SC-65

Accelerations & Angular Rate Time Responses for *Ares 1*-like LV to Wind Gust at Max-q

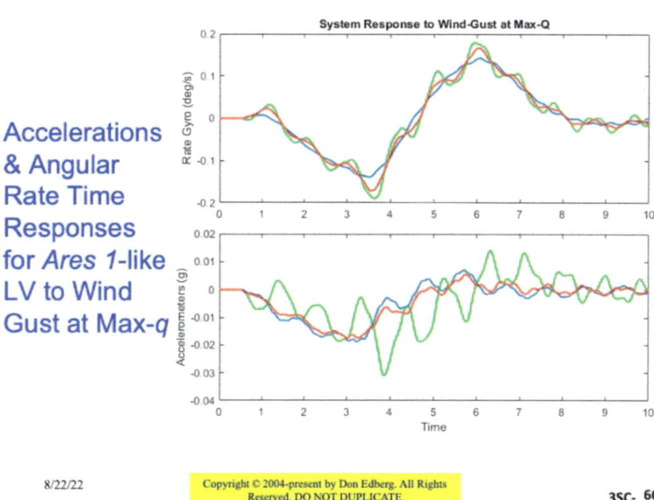

8/22/22

3SC- 66

Slosh of LOx Tank via Flow-3D®

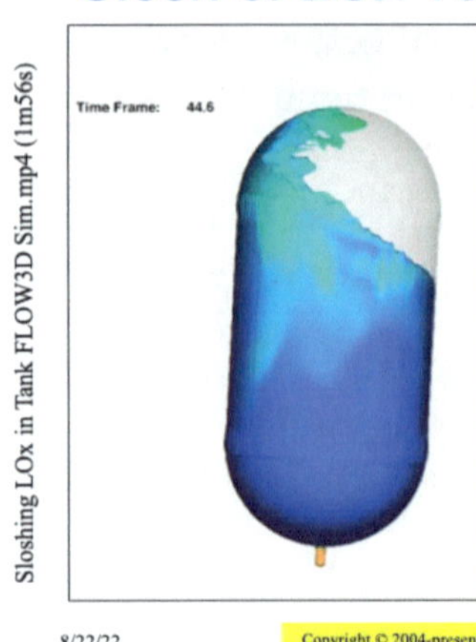

- 5 m ø, 10 m long LOx tank lateral shake at 0.1 g_0 , 3 Hz
- Drain @ 500 kg/s, ullage 1.02 bar
- Color scale represents velocity magnitude
- 58 s simulation courtesy Flow Science

8/22/22 3SC-67

Slosh System & Time Response

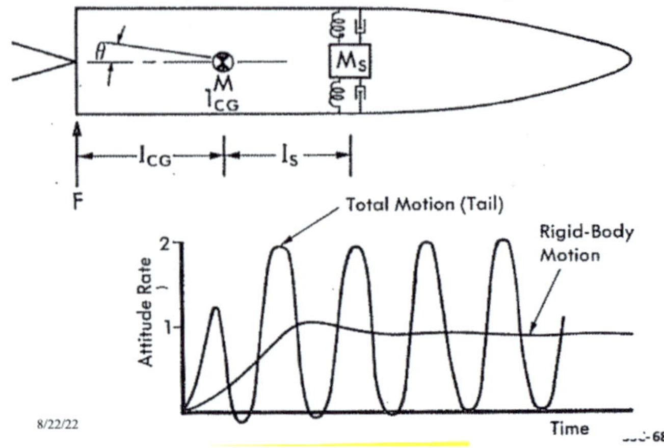

8/22/22 3SC-68

Max-*q* Acoustic Environment

- Flow of air makes boundary layer / wind noise that excites fairing & payload inside
- Acoustic insulation can attenuate noise transmission to payload

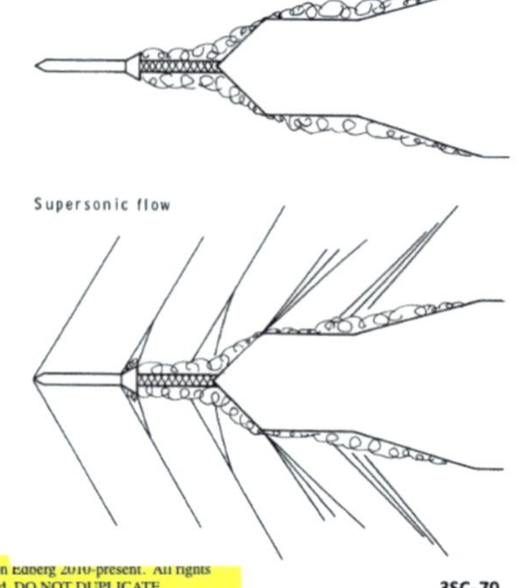

Courtesy: NASA

August 22, 2022 3SC-70

Launch Vehicle Aerodynamic Noise & Buffet Sources

- Depend on geometry (i.e. diameter changes), shock waves, boundary layers, or wakes from upstream objects
- q = dynamic pressure
- p_{rms} = effective pressure fluctuation
- TBL = turbulent boundary layer

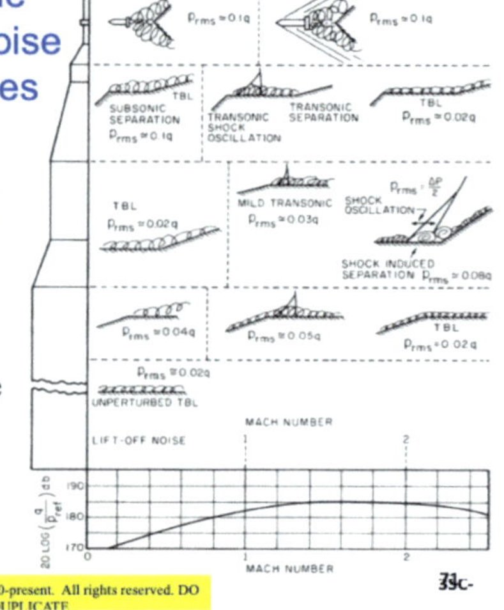

Courtesy NASA

August 22, 2022 3SC-71

Buffet Loads

- *Pressure oscillation* caused by turbulence or separated air flow
- Can excite both body-bending or local structural responses
- Greatest at transonic speeds where shock interactions augment pressure fluctuation
- Often greatest at locations where steady loads are max

August 22, 2022 3SC-72

Nose Shape Effect on Pressure Fluctuation Response

- Hammerhead vs. Conventional LV
- Identical mass and elastic properties, free-free suspension
- Hammerhead's higher response due to buffeting from the reverse taper

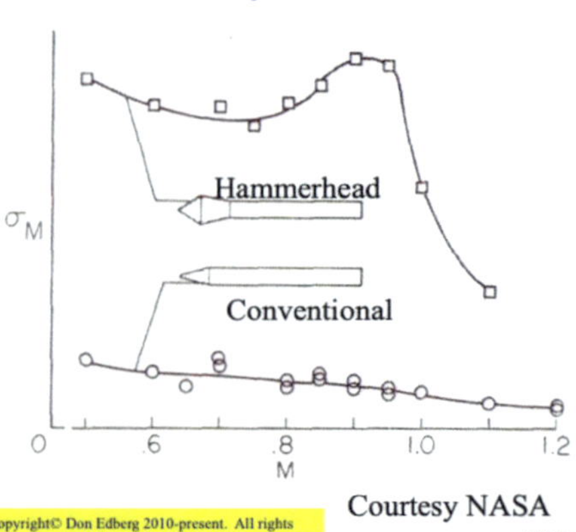

Courtesy NASA

August 22, 2022 3SC-74

Hammerhead Bending-Moment Buffet

- Hammerhead vs. "regular"
- Identical mass, elastic properties, & suspension
- Hammerhead's higher response due to buffeting from the reverse taper
- High peaks indicate low structural & aero damping
- 1st mode shows negative damping at higher Mach, could cause sustained oscillation

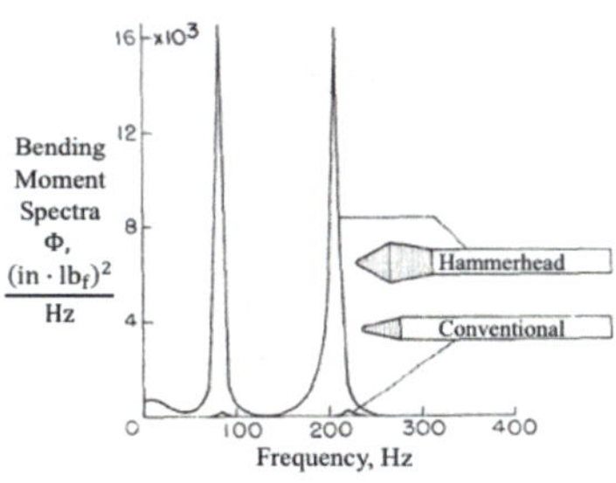

Hammerhead vs. regular bending moment power spectra for $M = 0.9$, $\alpha = 0°$

Courtesy NASA

August 22, 2022 3SC-75

"Hammerhead" Configuration Problem Solved for Atlas V + Starliner

- ULA & Boeing predicted issues during 1st stage flight sims, with loads from Starliner on Centaur
- Updated aerodynamic configuration revealed: *aeroskirt* incorporated aft of spacecraft
- New LV configuration:
  - Extends Starliner Service Module cylindrical surface
  - Improves aerodynamic characteristics of launch configuration
  - Reduces flight loads to acceptable levels
- Aeroskirt jettisoned during ascent; has provisions for venting if abort engines are fired

August 22, 2022 3SC-76

Buffet/Pressure Fluctuation Effects Due to Canted Booster Nose

(A) Sharp on-axis booster nose.

(B) Sharp canted booster nose.

(C) Canted ogive booster nose.

Courtesy NASA NESC

August 22, 2022 3SC-77

Pressure-Sensitive Paint on SLS Model in NASA Ames Wind Tunnel Shows Buffeting and Movement of Pressure Pulses

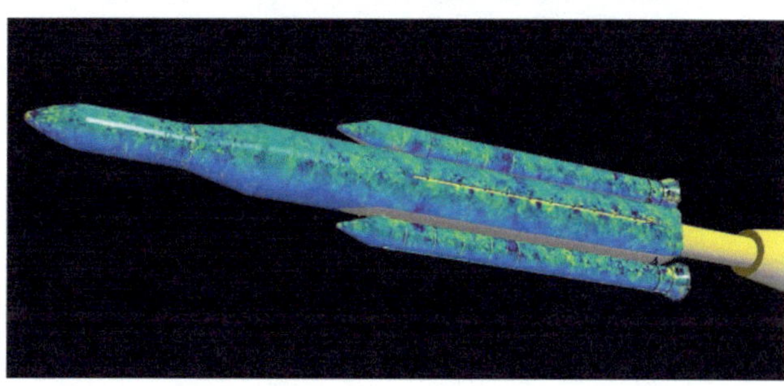

Courtesy NASA

August 22, 2022 3SC-78

Topics

- Transportation & launch loads cases
  - Mechanical transportation loads: air, ship, rail; typical load factors
- Anatomy of a launch vehicle (LV)
- Launch trajectory phases
- Flight environment (mechanical & acoustic)
- Load factors
- Flight events: mechanical & acoustic loads: ignition, liftoff, max-q, staging/separation, pogo, resonant burn oscillation
- **Separation systems: pyro & non-pyro**
- Spacecraft flight loads environment: quasi-static, vibration, shock, acoustic, thermal
  - Payload isolation

8/22/22 3SC-79

Delta II PLF & Spacecraft Separation

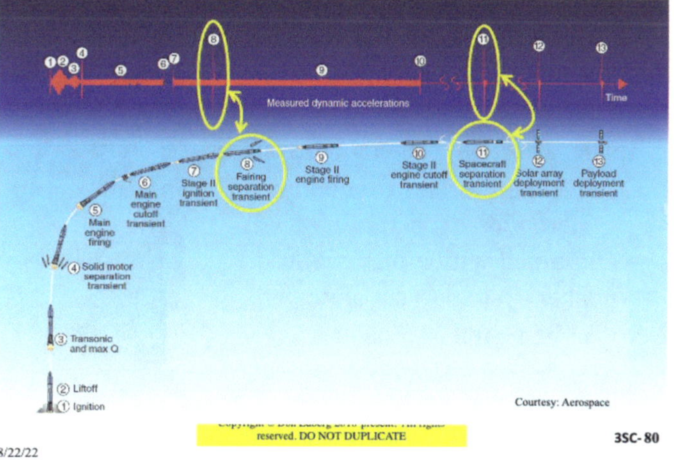

Courtesy: Aerospace

8/22/22 3SC-80

Stage Separation

- Multi-stage rockets "stage" for better performance
- Pyrotechnic stage separation a large shock event
- Non-pyro = less shock

Delta-II_2662ndstg.mp4

Courtesy: NASA, Space Online

Pyrotechnic Devices

- Small charges of explosive, a.k.a. "ordnance" or **pyro**technics used to cut through materials or activate devices
- Ordnance devices are really reliable ways to make things happen ***a single time***
  - Staging/separation mechanisms
  - Frangible bolts
  - Frangible nuts or pyro nuts

Pyro Shocks Can Be Large & Contaminating

Saturn V Stage Separation Test

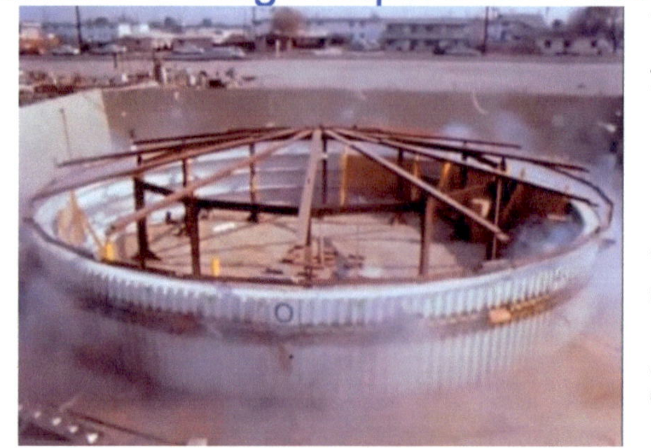

Saturn V stage separation test.!

Courtesy: NASA

Alternatives to Pyro Separation?

- Cannot test explosive system without replacing after test
- High shock levels

SpaceX *Falcon 9* interstage system

- Reusable mechanical clamping system with "separation collets" combined with a pneumatic pusher system to separate parts
- Can be tested & reused
- Lower shock levels

Image courtesy SpaceX

Payload Fairing ("Shroud") Separation

- Drop Payload Fairing (PLF) for more mass to orbit
- Separation involves use of linear explosive ordnance
- Considerable shock loading

(Atlas PLF separation shown)

Payload Fairing Separation Systems: No Payload Contamination, Please!

Systems that ***do not*** contain debris from explosive firings:

- Detonating cord ("det" cord)
- Linear shaped charges (LSC)

Systems that ***do*** contain their explosives/debris:

- "Super*Zip"
- "Sure-Sep"
- "Thrusting Joint"
- Hollow-Form Frangible Joint

*Super*Zip* Operation

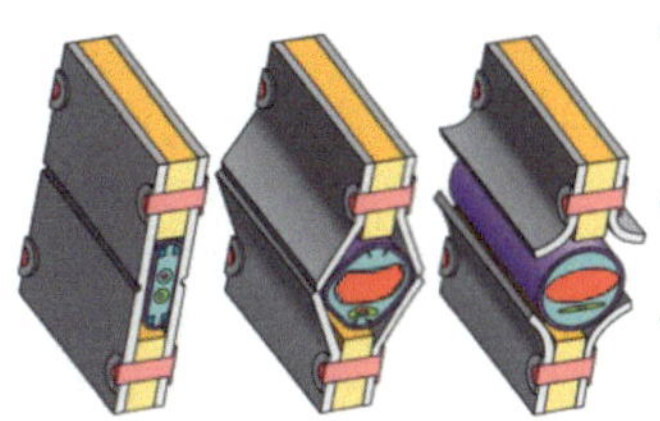

- Primacord linear explosive in an expandable sleeve
- Firing creates gases which expand sleeve
- Expanding sleeve fractures outer structure with pre-cut grooves
- Lockheed Martin

Computer-generated simulation of Super*Zip function. From left to right: 0, 40, & 85 μs (microseconds) after initiation of explosive

Courtesy Aerospace

*Super*Zip*

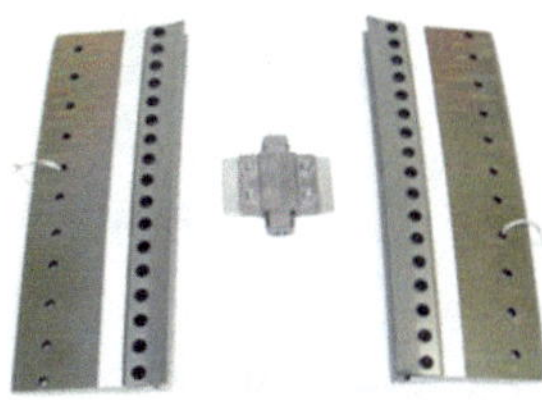

Severed segment of a separation ring with a sample of the Super*Zip expanding tube separation device, as it would have looked before separation, between the two halves.

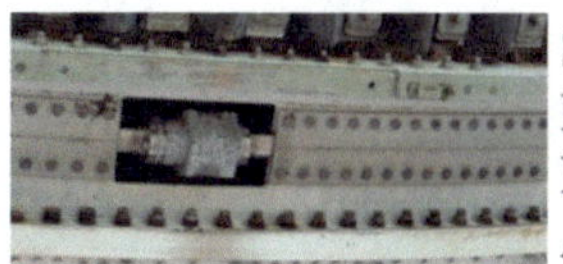

Super*Zip on Centaur G-Prime/Shuttle bay mount, MSFC (D. Edberg photo 2012 April)

Courtesy Aerospace

"Sure-Sep" Separation System

- Developed by McDonnell Douglas (now Boeing)
- Material fails in shear, reducing energy needed
- Simpler to machine

Courtesy Stockinger et al

"*Hollow-Form Frangible Joint*" (HFFJ) Separation System

- Developed by Orbital Sciences & Ensign-Bickford Aerospace
- Uses a simple aluminum extrusion
- Less expensive & simpler to machine

Falcon 9 PLF: Non-Pyrotechnic

- Still frame from video of FH
- Reusable pneumatic system like F9 stage separation
- PLF photo shows "hard points"
- F9 PUG says it has mechanical latches & pneumatic release / pusher system
- Provides benign shock environment
- Allows pre-flight testing
- Separation rails or other mechanisms not visible

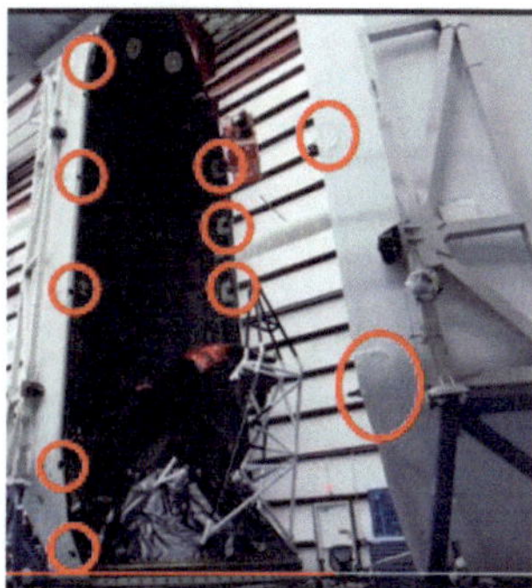

Courtesy SpaceX

PLF Separation Dynamics

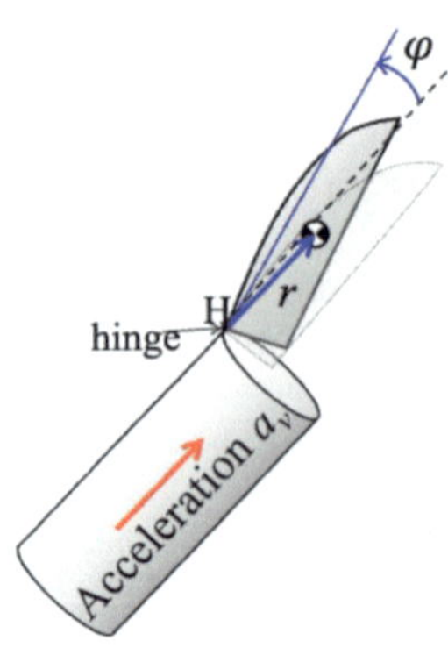

- Vehicle accelerating a_v in vacuum
- PLF of mass m separates
- First pivots at hinge H, then separates at angle φ_0 – long enough to provide 'outward' motion
- Don't want freed PLF parts to contact lower body or anything else

August 22, 2022 3SC-94

PLF Sep Dynamics

During hinged phase, PLF weight is ma_v

- PLF must split with enough rotational momentum for its CM to overcome effective weight's restoring force & "coast" past $\varphi = 0$, the forwards projection of hinge H
- $\sum$ Torques $T|_{\text{hinge } H} = ma_v(r \sin \varphi) = I_H \frac{d^2\varphi}{dt^2}$, where I_H is PLF MoI about hinge

Separation when $\varphi = \varphi_0$; effective PLF weight = 0

- PLF tumbles around its CM with rate $\frac{d\varphi}{dt}$
- CM follows a parabola with respect to LV axis

August 22, 2022 3SC-95

PLF Separation Ordnance

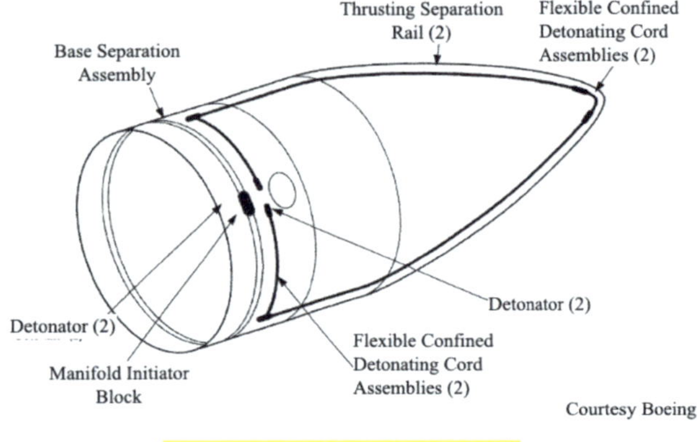

Courtesy Boeing

August 22, 2022 3SC-96

Delta IV Fairing Separation Test

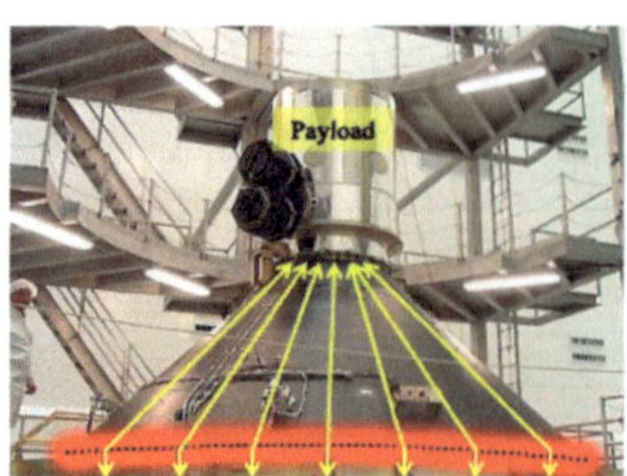

The sudden separation of the payload fairing exposes spacecraft components to the expected flight shock environment

Source: Boeing

August 23, 2022 3SC-97

Fairing Separation (or Lack)

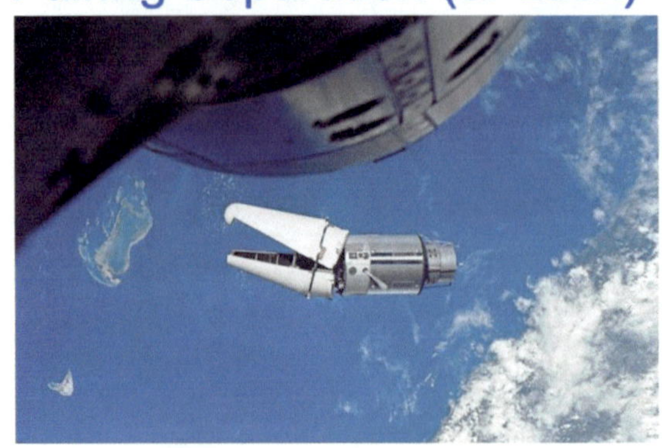

NASA photo

August 22, 2022 3SC-98

Atlas V PLF Sep Movie

Atlas V PLF SRB AEHF-6 Launch Highlights.mp4 courtesy ULA

August 22, 2022 3SC- 99

"Pogo" Vibrations

- Interaction between booster structural dynamics and *liquid fuel* propulsion system vibrations
- Low-frequency oscillation of entire LV in long axis, can be as high as 30 g_0 magnitude
- Similar to "water hammer," waves pass up and down plumbing causing pressure transients producing uneven burning which in turn produces thrust transients which shake structure and so on
- LV companies refer to it as "sustained oscillation" (*pogo* is a 'no-no' word for them)

8/22/22

3SC-100

Pogo

- On *Saturn V* (below), *Titan 2*, others
- Only occurs a certain time of flight

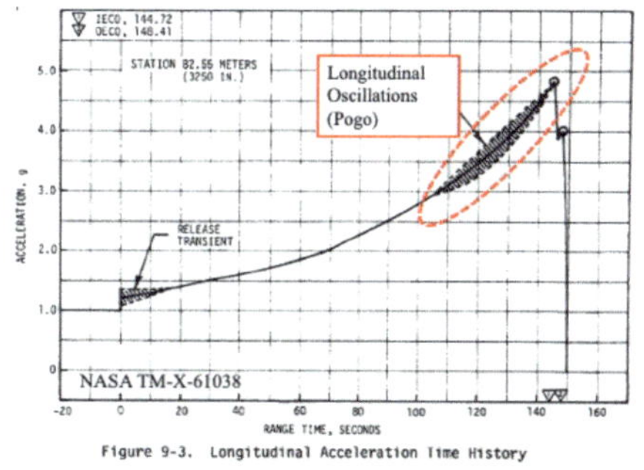

8/22/22

3SC-101

Pogo Instability Explained

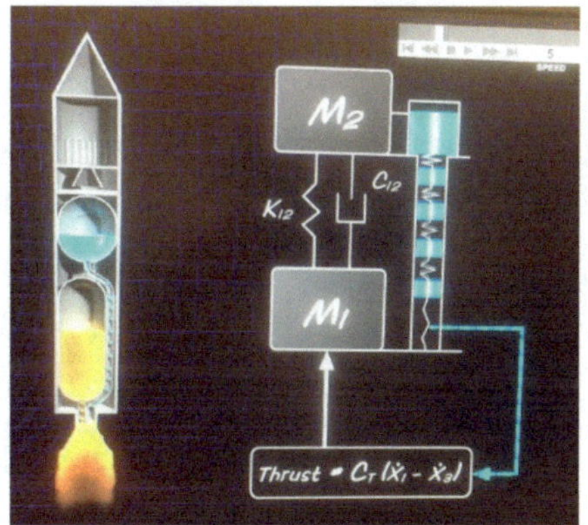

Caused by interaction of vehicle structural vibration mode with thrust oscillations:

- Vibrations cause flexing of the propellant-feed pipes
- … which induces thrust variation
- … which drives oscillations, causing more flex
- … causing more thrust variations, etc…
- Instability!

POGO system sim.MOV

8/22/22

3SC-102

Saturn V S-II Stage POGO Video

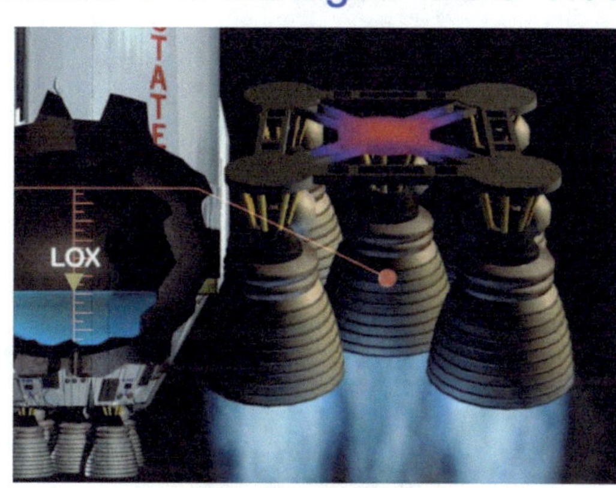

POGO Saturn S-II.m4v

8/22/22

3SC-103

AS-508 / Apollo 13 Pogo at Center Engine

- Acceleration at center S-II engine reached 34 g_0 at 15 Hz (estimated; full scale was 20 g_0)
- Caused large pressure oscillations that shut engine down at 160 s
- If oscillations had continued one more cycle, thrust structure would have failed

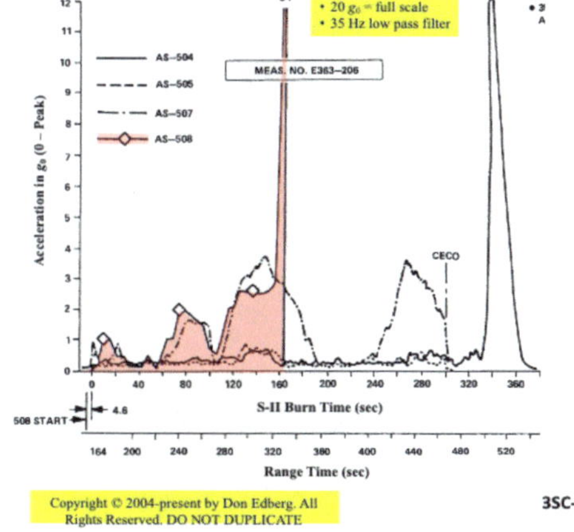

8/22/22

3SC-104

Pogo Similar to "Water Hammer"

Pogo occurs because of pressure pulses in liquid propellant system causing thrust oscillations that excite structural vibrations in LV structure that cause pressure pulses that cause thrust oscillations … and so on …

Water-Hammer-Action.mp4 (courtesy YouTube)

3SC-105

Get Rid of Water Hammer

- Need to add "compliance" to system
- Use a volume of gas attached to plumbing lines
- Pressure pulses compress gas rather than entire water system
- Under sink "accumulators" shown

D Edberg photo

8/22/22 3SC-106

Pogo Suppression

- Add gas-filled cavities — "accumulators" to propulsion system (called "pogo suppressors")
- Similar to adding capacitance to an electric circuit, the accumulators "absorb" or "filter" the pressure oscillations to minimize them
- Pogo oscillation discussion: NASA SP-4205

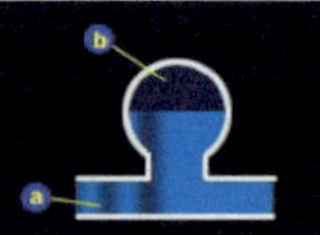

Pogo-induced pressure waves (a) are damped by the compression of a trapped gas (b), usually helium. Courtesy http://www.xmission.com/~jwindley/techs vpogo.html

8/22/22 3SC-107

Pogo Correction Devices

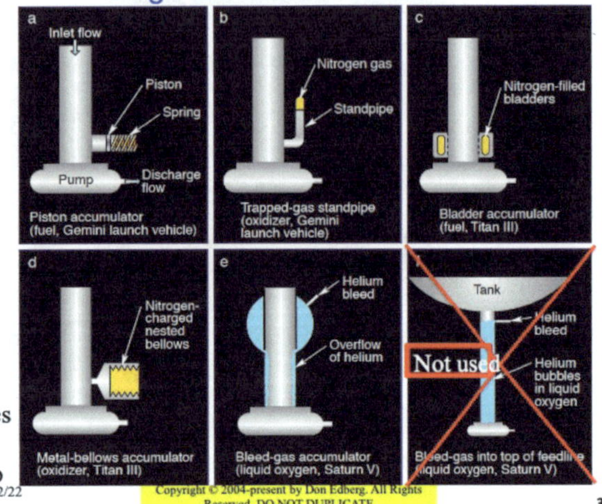

(courtesy Aerospace Corp.)

8/22/22 3SC-108

Pogo Model (courtesy Meyer, Aerospace)

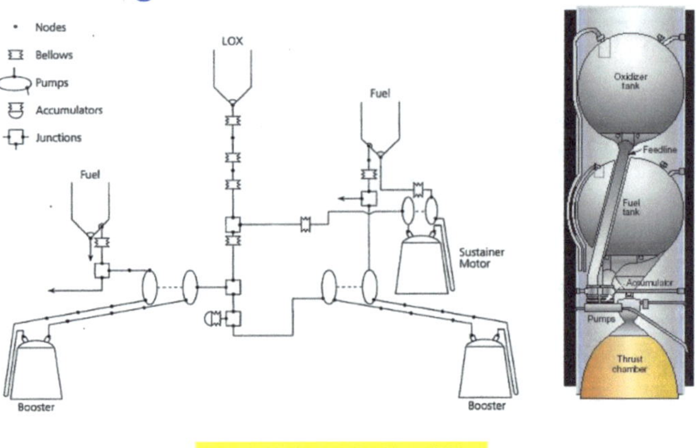

8/22/22 3SC-109

"Resonant Burn" Oscillations

- Oscillations like pogo, but occurs in solid motors
- Interaction between booster combustion chamber acoustics and solid propellant combustion
- Pressure waves pass up and down combustion chamber (think of an *organ pipe*) causing uneven burning which in turn produces thrust transients which shake structure and so on
- Imposes *significant* axial loads
- Was a BIG problem on *Ares I* (until its cancellation)
- No way to prevent resonant burn, must build LV and payload to survive (or isolate payload; see below)

8/22/22 3SC-110

Resonant Burn Due to Pressure Pulses (Like "Organ Pipe" Sounds)

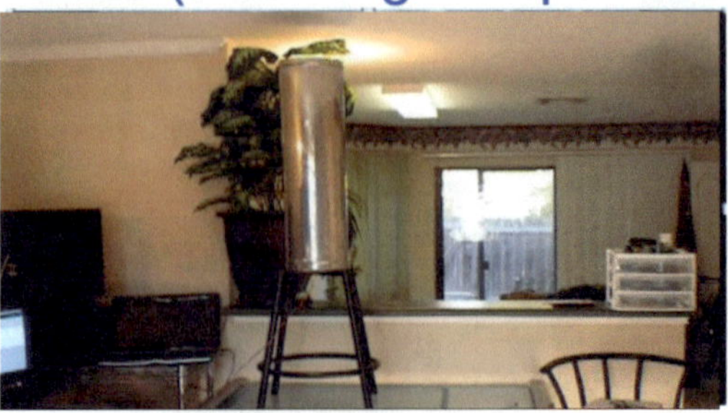

- Occurs due to pressure pulses ("sounds") in propellant exhaust coupling with structural vibrations in LV structure

Acoustic Resonance Tube.mp4 (courtesy youtube)

8/22/22 3SC-111

Resonance in A Cylinder

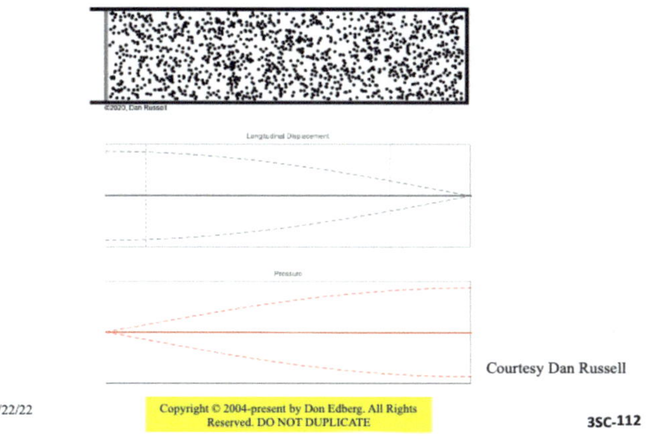

8/22/22

3SC-112

Normal & Unstable SRM Combustion (a.k.a. "Resonant Burn")

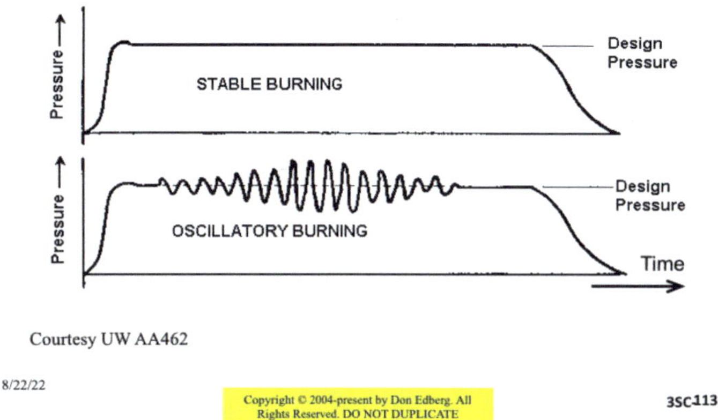

Courtesy UW AA462

8/22/22

3SC-113

Time History Showing Resonant Burn Oscillations (~190 s)

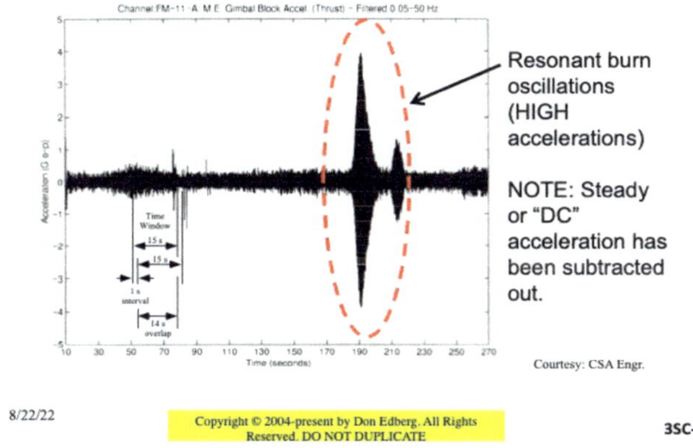

Courtesy: CSA Engr.

8/22/22

3SC-114

Separating Spacecraft From LV

Pyrotechnic or *pyro* devices utilize explosives & produce shocks:
- "Frangible" or intentionally-breakable pyrotechnic nuts or bolts (good for point attachments)
- Clamp band: good for circular shell structures or motor casings where you don't want point loads.

- Non-pyrotechnic release mechanisms: minimize shock:
  - Hold-down & release mechanisms from Ensign-Bickford, others
- More info in Lecture 13 SC03B Mechanisms

August 22, 2022

3SC-115

Pyro Shocks Are Energetic Events: Time History & SRS

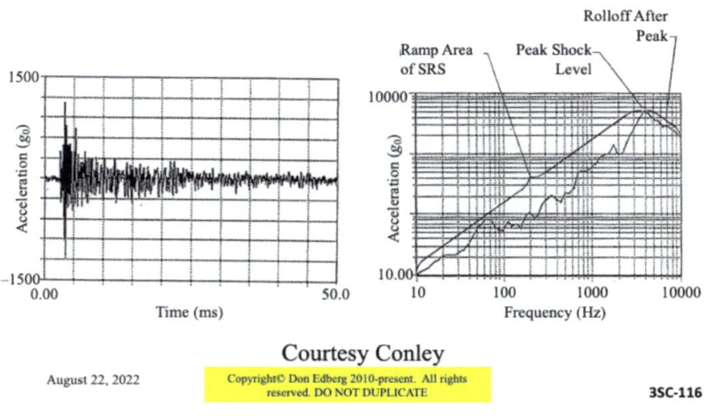

Courtesy Conley

August 22, 2022

3SC-116

Topics

- Transportation & launch loads cases
  - Mechanical transportation loads: air, ship, rail; typical load factors
- Anatomy of a launch vehicle (LV)
- Launch trajectory phases
- Flight environment (mechanical & acoustic)
- Load factors
- Flight events: mechanical & acoustic loads: ignition, liftoff, max-q, staging/separation, pogo, resonant burn oscillation
- Separation systems: pyro & non-pyro
- **Spacecraft flight loads environment: quasi-static, vibration, shock, acoustic, thermal**
  - Payload isolation

8/22/22

3SC-117

From the Point of View of the Spacecraft …

How does the LV affect the payload?

August 22, 2022

3SC-118

Quasi-Static Environment

- Slowly changing environments can be considered "static" (*g*-loading / quasi-steady or quasi-static accelerations)
- Design load factors are represented by Quasi-Static Loads (QSL), which are the most severe combination of dynamic & steady-state acceleration
- AXIAL COMPRESSIVE LOADS ARE LARGEST
- LATERAL LOADS BEND STRUCTURE & TEND TO BE WORSE THAN AXIAL

August 22, 2022

3SC-119

Strength Design Load Factor

- Initial design: use load factors based on LV acceleration
- Obtain from the Payload Planners' Guide
- Axial (thrust) direction and lateral direction factors
  - Typical values 6-8 g axial, 2-3 g lateral
- "Uncertainty factors" are applied as a margin until things become better understood (Coupled-loads analysis or CLA)

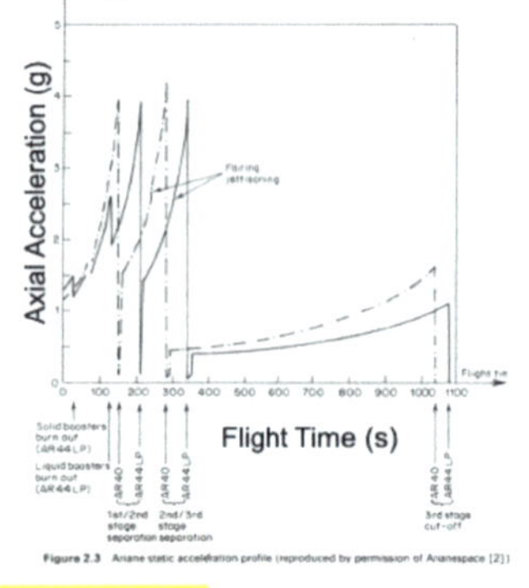

August 22, 2022

3SC-121

Loads on Spacecraft

- Loads also act on spacecraft, the LV's payload
- Spacecraft are attached to payload attach fitting (PAF) or LV Adapter (LVA)
- LV sets load conditions (*Atlas V* design factors below)

Atlas V Flight Condition	Lateral (+ = compression,	Axial – = tension)
Liftoff	**±2.0 g**	+1.8 ±2.0 g
Flight Winds	**±0.4 ±1.6 g**	+2.8 ±0.5 g
Strap-On Separation	±0.5 g	+3.3 ±0.5 g
BECO (max axial)	±0.5 g	**+5.5 ±1.0 g**
BECO (max lateral)	±1.5 g	+2.5 ±1.0 g
MECO (max axial)	±0.3 g	+4.5 ±1.0 g
MECO (max lateral)	±0.6 g	±2.0 g

August 22, 2022

3SC-122

Loads on Spacecraft

- Loads also act on spacecraft, the LV's payload
- Spacecraft are attached to payload attach fitting (PAF) or LV Adapter (LVA)
- LV sets load conditions (*Ariane V* QSL below)

Acceleration (g) / Critical flight events	Longitudinal		Lateral
	Static	Dynamic	Static + Dynamic
Liftoff	– 1.7	± 1.5	± 2
Max *q*	– 2.7	± 0.5	± 2
SRB thrust tail-off	– 4.55	± 1.45	± 1
Main core thrust tail-off	– 0.2	± 1.4	± 0.25
Max tension case: SRB jettison	+2.5		± 0.9

August 22, 2022

3SC-123

Atlas V Design Load Factors-Graphical

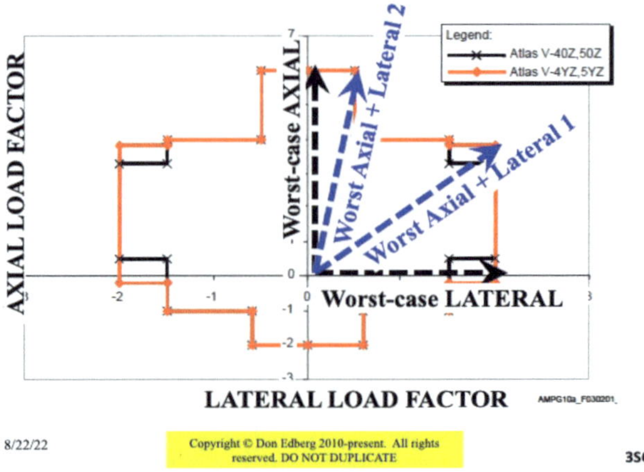

8/22/22

3SC-124

Different LV Vibration Environments

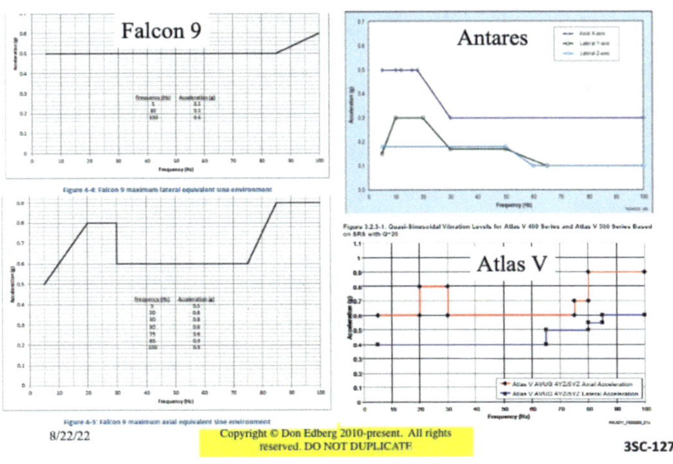

Different LV Shock Environments

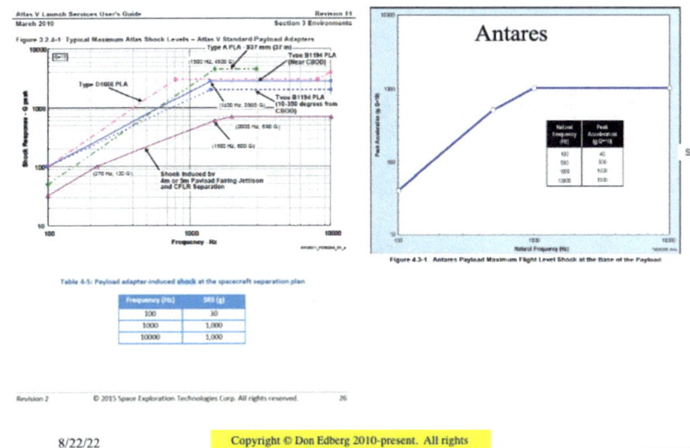

Stiffness Design Factors

- Each LV has its own frequency criteria
- Usual issues are bandwidth of control system vs. structural natural frequencies
- Selected to preclude control-structure interaction
- These are guidelines only
- Verify during coupled-loads analyses

Topics

- Transportation & launch loads cases
  - Mechanical transportation loads: air, ship, rail; typical load factors
- Anatomy of a launch vehicle (LV)
- Launch trajectory phases
- Flight environment (mechanical & acoustic)
- Load factors
- Flight events: mechanical & acoustic loads: ignition, liftoff, max-q, staging/separation, pogo, resonant burn oscillation
- Separation systems: pyro & non-pyro
- Spacecraft flight loads environment: quasi-static, vibration, shock, acoustic, thermal
  - Payload isolation

Payload Natural Frequency ≠ LV Freq.

- SC fundamental stiffness requirement is no resonant coupling with LV modes
- Dynamic decoupling needed: SC structure must exhibit natural frequency > LV freqs.

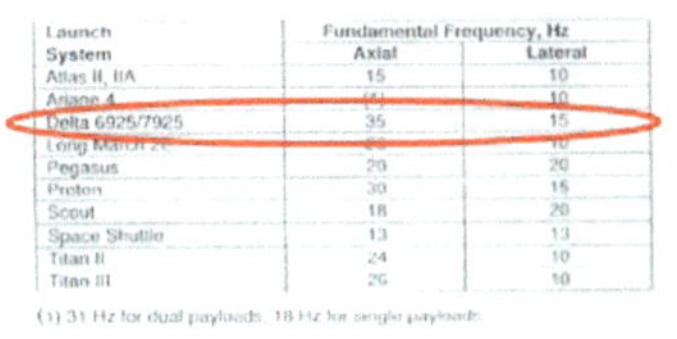

Launch System	Fundamental Frequency, Hz Axial	Lateral
Atlas II, IIA	15	10
Ariane 4	(1)	10
Delta 6925/7925	35	15
Long March 2E	[illegible]	10
Pegasus	20	20
Proton	30	15
Scout	18	20
Space Shuttle	13	13
Titan II	24	10
Titan III	26	10

(1) 31 Hz for dual payloads, 18 Hz for single payloads

Amplification ← → Attenuation (Frequency Isolation)

Transmissibility, TR

Frequency Ratio, Ω/ω_n

Payload Isolation: Helps With All Shock & Vibe

Replace the bolts that attach the avionics section to the payload cone with an isolation system

Courtesy CSA Engineering

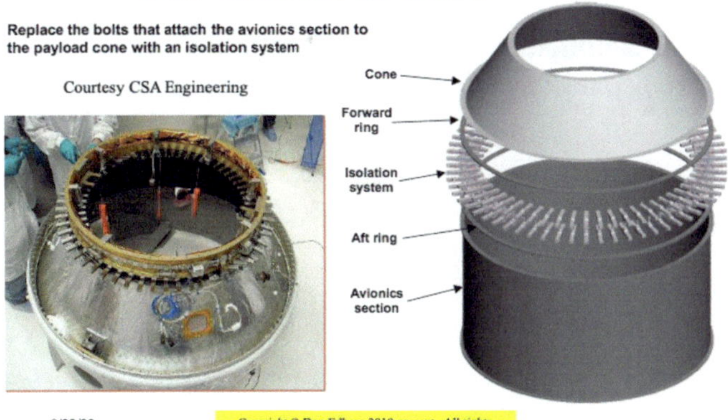

Payload Isolation Attenuates Vibes

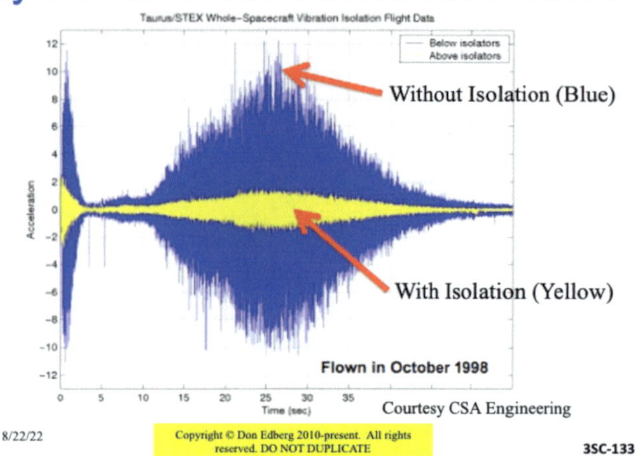

Courtesy CSA Engineering

8/22/22 3SC-133

Peak Lateral Acceleration, Isolated & Unisolated, "6-6" Isolation System on DII

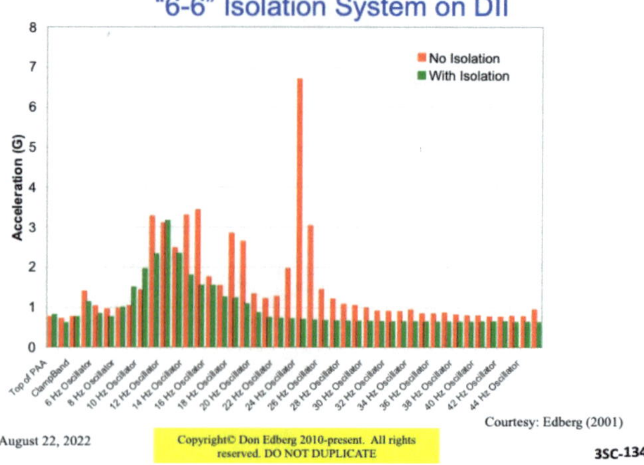

Courtesy: Edberg (2001)

August 22, 2022 3SC-134

Lateral SRS for 24 Hz Structure: Large Reduction in Levels >12 Hz, "6-6" Isolator/DII

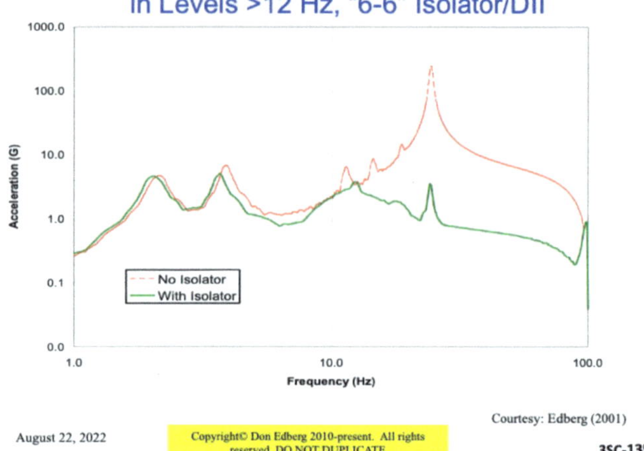

Courtesy: Edberg (2001)

August 22, 2022 3SC-135

Acoustic Loads

- Rockets produce *huge* noise
- Worst acoustic load periods:
  - Ignition/liftoff: noise can reflect off of ground near liftoff
  - Max-*q*: "boundary layer" noise excites fairing during climb
- Acoustic loads measured in dB or decibels

Noise Source (note: 3 dB ≈ factor of 2)	Sound Level
Hearing threshold	0
Dishwasher	50-53
Car @ 10 m	60-80
Jack hammer @ 1 m	100
Jet engine @ 100 m	110-140
Jet engine @ 30 m	150
Rocket launch	**165**
Theoretical limit	194

8/22/22 3SC-136

Payload Acoustic Environments

- Design curves are supplied by LV manufacturers (payload planner's guide)
- Provide maximum expected SPLs inside of payload fairing
- Levels vary depending on:
  - LV configuration
  - Payload geometry

August 22, 2022 3SC-137

Delta II Payload Acoustic Environment

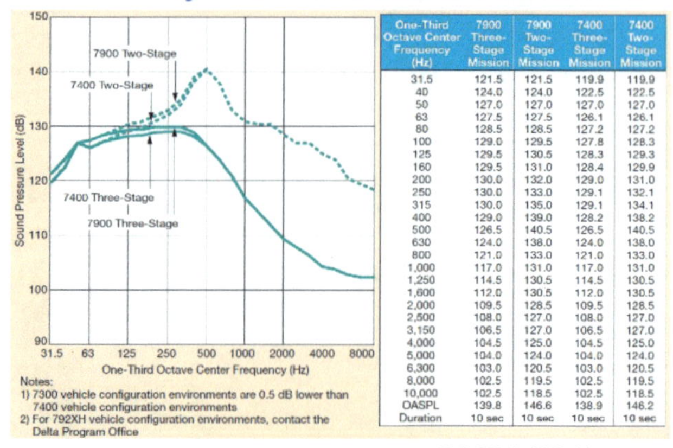

One-Third Octave Center Frequency (Hz)	7900 Three-Stage Mission	7900 Two-Stage Mission	7400 Three-Stage Mission	7400 Two-Stage Mission
31.5	121.5	121.5	119.9	119.9
40	124.0	124.0	122.5	122.5
50	127.0	127.0	127.0	127.0
63	127.5	127.5	126.1	126.1
80	128.5	128.5	127.2	127.2
100	129.0	129.5	127.8	128.3
125	129.5	130.5	128.3	129.3
160	129.5	131.0	128.4	129.9
200	130.0	132.0	129.0	131.0
250	130.0	133.0	129.1	132.1
315	130.0	135.0	129.1	134.1
400	129.0	139.0	128.2	138.2
500	126.5	140.5	126.5	140.5
630	124.0	138.0	124.0	138.0
800	121.0	133.0	121.0	133.0
1,000	117.0	131.0	117.0	131.0
1,250	114.5	130.5	114.5	130.5
1,600	112.0	130.5	112.0	130.5
2,000	109.5	128.5	109.5	128.5
2,500	108.0	127.0	108.0	127.0
3,150	106.5	127.0	106.5	127.0
4,000	104.5	125.0	104.5	125.0
5,000	104.0	124.0	104.0	124.0
6,300	103.0	120.5	103.0	120.5
8,000	102.5	119.5	102.5	119.5
10,000	102.5	118.5	102.5	118.5
OASPL	139.8	146.6	138.9	146.2
Duration	10 sec	10 sec	10 sec	10 sec

Notes:
1) 7300 vehicle configuration environments are 0.5 dB lower than 7400 vehicle configuration environments
2) For 792XH vehicle configuration environments, contact the Delta Program Office

D2 PPG (ULA)

August 22, 2022 3SC-138

Acoustic Attenuation in PLF: Passive Insulation

Acoustic Attenuation in PLF: Helmholtz Absorbers

Courtesy: NASA

Courtesy: USAF

External Acoustic Suppression

- Water acoustic suppression, on launch pad, also used to suppress noise
- LV exhaust ducted through tunnels to reduce radiation towards LV

Static Pressure During Launch

- Static pressure decreases during launch due to climb
- Trapped volumes need to vent

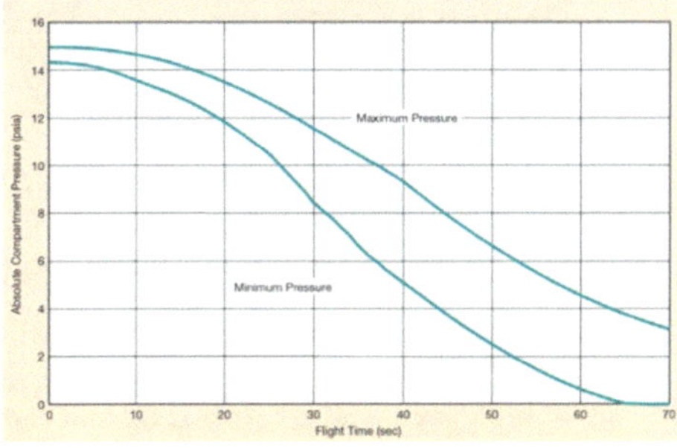

- Payload Planner's Guides (PPGs) provide pressure vs. time schedule: this courtesy D2 PPG (ULA)

Payload Thermal Loads

- Payload fairing internally air-conditioned before launch
- Loads typically managed by payload fairing during launch and ascent
- Payload "comfortable" while nose & skins warm from aero heating

Courtesy ULA

Delta II Heating Loads

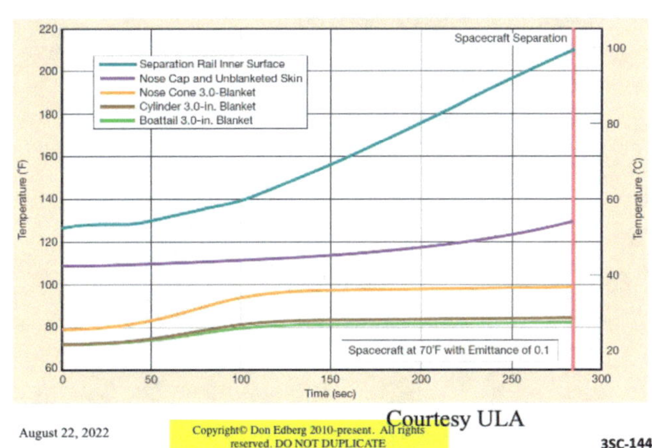

Courtesy ULA

Payload Fairing (PLF) Jettison: When?

- PLF present to protect payload during high q & high heat transfer situations
- PLF jettisoned once aero heating has dropped to a "small" value
  - Typically about solar radiation value ~1135 W/m^2 (0.1 BTU/ft^2/s) = ILS *Proton* spec.
- Mass reduction after jettison improves LV launch performance

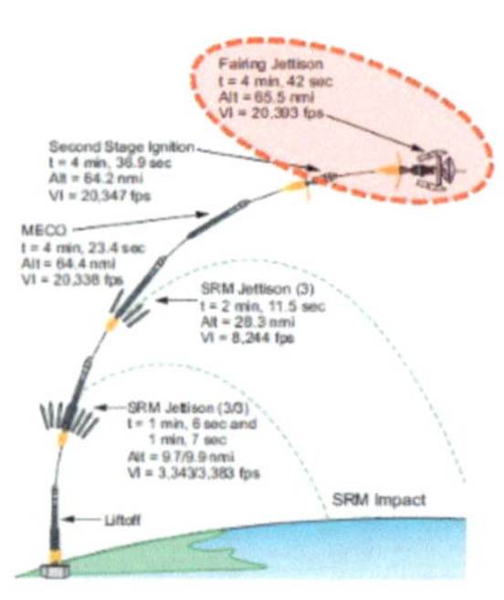

August 22, 2022 3SC-145

On-Orbit Thermal Environment (later lecture)

- SC environment dominated by received & transmitted radiation & internal heat sources
- Extremes are solar/reflected solar/planet IR heating and deep space cooling
- Molecular heating may occur in low orbit
- Detailed thermal analysis involves complex nodal networks accounting for heat inputs (internal & external), material properties, geometries, orbital considerations, etc...
- Entry heating dominated by radiation & convective heat transfer
  - Requires ablative or high temperature materials (e.g. ceramics) if survival of SV is required

August 22, 2022 3SC-146

References

- **Dynamics and Simulation of Flexible Rockets**. Barrows, T.M., and Orr, J.S., 2021, Elsevier, ISBN: 978-0-12-819994-7
- Abramson, H. N., "The Dynamic Behavior of Liquids in Moving Containers with Application to Space Vehicles," NASA SP-106, 1966
- Southwest Research Institute slosh testing article in **Aerospace Testing Magazine**, 2021: https://ati.mydigitalpublication.co.uk/publication/?m=63357&i=678037&p=22

8/22/22 3SC-147

Pyrotechnic Device Refs.

- Ensign Bickford
  https://www.ebad.com/launch-vehicle/
  https://www.ebad.com/separation-release-deployment/
- Frangible pyro nuts and bolts
  https://www.chemring.co.uk/what-we-do/countermeasures-and-energetics/space-initiators-and-release-mechanisms

August 22, 2022 3SC-148

For More LV Info, May We Suggest…

8/22/22 3SC-149

Orbital Mechanics Introduction

Orbits & Orbital Mechanics

- Why cover orbits *before* spacecraft design
  Answer: To look at how they affect design
- Orbital history & classification
- Circular & elliptical orbits
- Open orbits & escape trajectories
- Orbital Elements
- Special Orbits

Course Layout

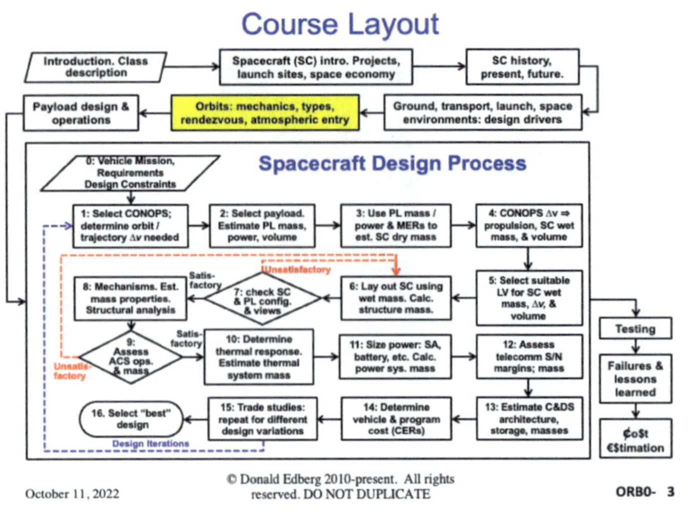
Introduction. Class description
Spacecraft (SC) intro. Projects, launch sites, space economy
SC history, present, future.
Payload design & operations
Orbits: mechanics, types, rendezvous, atmospheric entry
Ground, transport, launch, space environments: design drivers
0: Vehicle Mission, Requirements Design Constraints
Spacecraft Design Process
1: Select CONOPS; determine orbit / trajectory Δv needed
2: Select payload. Estimate PL mass, power, volume
3: Use PL mass / power & MERs to est. SC dry mass
4: CONOPS Δv ⇒ propulsion, SC wet mass, & volume
Unsatisfactory
Satis-factory
8: Mechanisms. Est. mass properties. Structural analysis
7: check SC & PL config. & views
6: Lay out SC using wet mass. Calc. structure mass.
5: Select suitable LV for SC wet mass, Δv, & volume
Satis-factory
9: Assess ACS ops. & mass
Unsatis-factory
10: Determine thermal response. Estimate thermal system mass
11: Size power: SA, battery, etc. Calc. power sys. mass
12: Assess telecomm S/N margins; mass
16. Select "best" design
15: Trade studies: repeat for different design variations
14: Determine vehicle & program cost (CERs)
13: Estimate C&DS architecture, storage, masses
Design Iterations
Testing
Failures & lessons learned
¢o$t €$timation

Orbits & Orbital Mechanics

- Why cover orbits *before* spacecraft design
  Answer: To look at how they affect design
- Orbital history & classification
- Circular & elliptical orbits
- Open orbits & escape trajectories
- Orbital Elements
- Special Orbits

How Does Orbit Affect Mission?

- Orbit → launch site/provider → payload mass & size
- Orbit → navigation, ground tracking, visibility, data transfer
- Orbit → eclipse times → sensor visibility, power generation, thermal control, radiation exposure
- Orbit maintenance → SC propulsion

Orbits & Orbital Mechanics

- Why cover orbits *before* spacecraft design
  Answer: To look at how they affect design
- Orbital history & classification
- Circular & elliptical orbits
- Open orbits & escape trajectories
- Orbital Elements
- Special Orbits

Kepler's Laws

1. The planets move in ellipses with the sun at one focus.
2. Areas swept out by the radius vector from the sun to a planet in equal times are equal.
3. The square of T, the period of revolution, is proportional to the cube of a, the semimajor axis. That is, T^2/a^3 = constant

Kepler's First Law

- Planets move in elliptical orbits with the Sun at one focus

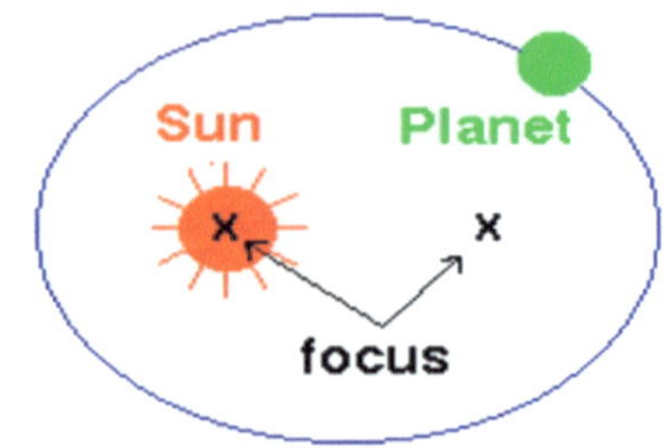
Sun
Planet
×
×
focus

Kepler's Second Law

A line joining a planet and the Sun sweeps out equal areas in equal time Kepler's 2nd law vdo.mp4

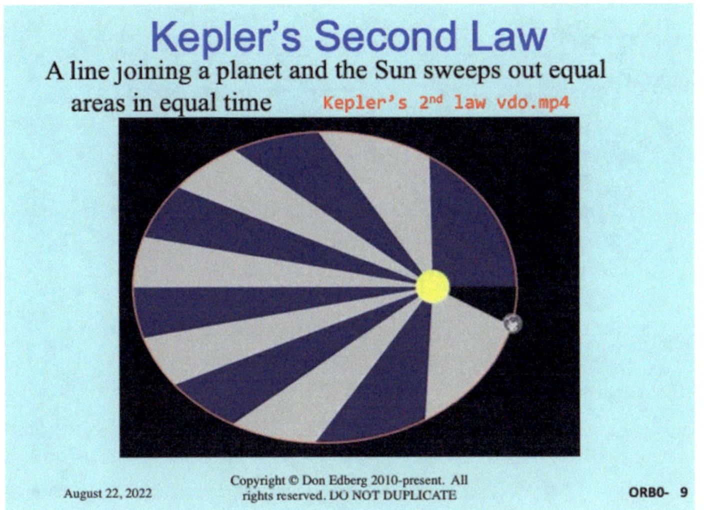

Conservation of Energy: The "*Vis-Viva*" Equation

$\frac{v^2}{2} - \frac{\mu}{r} = -\frac{\mu}{2a} = \varepsilon$ = specific energy = *constant*

Specific kinetic energy + specific potential energy = *constant*

Total energy is conserved. Here:

- v = speed
- r = distance from center of attracting mass
- a = semimajor axis
- μ = gravitational parameter GM

(*vis viva* = Latin for "live force")

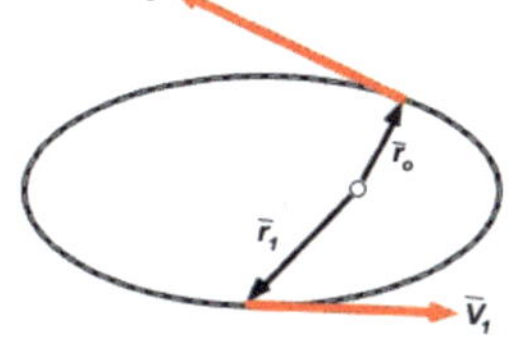

Conservation of Angular Momentum

$\mathbf{h} = \mathbf{r} \times \mathbf{v}$ = specific angular momentum (out of plane of screen/paper)

$|\mathbf{h}| = h = rv\sin\left(\frac{\pi}{2} - \gamma\right) = rv\cos\gamma$, where γ = flight path angle

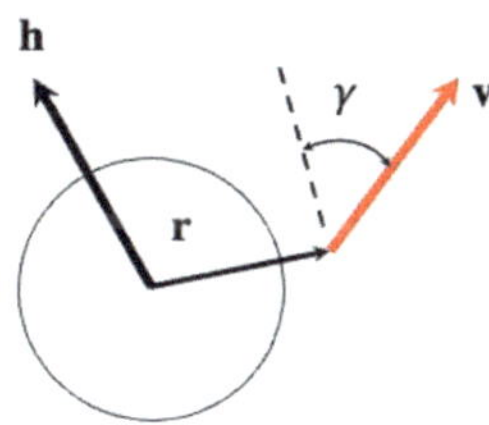

Kepler's Third Law

- The square of the period of a planetary orbit is proportional to the cube of its longest dimension

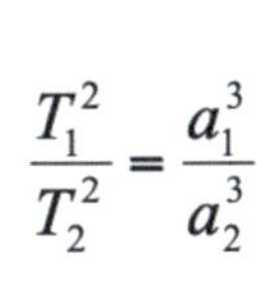

$$\frac{T_1^2}{T_2^2} = \frac{a_1^3}{a_2^3}$$

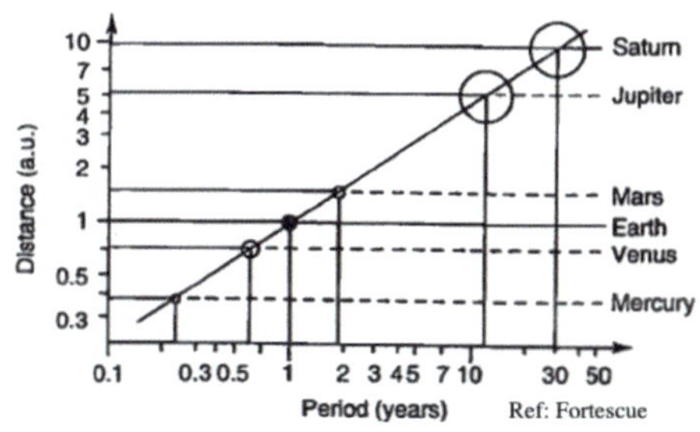

Ref: Fortescue

Universal Law of Gravitation

- Attractive force between two point masses in space:

$$\vec{F}_g = -\frac{Gm_1m_2}{r^2}\left(\frac{\vec{r}}{r}\right) = -\frac{Gm_1m_2}{r^2}\vec{e}_r \quad G = 6.67 \times 10^{-11}\frac{\text{Nm}^2}{\text{kg}^2}$$

= Gravitational Constant

Define $\mu = Gm_1$

$m_1 >> m_2$

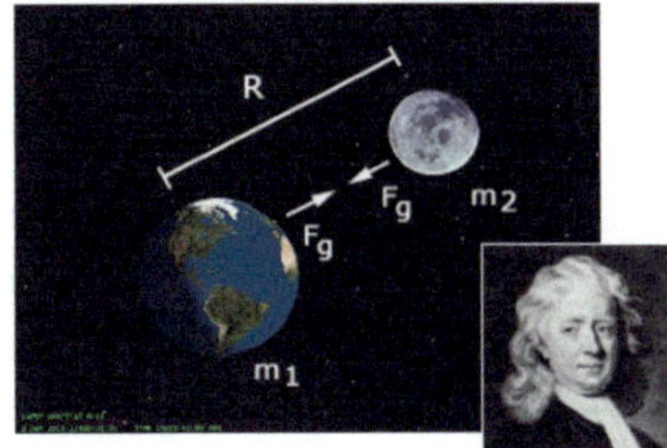

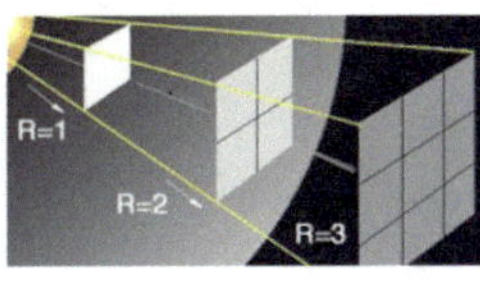

R=1
R=2
R=3

Orbits & Orbital Mechanics

- Why cover orbits *before* spacecraft design
  Answer: To look at how they affect design
- Orbital history & classification
- Circular & elliptical orbits
- Open orbits & escape trajectories
- Orbital Elements
- Special Orbits

Closed Orbits: Circles & Ellipses

- Flight path angle $\gamma = 0$ at both periapsis and apoapsis

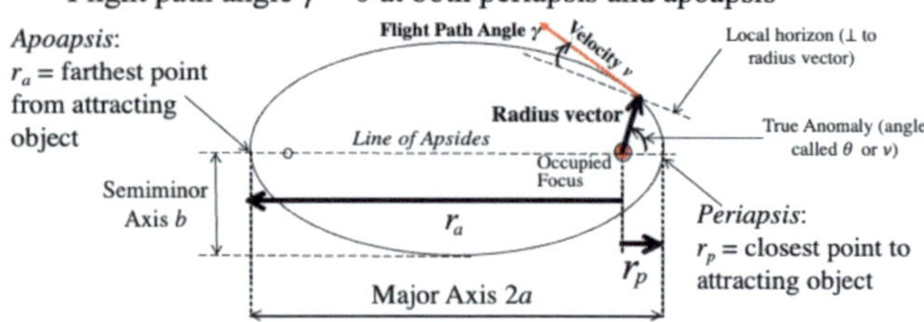

$$r_p = a(1-e);\; r_a = a(1+e)$$

$$e = \text{eccentricity} = \frac{r_a - r_p}{r_a + r_p}$$

Important Orbit Parameters (1 of 2)

- Periapsis radius r_p: distance from the *center* of the planet to the *closest* point in the orbit
  - when orbiting Earth, periapsis is often referred to as perigee
- Apoapsis radius r_a, the distance from the *center* of the planet to the *farthest* point in the orbit
  - when orbiting Earth, apoapsis is often referred to as apogee
- Semi-major axis $a = ½\,(r_p + r_a)$ measures the orbit's overall size. If circular, semimajor axis = orbit's radius.

Important Orbit Parameters (2 of 2)

- Periapsis altitude: $h_p = r_p - R$ = distance from the *surface* of the planet to the *closest* point in the orbit
- Apoapsis altitude: $h_a = r_a - R$ = distance from the *surface* of the planet to the *farthest* point in the orbit;
- Eccentricity $e = \frac{r_a - r_p}{r_a + r_p}$ indicates how "circular" the orbit is in shape.
  - For a circular orbit, $e = 0$.
  - For an elliptical orbit, $0 < e < 1$
  - Higher value of e: orbit is more "squashed"

Orbits are often described by their *altitude* or *height above the surface*, rather than their radius. (Ex: a 400 × 800 km orbit is providing above-surface heights

Orbital Period

$$T = P = 2\pi\sqrt{\frac{a^3}{\mu}}$$

- Earth orbits can be designed to have periods ranging from ~ 90 minutes to ~24 hours (geosynchronous) and more
- Be careful not to confuse radius with altitude ($r = R_{planet}$ + altitude h)
- For circular orbits, semimajor axis a = orbit radius r

Two-Body Orbital Mechanics

Two-body orbital mechanics requires three simplifying assumptions:

1. Point masses
2. Two bodies (rather than "n" bodies)
3. $M >> m$ (planet mass MUCH larger than spacecraft mass)

Simplifying Orbit Assumptions

Assumption 1: Bodies are Point Masses

- All mass can be collected at center
- Central force fields (spherical bodies).
  Note: *In reality, the Earth is not spherical but an oblate spheroid*

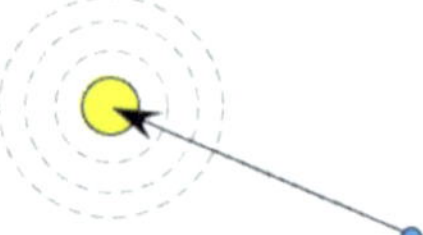

Assumption 2: Only *two* bodies

- We ignore outside influences (sun, moon, etc.) for now

Simplifying Orbit Assumptions (cont.)

Assumption 3: $m_1 >> m_2$, mass 1 >> mass 2

- One mass is MUCH greater than the other (i.e. Earth vs. satellite)
- This is like stating the center of mass of the system is at the center of the larger body (which is therefore stationary)
- *Not true* for Earth-moon system

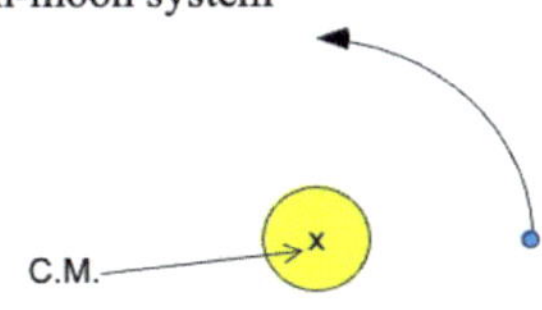

August 22, 2022 ORBO- 21

Solar System Data

Body	μ (km³/s²)	Radius (km)	Mean Solar Distance (km)
Sun	1.3271244E11	696,000	—
Mercury	22,032.1	2,439.7	57.9E6
Venus	324,858.8	6,051.8	108.2E6
Earth	398,600.4	6,378.14	149.6E6
Mars	42,828.3	3,397	228.0E6
Jupiter	126,712,000	71,492	778.4E6
Saturn	37,939,519.7	60,268	1433E6
Moon	4,902.8	1,737.4	—

August 22, 2022 ORBO- 22

All Trajectories Conic Sections

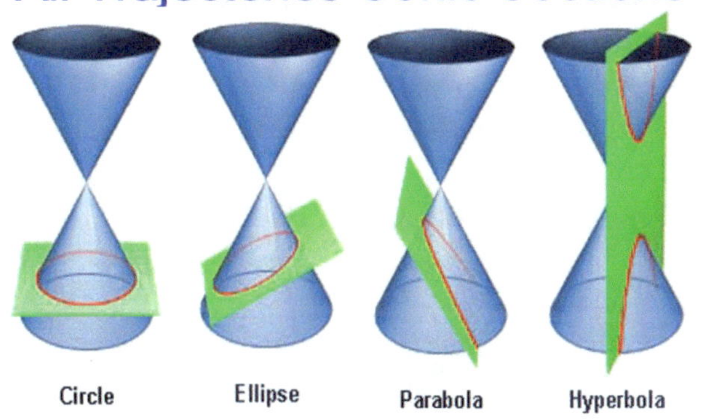

http://ccins.camosun.bc.ca/~jbritton/jbconics.htm

August 22, 2022 ORBO- 23

The Four Conic Sections: Closed & Open Orbits

Hyperbola e > 1

Parabola

Ellipse 0 < e < 1

Circle e = 0

Hyperbolic asymptotes

Courtesy Walter, **Astronautics**

August 22, 2022 ORBO- 24

Conic Section Equations

$$r = \frac{h^2/\mu}{1+e\cos\nu} = \frac{a(1-e^2)}{1+e\cos\nu} = \frac{p}{1+e\cos\nu}$$

$\nu =$ "true anomaly" angle (sometimes θ)

$$e = \text{eccentricity} = \frac{r_a - r_p}{r_a + r_p}$$

$p =$ parameter or "semilatus rectum"

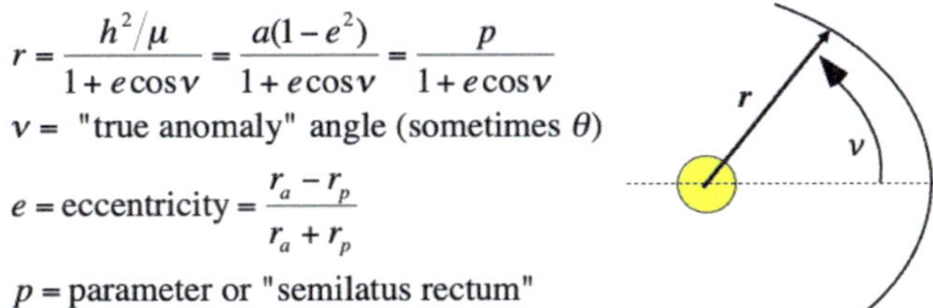

The term "semilatus rectum" means half the latus rectum, that is a translation from Greek used by Apollonius, whose work was used by Kepler in the development of his laws. It comes from a Latin Translation of *latus*, which means side, and *rectus*, which means straight or parallel to the directrix.
The Latin translation comes from the Greek translation of orthea pleaura, which is what Apollonius used to describe conic sections in his books. Through the evolution of languages the word rectum or rectos turned into "right," hence the term Right angle, and is the answer to the question asked as to why we have right triangles and not left ones. From this idea of vertical being right we also started to use the root of right as in good or proper, hence words like corRECT or RECKless.

August 22, 2022 ORBO- 25

Eccentricity *e* Defines Shape

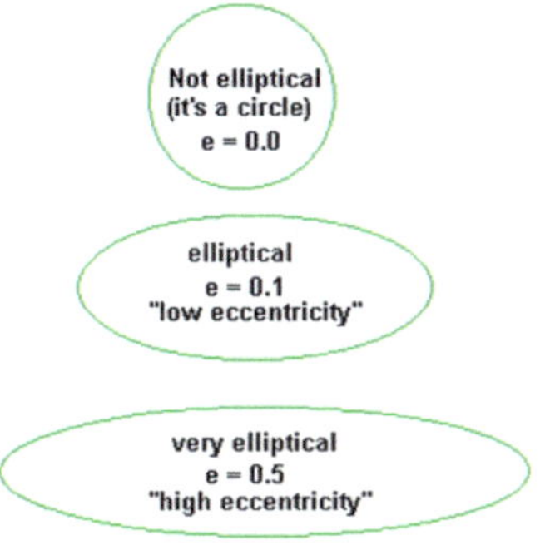

August 22, 2022 ORBO- 26

Four Kinds of Conics (note $\varepsilon \neq e$!)

Type	Eccentricity e	Specific Energy
Case 1: Circles	$e_{circle} = 0$	$\varepsilon_{circle} < 0$ (closed orbit)
Case 2: Ellipses	$0 < e_{ellipse} < 1$	$\varepsilon_{circle} < \varepsilon_{ellipse} < 0$ (closed orbit)
Case 3: Parabolas	$e_{parabola} = 1$	$\varepsilon_{parabola} = 0$ (escapes, no energy)
Case 4: Hyperbolas	$e_{hyperbola} > 1$	$\varepsilon_{hyperbola} > 0$ (escapes, $V > 0$)

"Gravity Well"

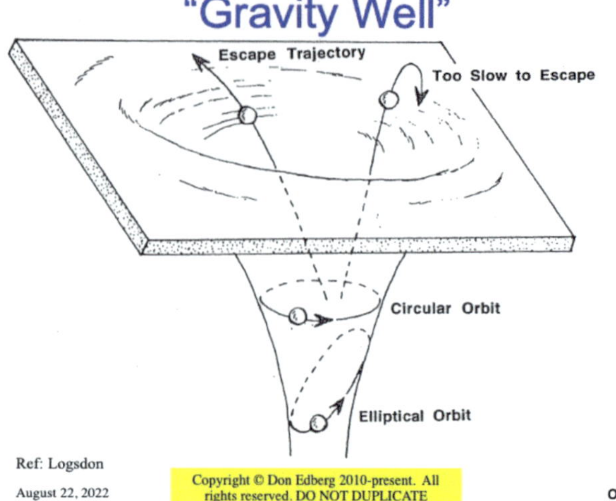

Ref: Logsdon

Four Orbit Types in Gravity Well

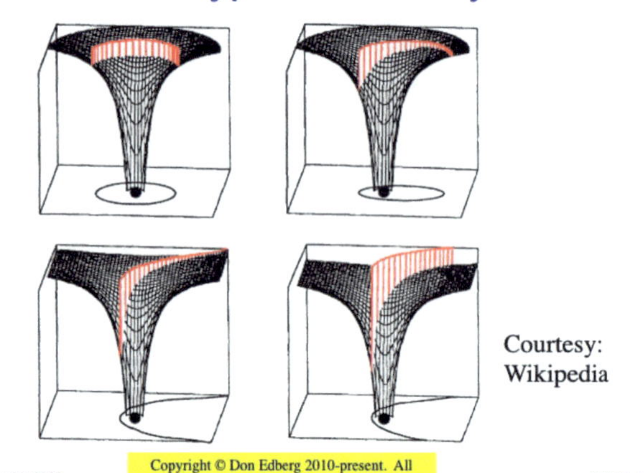

Courtesy: Wikipedia

LV Determines Energy ➔ Payload

- Maximum payload (spacecraft mass) determined by orbit energy requirements
- Interplanetary missions also have energy/time-of-flight/arrival speed constraints

Determine LV capabilities from:

1. LV Payload Planner's/User's Guides
2. Web sites such as NASA Launch services: http://elvperf.ksc.nasa.gov/elvMap/elvMap.ui.PerfGraph0?ReqType=Graph&OrbitType=C3&Contract=2&Vehicles=4

NASA LV Performance Curves

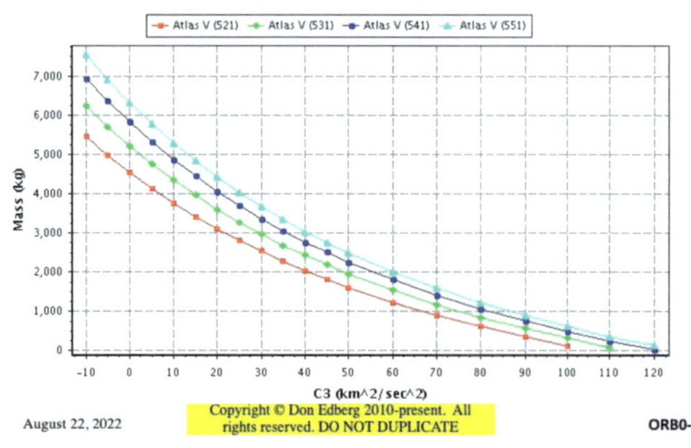

Conic Type 1: Circular Orbits

- Speed v_c is constant
- Speed is determined by orbital radius r (where $r = R_{planet} + h$) and attracting body's gravitational parameter μ:

$$v_c = \sqrt{\frac{\mu}{r}}$$

Circular Earth Orbital Period vs. Altitude

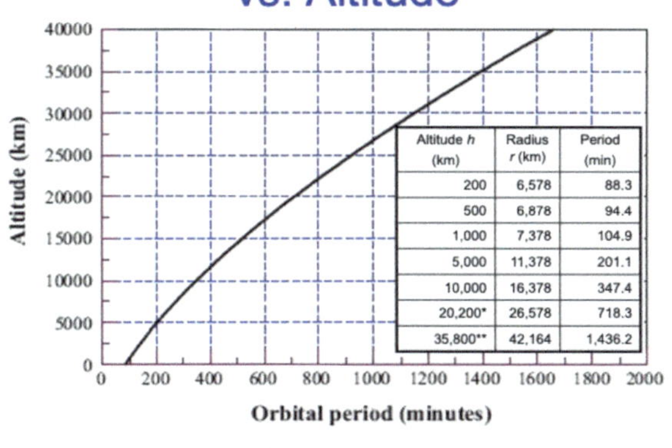

Altitude h (km)	Radius r (km)	Period (min)
200	6,578	88.3
500	6,878	94.4
1,000	7,378	104.9
5,000	11,378	201.1
10,000	16,378	347.4
20,200*	26,578	718.3
35,800**	42,164	1,436.2

Conic Type 2: Closed Orbit: Ellipse

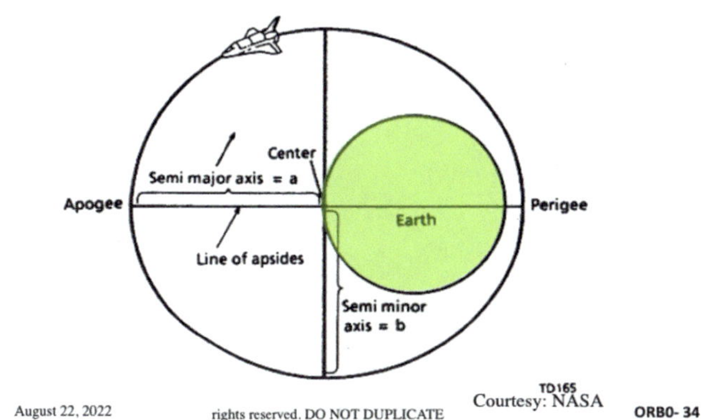

Courtesy: NASA

Orbits & Orbital Mechanics

- Why cover orbits *before* spacecraft design Answer: To look at how they affect design
- Orbital history & classification
- Circular & elliptical orbits
- Open orbits & escape trajectories
- Orbital Elements
- Special Orbits

Conic Type 4: Hyperbolic Trajectory

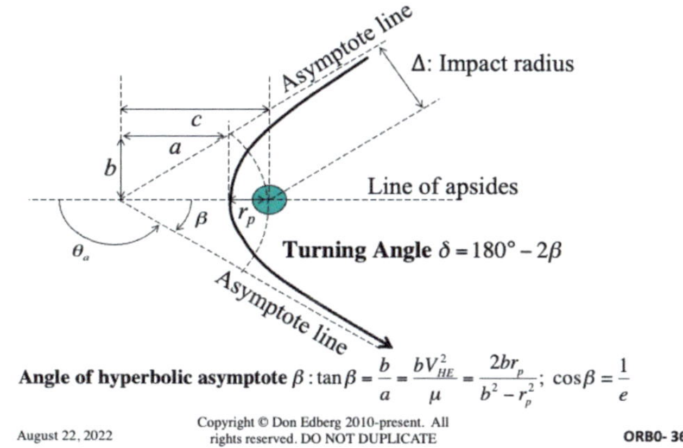

Angle of hyperbolic asymptote $\beta : \tan\beta = \frac{b}{a} = \frac{bV_{HE}^2}{\mu} = \frac{2br_p}{b^2 - r_p^2};\ \cos\beta = \frac{1}{e}$

"Pork Chop Plot," or Contours of *C3* (escape energy) for LV Selection

(Earth-to-Mars minimum C3 = 9 km²/s²; courtesy NASA)

Online calculator (2021): http://sdg.aero.upm.es/index.php/online-apps/porkchop-plot

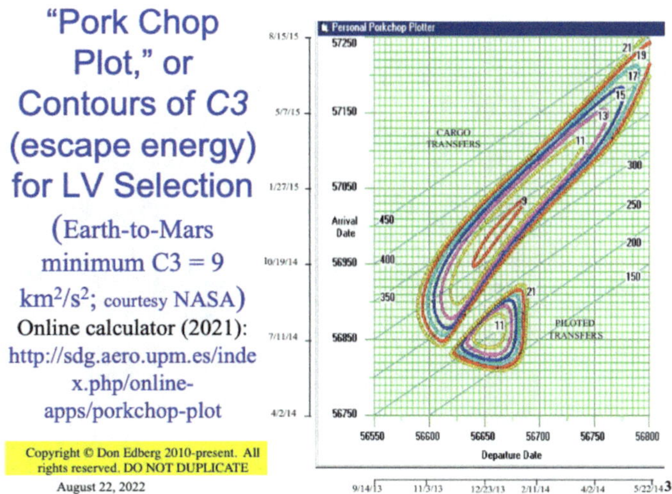

Earth-Centered Coordinate System

- Projection of Earth equator is *celestial equator*
- Earth's longitude (meridians) and latitude (parallels) give *right ascension* and *declination*
- North/South poles correspond to Earth's

north celestial pole
Right Ascension
Declination
celestial equator
south celestial pole
north celestial pole
star
DEC
RA
celestial equator
Vernal equinox (0h RA, 0° DEC)

Courtesy: Swinburne University

Sun-Centered Coordinate System & Reference Directions

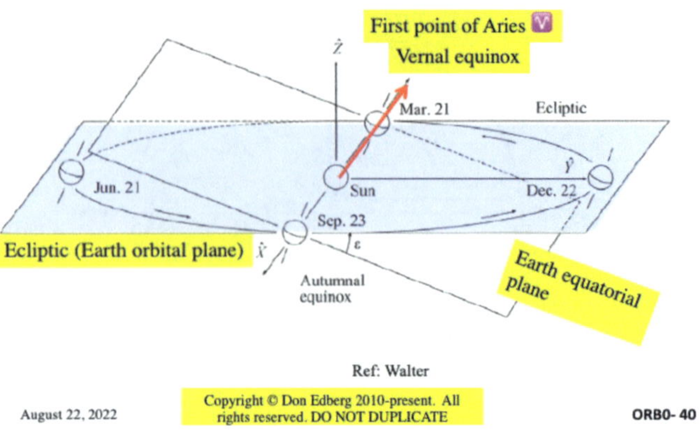

Ref: Walter

Orbits & Orbital Mechanics

- Why cover orbits *before* spacecraft design
  Answer: To look at how they affect design
- Orbital history & classification
- Circular & elliptical orbits
- Open orbits & escape trajectories
- Orbital Elements
- Special Orbits

Orbital Elements

For planar orbits, two elements determine the shape:

- Semimajor axis a
- Eccentricity e (don't confuse with energy ε).

To place a satellite within an orbit a third value is required: typically v (or θ), the "true anomaly" or angle from periapsis.

In general, six orbital elements are required to define an orbit

- Three added elements determine the orbital plane's orientation in space
- NOTE: orbital elements may be used in any coordinate system, including earth- and sun-centered

Orbit Geometry

Orbit's angular momentum vector **h**

i = inclination (orbit's tilt)

z = North

ν or θ = True anomaly: position in orbit (or time since periapsis passage)

Orbit

Periapsis

Orbital Plane

ω = Argument of Periapsis, or angle from Line of Nodes to periapsis line *in orbital plane*

Descending Node N_2

Line of Nodes

Line of Apsides

Plane of Equator

Line of Nodes

Ascending Node N_1

Ω = RAAN (Right Ascension or Longitude of Ascending Node), angle from Vernal Equinox *in equatorial plane*

Line of Apsides

Orbital Plane

Apoapsis

x ♈

Vernal Equinox (Reference dir.)

The Six Orbital Elements

Ref: Logsdon

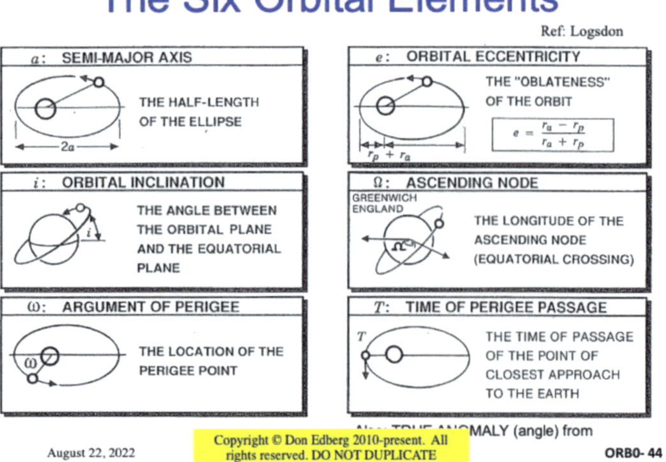

a: SEMI-MAJOR AXIS
THE HALF-LENGTH OF THE ELLIPSE
2a
e: ORBITAL ECCENTRICITY
THE "OBLATENESS" OF THE ORBIT
e = (r_a − r_p)/(r_a + r_p)
i: ORBITAL INCLINATION
THE ANGLE BETWEEN THE ORBITAL PLANE AND THE EQUATORIAL PLANE
Ω: ASCENDING NODE
GREENWICH ENGLAND
THE LONGITUDE OF THE ASCENDING NODE (EQUATORIAL CROSSING)
ω: ARGUMENT OF PERIGEE
THE LOCATION OF THE PERIGEE POINT
T: TIME OF PERIGEE PASSAGE
THE TIME OF PASSAGE OF THE POINT OF CLOSEST APPROACH TO THE EARTH

...MALY (angle) from

Orbital Elements – Movie

Orbital elements.mp4

Inclination

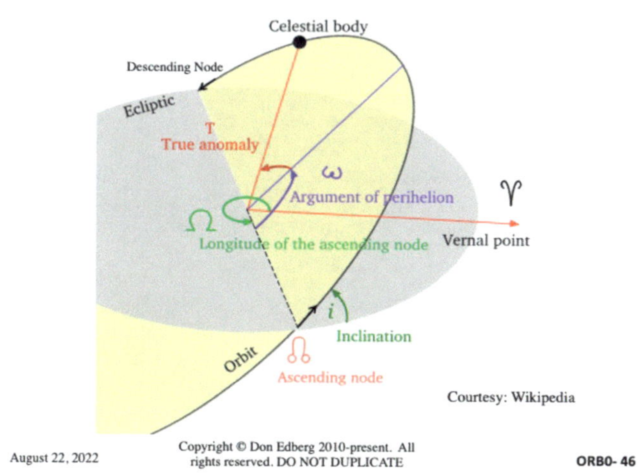

August 22, 2022 ORBO- 46

Orbits & Orbital Mechanics

- Why cover orbits *before* spacecraft design
  Answer: To look at how they affect design
- Orbital history & classification
- Circular & elliptical orbits
- Open orbits & escape trajectories
- Orbital Elements
- Special Orbits

August 22, 2022 ORBO- 47

Orbit Types: Described by Altitude, Eccentricity, & Inclination

LEO	Low Earth Orbit	Altitude < ~1000 km; any inclination, nearly circular	Space Shuttle, ISS, CubeSats
MEO	Medium Earth Orbit	Between LEO & GEO	GPS, comm sats
GEO	Geostationary Earth Orbit	Orbit period = Earth's, $i = 0°$. $R = 42164$ km, $h = 35786$ km.	TDRSS, satellite TV & radio
GTO	GEO Transfer Orbit	Elliptical orbit between LEO and GEO	Used by all GEO sats
HEO	High Earth Orbit	Higher than GEO	Chandra X-Ray Observatory
PEO	Polar Earth Orbit	90° Inclination, usually low	Spy sats
SSO	Sun-Synchronous Orbit	Between 90 – 100° inclination, usually low	Weather sats Spy sats
Escape	Hyperbolic orbit	Escapes from Earth's sphere of reference, orbits Sun	Interplanetary spacecraft
SEO	Subterranean Earth Orbit	Orbit which intersects Earth's surface, "lithobraking"	Failed launch, or atmosphere entry

August 22, 2022 ORBO- 48

Several Orbit Inclination Angles i

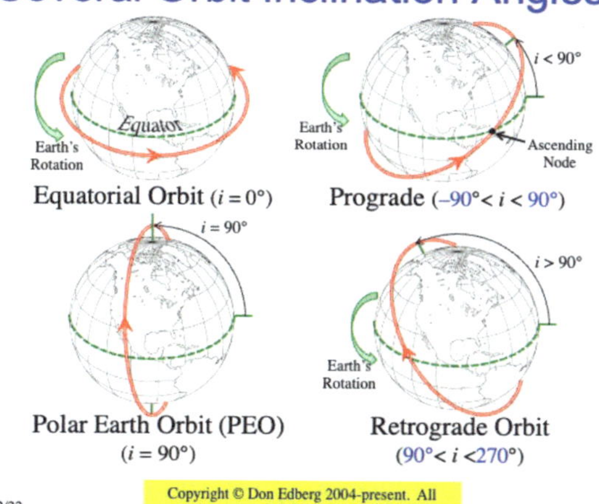

8/22/22 ORBO- 49

Movie: Prograde, Retrograde, & Sun Sync Orbits

pro-retro-sunsync.mp4

August 22, 2022 ORBO- 50

Alternate Orbital Descriptions

- Needed for "degenerate" orbital situations such as:
- Zero eccentricity (circular orbit)
- Zero inclination
- Kinds of elements
  - NORAD Two-Line-Elements (TLEs), Classical Elements with a particular way of interpreting perturbations
  - Latitude, Longitude, Altitude and Velocity
- Mathematical conversion possible between any of these

August 22, 2022 ORBO- 51

"Two-Line" Elements

- US Strategic Command (USSTRATCOM) maintains database of orbiting objects
- Objects described by *Two-Line elements* containing info on orbital elements and rates of change
- Software calculates position to a few km
- Sample:

```
Zarya module
123456789012345678901234567890123456789012345678901234567890123456789
---------|---------|---------|---------|---------|---------|---------
1 25544U 98067A   06287.64456019  .00008182  00000-0  53355-4 0  8609
2 25544 051.6354 292.0281 0013277 096.0881 057.9543 15.76874518452008
---------|---------|---------|---------|---------|---------|---------
123456789012345678901234567890123456789012345678901234567890123456789
```

Ref: Handbook of Space Flight

Two-Line Element Code (1 of 2)

Row	Column	Description	Example	Meaning
1	1	Line number identification (= 1)	1	First line
	3–7	NORAD catalog number (Example 16609)	25554	ISS (Zarya Module)
	8	Security classification	U	Not classified
	10–17	International COSPAR satellite identification (yynnnaaa), consisting of the year (yy), launch number (nnn) and piece letter (aaa)	98067A	First catalog object of the 67th launch of the year 1998
	19–20	Epoch of the orbital element (year)	06	2006
	21–32	Epoch day and fraction of 24-hour day (UTC)	287.64456019	October 14, 15:28:10
	34–43	First time derivative of the mean anomaly (in [rev/d²]) or ballistic coefficient *B*	.00008182	0.00008182
	45–52	Second time derivative of the mean anomaly (in [rev/d³]) (Decimal between columns 45 and 46; exponent in columns 51–52)	00000-0	0.0
	54–61	Bstar/drag term *B** (in [1/R_E]); (Decimal between columns 54 and 55; exponent in columns 60–61)	53355-4	$0.53355 \cdot 10^{-4}$
	63	Ephemeris type	0	SGP4 model
	65–68	Element number	860	860
	69	Check sum (modulo 10)	9	9

Two-Line Element Code (2 of 2)

Row	Column	Description	Example	Meaning
2	1	Line number identification (= 2)	2	Second line
	3–7	NORAD catalog number (example 16609)	25554	ISS (Zarya Module)
	9–16	Inclination (in [°])	051.6354	51.6354°
	18–25	Right ascension of ascending node (in [°])	292.0281	292.0281°
	27–33	Eccentricity with assumed leading decimal (between columns 26 and 27)	0013277	0.0013277
	35–42	Argument of the perigee (in [°])	096.0881	96.0881°
	44–51	Mean anomaly (in [°])	057.9543	57.9543°
	53–63	Mean motion (in [rev/d])	15.76874518	15.76874518 [rev/d]
	64–68	Revolution number at epoch	45200	45200
	69	Check sum (modulo 10)	8	8

Ref: Handbook of Space Flight

Kepler's Equation

- Used to calculate orbits numerically
- Variables are
  - true anomaly ν or θ
  - "mean anomaly" M
  - "eccentric anomaly" E
  - eccentricity e
- Solved numerically to provide time since periapsis
- $M = E - e \sin E$

Geostationary Orbit

- Period = 1 sidereal day (23 h 56 m 4 s)
- Must have zero inclination (equatorial)

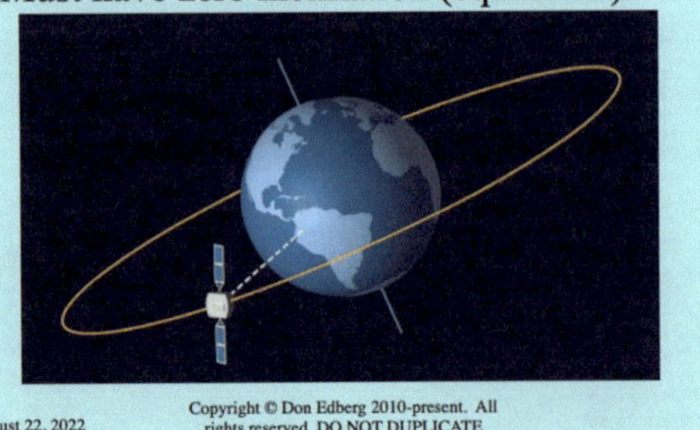

geostationary2.gif

Solar vs. Sidereal Day

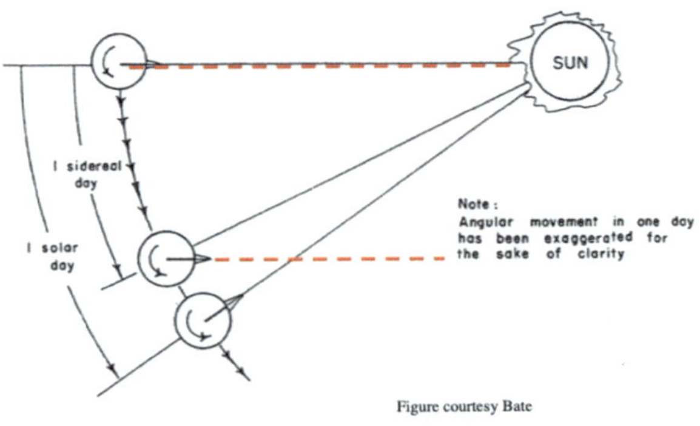

Figure courtesy Bate

Geostationary is Prime Real Estate

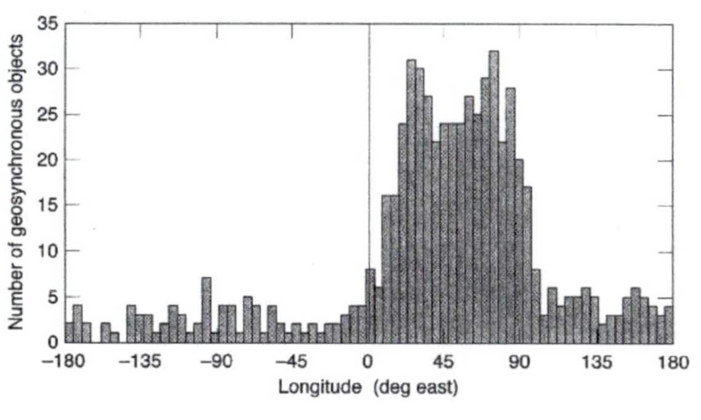

August 22, 2022 Ref: Rainey ORBO- 63

A Three-Spacecraft "Constellation: …

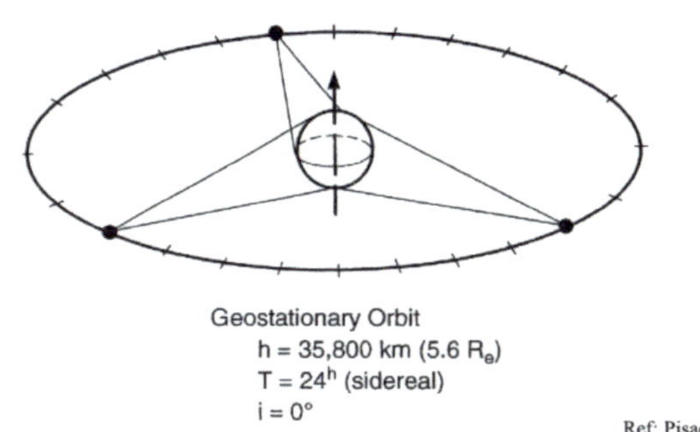

Geostationary Orbit
h = 35,800 km (5.6 R_e)
T = 24^h (sidereal)
i = 0°

Ref: Pisacane

August 22, 2022 ORBO- 64

… Covers Most of Earth

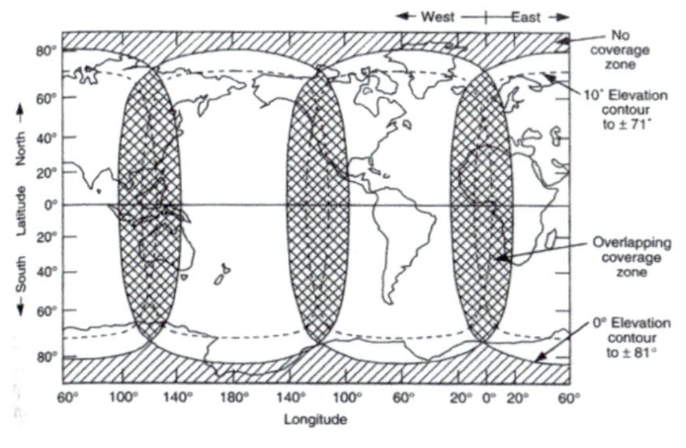

Ref: Pisacane

August 22, 2022 ORBO- 65

The Global Positioning System (GPS) Constellation

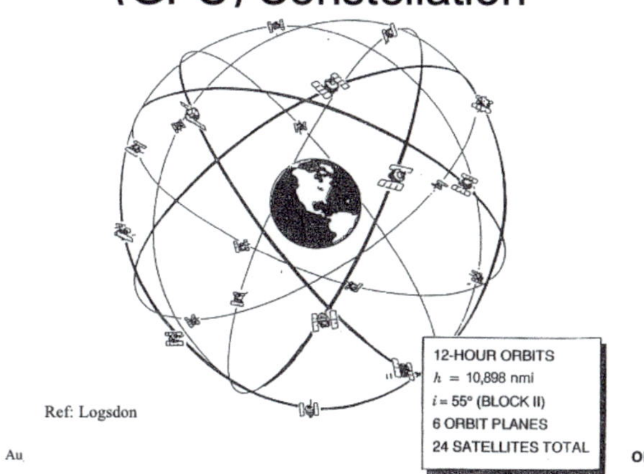

Ref: Logsdon

Au ORBO- 66

Blank slide

August 22, 2022 ORBO- 67

The Space Environment

August 22, 2022

ORB1A- #1

Space Environment Topics

- Non-uniform gravity fields (oblateness, "masscons")
- Orbital disturbances
  - Nodal regression
  - Periapsis advance
  - Station-keeping issues for GEO orbits
- Special orbits take advantage of oblate effects
  - Molniya orbits
  - Sun-synchronous orbits
- Lagrange points & restricted 3-body problem
- Space radiation: sources and effects
  - Types of radiation
  - The sun
  - Earth's radiation belts
- Summary

August 22, 2022

ORB1A- 2

Space Environment

Two effects:

- **Affects the orbits, ground tracks, & health of ALL spacecraft (independent of spacecraft) — *this* section**
- Effects on a particular spacecraft, the health & stability of the spacecraft itself & its systems: next section

Course Layout

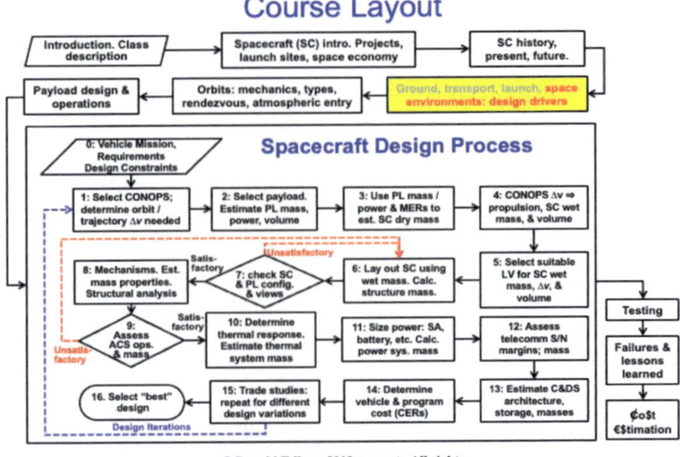

Introduction. Class description
Spacecraft (SC) intro. Projects, launch sites, space economy
SC history, present, future.
Payload design & operations
Orbits: mechanics, types, rendezvous, atmospheric entry
Ground, transport, launch, space environments: design drivers
Spacecraft Design Process
0: Vehicle Mission, Requirements Design Constraints
1: Select CONOPS; determine orbit / trajectory Δv needed
2: Select payload. Estimate PL mass, power, volume
3: Use PL mass / power & MERs to est. SC dry mass
4: CONOPS Δv ⇒ propulsion, SC wet mass, & volume
Unsatisfactory
Satis-factory
8: Mechanisms. Est. mass properties. Structural analysis
7: check SC & PL config. & views
6: Lay out SC using wet mass. Calc. structure mass.
5: Select suitable LV for SC wet mass, Δv, & volume
9: Assess ACS ops. & mass
10: Determine thermal response. Estimate thermal system mass
11: Size power: SA, battery, etc. Calc. power sys. mass
12: Assess telecomm S/N margins; mass
Testing
Failures & lessons learned
16. Select "best" design
15: Trade studies: repeat for different design variations
14: Determine vehicle & program cost (CERs)
13: Estimate C&DS architecture, storage, masses
Design Iterations
¢o$t €$timation

Space Environment Topics

- Non-uniform gravity fields (oblateness, "masscons")
- Orbital disturbances
  - Nodal regression
  - Periapsis advance
  - Station-keeping issues for GEO orbits
- Special orbits take advantage of oblate effects
  - Molniya orbits
  - Sun-synchronous orbits
- Lagrange points & restricted 3-body problem
- Space radiation: sources and effects
  - Types of radiation
  - The sun
  - Earth's radiation belts
- Summary

Gravitational Perturbations: Three Sources

1. "Bulge" at equator or *oblateness* or *flatness f* of spinning planet.
2. Earth density is not uniform (land vs. ocean, mountains, etc.)

These two effects are described by planet's gravitational potential equation

Also …

3. Added perturbations from sun, moon, other nearby planets

Oblateness or Equatorial Bulge

- Planet's spin creates "bulge" at equator or *oblateness* or *flatness f*

  Note: *flatness* $f = \frac{R_{polar}}{R_{equator}}$

  $f_{Earth} = 0.003353$

  $f_{Mars} = 0.00648$

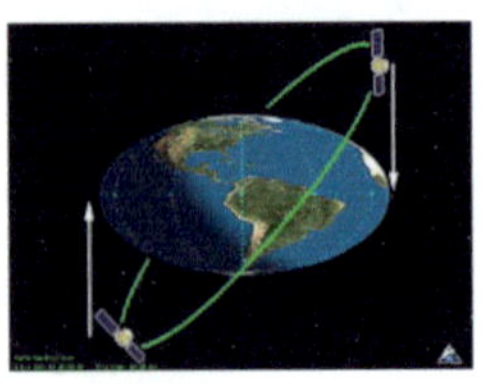

- Bulge attracts SC; torque on orbit creates two effects:
  - Nodal regression
  - Rotation of the line of apsides

Gravitational Potential Described in Power Series

$U(r, \Phi, \Lambda) =$ Gravitational Potential (radius, lat., long.)

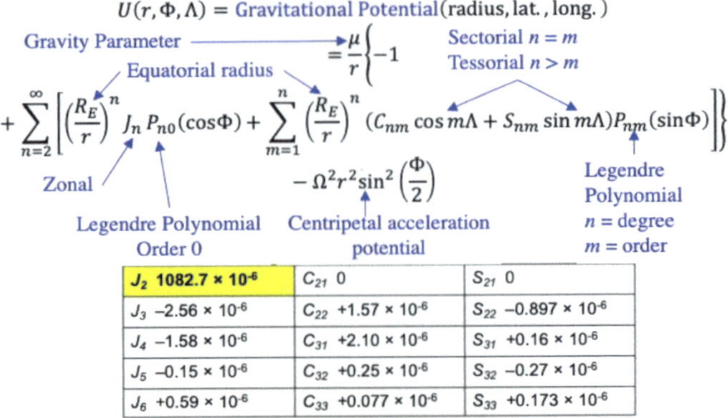

$$= \frac{\mu}{r}\left\{-1 + \sum_{n=2}^{\infty}\left[\left(\frac{R_E}{r}\right)^n J_n P_{n0}(\cos\Phi) + \sum_{m=1}^{n}\left(\frac{R_E}{r}\right)^n (C_{nm}\cos m\Lambda + S_{nm}\sin m\Lambda)P_{nm}(\sin\Phi)\right]\right\} - \Omega^2 r^2 \sin^2\left(\frac{\Phi}{2}\right)$$

Gravity Parameter
Equatorial radius
Sectorial $n = m$
Tessorial $n > m$
Zonal
Legendre Polynomial Order 0
Centripetal acceleration potential
Legendre Polynomial
n = degree
m = order

J_2 1082.7 × 10^{-6}	C_{21} 0	S_{21} 0
J_3 −2.56 × 10^{-6}	C_{22} +1.57 × 10^{-6}	S_{22} −0.897 × 10^{-6}
J_4 −1.58 × 10^{-6}	C_{31} +2.10 × 10^{-6}	S_{31} +0.16 × 10^{-6}
J_5 −0.15 × 10^{-6}	C_{32} +0.25 × 10^{-6}	S_{32} −0.27 × 10^{-6}
J_6 +0.59 × 10^{-6}	C_{33} +0.077 × 10^{-6}	S_{33} +0.173 × 10^{-6}

Types of Shape Functions

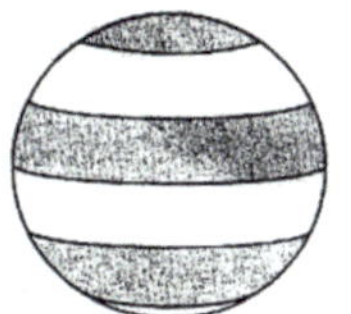
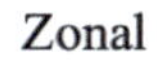
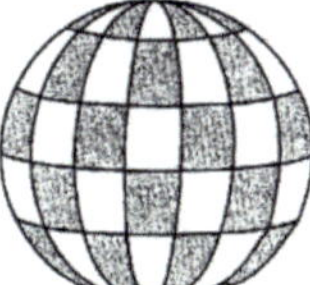
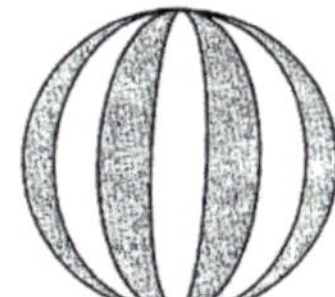

Zonal	Tesseral	Sectoral
Depend only on latitude	Depend on latitude & longitude	Depend only on longitude

Source: Pisacane

August 22, 2022

ORB1A- 9

J_2 — Zonal Coefficient (Due to Oblateness)

J_2: different from oblateness/flatness f

Planet	J2
Earth	0.00108263
Mars	0.00196
Jupiter	0.0146964
Saturn	0.01629071

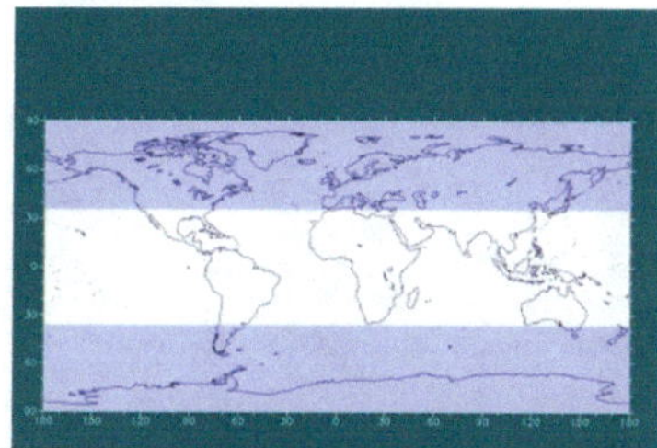

Data on Outer Planet Gravity Fields are available online: ssd.jpl.nasa.gov/?gravity_fields_op [Accessed 10/14/13]

August 22, 2022

ORB1A- 10

Space Environment Topics

- Non-uniform gravity fields (oblateness, "masscons")
- Orbital disturbances
  - Nodal regression
  - Periapsis advance
  - Station-keeping issues for GEO orbits
- Special orbits take advantage of oblate effects
  - Molniya orbits
  - Sun-synchronous orbits
- Lagrange points & restricted 3-body problem
- Space radiation: sources and effects
  - Types of radiation
  - The sun
  - Earth's radiation belts
- Summary

August 22, 2022

ORB1A- 11

Regression of Nodes Due to Torque From Equatorial Bulge

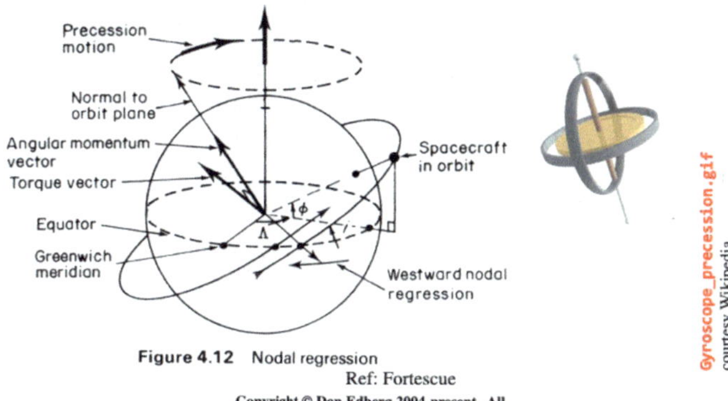

Figure 4.12 Nodal regression

Ref: Fortescue

August 22, 2022

ORB1A- 12

Nodal Regression Defined By

$$\dot{\Omega} = \frac{d\Omega}{dt} = -\frac{3nJ_2R_0^2\cos i}{2a^2(1-e^2)^2} = -\frac{3\mu^{1/2}J_2R_0^2\cos i}{2a^{7/2}(1-e^2)^2}$$

- Ω = place where orbit "cuts through" the equatorial plane relative to vernal equinox
- n = orbital rate = $(\mu/a^3)^{1/2}$
- μ = gravitational parameter
- J_2 = zonal coefficient describing "bulge"
- R_0 = Earth mean radius
- i = orbital inclination (orbit tilt relative to equator)
- a = orbit semimajor axis
- e = orbit eccentricity (how "flat" orbit is)

Earth: $\dot{\Omega} = -2.06474\times10^{14}\frac{\cos i}{2a^{3.5}(1-e^2)^2}\frac{\text{deg}}{\text{day}}$ (a in km)

August 22, 2022

ORB1A- 13

Nodal Regression: *Earth* Orbit

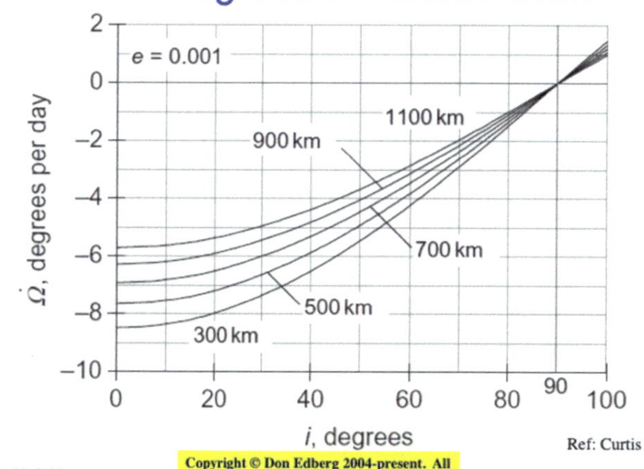

Ref: Curtis

August 22, 2022

ORB1A- 14

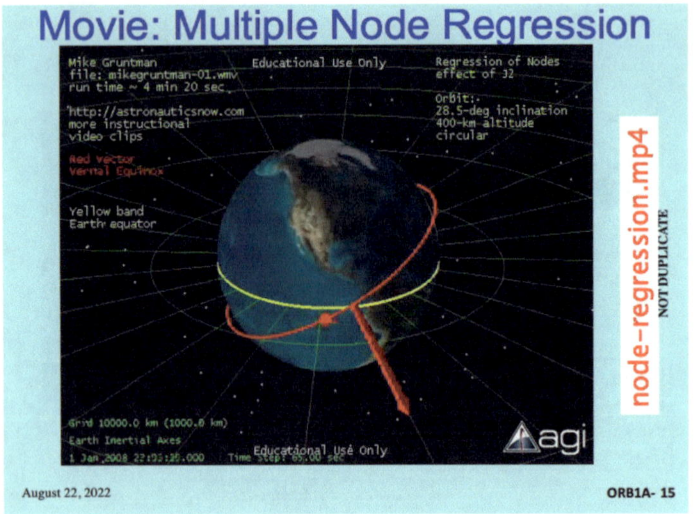

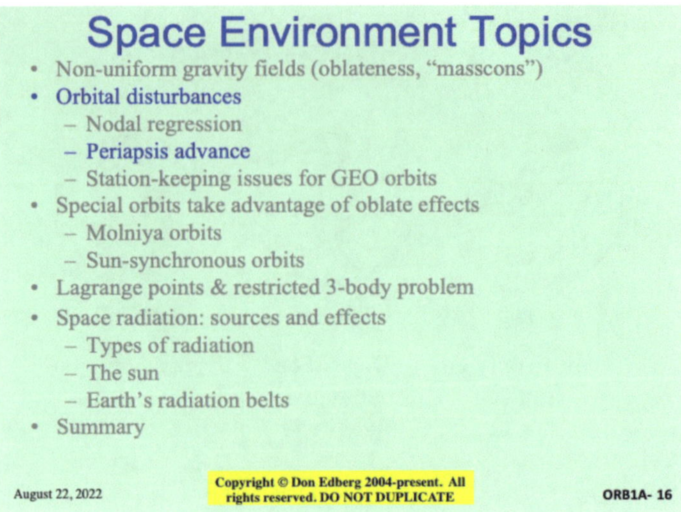

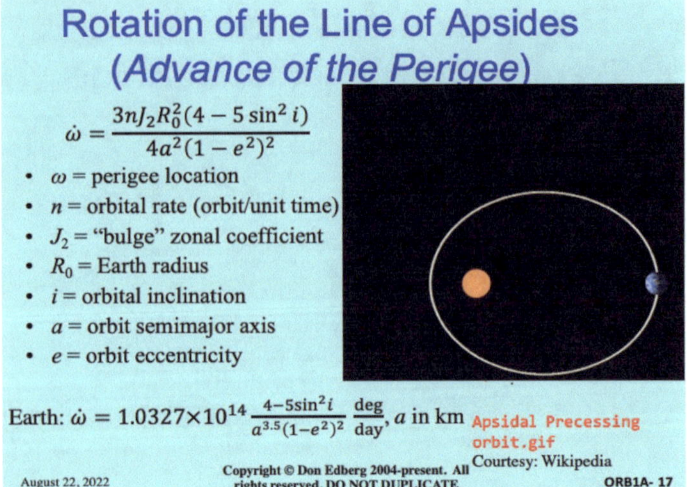

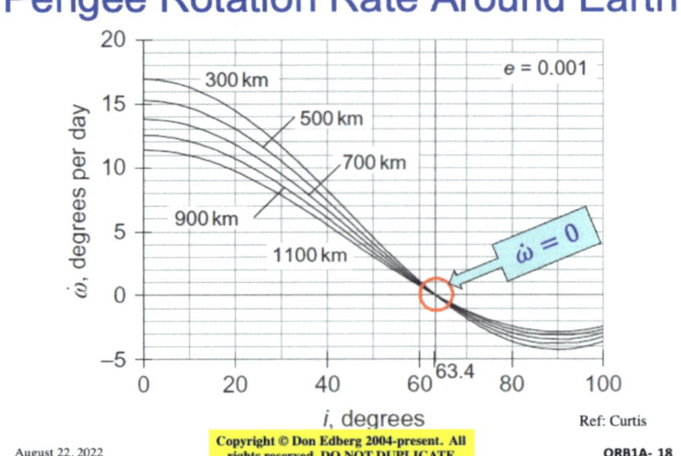

Simplified Regression Equations

$$\dot{\Omega} = \frac{d\Omega}{dt} = -A\cos i$$

$$\dot{\omega} = \frac{A}{2}(4 - 5\sin^2 i)$$

$$\text{where } A \equiv 3J_2\left(\frac{R_{planet}}{2r_ar_p}\right)^2\sqrt{2\mu(r_a+r_p)}$$

August 22, 2022

ORB1A- 19

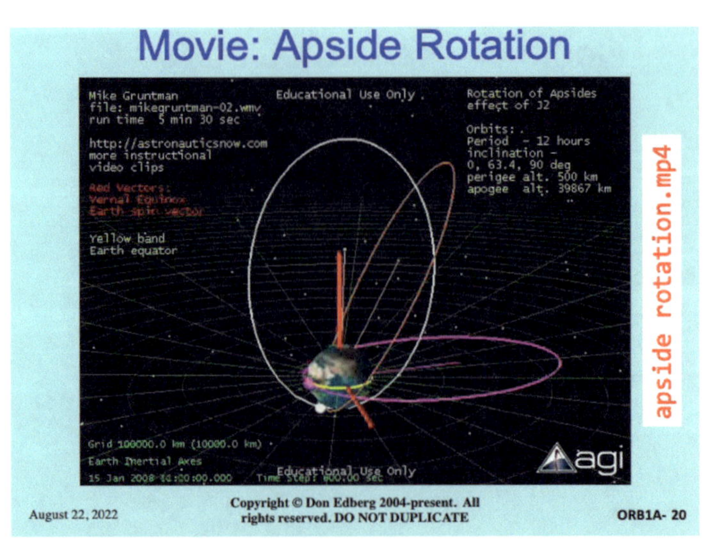

Space Environment Topics

- Non-uniform gravity fields (oblateness, "masscons")
- **Orbital disturbances**
  - Nodal regression
  - Periapsis advance
  - **Station-keeping issues for GEO orbits**
- Special orbits take advantage of oblate effects
  - Molniya orbits
  - Sun-synchronous orbits
- Lagrange points & restricted 3-body problem
- Space radiation: sources and effects
  - Types of radiation
  - The sun
  - Earth's radiation belts
- Summary

Equatorial Ellipticity of Earth Affects East-West Station-Keeping

- Earth uneven along equator. Actual shape of equator is not circular
- Similar to J_2 bulge, attracts SC longitudinally
- Causes station-keeping issues with equatorial-orbiting SC (drift in longitude)
- Requires station-keeping propellant

11.4° W Metastable
Acceleration
Drift
λ_s = 104.9° W Stable
λ_s = 75.1° E Stable
Metastable 162.1° E

Ref: Walter

Equatorial Longitudinal Acceleration

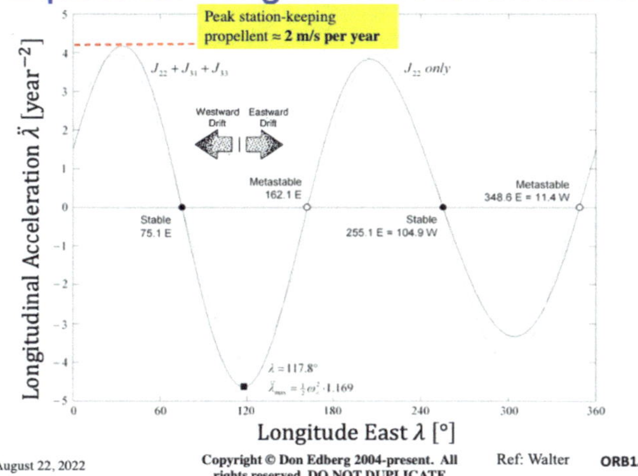

Ref: Walter

Multi-Body Effects

- Gravity from additional objects complicates matters greatly
- Effects on Earth-orbiters are primarily due to the Sun and Moon
- North-South station-keeping fuel needed to provide $\sim \frac{\mathbf{48\ m/s}}{\mathbf{year}}$ to maintain 0° GEO due to sun & moon gravity (per Logsdon)

Gravitational Perturbations: Reality is More Complicated Than Two Body Motion

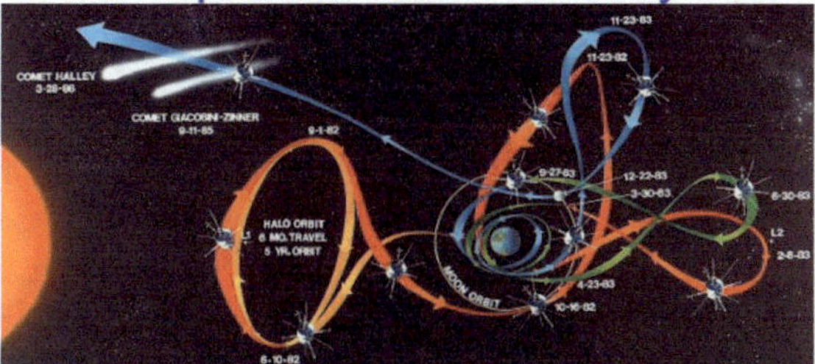

- No explicit solution exists for the three-body problem (or more bodies)
- Solution is to use numerical integration

Space Environment Topics

- Non-uniform gravity fields (oblateness, "masscons")
- Orbital disturbances
  - Nodal regression
  - Periapsis advance
  - Station-keeping issues for GEO orbits
- **Special orbits take advantage of oblate effects**
  - **Molniya orbits**
  - Sun-synchronous orbits
- Lagrange points & restricted 3-body problem
- Space radiation: sources and effects
  - Types of radiation
  - The sun
  - Earth's radiation belts
- Summary

Special Orbit #1: Stay Still With Respect to the Equator

- Say you want orbit to *stay put* relative to equator, so periapsis and apoapsis do not move (normally the apsides rotate, right?)
- This is an orbit that has *zero* apsidal rotation
- Choose inclination i to get zero apsidal rotation: $4 - 5\sin^2 i = 0$
- Solutions: $i = 63.4°$, or $116.6°$ (retrograde)

August 22, 2022

ORB1A- 27

Special Orbit: *Molniya* Orbit

- 12 h period elliptical orbit
- $i = 63.4°$ or $116.6°$ (no apsidal rotation, so fixed with respect to Earth)
- Good coverage for higher latitudes (i.e. Russia)
- Only small plane change required for Russia (57° launch site latitude)

August 22, 2022

ORB1A- 28

More on Molniya

- 12 h period elliptical orbit
- Good coverage of higher latitudes
- Constellation of two or three at various longitudes needed for continuous coverage

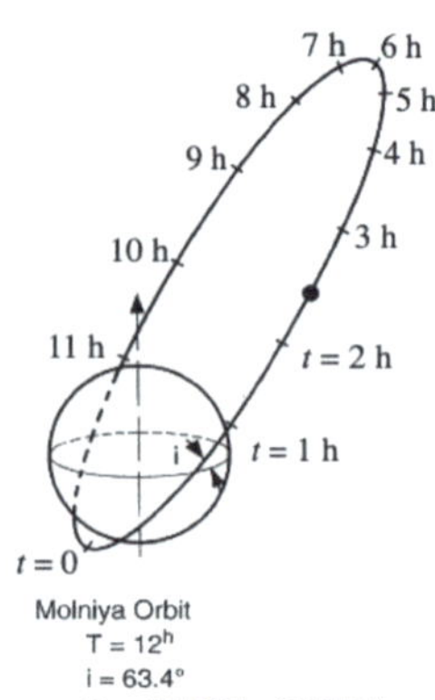

Ref: Pisacane

August 22, 2022

ORB1A- 29

Movie: Molniya Orbit

molniya-orbit.mp4

August 22, 2022

ORB1A- 30

Molniya Ground Track

~ 8 hr coverage of high-latitude areas per orbit

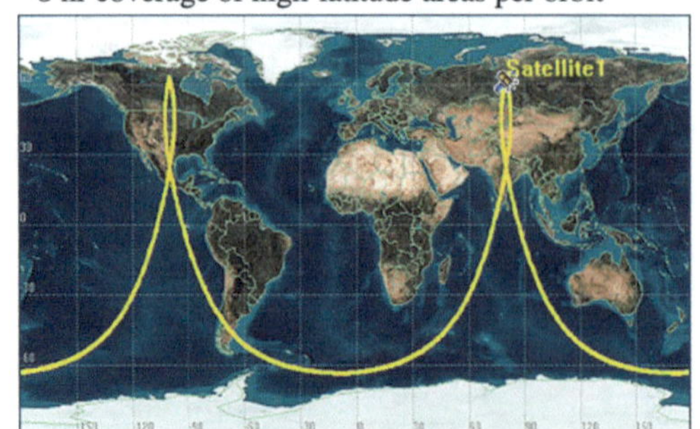

August 22, 2022

ORB1A- 31

Problem With Molniya Orbit

- Apogee remains over N hemisphere
- SV reaches apogee different times of day
- Solved with ACE orbit

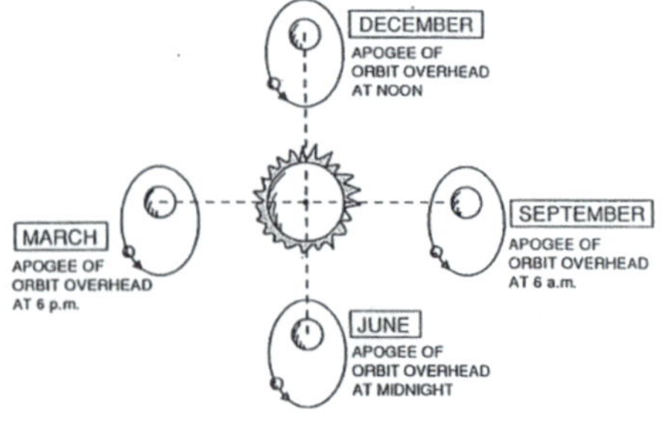

Courtesy: Logsdon

August 22, 2022

ORB1A- 32

ACE (Apogee at Constant time-of-day Equatorial) Orbit

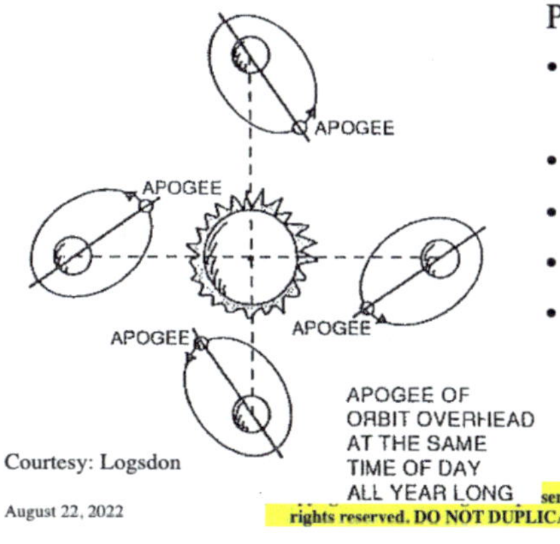

Courtesy: Logsdon

Parameters

- Period = 4.8 hours (typical)
- Inclination = 0°
- Perigee = 557 NM
- Apogee = 8,154 NM
- Apsidal rotation rate = 360°/year (sun-synchronous apsidal rotation)

Space Environment Topics

- Non-uniform gravity fields (oblateness, "masscons")
- Orbital disturbances
  - Nodal regression
  - Periapsis advance
  - Station-keeping issues for GEO orbits
- Special orbits take advantage of oblate effects
  - Molniya orbits
  - Sun-synchronous orbits
- Lagrange points & restricted 3-body problem
- Space radiation: sources and effects
  - Types of radiation
  - The sun
  - Earth's radiation belts
- Summary

Special Orbit 2: *Sun-Synchronous*

- Want to fly over a specific location at the same time of day each day
- *Nodal regression rate* must equal rate planet travels around sun
- Earth moves 360° in 365 days, or ~1°/day (actually 0.9856°/day). Adjust inclination i to provide equal nodal regression
- *Retrograde* orbit ($i > 90°$) needed
- Spacecraft flies over a given point at the same time, same sun angle, each day
- Good for earth photography: weather, spy photos, sensing ground construction, etc. Est: *N-1* 102 m high, actual 105 m.

Soviet *N-1* Courtesy: NRO

Sun-Synchronous Orbit

Inclination $i > 90°$ chosen so orbital drift matches change in sun angle each day: 360°/365.25 days = 0.9856°/day for Earth

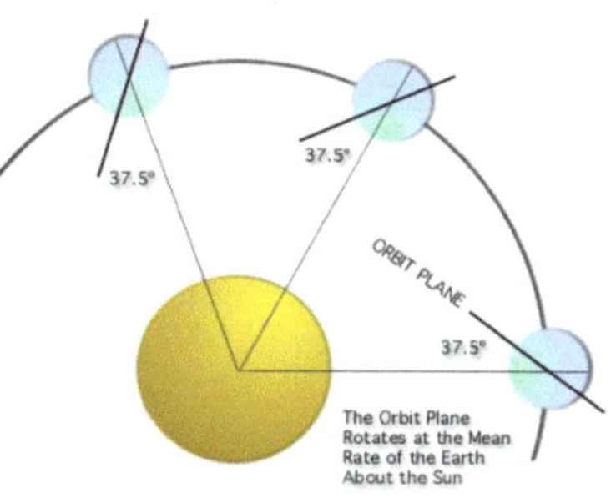

Sun-Synchronous Earth Orbits

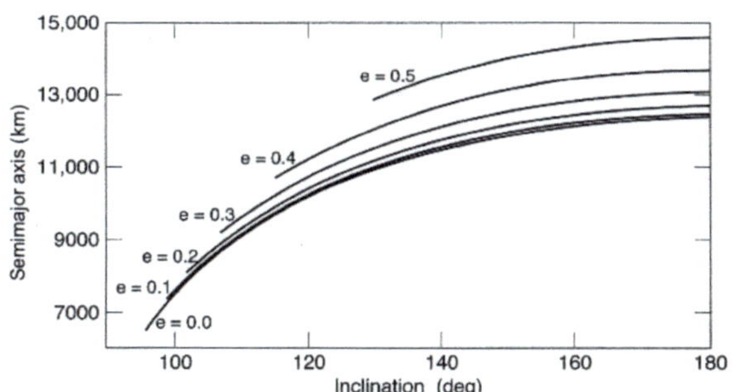

Fig. 12.3. Solution space for sun-synchronous orbits.

Ref: Rainey

Sun-Sync Orbits Referred by Time:

LST = Local Solar Time
LTDN = Local Time of Descending Node
(courtesy Aviation Week)

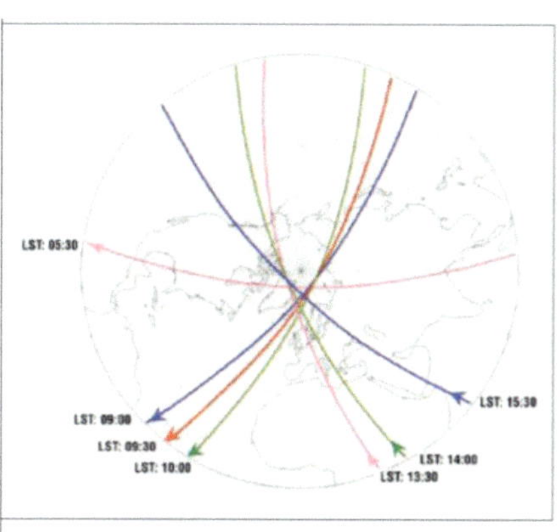

The World Meterological Organization's polar LEO coverage plan includes seven orbits: DWSS and the NOAA/NASA Joint Polar Satellite System (pink) in the 05:30 and 13:30 slots, respectively; Russia's Meteor (blue) in the 09:00 and 15:30 slots; Europe's EPS-SG (red) in the 09:30 slot; and China's FY-3 spacecraft (green) in the 10:00 and 14:00 slots.

Special "*Full-Sun*" Sun-Sync Orbit: Always Illuminated!

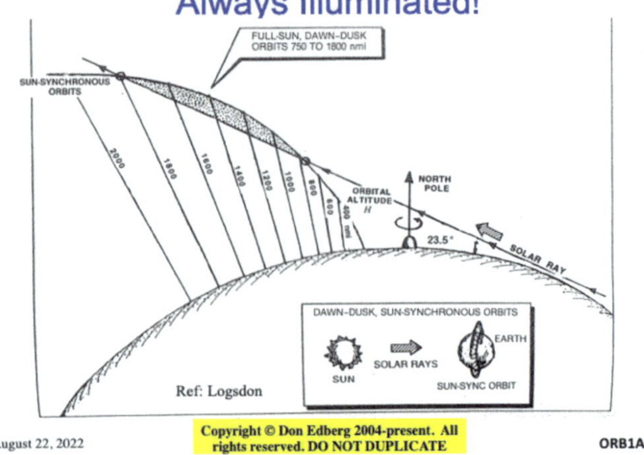

Ref: Logsdon

Special Orbits: Summary

Courtesy: Logsdon

Space Environment Topics

- Non-uniform gravity fields (oblateness, "masscons")
- Orbital disturbances
  - Nodal regression
  - Periapsis advance
  - Station-keeping issues for GEO orbits
- Special orbits take advantage of oblate effects
  - Molniya orbits
  - Sun-synchronous orbits
- Lagrange points & restricted 3-body problem
- Space radiation: sources and effects
  - Types of radiation
  - The sun
  - Earth's radiation belts
- Summary

Lagrange Points and the Restricted Three-Body Problem

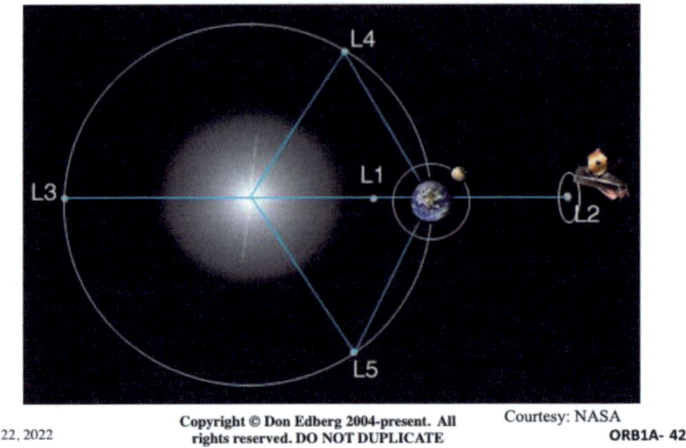

Courtesy: NASA

Lagrange Points

- Locations where the *combined gravitational pull* of the two large masses provides the exact centripetal force required to orbit with them
- Sun-Earth *L1* has been the destination for several Sun-science missions (*ISEE-3* in 1980s, *SOHO*, *Genesis*, others planned). JWST will be in *L2* (previous slide)
- Note that all Lagrange points are in motion, "revolving in formation" with main attracting bodies
- Scott Manley 13m30s explaining video at `https://www.youtube.com/watch?v=7PHvDj4TDfM`

Lagrange Points Rotate with System

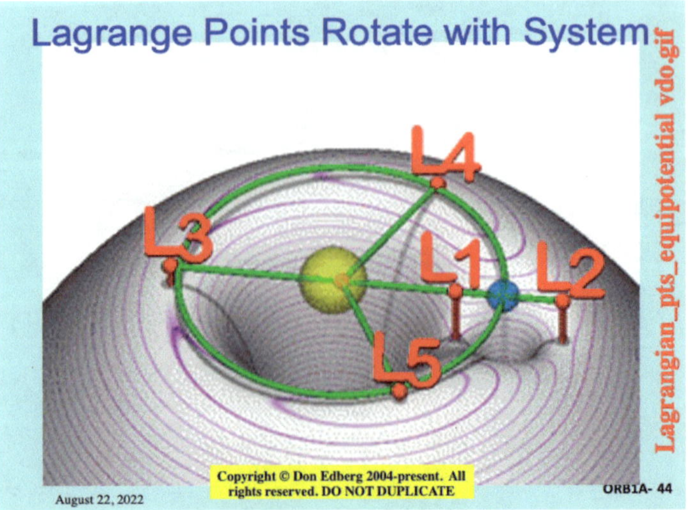

Stability of Lagrange Points

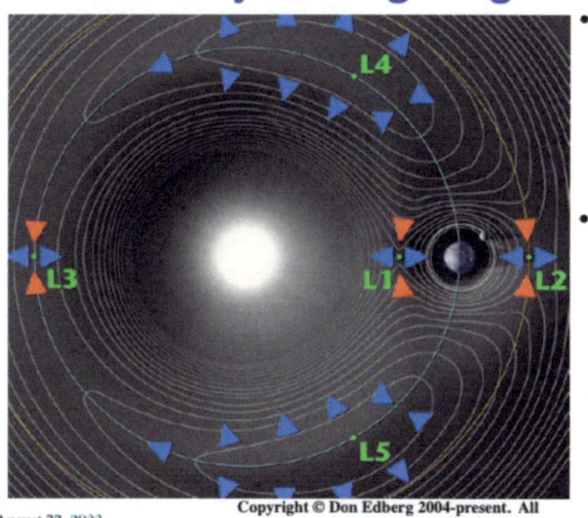

L4
L3
L1
L2
L5

- All "in-line" Lagrange points (L1-3) are unstable (red arrows)
- All others (L4, L5) are 'stable' (small motions only)

August 22, 2022 ORB1A- 45

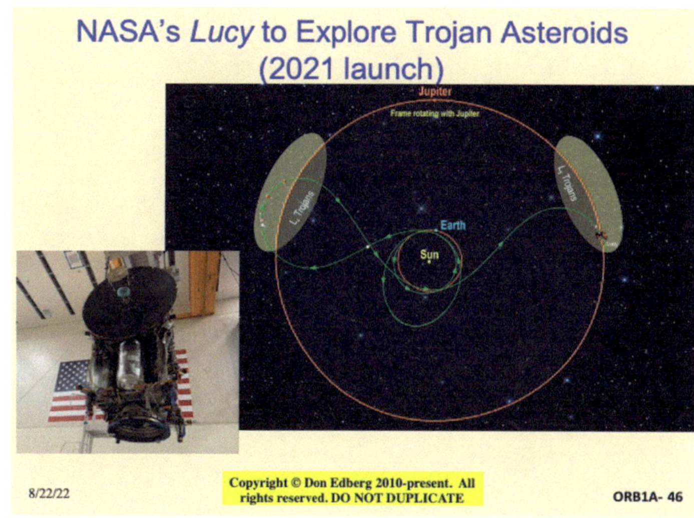

NASA's *Lucy* to Explore Trojan Asteroids (2021 launch)
Jupiter
Frame rotating with Jupiter
L₄ Trojans
L₅ Trojans
Earth
Sun

8/22/22 ORB1A- 46

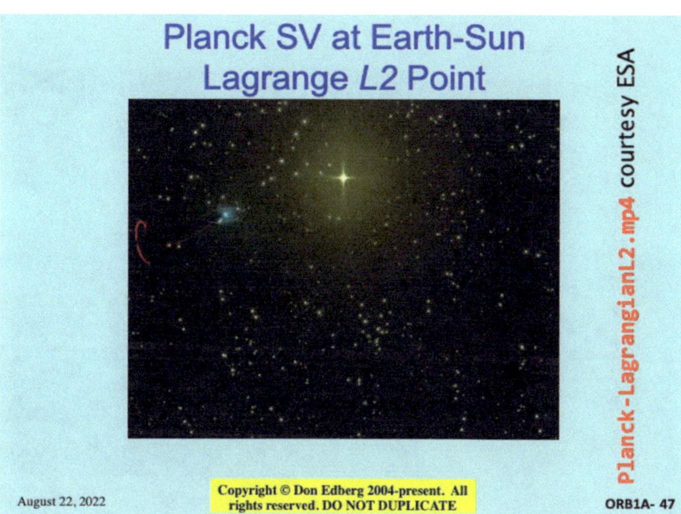

Planck SV at Earth-Sun Lagrange *L2* Point
Planck-LagrangianL2.mp4 courtesy ESA

August 22, 2022 ORB1A- 47

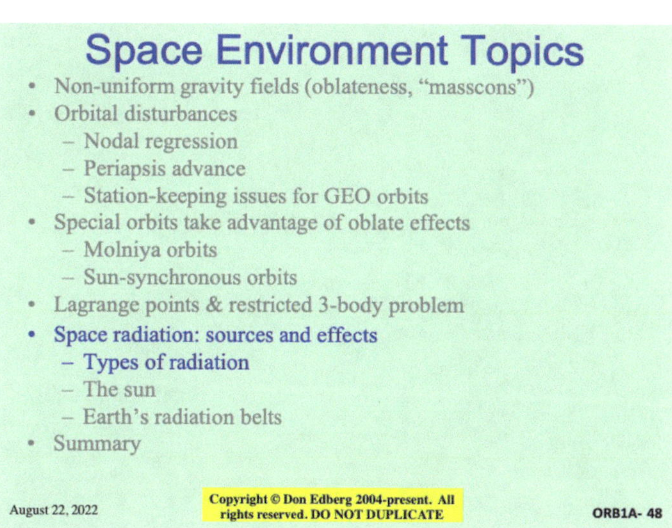

Space Environment Topics
• Non-uniform gravity fields (oblateness, "masscons")
• Orbital disturbances
– Nodal regression
– Periapsis advance
– Station-keeping issues for GEO orbits
• Special orbits take advantage of oblate effects
– Molniya orbits
– Sun-synchronous orbits
• Lagrange points & restricted 3-body problem
• Space radiation: sources and effects
– Types of radiation
– The sun
– Earth's radiation belts
• Summary

August 22, 2022 ORB1A- 48

Space Radiation

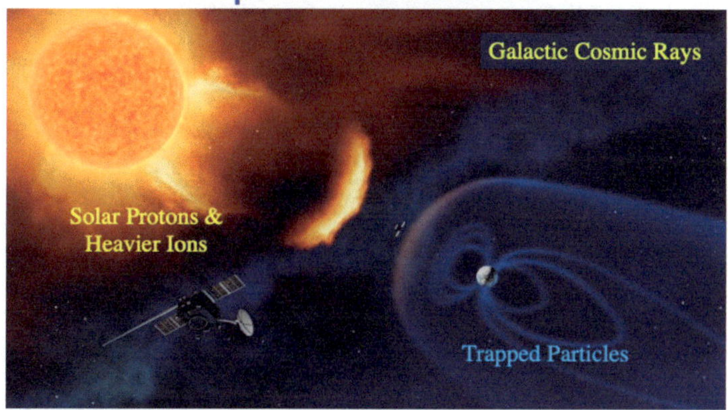

Galactic Cosmic Rays
Solar Protons & Heavier Ions
Trapped Particles

August 22, 2022 ORB1A- 49

Radiation Types

- Galactic cosmic rays (GCR)
- Solar flares
- Solar coronal mass ejections
- Protons
- Electrons

TRACE/NASA

Extreme UV image courtesy NASA's Solar Dynamics Observatory

August 22, 2022 ORB1A- 50

Cosmic Ray Summary

- Cosmic ray = ionized (+ or – charged) atom: 83% H, 13% He, 1% heavy elements, 3% electrons
- Heavy elements more damaging
- Highly variable – big events rare but very dangerous
- Earth's magnetic field is an effective shield (but flux over poles same as in deep space)
- Astronaut in spacesuit during EVA could not survive dose from a single major flare
- Hiding on the back side of the moon does not help
- GCR flux lower during solar max (no explanation)

Courtesy Karl Pfitzer, Boeing

August 22, 2022 ORB1A- 51

Proton/Heavy Ion Radiation Interactions

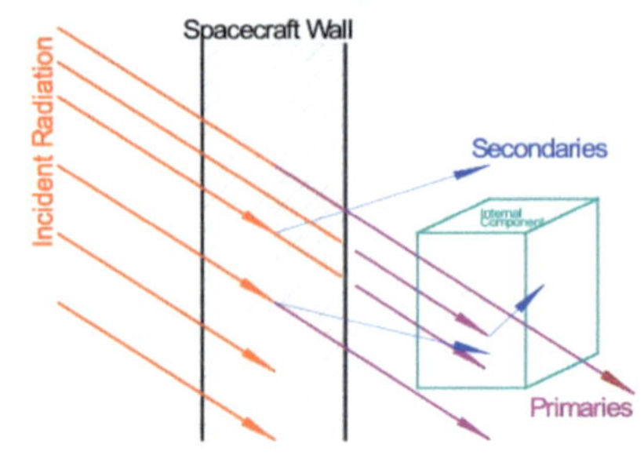

- Use low Z (atomic #) material for shielding – do not use lead
- Hydrogen is the most effective shield
- Secondaries may actually increase dose inside shield

Courtesy Karl Pfitzer, Boeing

August 22, 2022 ORB1A- 52

Space Environment Topics

- Non-uniform gravity fields (oblateness, "masscons")
- Orbital disturbances
  - Nodal regression
  - Periapsis advance
  - Station-keeping issues for GEO orbits
- Special orbits take advantage of oblate effects
  - Molniya orbits
  - Sun-synchronous orbits
- Lagrange points & restricted 3-body problem
- Space radiation: sources and effects
  - Types of radiation
  - The sun
  - Earth's radiation belts
- Summary

August 22, 2022 ORB1A- 53

The Active Sun

- Solar storms
  - Solar Flares
  - Coronal Mass Ejections (CMEs)
- GCR >> CME > solar flares

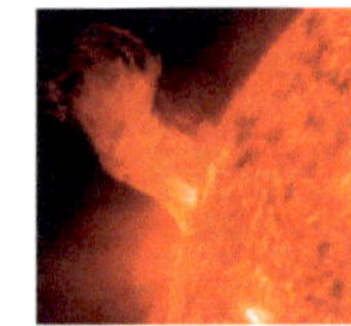

August 22, 2022 ORB1A- 54

Sun: 11 Year Activity Cycle

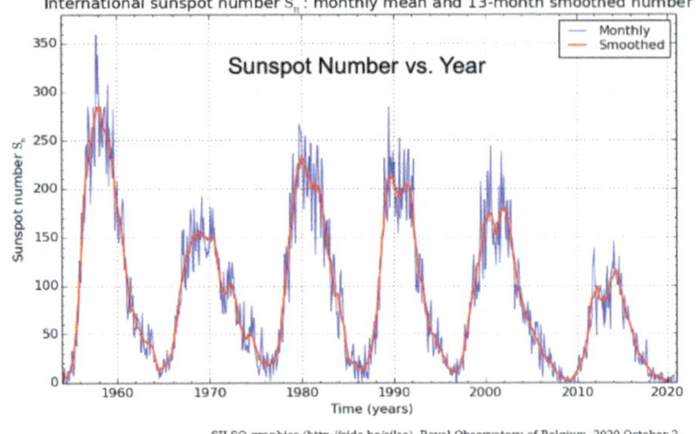

August 22, 2022 ORB1A- 55

Solar Flare: Multiple Effects

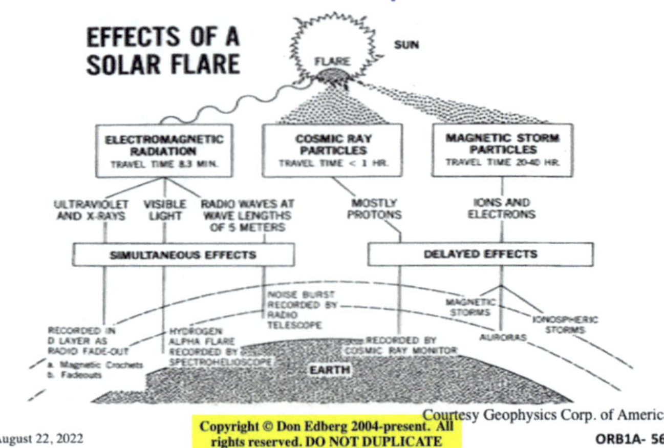

Courtesy Geophysics Corp. of America

August 22, 2022 ORB1A- 56

Solar Proton Events

Enough energy to penetrate spacecraft & affect devices

- May last hours or days
- Intensity varies from event to event, orders of mag. in minutes

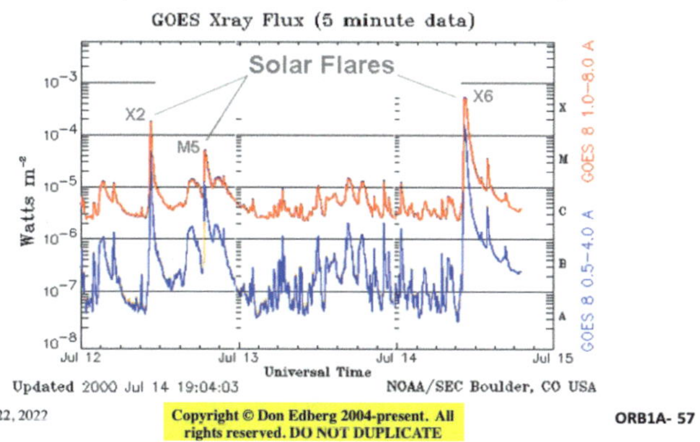

August 22, 2022

ORB1A- 57

Solar Proton Event Signature

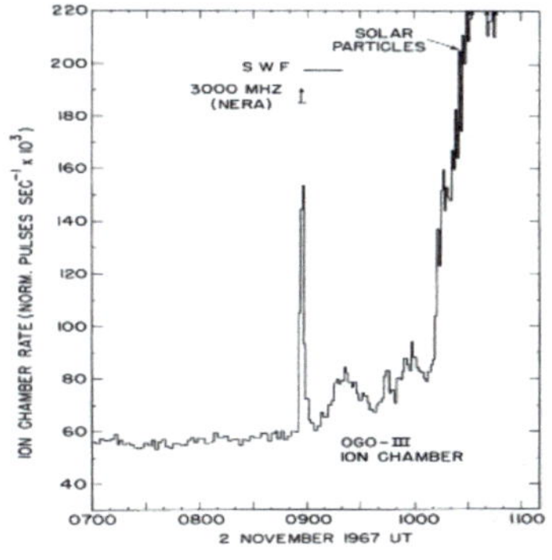

Event onset observed via visual flare, radio & x-ray burst arriving at light speed

Protons come after some delay. Delay varies, depends on high energy content of flare (more energetic protons arrive 1st).

Delay and rise rate of the onset used to predict event intensity.

Courtesy Karl Pfitzer, Boeing

August 22, 2022

ORB1A- 58

Space Weather @ NOAA site: www.swpc.noaa.gov

- Monitors solar winds, solar flares, CMEs, auroras, etc.

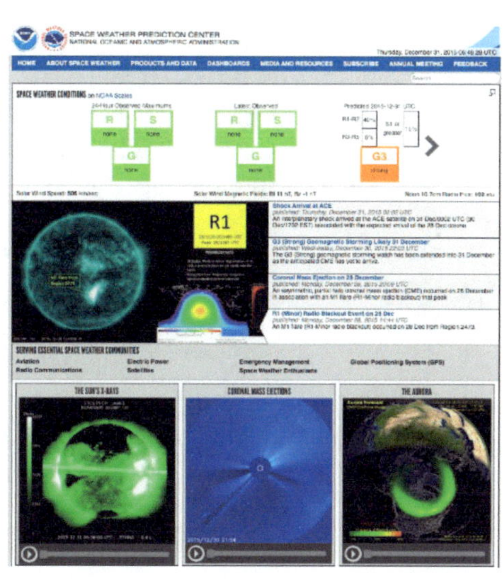

August 22, 2022

Solar Activity & Atmosphere

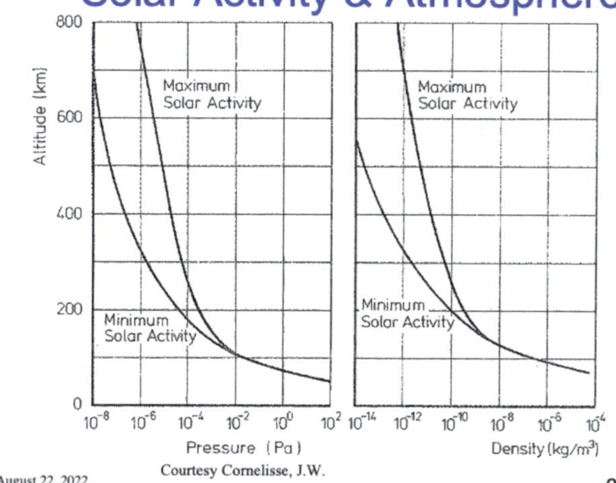

Courtesy Cornelisse, J.W.

August 22, 2022

ORB1A- 60

Absorbed Radiation Dose

- 1 Gy (Gray) = a dose of 1 joule of energy per kilogram of mass (1 J/kg)

  $\text{Gy} = 1\ \text{J/kg} = 6.2415 \times 10^{12}\ \text{MeV/kg}$
- rad = 0.01 Gy
- *Sievert* unit: a dose of ionizing radiation of 1 J/kg of recipient *biological* mass

Damage caused is a function of the absorbing material (semiconductor, tissue, etc)

August 22, 2022

ORB1A- 61

Space Environment Topics

- Non-uniform gravity fields (oblateness, "masscons")
- Orbital disturbances
  - Nodal regression
  - Periapsis advance
  - Station-keeping issues for GEO orbits
- Special orbits take advantage of oblate effects
  - Molniya orbits
  - Sun-synchronous orbits
- Lagrange points & restricted 3-body problem
- Space radiation: sources and effects
  - Types of radiation
  - The sun
  - Earth's radiation belts
- Summary

August 22, 2022

ORB1A- 62

Radiation Belts (AKA Van Allen Belts)

Outer Electron Belt: max density ~17,000 km alt. (4 – 6.5 R_E)

GPS Spacecraft 20,100 km

South Atlantic Anomaly

GEO Orbit NASA Solar Dyn. Observatory 35,400 km

Inner Proton Belt: max density ~3,000 km alt. (1.5 – 2 R_E)

Low Earth Orbit (LEO) International Space Station 370 km

Van Allen Probe-A

The "Slot": low electron density ~6,000 – 12,000 km alt. (2 – 4 R_E)

Van Allen Probe-B

Courtesy: spacesafetymagazine.com

Electron Belt Characteristics

- Inner belt: electron energy between 0.04 & 4.5 MeV
  - Effective shielding possible at these energies
- "Slot" between inner and outer belt
  - During active periods, slot fills with electrons E > 0.4 MeV and decays over weeks
- Outer belt: electron energy up to 7 MeV
  - Shielding more difficult at this energy level

Radiation Belts

- Extend from low Earth orbit to about 10 Earth radii
- Influenced by Earth magnetic field & sun activity: CMEs can push magnetosphere from normal 12 R_E position to less than 6 R_E
- Belts move higher as atmospheric density increases
- Preferred orbit height:
  - High enough to minimize atmospheric drag, but below the heart of the radiation belts

Courtesy Karl Pfitzer, Boeing

Proton Radiation Belt

- Single belt structure
- Range from 1.3 – 3.5 RE
- Most energetic protons (400 MeV) at lower altitudes
- Least energy higher altitudes
- Proton belt dips to low altitude at South Atlantic Anomaly (SAA)
- Many SEUs (single-event upsets) in SAA

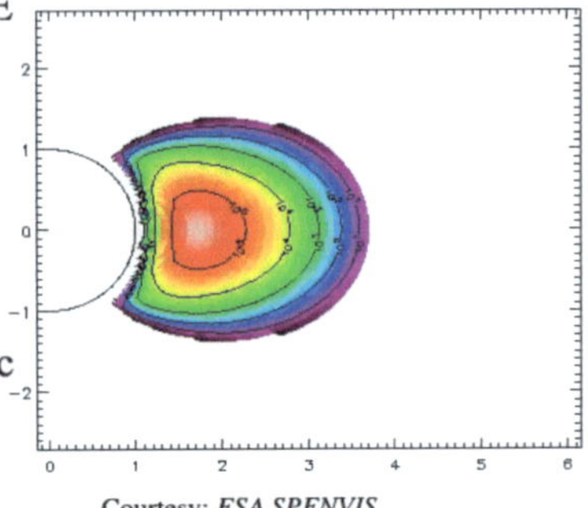

Courtesy: *ESA SPENVIS*

South Atlantic Anomaly (SAA)

- SAA = localized decrease in dipole term of Earth's magnetic field causes west & south drift of the ground-level local minimum in magnetic field
- Particles penetrate to lower altitudes over SAA

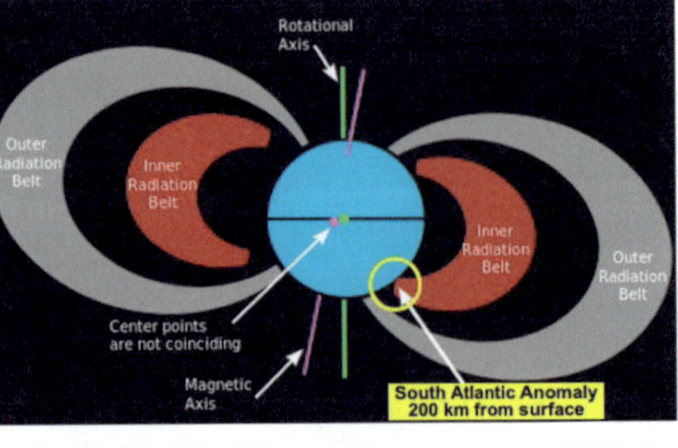

Courtesy Wikipedia

Earth Magnetic Field

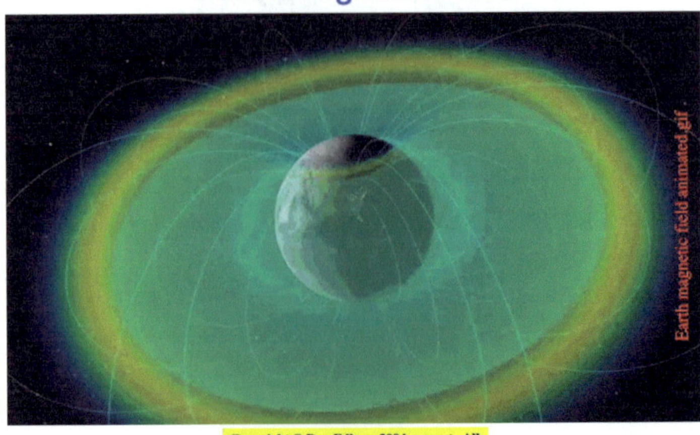

Why IS There a South American Anomaly?

- Earth's ferromagnetic liquid core generates a magnetic field whose N-S axis is *tilted* about 16° from spin axis
- North magnetic pole is ~7° away from the North Pole, but South magnetic pole is ~25° from the South Pole
- A line drawn from North to South *magnetic* poles *does not pass through the center of the Earth*
- Earth magnetic field is torus-shaped (donut), but its tilted & offset from Earth's center
- One inner surface of the donut is more squished up against the side of the earth than the other
- The *offset* causes the South Atlantic Anomaly

Spatial Distribution of SAA Protons

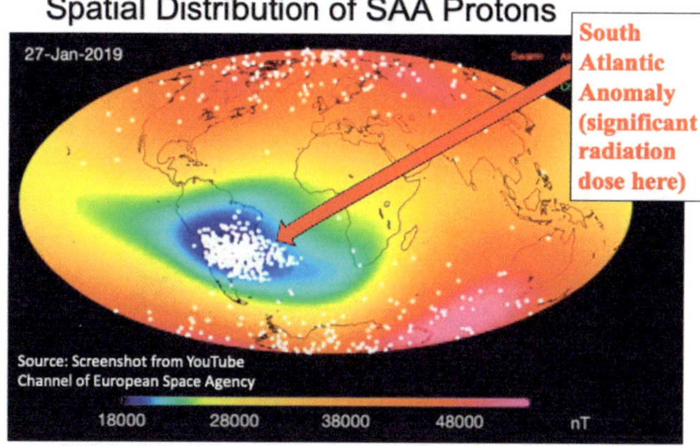

SAA dominates other radiation sources and is a localized phenomenon

Space Environment Topics

- Non-uniform gravity fields (oblateness, "masscons")
- Orbital disturbances
  - Nodal regression
  - Periapsis advance
  - Station-keeping issues for GEO orbits
- Special orbits take advantage of oblate effects
  - Molniya orbits
  - Sun-synchronous orbits
- Lagrange points & restricted 3-body problem
- Space radiation: sources and effects
  - Types of radiation
  - The sun
  - Earth's radiation belts
- **Summary**

Orbit Perturbations

- Caused by non-spherical shape & density variations
  - Nodal regression
  - Advance of periapsis
  - Use for special orbits: Molniya & sun-sync orbits
- Density variations on Earth and other bodies (moon, sun, etc.) also cause perturbations
  - Important for geosynchronous earth orbit (GEO) station-keeping, additional propulsion needed to overcome
- Three-body problem solved numerically
  - Used for spacecraft such as James Webb Space Telescope

Radiation Comments

- Space environment very complex & very hostile
- Modern electronics more sensitive than ever to space environment. Cosmic rays, solar protons, trapped radiation belts produce variety of single-event hazards
- Environments in LEO & GEO are as different as weather in Sahara and Amazon jungle. Each region has its own "gotchas," summary previous
- Before designing something for space, consult with someone who really understands it. What may work for one region of space may not work in another region: there are lots of ways to do it wrong.

Courtesy Karl Pfitzer, retired Boeing Technical Fellow

How The Space Environment Affects Spacecraft

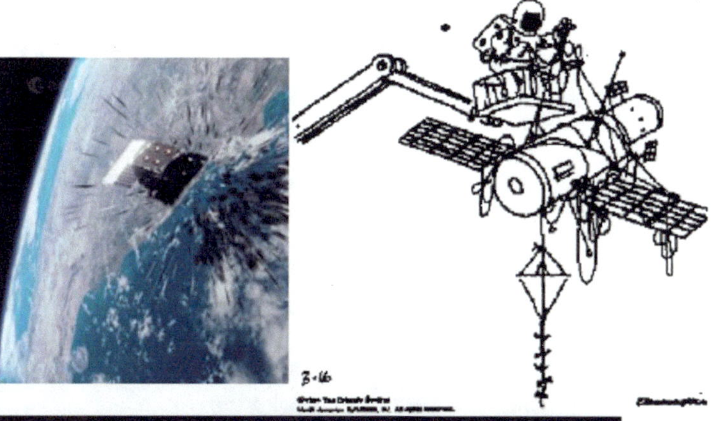

8/22/22

ORB1B- 1

Space Environment Effect on SC Systems: Topics

- Aerodynamic drag
- Gravity gradient
- Magnetism
- Solar pressure
- Radiation & particles
- Static buildup & electrical discharges
- Atomic oxygen
- Orbital debris
- Outgassing, cold welding, lubrication
- Internal disturbances

8/22/22

ORB1B- 2

Space Environment

Two effects:

- Affects orbits & health of SC, independent of type = previous section
- **Affects the health & stability of the individual spacecraft itself & its systems: *this* section**

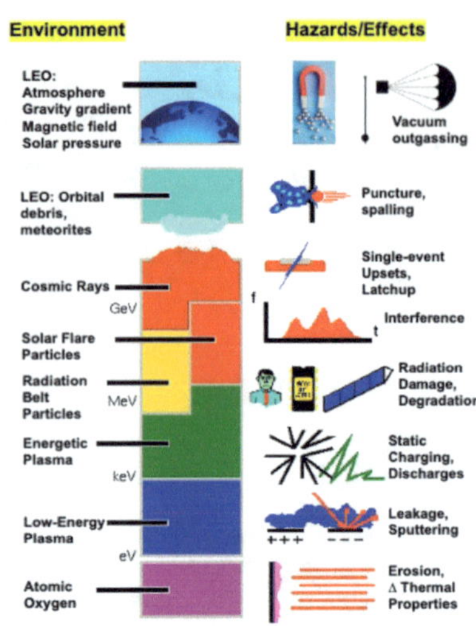

Environment
Hazards/Effects
LEO: Atmosphere Gravity gradient Magnetic field Solar pressure
Vacuum outgassing
LEO: Orbital debris, meteorites
Puncture, spalling
Cosmic Rays
Single-event Upsets, Latchup
Interference
Solar Flare Particles
Radiation Belt Particles
Radiation Damage, Degradation
Energetic Plasma
Static Charging, Discharges
Low-Energy Plasma
Leakage, Sputtering
Atomic Oxygen
Erosion, Δ Thermal Properties

Courtesy ESA/ESTEC

Course Layout

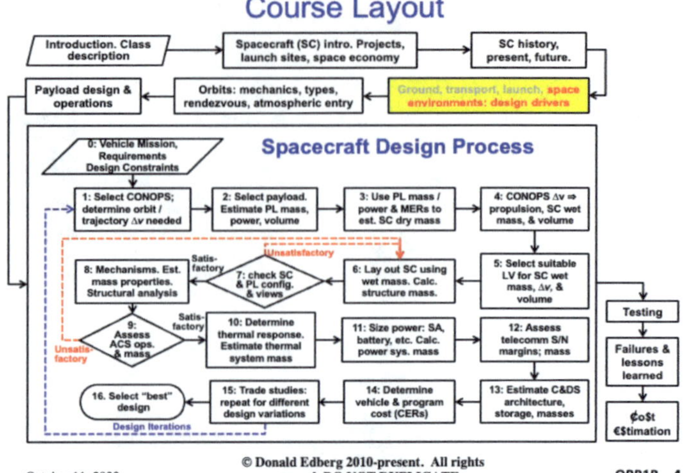

Introduction. Class description
Spacecraft (SC) intro. Projects, launch sites, space economy
SC history, present, future.
Payload design & operations
Orbits: mechanics, types, rendezvous, atmospheric entry
Ground, transport, launch, space environments: design drivers
Spacecraft Design Process
0: Vehicle Mission, Requirements Design Constraints
1: Select CONOPS; determine orbit / trajectory Δv needed
2: Select payload. Estimate PL mass, power, volume
3: Use PL mass / power & MERs to est. SC dry mass
4: CONOPS Δv ⇒ propulsion, SC wet mass, & volume
5: Select suitable LV for SC wet mass, Δv, & volume
6: Lay out SC using wet mass. Calc. structure mass.
7: check SC & PL config. & views
8: Mechanisms. Est. mass properties. Structural analysis
9: Assess ACS ops. & mass
10: Determine thermal response. Estimate thermal system mass
11: Size power: SA, battery, etc. Calc. power sys. mass
12: Assess telecomm S/N margins; mass
13: Estimate C&DS architecture, storage, masses
14: Determine vehicle & program cost (CERs)
15: Trade studies: repeat for different design variations
16: Select "best" design
Satisfactory
Unsatisfactory
Design Iterations
Testing
Failures & lessons learned
¢o$t €$timation

Disturbances to Spacecraft

- Aerodynamic drag
- Gravity gradient
- Magnetism
- Solar pressure
- Radiation & particles
- Static buildup & electrical discharges
- Atomic oxygen
- Orbital debris
- Outgassing, cold welding, lubrication
- Internal disturbances

Aerodynamic Drag

- Calculate drag (D) from classic definition:
  $D = C_D A \frac{\rho}{2}(v - \Omega_E r \cos i)^2 = \frac{C_D A \rho v_{rel}^2}{2} = C_D A q,$
  q = dynamic pressure
  velocity correction for earth's rotation Ω_E
  C_D = drag coeff.; A = cross sectional area
- Density ρ is small but varies a lot.
  Velocity v is LARGE so:
- Drag can be significant, i.e. *Skylab* loss
- Deceleration $a = \frac{D}{m} = \frac{C_D A}{m} \cdot \frac{\rho v_{rel}^2}{2} = \beta q$
  β = ballistic coefficient

Aerodynamic Drag Pressure vs. Altitude & Temp.

Almost two orders of magnitude change with sun behavior!
Note:
1 dyne/cm² = 0.1 Pa = 0.00209 lb/ft²

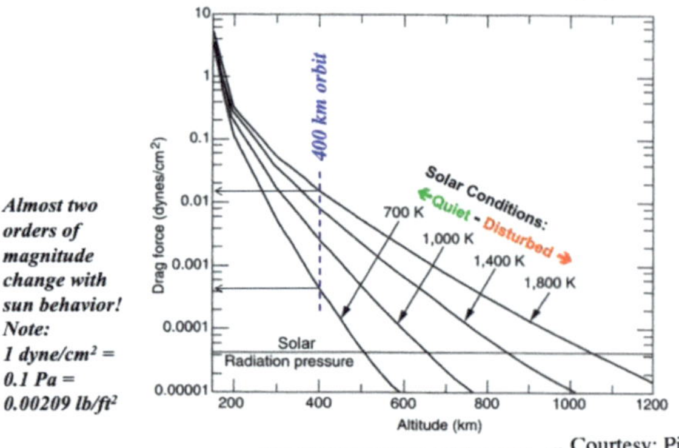

Drag force (dynes/cm²)
10
1
0.1
0.01
0.001
0.0001
0.00001
400 km orbit
Solar Conditions:
Quiet - Disturbed
700 K
1,000 K
1,400 K
1,800 K
Solar
Radiation pressure
200
400
600
800
1000
1200
Altitude (km)

Courtesy: Pisacane

Solar Flare Destroys 40 SpaceX Starlink Satellites (2020 Feb. 03)

- 1 day after launch, solar flare expands Earth atmosphere, creating increased drag
- Drag lowers orbits; SC put into safe mode edgewise to velocity vector
- Drag makes SC unable to leave safe mode for orbit-raising, 40 of 49 SC burn up in atmosphere

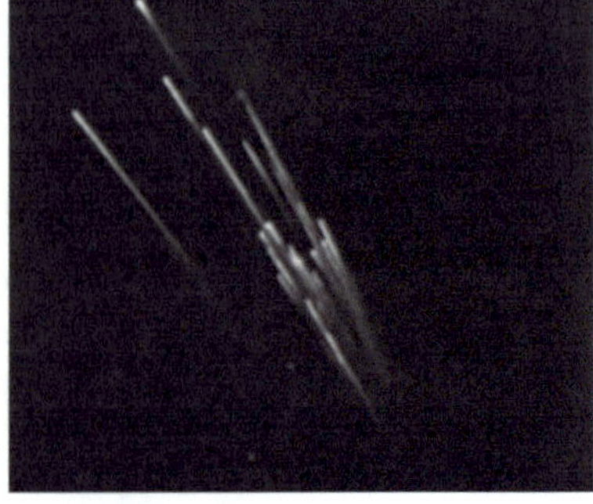

AvWeek 2022FEB21 p. 34

Courtesy: Skywatch Media News

Drag Coefficients

- C_D difficult to estimate for spacecraft
- $C_D \approx 2.2$ commonly used for orbiting objects (based on Newtonian particle physics of free molecular flow)

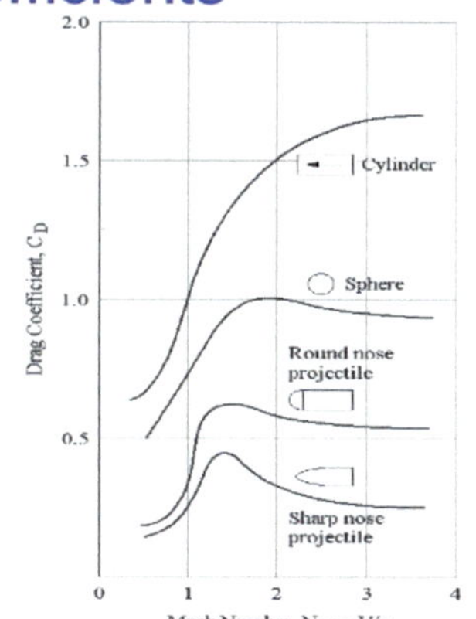

8/22/22

ISS Orbital Decay 2009-10

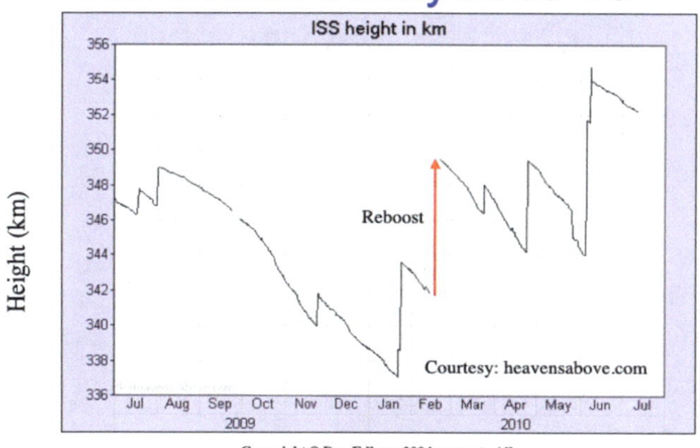

8/22/22

Aero Drag Also Creates Torques

- Center of pressure (CP) where drag acts and center of gravity (CG) which body likes to rotate about are likely separated
- Separation = moment arm d
- Torque = drag force × moment arm = $\mathbf{D} \times \mathbf{d}$
- Thrusters or reaction wheels needed to absorb (desaturate)

8/22/22

Example: Aerodynamic Torque Calculation

- Frontal area 5 m^2, $C_D = 2.0$.
- 400 km orbit, density = 4×10^{-12} kg/m^3
- Aerodynamic force $D = 1.2 \times 10^{-3}$ N
- For a 0.5-meter separation d from drag area center to CG:
  $T = |D \times d| = 0.6 \times 10^{-3}$ N·m

Ref: Griffin

8/22/22

Disturbances to Spacecraft

- Aerodynamic drag
- Gravity gradient
- Magnetism
- Solar pressure
- Radiation & particles
- Static buildup & electrical discharges
- Atomic oxygen
- Orbital debris
- Outgassing, cold welding, lubrication
- Internal disturbances

8/22/22

Gravity Gradient ("GG")

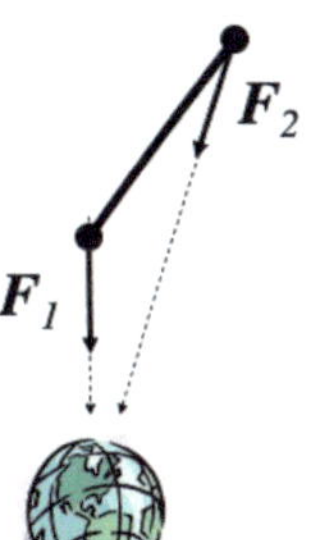

- Gravitational force $F_1 > F_2$, since $r_1 < r_2$
- A torque is generated
- Object tries to align with planet center

8/22/22

More GG Comments

- GG creates a stretching force on long spacecraft
- There are forces pulling up and down
- Acting on the earth we call these "tidal forces"
- On a quasi-static spacecraft, only a small portion will have "micro-gravity"

ISS DAC-9 Microgravity Plots

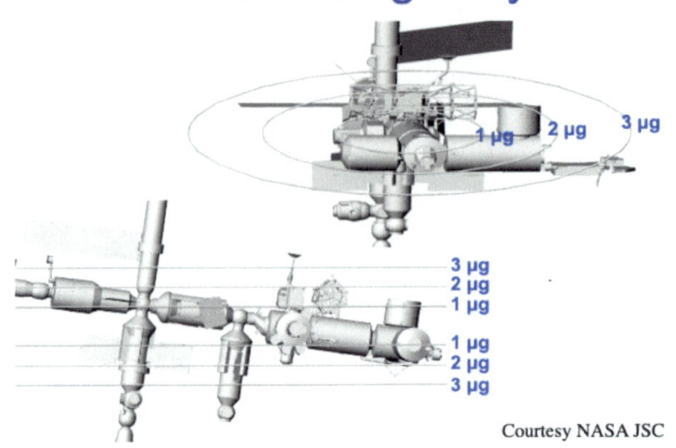

Courtesy NASA JSC

Gravity Gradient Torques

- For small angles:

$$T_x = 3n^2\,[(I_z - I_y)\varphi - I_{xy}\theta + I_{yz}]$$
$$T_y = 3n^2\,[(I_z - I_x)\theta - I_{xy}\varphi + I_{xz}]$$
$$T_z = 3n^2\,[I_{xz}\varphi + I_{yz}\,\theta]$$

where $n^2 = \frac{\mu}{r^3}$ or

$$n = \sqrt{\frac{\mu}{r^3}} = \text{orbital rate}$$

Courtesy Northrop Grumman Space Hdbk

Calculation of GG Torque

- GG torque depends on orbit height, inertia properties
- I_x, I_y, I_z are moments of inertia about *x, y,* and *z* axes respectively
- I_{ij} are products of inertia (zero if principal directions)
- θ is roll angle, φ is pitch angle
- Torque is zero if aligned or perpendicular to nadir
- For low orbiting SC with 1,000 kg·m^2 inertia moment difference, $T = 6.7 \times 10^{-5}$ N·m/deg

GG Comments

- CG ≠ CM!!!
- GG torques are *small* but NOT negligible: Skylab $T_{gg} \approx 3$ N·m led to saturation of CMGs
- There is no damping so oscillations may continue for a *long* time
- Oscillations called "libration"
- GG libration has frequency $f = 0.276\ n$ (very slow, <1 cycle per orbit)

Disturbances to Spacecraft

- Aerodynamic drag
- Gravity gradient
- Magnetism
- Solar pressure
- Radiation & particles
- Static buildup & electrical discharges
- Atomic oxygen
- Orbital debris
- Outgassing, cold welding, lubrication
- Internal disturbances

Planet Magnetic Fields Torque SC (1 of 4)

- B field of Earth tilted 11° relative to spin axis & centered ~400 km from geometric center, varies as $1/r^3$.
- Earth magnetic field strength calculated from

$$B = B_0(r_0/r)^3\sqrt{3\sin^2\lambda}$$

where $B_0 = 3 \times 10^{-5}$ tesla

r = orbital radius

- r_0 = Earth radius 6,378 km

λ = magnetosphere latitude (can use geographic lat. for approximate calculations)

Fig. 7.23 Tilted-centered dipole model for Earth's magnetic field.

Magnetic Field Torque SC (2 of 4)

- A current-carrying coil experiences a torque

$T_m = niAB\sin\theta$, where

T_m = magnetic torque (Nm)

n = number of loops in coil

i = loop current (A or amps)

A = coil area (m²)

B = magnetic field strength (units T or tesla; note 1 T = 10^4 G or gauss)

θ = angle between magnetic field lines & coil's normal vector

  - Don't confuse torque OR thrust T (italic) with magnetic field strength units T = tesla

Magnetic Torque (3 of 4)

- Any given SC has a *residual magnetic field* or *dipole* strength $M_{mag} = niA + M_0$ resulting from
  - Internal current flows through wire coils niA &
  - Residual magnetism M_0 from magnetized metallic parts
- Value of residual dipole strength obtained by test, may range from 0.2 – 20 A·m² (according to Brown 5.2.2)
- torque $T_m = M_{mag}B\sin\theta$

Magnetic Torque (4 of 4)

- From Earth field strength eq.

$B = B_0(r_0/r)^3\sqrt{3\sin^2\lambda + 1}$,

can see that field strength at poles ($\lambda = \pm 90°$) is twice that at equator ($\lambda = 0°$)

- In solar system, Venus & Mars have negligible magnetic fields
- Jupiter has *very large* field:

4.3×10^{-4} T at radius 71,372 km

- Compare to Earth 3.1×10^{-5} T at radius 6,378 km

Example SC Magnetic Torque Calculation

- SC has residual dipole M_{mag} = 2A·m² in a circular equatorial orbit at 400 km. What is mag torque strength?
- r = 6,378 + 400 km = 6,778 km
- Here Earth's mag field is $B = B_0(r_0/r)^3\sqrt{3\sin^2\lambda + 1}$

$$= 3\times10^{-5}\,\text{T}(6{,}378/6{,}778)^3\sqrt{1}$$

$$= 2.50\times10^{-5}\,\text{T}$$

- Typical mag torque value (max at θ= 90°):

$$T_m = M_{mag}B\sin\theta$$

$$T_m = (2.50\times10^{-5}\,\text{T})(2\,\text{Am}^2)$$

$$= 5\times10^{-5}\,\text{Nm}$$

Mag Torque = Control Mechanism

- Add "torque rods" or "magnetorquers" = electromagnets that can be turned off and on based on spacecraft orientation
- Can control SC attitude using local knowledge of Earth's magnetic field
- Common for small spacecraft (CubeSats, nanosats)
- See "Spacecraft Control" section for more detail

Disturbances to Spacecraft

- Aerodynamic drag
- Gravity gradient
- Magnetism
- Solar pressure
- Radiation & particles
- Static buildup & electrical discharges
- Atomic oxygen
- Orbital debris
- Outgassing, cold welding, lubrication
- Internal disturbances

8/22/22 ORB1B-27

Solar Force & Torque

Photons in light have momentum, create pressure when absorbed (more if reflected)

Solar force $F_s = P_s A(1 + \rho)$ where:

- P_s is solar pressure, 4.6×10^{-6} Pa/$(r/1\text{ AU})^2$, where r = distance from sun; 1 AU = mean Earth-sun dist. Varies $\sim \frac{1}{r^2}$ (inverse square)
- A = cross sectional area
- $\rho \approx$ *reflectance* (don't confuse with density)
  $0 \le \rho \le 1$ ($\rho = 1$ for a mirror)
  For spacecraft bodies: $\rho \approx 0.5$
  For solar arrays: $\rho \approx 0.3$
- Torque = Force $\mathbf{F_s} \times \mathbf{r}$ (moment arm)

R=1 R=2 R=3

8/22/22 ORB1B-28

Solar Torque Example

Earth Orbit Spacecraft

- SC solar array area $A = 5$ m^2
- Offset from center $L = 0.1$ m
- Reflectivity $\rho = 0.5$
- Solar torque $T_s = F_s r = P_s A(1 + \rho)r$
  $= (4.6 \times 10^{-6}$ Pa$)(5$ m$^2)(1 + 0.5)(0.1$ m$)$
  $T_s = 3.5 \times 10^{-6}$ N·m

r
Solar Panel
solar force F_s
Torque T_s

8/22/22 ORB1B-29

Solar Torque (cont.)

- Big driver in GEO spacecraft
- Important above 1000 km from Earth
- Actually used for roll control on *Mariner 10*!
- Can use for free *propulsion*: solar sails are feasible but control & deployment can be issues

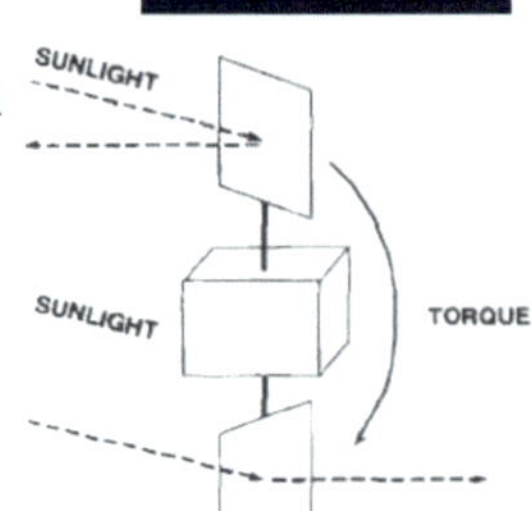

Courtesy **Introduction to Space Sciences**, Campbell & McCandless, 1996.

8/22/22 ORB1B-30

Solar Pressure Makes GEO into Ellipse, fix at Δv Points

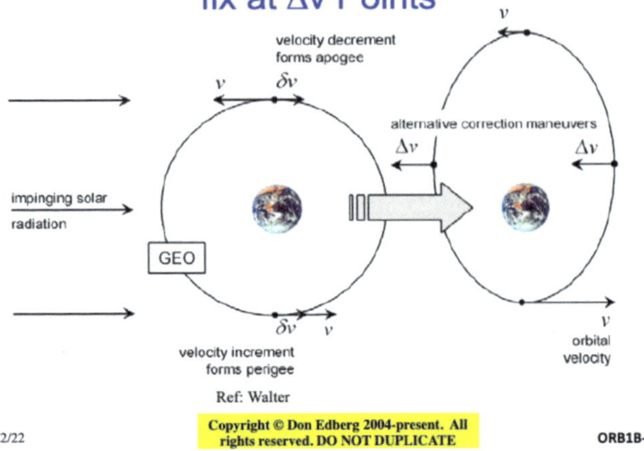

Ref: Walter

8/22/22 ORB1B-31

Solar Pressure Affects GEO SV *e*:

Depends on *m*/*A*, increases gradually over 6 mo., then decreases.

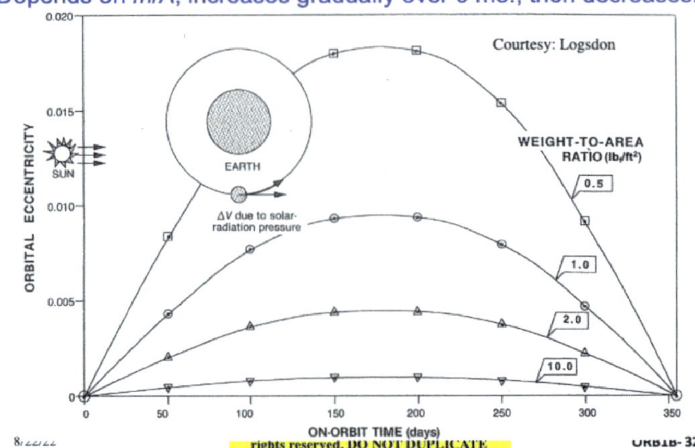

8/22/22 ORB1B-32

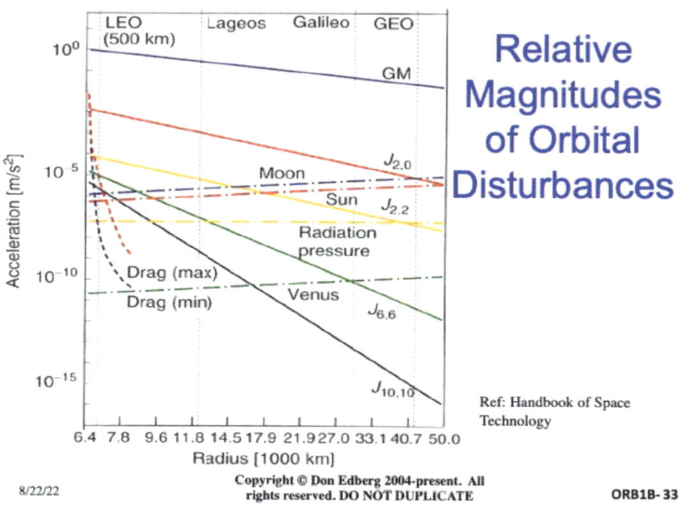

Relative Magnitudes of Orbital Disturbances

Ref: Handbook of Space Technology

8/22/22

ORB1B- 33

Disturbances to Spacecraft

- Aerodynamic drag
- Gravity gradient
- Magnetism
- Solar pressure
- **Radiation & particles**
- Static buildup & electrical discharges
- Atomic oxygen
- Orbital debris
- Outgassing, cold welding, lubrication
- Internal disturbances

8/22/22

ORB1B- 35

Radiation & Single-Event Effects

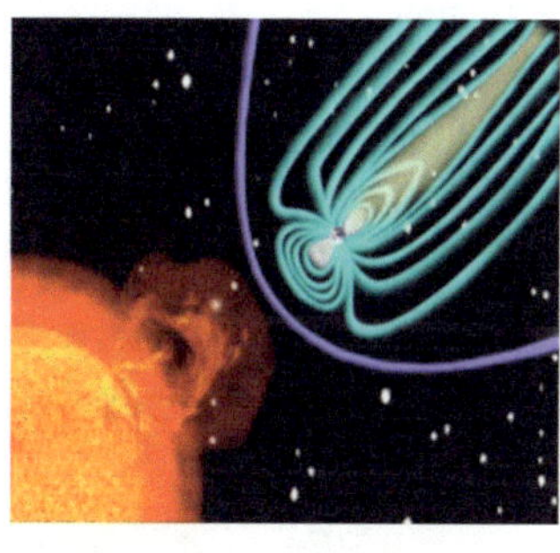

Courtesy: ESA

8/22/22

ORB1B- 36

Mars Rover Exhibits "Odd" Behavior

NY Times (1/29/2009): "Mission managers at JPL … reported that Spirit had behaved oddly," when it "acknowledged receiving its driving directions from Earth, but it did not move." The rover also had "no memory" of the day's activities." Another rover system recorded power being drawn from the batteries for 90 min ... But before-and-after images showed that the Spirit ended the day exactly where it started." According to the article, "One hypothesis is that **a cosmic ray hit the electronics and scrambled the rover's memory**." The rover is now in 'good health' and is responding to commands.

8/22/22

ORB1B- 37

Single Event Effect (SEE)

- An energetic particle passes through a semiconductor
- It creates a trail of ionized particles in the vicinity of a reverse-biased PN junction
- The sudden flux in ionized particles can cause a swing in bias across the junction
- This change may alter the state of the device

Courtesy Tribble

8/22/22

ORB1B- 38

Single Event Effect Illustration: Upset & Latchup

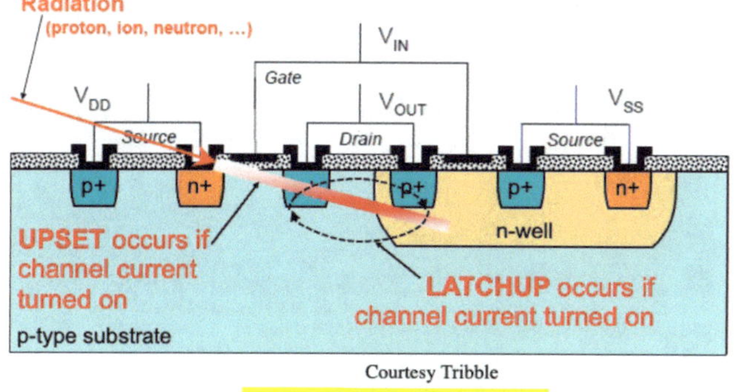

Courtesy Tribble

8/22/22

ORB1B- 39

Single-Event Phenomena / Effects (1)

- Single Event *Upset*
  - Particle causes "bit flip" in computer memory altering logic or data
  - System-level manifestations depend on application, can be system crashes or invalid data
  - Reboot system to eliminate effects

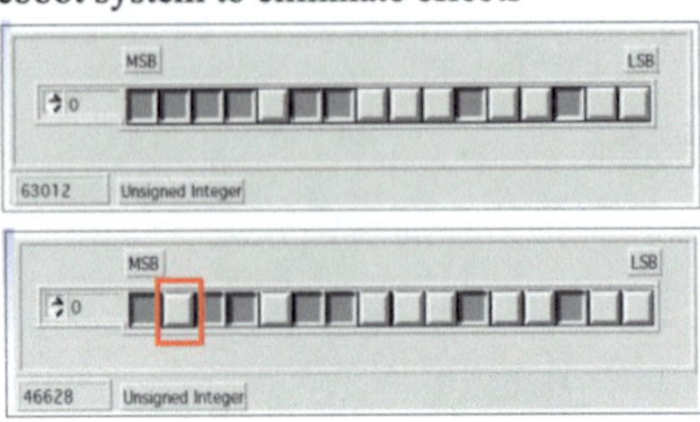

Courtesy Tribble

8/22/22

Single-Event Phenomena / Effects (2)

- Single Event *Latchup*
  - Low-resistance path develops between power & ground through the device, usually destructive
  - Sometimes observe "mini-latch" behavior, must power-down to reset
- Single Event *Functional Interrupt*
  - Upset which places a device in an ill-defined condition
  - Causes system lock up or jump into unknown configuration
- Single Event *Transient*
  - Spurious voltage spike that can cause system-level effects
  - Increased noise in the system

Courtesy Tribble

8/22/22

Single-Event Phenomena/Effects (3)

- Single Event *Burnout*
  - Localized short through power MOSFET
  - Permanently damages the part; device fails
- Single Event *Gate Rupture*
  - Localized short through drain-to-oxide interface in a power MOSFET
  - Permanently increases gate leakage
- Single Event *Dielectric Rupture*
  - Oxide damage in non-volatile elements or antifuse type FPGAs

Courtesy Tribble

8/22/22

Typical Electronics Radiation Failure Doses

Technology	Failure Level (krad)
Linear ICs	2 – 50
Mixed Signal ICs	2 – 30
Flash Memories	5 – 15
DRAMs	15 – 50
Microprocessors	15 – 70

8/22/22

Shuttle Single-Event Upsets (NASA)

This figure appears to be a compilation of all "bit flips" that were downlinked to JSC after the shuttle switched to modern static random access memory. SEUs are yellow squares; multi-bit SEUs (2-8 bits upset by a single particle) are red triangles. From NASA SP-2010-3049, "Wings in Orbit."

8/22/22

Proton/Heavy Ion Radiation Interactions with Materials

Courtesy Karl Pfitzer, Boeing

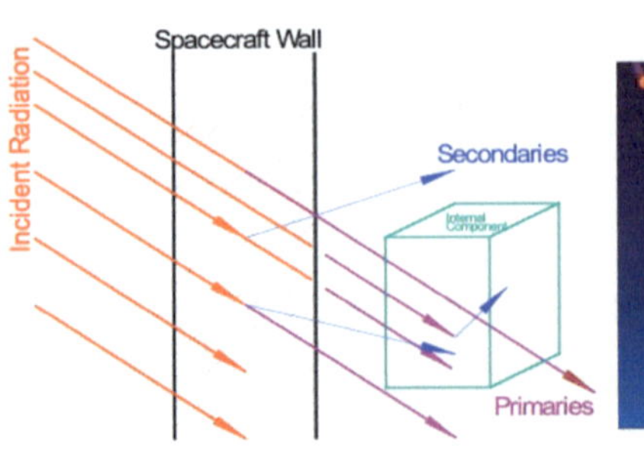

Courtesy NASA

8/22/22

Preferred Shielding Materials

- Low Z (atomic #) material preferred
  - DO NOT use lead
  - Secondaries may actually increase dose inside shield
- Hydrogen is the most effective shield
  - Water is a good material with multiple space uses!
- Polyethylene & plastics are also effective

8/22/22 ORB1B- 46

Shielding

- Typical shielding = Aluminum; standard boxes provide decent shielding
- Al shielding effectiveness in LEO shown right
- Multiple-layer shielding = [Low Z / High Z / Low Z] composite

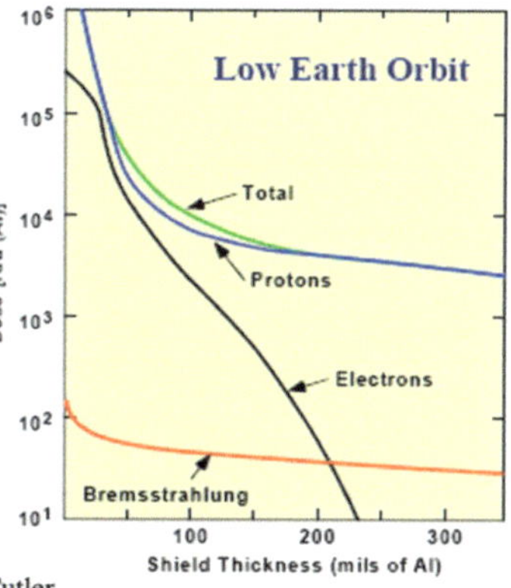

Courtesy James Cutler

8/22/22 ORB1B- 47

Juno Radiation Vault

- Compartment inside houses much of the probe's electronics & computers to offer increased protection of radiation at Jupiter
- Roughly a cube, walls made of 1 cm thick titanium, each side ~1 m^2, mass ~ 200 kg
- Inside C&DH & power control
- Should reduce radiation exposure to 20,000 krad by 1/800 = 25 rad
- Does not stop all radiation – significantly reduces it to limit damage to SC's electronics (Wikipedia)

8/22/22 ORB1B- 48

Low Molecular Weight Shielding

2011: ILC-Dover has new flexible lightweight borated polyethylene material for shielding against GCR and high-energy protons called *ArmorFlex*

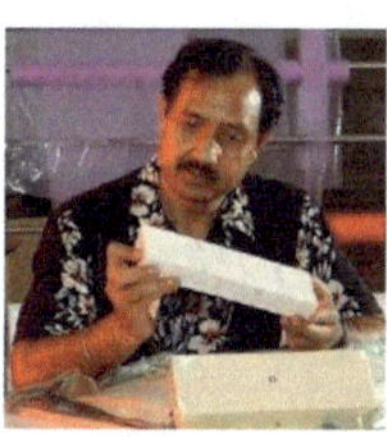

2006: NASA has flexible plastic material for shielding called *RXF1*: 3 times the tensile strength of aluminum, yet is 2.6 times lighter

8/22/22 ORB1B- 49

To Combat Single-Event Phenomena

Utilize shielding

- Adds weight and volume
- Aluminum cases somewhat effective
- Low molecular weight materials such as polyethylene
- Not effective on very energetic (GeV) charged particles

Part selection

- Use rad-hard devices that can withstand the total dose environment anticipated, or are immune or resistant to SEE

8/22/22 ORB1B- 50

Commercial vs. Rad-Hard Memory

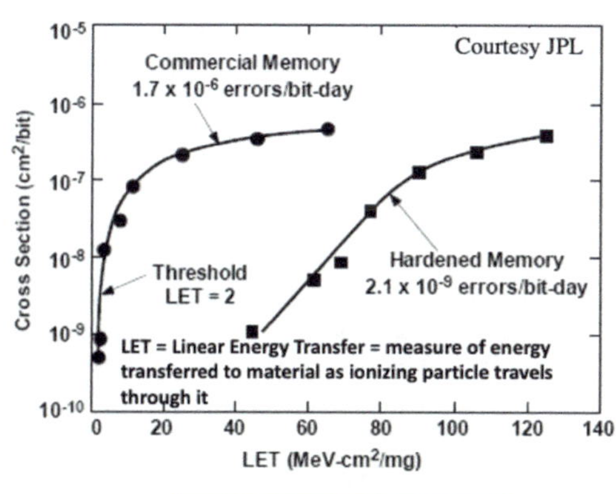

8/22/22 ORB1B- 51

To Combat SE Phenomena (cont.)

- Hardware
  - Make fault-tolerant
  - Use redundancy, parallel devices
  - Install power control circuitry: power cycling can recover many errors (complicates power system)
  - Install current monitoring hardware: detection circuitry can catch errors & cycle power
- Software
  - Provide automatic error detection & correction
  - "Majority" voting logic to override faults
  - Reset (reboot) on-board computer at regular intervals
  - Error detection & correction (= EDAC), Hamming codes, etc.

Mars Odyssey's Computer Systems Needs A Reboot

Space.com (3/5/2009): "NASA's Mars Odyssey orbiter, circling the red planet for nearly eight years, needs a risky reboot to address a long-known, potential vulnerability in its memory system … The chief concern about the potential memory vulnerability stems from the length of time that the spacecraft has been exposed to the accumulated effects of the space radiation environment since the last reboot" in 2003. This will "demonstrate whether Odyssey's onboard backup systems will be available should they ever be required."

Disturbances to Spacecraft

- Aerodynamic drag
- Gravity gradient
- Magnetism
- Solar pressure
- Radiation & particles
- Static buildup & electrical discharges
- Atomic oxygen
- Orbital debris
- Outgassing, cold welding, lubrication
- Internal disturbances

Internal Electrostatic Discharges (ESDs)

- MeV electrons penetrate into spacecraft interior – no high energy protons to equalize
- Sun energizes surface electrons via photoelectric effect
- Slow process, takes days to build up charge
- Dielectrics & floating pieces of conductors get charged

Harwood, **SpaceFlight Now**, 9/25/2008: "Russian troubleshooters believe **electrical arcing** between the space environment and ISS most likely caused recent problems with explosive bolts used to separate Soyuz re-entry vehicles." ISS program manager Michael Suffredini said "the TMA-13 spacecraft," the next Soyuz to carry a crew to the station, "has been modified to **minimize the threat of arcing**."

Courtesy Karl Pfitzer, Boeing

NASA External ESD Design Guidelines:

- All structural & mechanical parts, electronic boxes, etc. must be electrically bonded. All grounds must be tied together.
- All cables exiting the Faraday cage **must** be shielded; shields must be terminated outside on the connector shell and not via a connector pin to the interior
- Use "leaky" dielectrics to automatically dissipate slowly
- Cover insulating surfaces, thermal control coatings with a vacuum-deposited metallic film such as Indium-Tin-Oxide. All externally visible surfaces must be conducting in an ESD sense (resistance of any point to ground < 10^3 MΩ)
- Test in vacuum chambers with electron guns

Courtesy Karl Pfitzer, Boeing

Single-Point Grounding Scheme

- ✓ Single 0-V reference point on entire SC (disadvantage: numerous long ground leads make harness heavy, can cause interference at high frequencies)
- ✓ Power fed by twisted-pairs
- ✓ No currents thru SC structure (reduce interference)

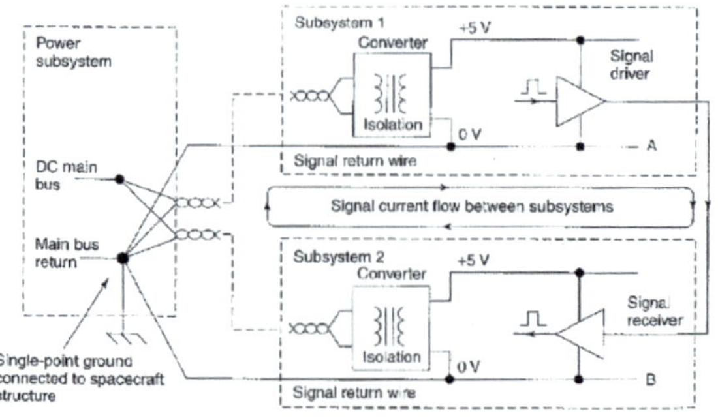

Courtesy Fortescue

Multi-Point Grounding Scheme

- ✓ Zero-volt reference low-inductance conductive surface ground plane, plated Cu or Al honeycomb, or conductive strips on CF
- ✓ Effect of many local grounds negligible since low inductance

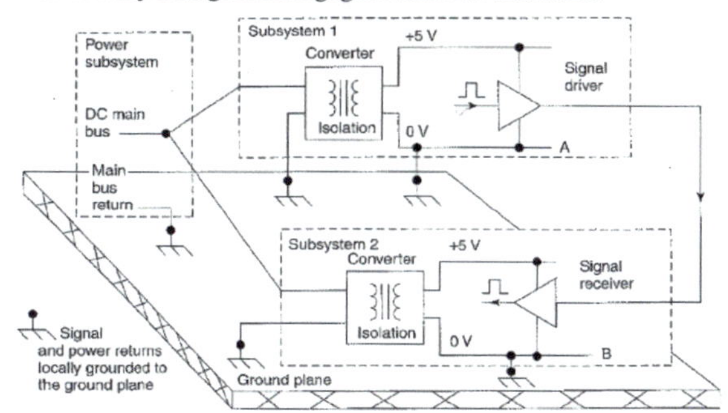

8/22/22 Courtesy Fortescue ORB1B- 58

Generic Spacecraft Hardening

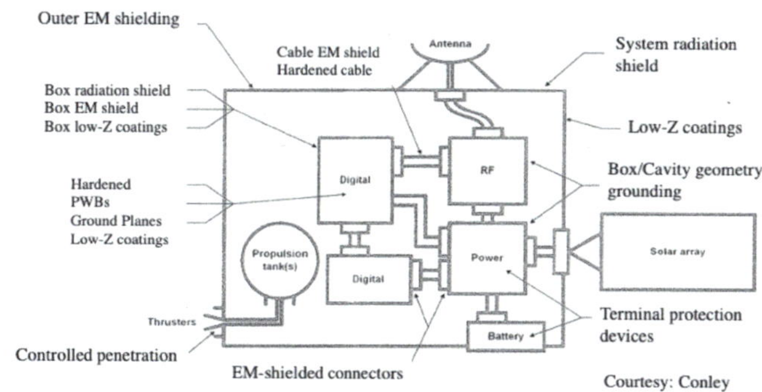

8/22/22 ORB1B- 59

"Whisker" Failures

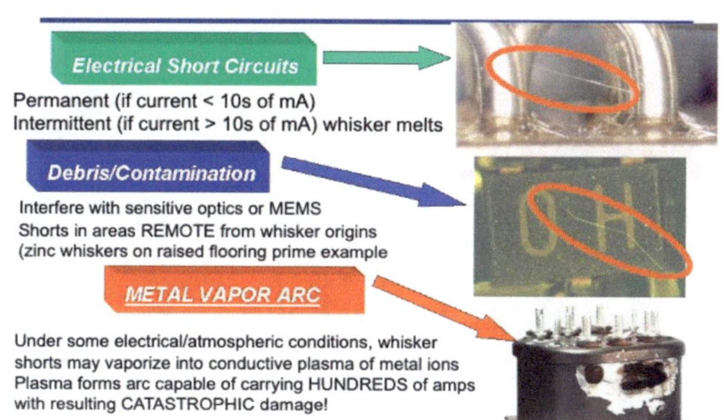

8/22/22 Courtesy NASA ORB1B- 60

Disturbances to Spacecraft

- Aerodynamic drag
- Gravity gradient
- Magnetism
- Solar pressure
- Radiation & particles
- Static buildup & electrical discharges
- **Atomic oxygen**
- Orbital debris
- Outgassing, cold welding, lubrication
- Internal disturbances

8/22/22 ORB1B- 61

Atomic Oxygen

- Atomic oxygen (AO) created when charged particles hit oxygen molecules:
  Molecule O_2 + energy = 2 O atoms
- AO is STRONG oxidizer
- AO causes changes to surfaces
  - Changed thermal characteristics affect (absolute) equilibrium temperature: can show
  - $\Delta T = \frac{T}{4} \cdot \frac{\Delta(\alpha/\varepsilon)}{(\alpha/\varepsilon)} = \frac{T}{400} \cdot \left[\% \text{ change in } \left(\frac{\alpha}{\varepsilon}\right)\right]$
  - Degrades sensor performance
  - Pits solar arrays, hurting performance
  - Weakens materials

8/22/22 ORB1B- 62

Atomic Oxygen Effects: 3.95 year exposure

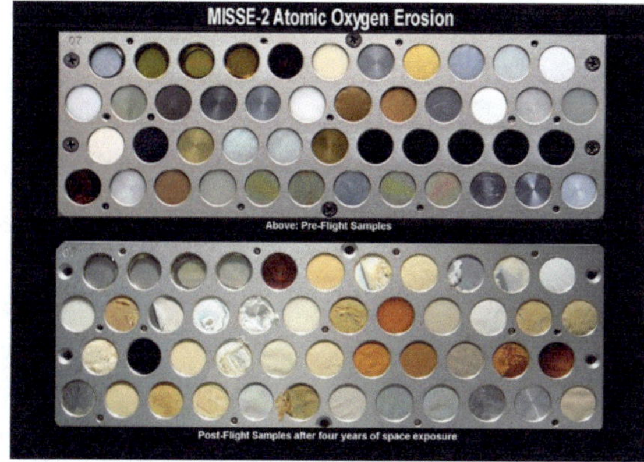

Courtesy NASA/TM-2010-216903 November 2010

8/22/22 ORB1B- 63

Disturbances to Spacecraft

- Aerodynamic drag
- Gravity gradient
- Magnetism
- Solar pressure
- Radiation & particles
- Static buildup & electrical discharges
- Atomic oxygen
- **Orbital debris**
- Outgassing, cold welding, lubrication
- Internal disturbances

8/22/22 ORB1B- 64

Royal Aeronautical Society

8/22/22 ORB1B- 65

Orbital Debris (MicroMeteoroid and Orbital Debris = MMOD)

Distribution of the Catalogued objects (1957 - 2000)

ESA orbital debris.qt

8/22/22 ORB1B- 66

ISS Fires Engines to Avoid Orbital Debris

- At 10:36 p.m. EDT Sat, 2 Apr 2011, ground controllers moved ISS away from a piece of orbital debris. The object is a relic from a collision between the *COSMOS 2251* and *Iridium 33* satellites in February 2009 and had been close to the station's orbit prior to the debris avoidance maneuver (DAM).
- The DAM, performed during the Expedition 27 crew sleep period, used thrusters from three spacecraft, the European Space Agency's *Johannes Kepler* Automated Transfer Vehicle 2 (ATV2), the *Zvezda* service module and Progress 41P.
- Mission Control Center had been monitoring a series of conjunctions between the International Space Station and the orbital debris. The Expedition 27 crew was informed of the possible conjunction and planned maneuver.

8/22/22 ORB1B- 67

Quantity of In-Orbit Objects Increasing…

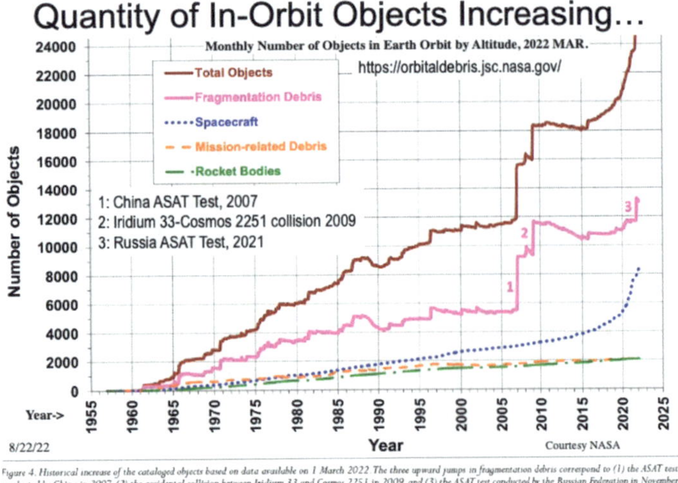

8/22/22 Courtesy NASA

Figure 4. Historical increase of the cataloged objects based on data available on 1 March 2022. The three upward jumps in fragmentation debris correspond to (1) the ASAT test conducted by China in 2007, (2) the accidental collision between Iridium 33 and Cosmos 2251 in 2009, and (3) the ASAT test conducted by the Russian Federation in November 2021. More Cosmos 1408 fragments are expected to the added to the catalog in the coming weeks and months.

68

MASS of In-Orbit Objects Increasing…

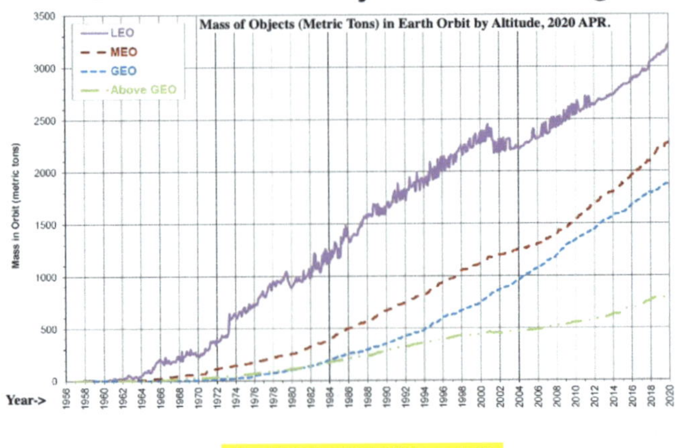

8/22/22 ORB1B- 69

Debris Maps, LEO & GEO (2009)

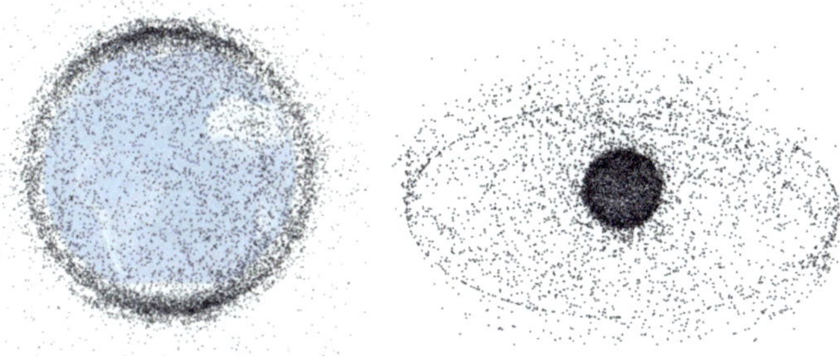

Courtesy: Each dot represents a bit of known space junk that's at least 4 inches (10 cm). 19,000 manmade objects this size or bigger orbit Earth as of July 2009; most are in low-Earth orbit. *Countless smaller objects are also circling the planet*. Credit: NASA/Orbital Debris Program Office.

8/22/22 ORB1B- 70

Debris Maps in LEO and GEO

Courtesy: Each dot represents a bit of known space junk that's at least 1 mm. 25,182+ manmade objects this size or bigger in orbit Earth as of 03/2022; most are in low-Earth orbit. *Countless smaller objects are also circling the planet*. Credit: ESA

8/22/22 ORB1B- 71

Orbital Debris Plot

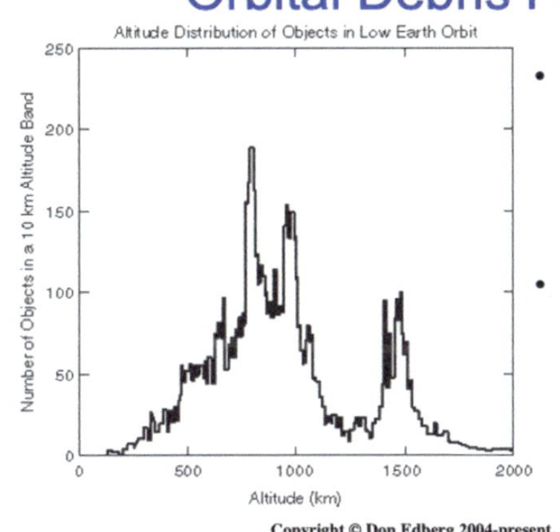

- Plot shows # of objects with diameter > 10 cm in a 10-km band vs. altitude
- Any of these could readily disable a spacecraft

8/22/22 ORB1B- 72

USAF Space Collision Movie

USAF Space collision.mp4

8/22/22 ORB1B- 73

Debris: Serious Problem

- *Cerise* spacecraft gravity gradient boom severed by *Ariane* fragment
- *Hubble Space Telescope* antenna dish penetrated by fragment (NASA)
- *Hubble Space Telescope* solar array penetrated by fragment (source: ESA)
- *Challenger* window hit by paint fleck with energy of a rifle bullet, did not fully penetrate (STS-7, 1983)
- Odds of hit by object > 1 mm diameter: estimated 1:1000. Over 1000 hits to shuttle, 100 windows replaced.

http://www.spacesafetymagazine.com

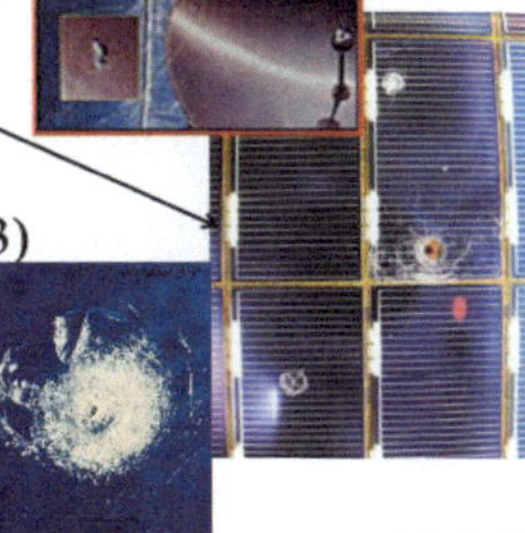

8/22/22 ORB1B- 74

More Debris

- Radiator: STS-118, Space Shuttle Endeavor
- 2.2 mm object moving at several thousand km/s
- Went completely through; came out back side with a 5 mm hole

8/22/22 ORB1B- 75

SC to SC Collisions Not So Rare

1991	Inactive *Cosmos 1934* hit by cataloged debris from *Cosmos 296*
1996	Active *Cerise* hit by cataloged debris from Ariane rocket stage
1997	Inactive *NOAA 7* hit by uncataloged debris large enough to change its orbit and create additional debris
2002	Inactive *Cosmos 539* hit by uncataloged debris large enough to change its orbit and create additional debris
2005	U.S. rocket body hit by cataloged debris from Chinese rocket stage
2007	Active *Meteosat 8* hit by uncataloged debris large enough to change its orbit
2007	Inactive NASA *UARS* believed hit by uncataloged debris large enough to create additional debris
2009	Active *Iridium* hit by inactive *Cosmos 2251*

Courtesy: **The Space Review**

Space Debris Movie

Movie courtesy ESA: space_debris_ESA.mp4

Debris Shield ("Whipple" Shield)

- Add an outside "bumper"
- Object impacts with bumper and splits into smaller objects with lower energy
- Mass and volume penalty

Illustration courtesy Douglas Hdbk

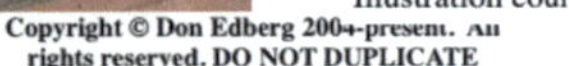

Orbital Good Citizenship

- Plan on de-orbiting before running out of propellant or control
  - Extra propellant required to lower orbit
  - Add drag device to hasten orbital decay
- GEO spacecraft moved to higher or lower orbits for retirement

USSF Guidelines for Disposal Orbits

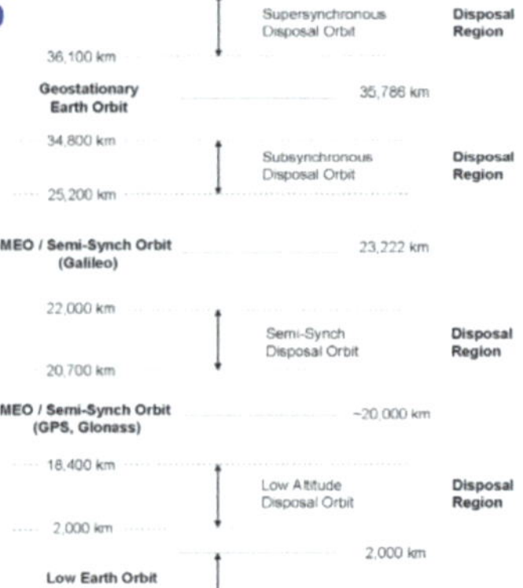

Disturbances to Spacecraft

- Aerodynamic drag
- Gravity gradient
- Magnetism
- Solar pressure
- Radiation & particles
- Static buildup & electrical discharges
- Atomic oxygen
- Orbital debris
- **Outgassing, cold welding, lubrication**
- Internal disturbances

Vacuum Causes Outgassing

- Vacuum sucks volatile chemicals out of materials
  - New car windshields "fog up" etc.
- Also may have depositions from thruster firings
- May affect structural characteristics
- Can cause electrical arcing
- Possibility of vapor condensation on coolest surfaces around … usually optics/mirrors!
- BSS-702 SC had (cold) solar concentrators which collected outgas, causing reduction of available power (p. 230 **Space Systems Failures**)

Cassini & *STARDUST* Contamination

Cassini contamination: repeated, months-long heating to +4° C fixed it (normal temp –90° C)

Courtesy http://photojournal.jpl.nasa.gov/jpeg/PIA03477.jpg

  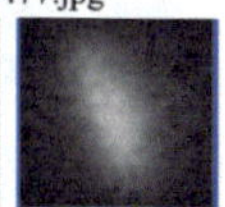

STARDUST: (L): Pre-launch image of internal calibration lamp filament. (C): Degraded image. (R): After heating.

Courtesy JPL

Outgassing Solutions

1. Avoid materials with high fractions of condensates that will outgas (RTV, etc.) Table follows.
2. Bake-out contaminants in vacuum chamber before flight
3. Burn-off heaters during mission

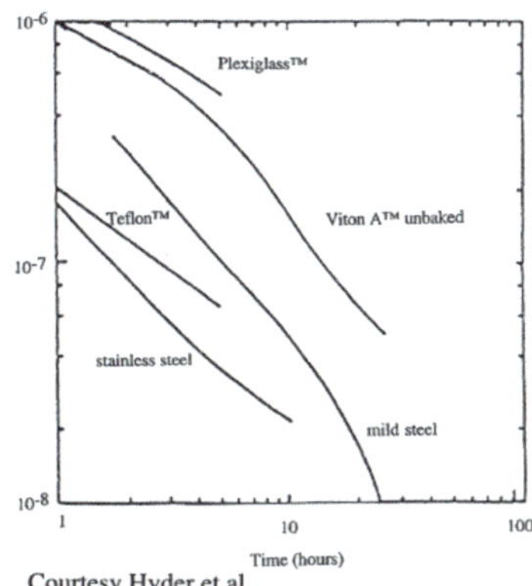

Courtesy Hyder et al.

Outgassing of Materials (1 of 2)

CVCM = Condensed Contaminants, TML = Total Mass Loss

Material Type	Name	Use	TML (%)	CVCM (%)
Silicone	PCB-Z	Conductive white paint	0.60	0.10
Urethane	A276	White paint	0.99	0.08
Urethane	Thermofit RT876	Wire insulation sleeve	0.8	0.08
Silicone Rubber	Eccoshield SV-R	Conductive seals	0.3	0.08
Epoxy	Araldite AV100/SV100	Adhesive	1.1	0.07
Grease	Braycote 602	Lubricant	0.15	0.06
Epoxy	Stycast 1090/9	Potting foam	0.55	0.04
PETP	Gude Space DPTH	Harness tape	0.5	0.04
Urethane	Aeroglaze Z306	Black paint	0.92	0.03
Polyester	Scotch 850 Silver	Thermal tape	0.6	0.03
Epoxy	Scotchweld 1838	Adhesive	0.65	0.03
Silicone	RTV566	Sealant, adhesive	0.27	0.03

From **Principles of Space Instrument Design**, Cruise, et al

Outgassing of Materials (2 of 2)

CVCM = Condensed Contaminants, TML = Total Mass Loss

Material Type	Name	Use	TML (%)	CVCM (%)
Polyimide	Kapton H	Thermal film	1.03	0.02
Fluorocarbon	Viton B910	Rubber seals	0.5	0.02
Polyacetyl	Delrin 550	Plastic parts	0.39	0.02
Urethane	Torrseal	Sealing resin	1.0	0.015
PETP	Mylar A	Thermal film	0.25	0.015
Epoxy-Carbon	Cycon C89/HM-S 40/60	Structure composite	0.6	0.01
Urethane	Solithane 113	Potting resin	0.37	0.01
Fluoralkylether	Fomblin Z25	Lubricating oil	0.06	0.01
Polyimide	Vespel SP-3	Machined insulators	1.08	0
DAP		Connector Bodies	0.44	0
PTFE/glass/MoS_2	Duroid 5813	Bearings, composite	0.08	0
Glass, Woven	Betacloth	Thermal Blanket	0.03	0

From **Principles of Space Instrument Design**, Cruise, et al

 Courtesy Cruise, et al

Vacuum Bake-Out

- Bake all parts in vacuum chamber before flight

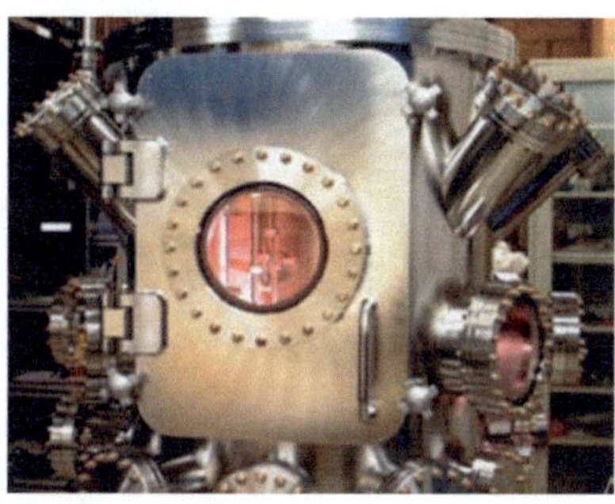

Courtesy Aerospace

Cold Welding & Vacuum Lubricants

- Vacuum: tiny spaces between metals disappear; Two parts can weld together
- Avoid moving parts where possible
- Smooth, polished surfaces preferred
- Vacuum lubricants that do not outgas: graphite, Molybdenum Disulphide (MoS_2) grease
- Wet lubrication preferred due to lower friction. Avoid MoS_2 *solid* lubricants. Lead or other soft metal coatings may be applied by ion plating (caution: lead oxidizes rapidly)
- *Cause of Galileo antenna jam?*

Disturbances to Spacecraft

- Aerodynamic drag
- Gravity gradient
- Magnetism
- Solar pressure
- Radiation & particles
- Static buildup & electrical discharges
- Atomic oxygen
- Orbital debris
- Outgassing, cold welding, lubrication
- Internal disturbances

8/22/22 ORB1B- 88

Internal Mechanical Disturbances

- Articulating mechanisms (solar arrays, camera platforms)
- Jettisons, including venting and waste (STS)
- Propellant sloshing
- Flexible structural vibration
- Imbalances in spinning items: RWAs, CMGs, bearing flaws
- Crew activities (push-offs, treadmill, sneezes, etc.)

Courtesy: NASA

8/22/22 ORB1B- 89

ISS Vibration Environment

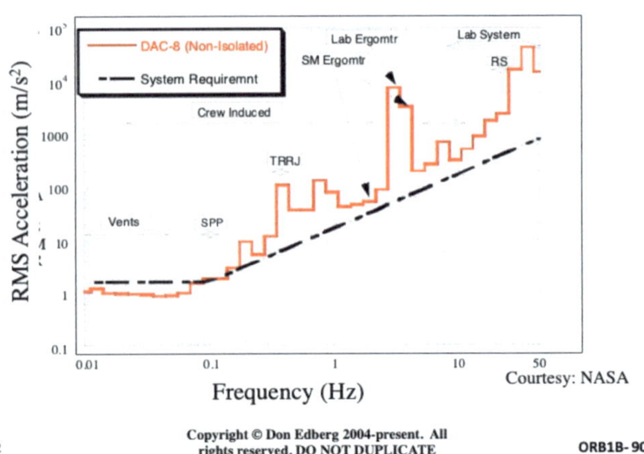

Courtesy: NASA

8/22/22 ORB1B- 90

STABLE Vibration Isolation System

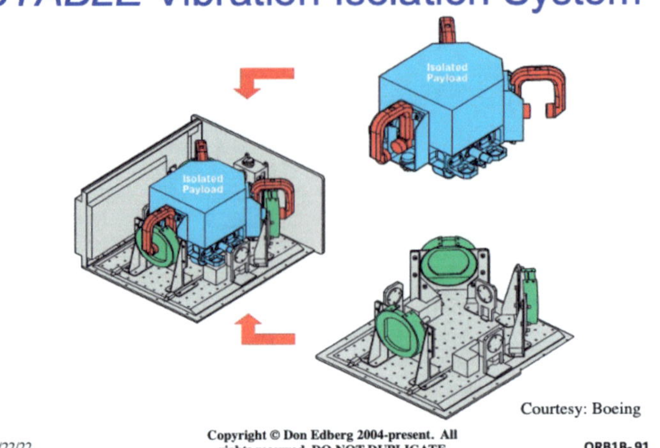

Courtesy: Boeing

8/22/22 ORB1B- 91

STABLE in USML-02, 1995

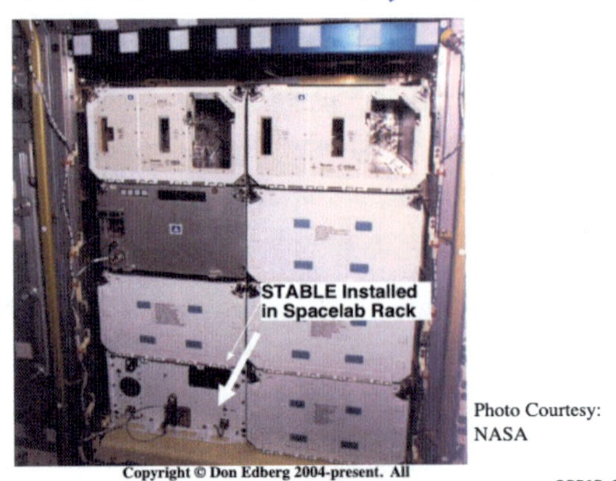

Photo Courtesy: NASA

8/22/22 ORB1B- 92

STABLE Flight Results (STS-73, 1995)

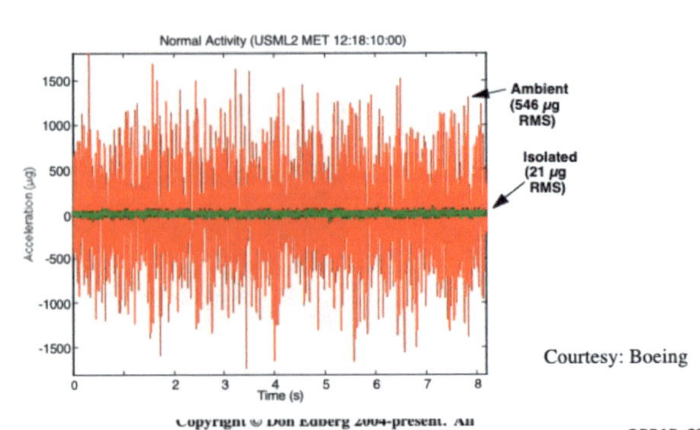

Courtesy: Boeing

8/22/22 ORB1B- 93

More Orbital Mechanics: Changing Orbits, Orbital Transfers & Trajectories

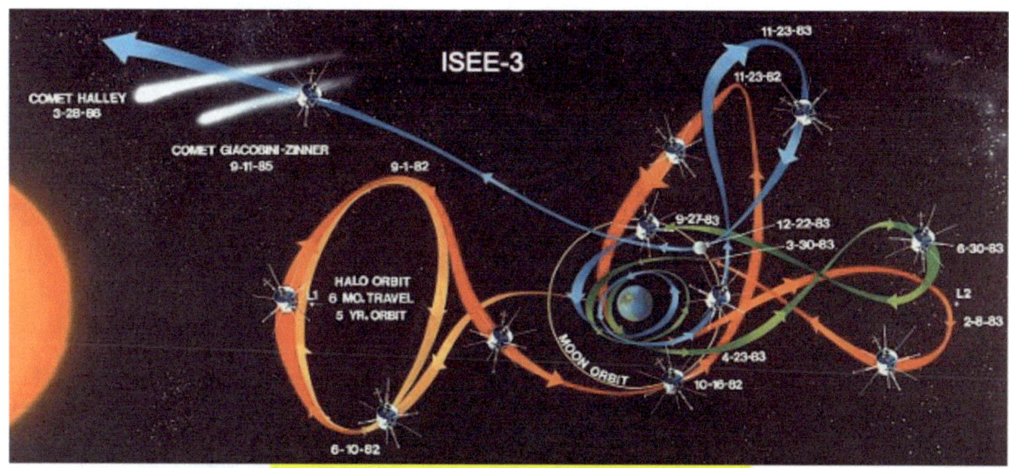

August 22, 2022

(53)

Topics: Changing Orbits

- In-plane orbit shape changes
  - Effect of speed changes
  - Transfer orbits, Hohmann transfers
- Orbital plane (inclination) changes
- Combined in-plane & inclination changes
- Interplanetary transfers / patched conics
- Gravity assists

August 22, 2022

Course Layout

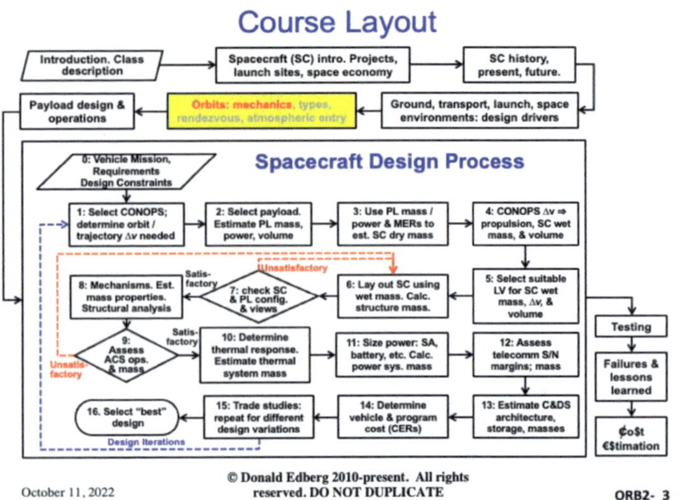

Introduction. Class description
Spacecraft (SC) intro. Projects, launch sites, space economy
SC history, present, future.
Payload design & operations
Orbits: mechanics, types, rendezvous, atmospheric entry
Ground, transport, launch, space environments: design drivers
0: Vehicle Mission, Requirements Design Constraints
Spacecraft Design Process
1: Select CONOPS; determine orbit / trajectory Δv needed
2: Select payload. Estimate PL mass, power, volume
3: Use PL mass / power & MERs to est. SC dry mass
4: CONOPS Δv ⇒ propulsion, SC wet mass, & volume
Unsatisfactory
8: Mechanisms. Est. mass properties. Structural analysis
Satis-factory
7: check SC & PL config. & views
6: Lay out SC using wet mass. Calc. structure mass.
5: Select suitable LV for SC wet mass, Δv, & volume
Testing
9: Assess ACS ops. & mass
Unsatis-factory
Satis-factory
10: Determine thermal response. Estimate thermal system mass
11: Size power: SA, battery, etc. Calc. power sys. mass
12: Assess telecomm S/N margins; mass
Failures & lessons learned
16. Select "best" design
15: Trade studies: repeat for different design variations
14: Determine vehicle & program cost (CERs)
13: Estimate C&DS architecture, storage, masses
¢o$t €$timation
Design iterations

Topics: Changing Orbits

- In-plane orbit shape changes
  - Effect of speed changes
  - Transfer orbits, Hohmann transfers
- Orbital plane (inclination) changes
- Combined in-plane & inclination changes
- Interplanetary transfers / patched conics
- Gravity assists

Changing Orbit Shapes

- Change between one orbit and another having same periapsis or apoapsis (higher or lower ellipse, for example)

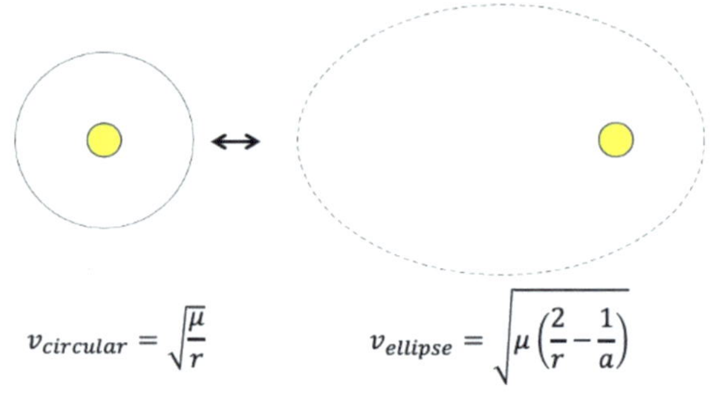

$$v_{circular} = \sqrt{\frac{\mu}{r}} \qquad v_{ellipse} = \sqrt{\mu\left(\frac{2}{r} - \frac{1}{a}\right)}$$

Speed Change (Δv) Parallel to Velocity Vector Alters Orbit Shape

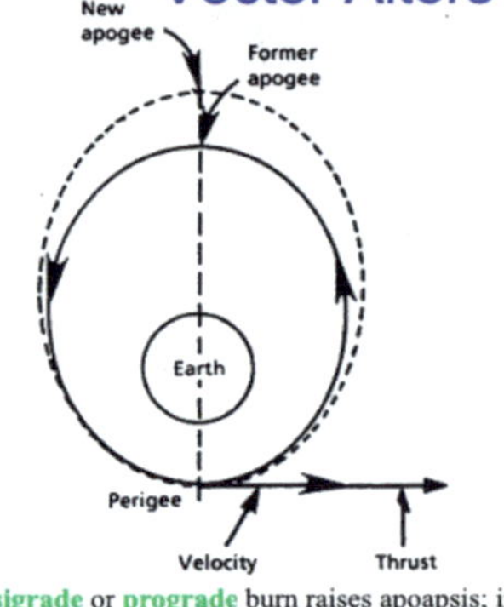

New apogee
Former apogee
Earth
Perigee
Velocity
Thrust

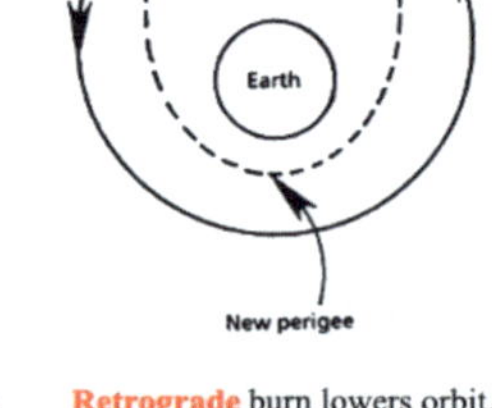

New apogee
Retrograde thrust vector
Earth
New perigee

Posigrade or **prograde** burn raises apoapsis; in LEO, 1 m/s raises ~3 km! Burn location *does not change.*

Retrograde burn lowers orbit. Burn location altitude *does not change.*

Δv *Parallel* to Velocity Direction

- Burns change orbit at *opposite* side
- Burns do not change altitude where made
- Burn with/against velocity direction raises/lowers orbit opposite side
- Speeding up slows you down…And vice versa
- A Δv of 1 m/s raises orbit about *3 km* at apoapsis (LEO)

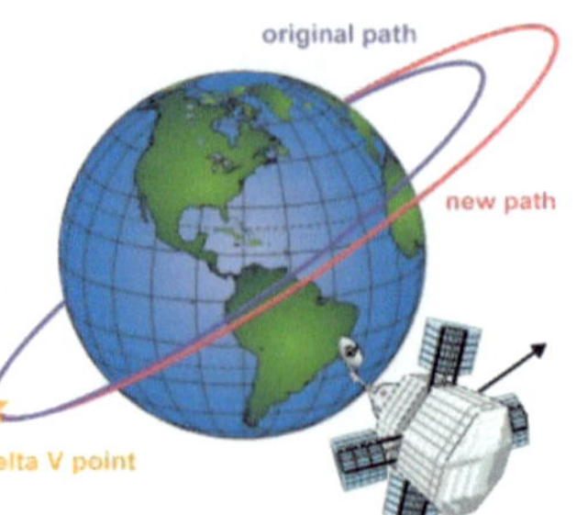

original path
new path
Delta V point

Δv Radially Away From Attracting Object Alters *Eccentricity*

- What happens if the SC thrusts *upwards?*
- Changes *eccentricity*
- 1 m/s speed changes altitude by ±800 m

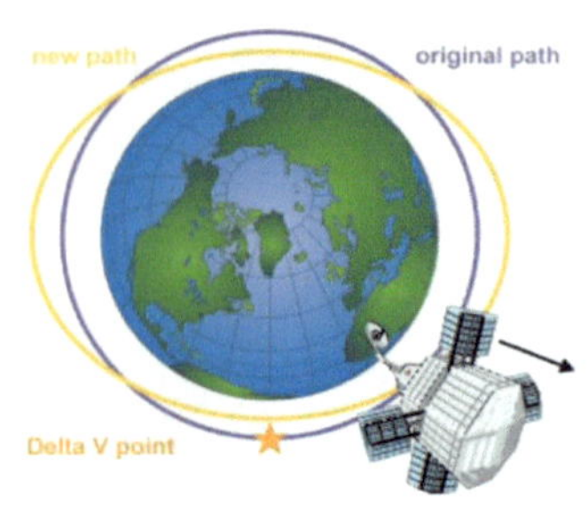

new path
original path
Delta V point

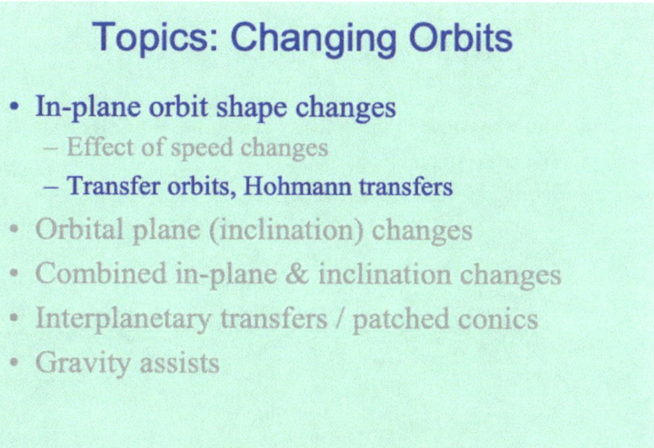

Hohmann Transfer

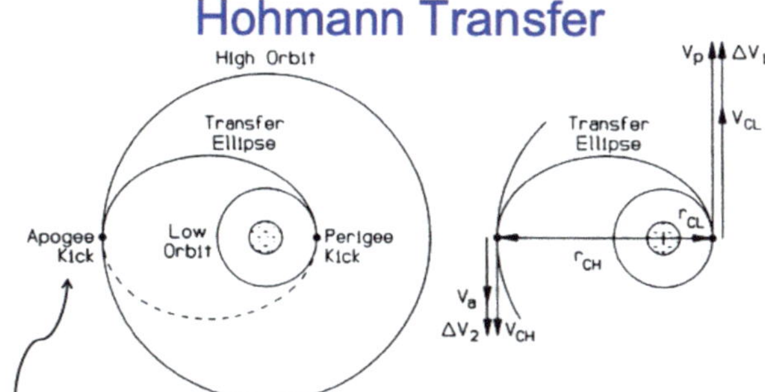

Apogee Boost Motor: ABM, Apogee Kick Motor: AKM

Required total speed is:

$$\Delta v_{Hohmann} = \sqrt{\frac{\mu}{r_{CL}}}\left[\frac{1}{\sqrt{\alpha}} - \frac{(1-\alpha)\sqrt{2}}{\sqrt{\alpha(1+\alpha)}} - 1\right], \text{where } \alpha = \frac{r_{CH}}{r_{CL}}$$

Ref: Curtis

Hohmann Transfer

- Use to transfer between two orbits that are not touching
- In many cases (but not all), the two impulse Hohmann transfer ellipse between an inner and outer circle will require minimum Δv
- Note: Earth & Mars orbits are not coplanar nor circular, transfer is 'Hohmann-like'

Mars orbit around sun
Earth orbit around sun
Transfer orbit

Disposal Orbits

IADC recommends fulfilling two conditions for disposal to an orbit above the GEO protected region:

1. A minimum *increase* in perigee altitude of: $\Delta h_p = 235$ km + 1000 × CR × A/m, where
   - CR = 1.02 ~ 1.07, solar radiation pressure coefficient
   - A/m = aspect area to dry mass ratio (m²/kg)
   - 235 km = upper altitude of the GSO protected region (200 km) + maximum descent of a re-orbited spacecraft due to luni-solar & geopotential perturbations (35 km)
2. Eccentricity $e \leq 0.003$

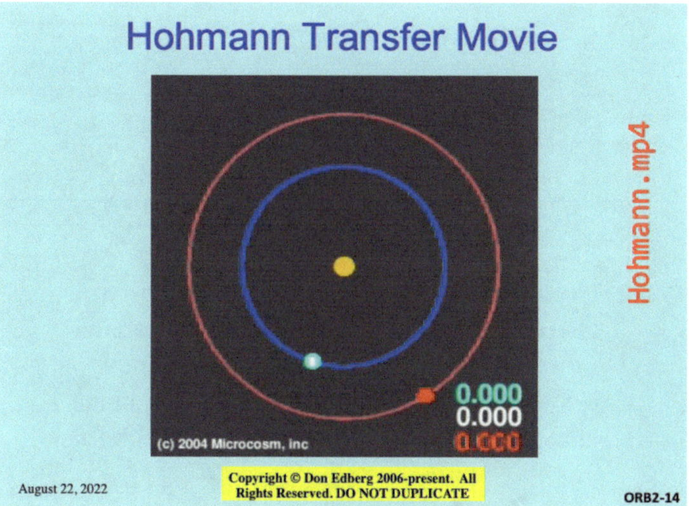

Topics: Changing Orbits

- In-plane orbit shape changes
  - Effect of speed changes
  - Transfer orbits, Hohmann transfers
- **Orbital plane (inclination) changes**
- Combined in-plane & inclination changes
- Interplanetary transfers / patched conics
- Gravity assists

Hohmann Transfer Including Plane Change

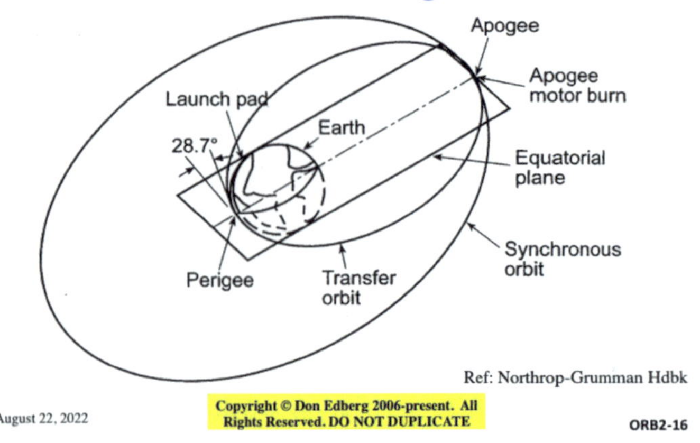

Ref: Northrop-Grumman Hdbk

August 22, 2022

ORB2-16

Inclination Change Only

- The *magnitude* of the velocity vector doesn't change
- The angle (inclination) changes

Δv required for inclination change Δi is: $\Delta v = 2v\sin\left(\frac{\Delta i}{2}\right)$

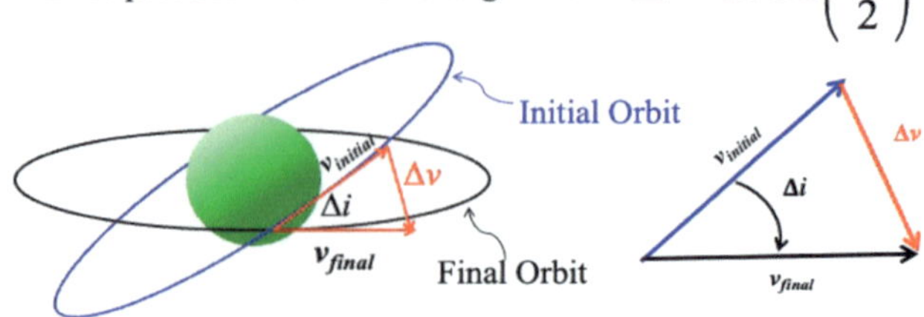

August 22, 2022

ORB2-18

Plane Changes

Launch Site

Equator

(a)

(b)

- Plane changes ***very expensive*** in terms of Δv
- Choices:
  - Choose launch latitude if possible
  - During ascent ("dogleg" maneuver, let booster do it)
  - At apoapsis of an ellipse

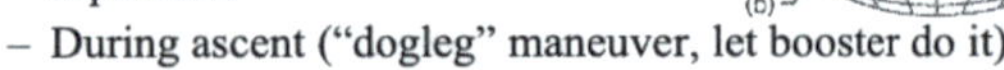

- Usually best to perform when speed is *lowest*: apoapsis or further.
- If plane change ***and*** orbit shaping required, often best to combine into a single Δv (but not always)

August 22, 2022

ORB2-19

Topics: Changing Orbits

- In-plane orbit shape changes
  - Effect of speed changes
  - Transfer orbits, Hohmann transfers
- Orbital plane (inclination) changes
- Combined in-plane & inclination changes
- Interplanetary transfers / patched conics
- Gravity assists

August 22, 2022

ORB2-20

Combined Changes

Use to change the magnitude of *a* as well as the inclination *i*

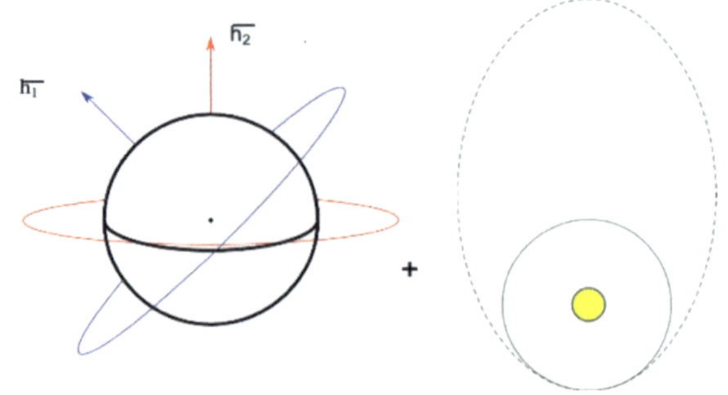

August 22, 2022

ORB2-21

Combined Plane Changes

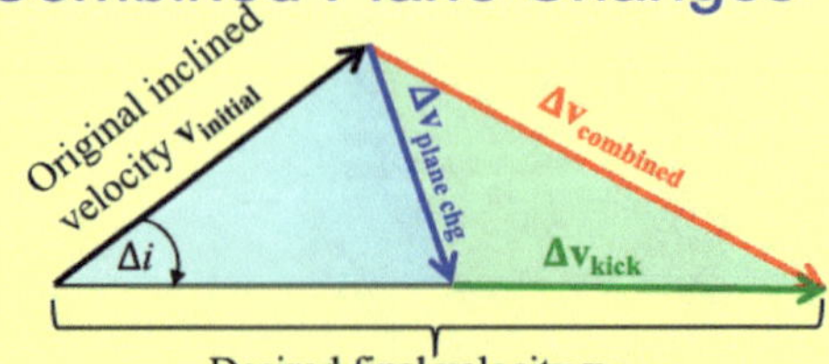

- Usually want to transfer to higher orbit and change inclination to zero at top of transfer
- *Law of Cosines* used to obtain combined orbit shaping/plane change $\Delta v_{combined}$, since vector sum is less than two sequential burns

August 22, 2022

ORB2-22

Transfer to Higher Orbit, Less *i*

- Vector quantities **bold-face**, not italicized; magnitudes of vectors *italicized*
- $\mathbf{v}_{\mathbf{initial}}$ = original inclined velocity vector
- $\mathbf{v}_{\mathbf{final}}$ = desired final velocity vector
- $\Delta\mathbf{v}_{\mathbf{plane\ change}}$ = $\Delta\mathbf{v}$ needed to change orbit inclination by amount $\Delta i = i_f - i_i$
- $\Delta\mathbf{v}_{\mathbf{kick}}$ = speed change needed to shape orbit
- $\Delta v_{combined} = \sqrt{v_{initial}^2 + v_{final}^2 - 2v_{initial}v_{final}\cos\left(i_f - i_i\right)}$
- $v_{initial} = v_{apoapsis}$ of transfer orbit (inclination i_i),
- $v_{final} = v_{circular}$ of final orbit, inclination i_f (0 for GEO)

Split Plane Change (cont.)

$$\frac{\partial \Delta v_{total}}{\partial(\Delta i)} = 0 = \frac{v_p v_{CL}\sin(\Delta i)}{\sqrt{v_p^2 + v_{CL}^2 - 2v_p v_{CL}\cos(\Delta i)}} - \frac{v_a v_{CH}\sin(i - \Delta i)}{\sqrt{v_a^2 + v_{CH}^2 - 2v_a v_{CH}\cos(i - \Delta i)}}$$

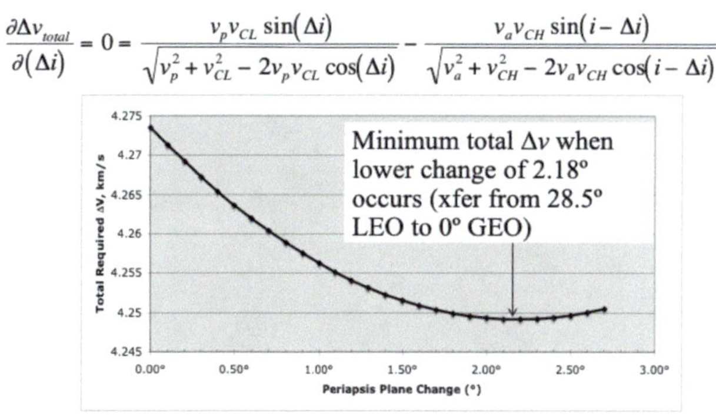

Optimal Two-Burn 28.5° LEO to 0° GEO

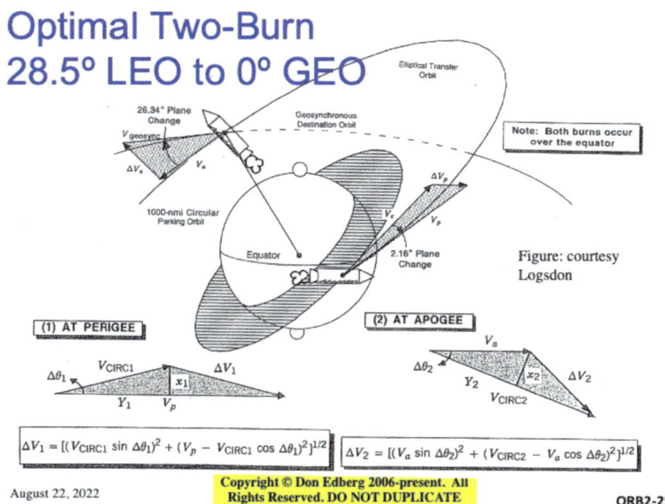

Figure: courtesy Logsdon

Combined Plane Change Example: 185 km × 28.5° Earth parking orbit to 42160 km (radius) × 0° GEO orbit

Case 1: First do –28.5° inclination change. Then do Hohmann xfer Δv_1 & Δv_2	Case 2: 1st do Hohmann xfer Δv_1 & Δv_2. Then change inclination – 28.5° (simple).	Case 3: First do combined plane & inclination change, then do Hohmann xfer Δv_2	Case 4: First do Hohmann xfer Δv_1. Then do combined change at apogee.	Case 5: 1st, do Hohmann xfer Δv_1 & 2.18° plane change. 2nd 26.32° Δi & Δv_2 xfer burn (**optimum**).
Δv_{simple} = 3.837 km/s (in orbit 1)	$\Delta v_{Hohmann}$ = 3.938 km/s	$\Delta v_{combined}$ = 5.041 km/s (at perigee)	Δv_1 = 2.459 km/s	Δv_1 = 2.483 km/s
$\Delta v_{Hohmann}$ = 3.938 km/s	Δv_{simple} = 1.514 km/s (in orbit 2)	Δv_2 = 1.479 km/s	$\Delta v_{combined}$ = 1.838 km/s (at apogee)	$\Delta v_{combined}$ = 1.79 km/s (at apogee)
Δv_{total} = 7.77 km/s	Δv_{total} = 5.452 km/s	Δv_{total} = 6.52 km/s	**Δv_{total} = 4.297 km/s**	**Δv_{total} = 4.273 km/s (–0.56%)**

Optimal Transfer Not Necessarily Hohmann

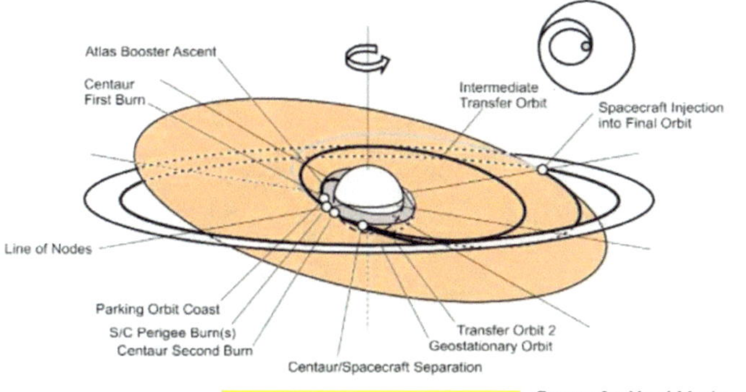

Courtesy Lockheed-Martin

Bi-elliptic Transfer: Go Out Further

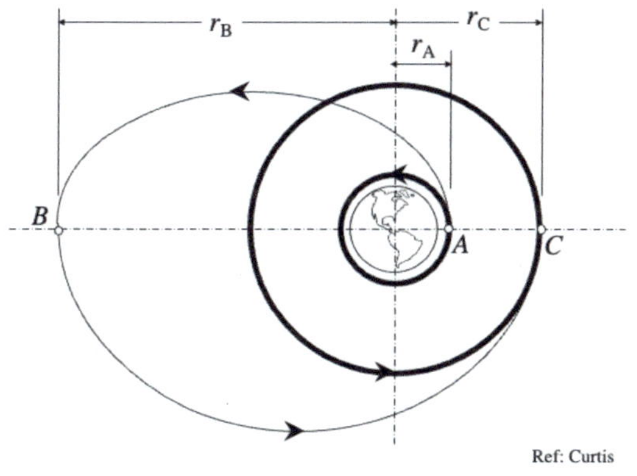

Ref: Curtis

Bi-Elliptic More Efficient? Depends.

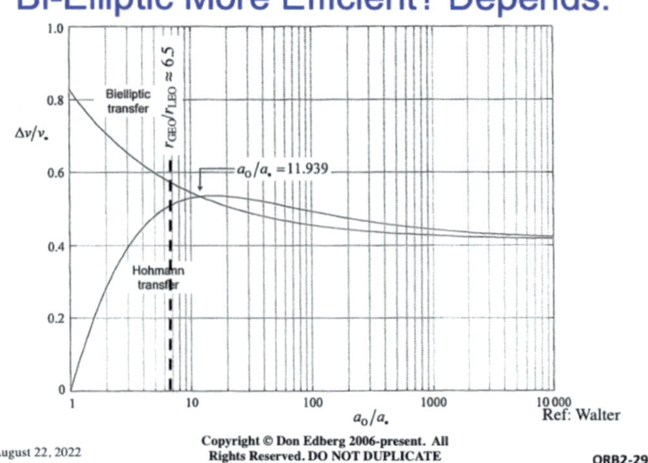

Ref: Walter

Advantages & Disadvantages

- If $r_C < 11.94\, r_A$, Hohmann better
- If $r_C > 11.94\, r_A$, bi-elliptic better
- If there is an advantage, it's not huge
- *Three* burns required (vs. *two* for Hohmann)
- Bi-elliptic time-of-flight >> Hohmann
- Required speed change is $\left(\alpha = \frac{r_c}{r_a}, \beta = \frac{r_b}{r_a}\right)$:

$$\Delta v_{bi-elliptic} = \left[\sqrt{\frac{2(\alpha+\beta)}{\alpha\beta}} - \frac{1+\sqrt{\alpha}}{\sqrt{\alpha}} - \sqrt{\frac{2}{\beta(1+\beta)}}(1-\beta)\right]\sqrt{\frac{\mu}{r_A}}$$

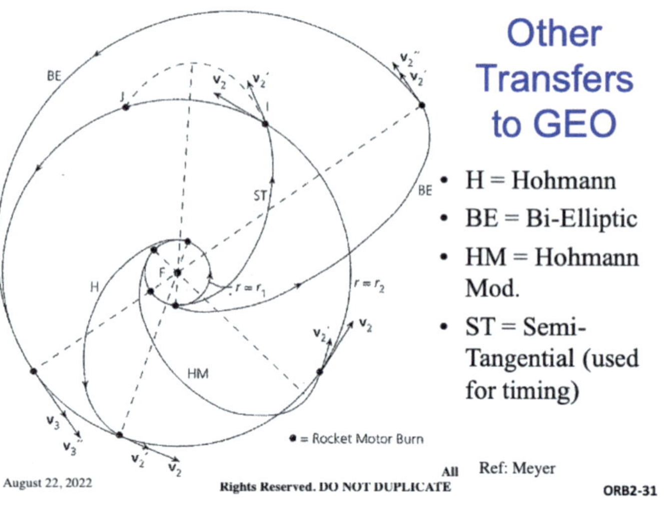

Other Transfers to GEO

- H = Hohmann
- BE = Bi-Elliptic
- HM = Hohmann Mod.
- ST = Semi-Tangential (used for timing)

Ref: Meyer

General Transfers

- Can be Hohmann, Fast, or Low-Thrust spiral

Hohmann   Direct   Spiral

Fast Transfer

- Faster than Hohmann
- More "expensive" since more Δv required
- Often used for quicker planetary transfers
- Elliptical or other conics possible

Type I	Type II	Type III
Parallel at periapsis	Parallel at apoapsis	Not parallel at periapsis or apoapsis (general case)

Topics: Changing Orbits

- In-plane orbit shape changes
  - Effect of speed changes
  - Transfer orbits, Hohmann transfers
- Orbital plane (inclination) changes
- Combined in-plane & inclination changes
- Interplanetary transfers / patched conics
- Gravity assists

Interplanetary Trajectories: Transfers in Solar System

- Complex problem: motion affected by departing planet, arriving planet, sun, other planets
- Multi-body problem not solvable analytically, computers find solutions
- Departure planet orbit typically on different plane than arrival planet
- Approximations made to solve by hand

Patched Conic Method

- Assume the transfer can be divided into three separate "phases"
- In each phase, the two-body equations of motion can be applied separately
- A "sphere of influence" defines transitions from one phase to the next

II

III

I

Simplified Solution: "Method of Patched Conics"

Split into three conics, "patch" them together

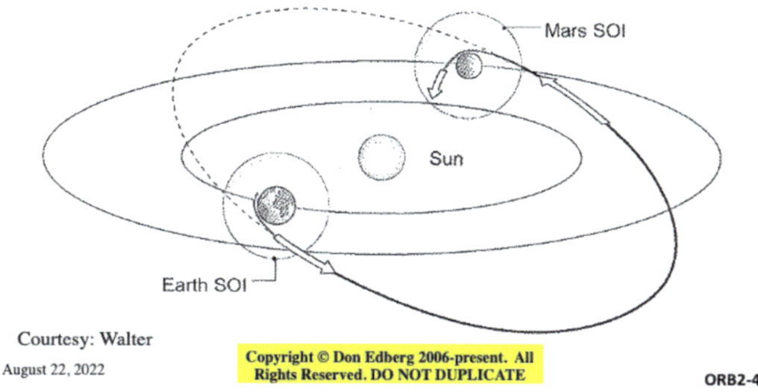

Courtesy: Walter

Three Phases for Solar System Travel

- I: Earth escape (hyperbolic excess velocity)
- II: Heliocentric Hohmann transfer (ellipse coast around sun, need planetary locations)
- III: Planetary arrival/capture (hyperbolic, then establish orbit)

Example: transfer from Earth to Mars

- Assumptions:
  - Circular orbits
  - Co-planar orbits: no plane changes

Earth-Mars Patched Conic

- Solve Phase II first to get end point requirements.

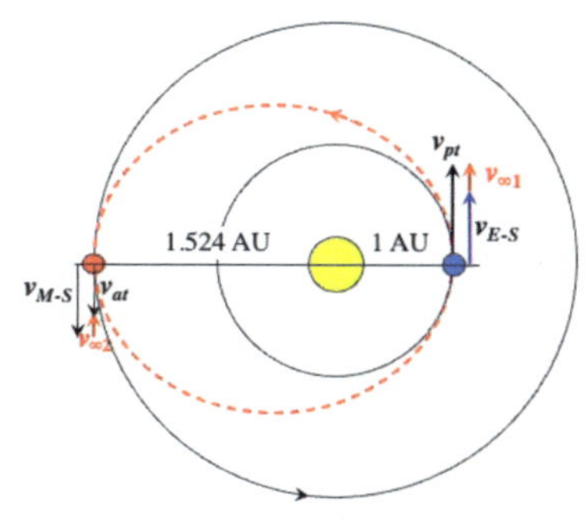

$v_{pt} = 32.73$ km/s

$v_{at} = 21.476$ km/s

$\varepsilon_t = -3.514 \times 10^8$ m²/s²

$v_E = 29.79$ km/s

$v_M = 24.14$ km/s

Phase I: Earth Departure within *SOI*

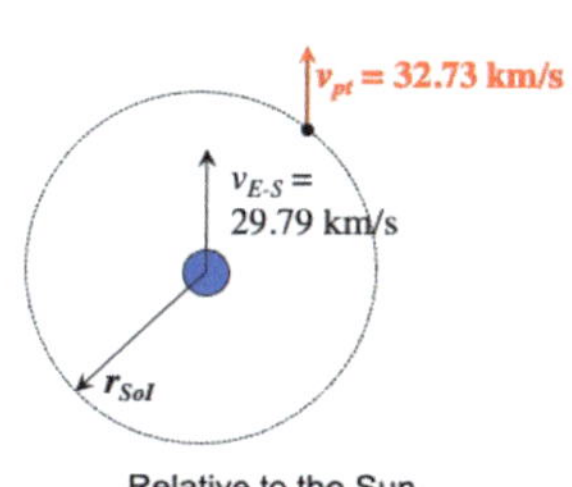

Relative to the Sun

v_∞ = 32.73 – 29.79 km/s = 2.94 km/s

$(v_{E\text{-}S})$

r_{SoI}

Relative to the Earth

Sphere of influence (SOI): a transitional region, inside of which the *planet's* gravitational influence is *more* important than the *sun's*.

Phase I: Earth Departure (cont.)

The sphere of influence is defined as:

$$r_{SOI} = r_1 \left(\frac{m_p}{m_{Sun}} \right)^{0.4}$$

- Outside of the planet's SOI, the sun can be assumed to be the major gravity source.
- Some values:

$r_{SOI\ Earth} = 924\ 500$ km $= 502\ 500$ nm

$r_{SOI\ Mars} = 578\ 100$ km $= 314\ 200$ nm

August 22, 2022 ORB2-44

Phase I: Earth Departure (cont.)

- For the patched conic approach, assume:

$v_{SOI} = v_\infty$

$v_{\infty 1} = v_{pt} - v_E = 32.73 - 29.79$ km/s $= 2.94$ km/s

$C3 = 2\varepsilon = {v_\infty}^2 = 8.64$ km²/s²

C3 measures LV interplanetary launch capacity: larger *C3* indicates higher escape velocity

Note: C3 < 0 means closed orbit; $C3 = v^2 - 2\mu/r$

Given the required v_∞ and orbital height, we can calculate Δv required to obtain v_∞.

Common to enter parking orbit first, but may do direct escape instead

August 22, 2022 ORB2-45

Escape After Parking Orbit

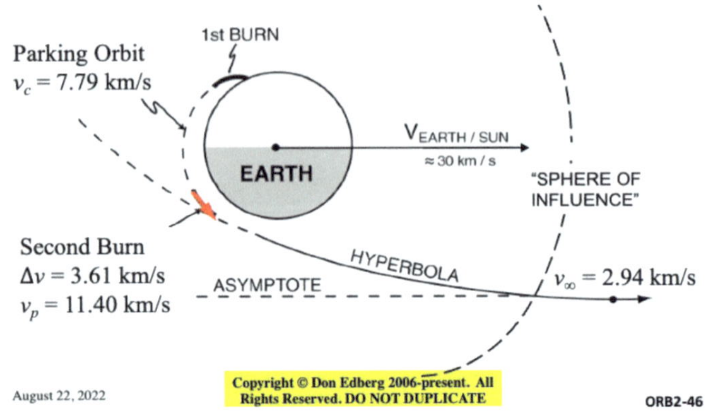

August 22, 2022 ORB2-46

Escape From Any Inclination (No Plane Change Needed)

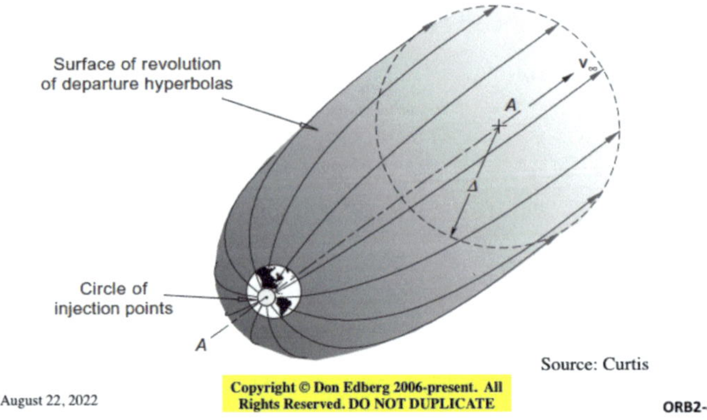

Source: Curtis

August 22, 2022 ORB2-47

Min. Planetary Transfer Speeds

MINIMUM HYPERBOLIC V_∞ (km/s)

SUN*	MERCURY*	VENUS*	MARS	JUPITER	SATURN	URANUS	NEPTUNE	PLUTO	SOLAR ESCAPE
29.9	7.6	2.5	3.0	8.8	10.3	11.3	11.7	11.9	12.4

* must launch ***opposite*** to Earth's motion around the Sun

LAUNCH SPEED (1,000 FPS)
VENUS
MARS
MERCURY
JUPITER
SATURN
SOLAR SYSTEM ESCAPE SPEED
DISTANCE FROM SUN (ASTRONOMICAL UNITS)

- These are the minima needed for transfer *without* gravity assists
- v_∞ or *C3* must be delivered by LV

Courtesy: Douglas

August 22, 2022 ORB2-48

During Phase II: Mid-Course Corrections (Trajectory Correction Maneuvers)

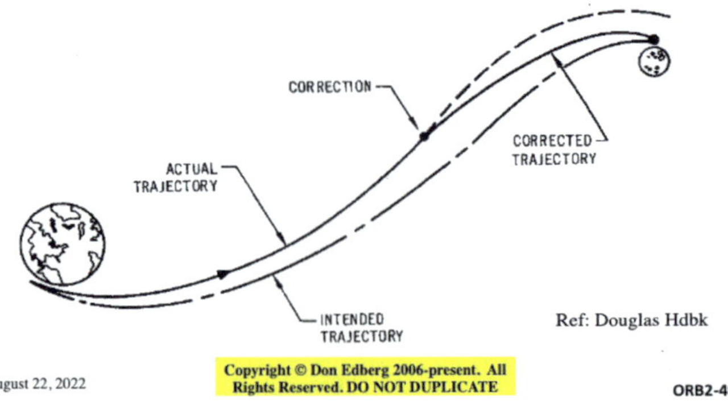

Ref: Douglas Hdbk

August 22, 2022 ORB2-49

Launch Error Sensitivity Analysis Example: *Earth-Mars Transfer*

- Aphelion radius: $R_{sun\text{-}Mars} = h^2/\mu_{sun}(1-e)$
- Take variation in radius r with respect to burnout position δr_p and burnout speed δv_p:

δr & δv are errors

$$\frac{\delta r}{R_{sun-Mars}} = \frac{2}{1 - R_{sun-Earth} v_D^2 / 2\mu_{sun}} \left[\frac{\mu_{Earth}}{v_D v_\infty r_p} \left(\frac{\delta r_p}{r_p} \right) + \frac{v_\infty^2 r_p + 2\mu_{Earth}}{v_D v_\infty r_p} \left(\frac{\delta v_p}{v_p} \right) \right]$$

For a mission to Mars, leaving Earth from a 300 km orbit,
v_D = departure speed = 32.73 km/s (relative to sun),
r_p = 6678 km, Δv = 3.59 km/s (escape hyperbola @ 300 km)
v_p = 11.32 km/s, v_∞ = 2.943 km/s (escaping from S.o.I.)
$R_{sun\text{-}Earth} = 149.6 \times 10^6$ km, $R_{sun\text{-}Mars} = 227.9 \times 10^6$ km
$\mu_{earth} = 398600.4$ km^3/s^2, $\mu_{sun} = 1.327 \times 10^{11}$ km^3/s^2

August 22, 2022

Ref: Curtis, Ch. 8

ORB2-50

Burnout Error Results

- We find: $$\delta r = R_{sun-Mars} \times \left[3.127 \left(\frac{\delta r_p}{r_p} \right) + 6.708 \left(\frac{\delta v_p}{v_p} \right) \right]$$
- A 0.01% or 1 part in 10,000 error (11.32 km/s ÷ 10000 = **1.13 m/s**) in burnout speed v_p changes target radius by 6.708 ÷ 10000 = 0.067% × ($R_{sun\text{-}Mars}$) = **153,000 km**!
- A 0.01% or 1 part in 10,000 error (6678 km ÷ 10000 = **668 m** ~ 2200 ft) error in burnout altitude r_p changes target radius by 0.031% × ($R_{sun\text{-}Mars}$) = **71,260 km**!
- Note that R_{Mars} = 3396 km: these are **HUGE** misses
- These small burnout errors *must be corrected* by midcourse maneuvers ("trajectory correction maneuvers" or TCMs)

Ref: Curtis, Ch. 8

August 22, 2022

ORB2-51

New Horizons Fires Engine To Aim For Kuiper Belt's 2014 MU69

SPACE (2/2/17) reports that on Wednesday, NASA's *New Horizons* spacecraft briefly fired its engine to refine its course for its next flyby target: Kuiper belt object 2014 MU69, which orbits a billion miles beyond the probe's last target, Pluto. The adjustment increased the spacecraft's speed by only one mile per hour, but the mission's principal investigator said that "will add up to an aim point refinement of almost 6,000 miles." The adjustment was made in response to recent observations of 2014 MU69's orbit by NASA's *Hubble Space Telescope*. *New Horizons* is on course to reach the object on January 1, 2019.

August 22, 2022

ORB2-52

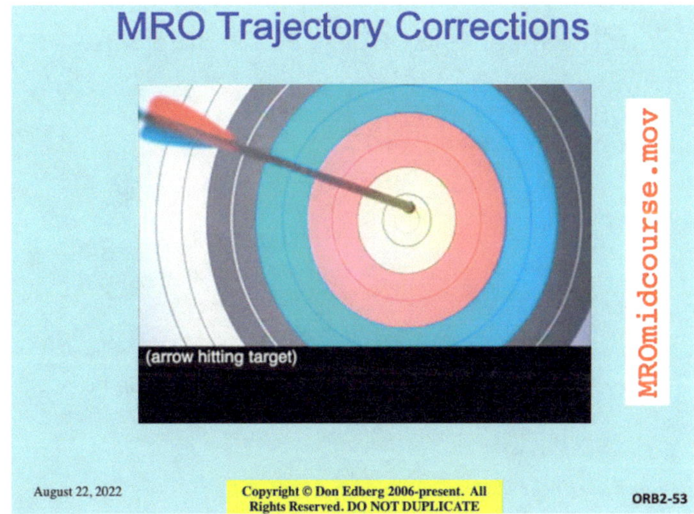

MRO Trajectory Corrections
(arrow hitting target)
MROmidcourse.mov

August 22, 2022

ORB2-53

Phase III: Planetary Arrival

- For arrival at the planet: plan to get 'rear-ended,' since transfer speed $v_{at} < v_{M\text{-}S}$ (Mars speed)!

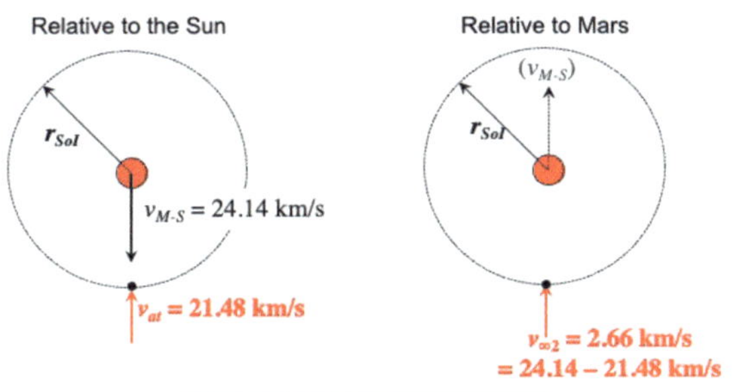

August 22, 2022

ORB2-54

Arrival (cont.)

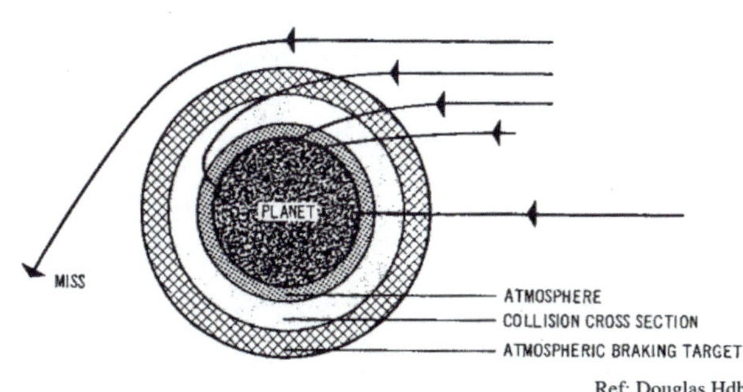

Ref: Douglas Hdbk

August 22, 2022

ORB2-55

Phase III: Planetary Arrival (cont.)

- Approach is important!

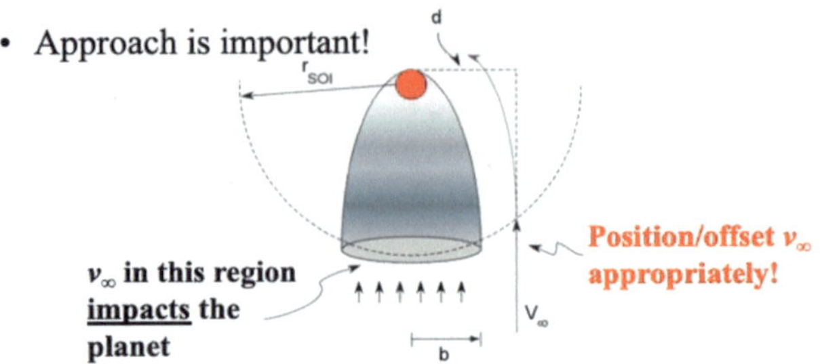

$v_{at} = 21.476$ km/s, $v_{mars} = 24.14$ km/s, so $v_\infty = \mathbf{2.664\ km/s}$

August 22, 2022

ORB2-56

Phase 3: Planetary Arrival (cont.)

- From Hale, **Introduction to Space Flight**:

$$d = r_{SOI}\cos(\phi_{SOI}) = \frac{h}{v_\infty}$$

= distance from asymptote to planet center

$$b = r_{planet}\sqrt{1 + \frac{v_{esc_{planet}}^2}{v_\infty^2}} = \text{Impact radius}$$

$$v_{esc_{planet}} = \sqrt{\frac{2\mu_{planet}}{r_{planet}}} = \text{escape speed}$$

August 22, 2022

ORB2-57

Phase 3: Planetary Arrival (cont.)

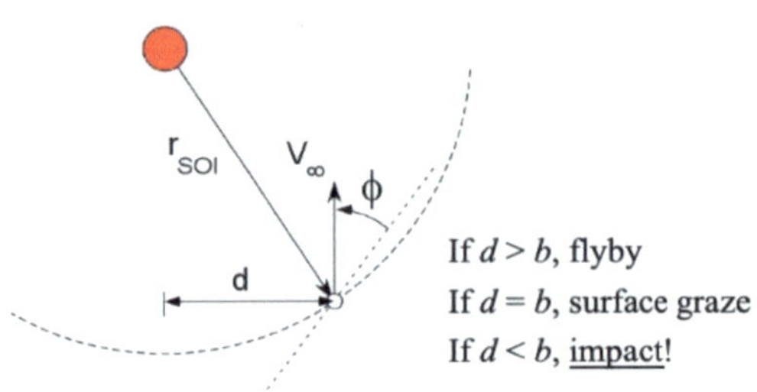

If $d > b$, flyby

If $d = b$, surface graze

If $d < b$, impact!

Alternatively, just find r_p for a given r_{SOI}, ϕ_{SOI}, v_∞

August 22, 2022

ORB2-58

Targeting: "*B-Plane*"

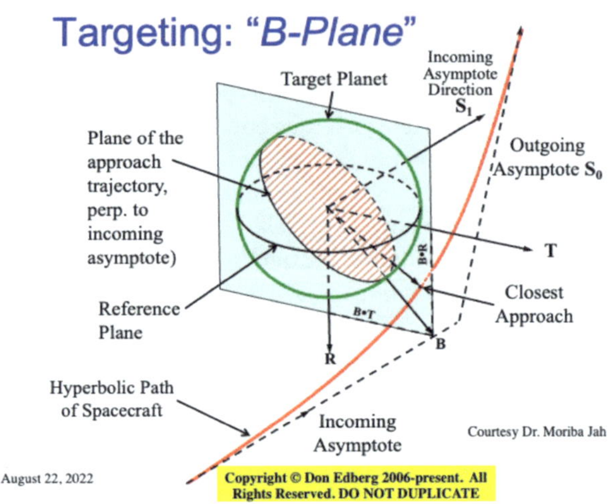

Courtesy Dr. Moriba Jah

August 22, 2022

ORB2-59

MRO Capture Burn @ Mars

The Challenges of Getting to Mars

Burn and Capture

MRO-Burn&Capture.mp4

August 22, 2022

ORB2-60

Inner Planet Transfers

- Launches to inner planets still have a positive v_∞ with respect to the Earth, just in the *opposite* direction of v_e ("leave out the back side")
  - Still move in the directions of the planets; no retrograde!

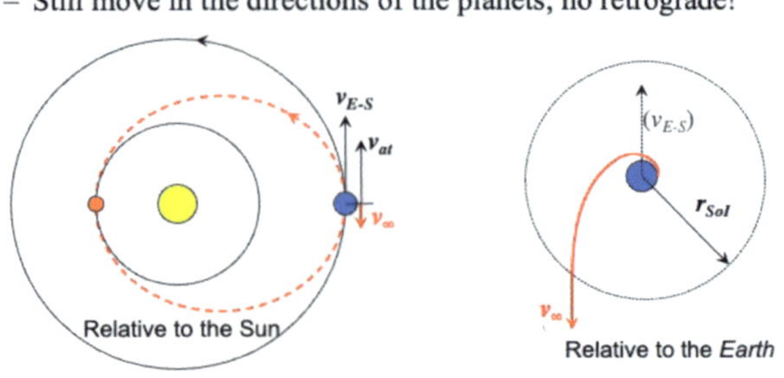

August 22, 2022

ORB2-61

Summary of Earth Exit Strategies

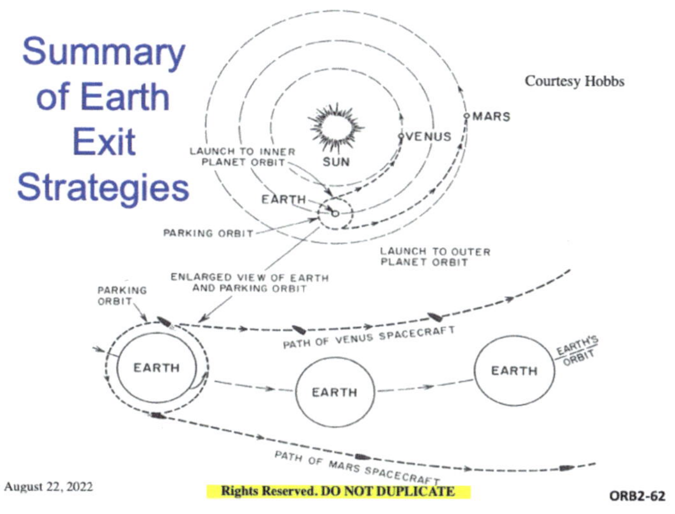

Real Interplanetary Transfers

- Planetary orbits are NOT coplanar & NOT circular
  - Transfer orbit requires a plane change
- Heliocentric transfer orbit typically NOT a Hohmann transfer
  - Could be a faster ellipse
  - Or a hyperbolic transfer
- Must account for ignored influences
- Computer solution:
  - "Trajectory Optimization tool" www.orbithangar.com/searchid.php?ID=5418 , or
  - http://sdg.aero.upm.es/index.php/online-apps/porkchop-plot

TWO Main Trajectory Considerations:

Departure *C3 defines* LV exit performance, SC Time of Flight (AKA "pork chop" plot)

Arrival speed: Capture requirements may *define* SC propulsion requirements

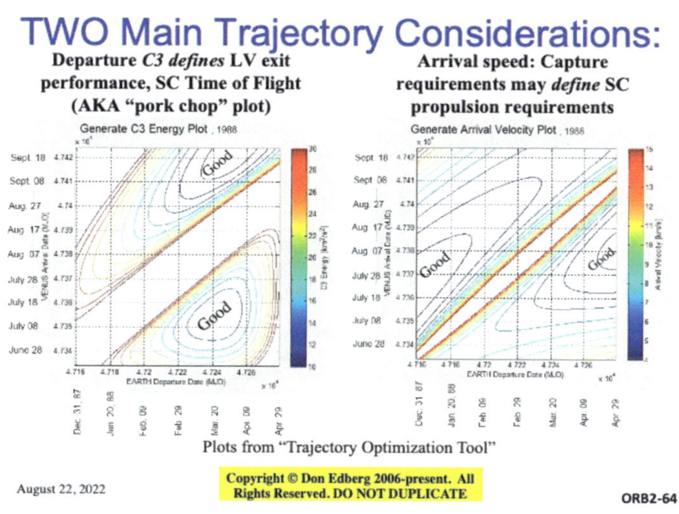

Plots from "Trajectory Optimization Tool"

NASA Mission Design Ctr. Trajectory Browser

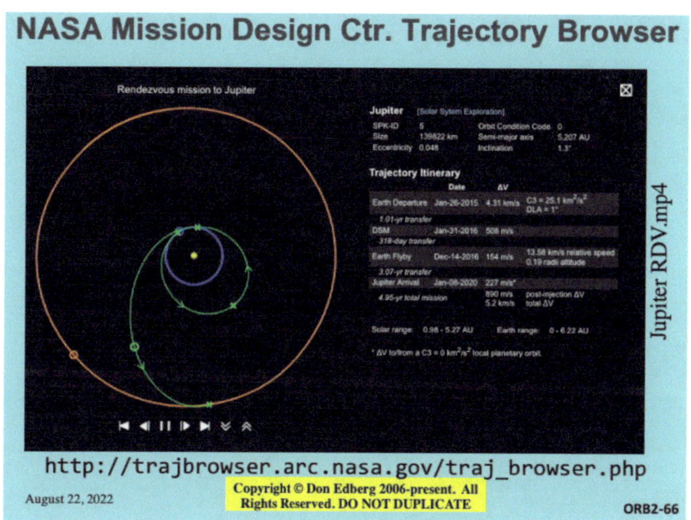

http://trajbrowser.arc.nasa.gov/traj_browser.php

Topics: Changing Orbits

- In-plane orbit shape changes
  - Effect of speed changes
  - Transfer orbits, Hohmann transfers
- Orbital plane (inclination) changes
- Combined in-plane & inclination changes
- Interplanetary transfers / patched conics
- **Gravity assists**

Gravity Assists: Δv for Free!

Planet's Direction of Motion

a – Arrival
d – Departure

Planet

Planet's Sphere of Influence

Gravity Assist From Different Frames

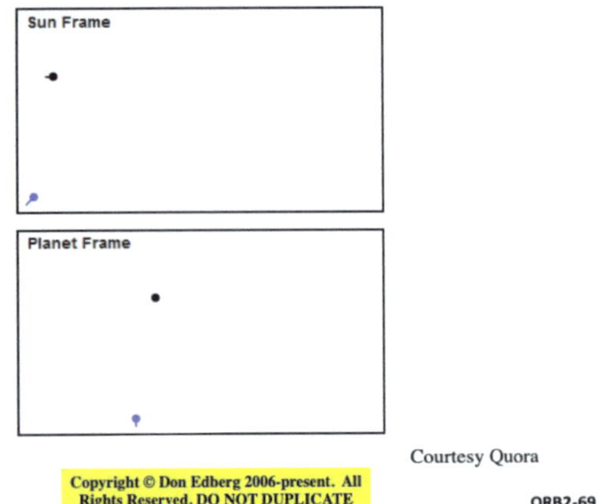

Courtesy Quora

Jupiter Swingby

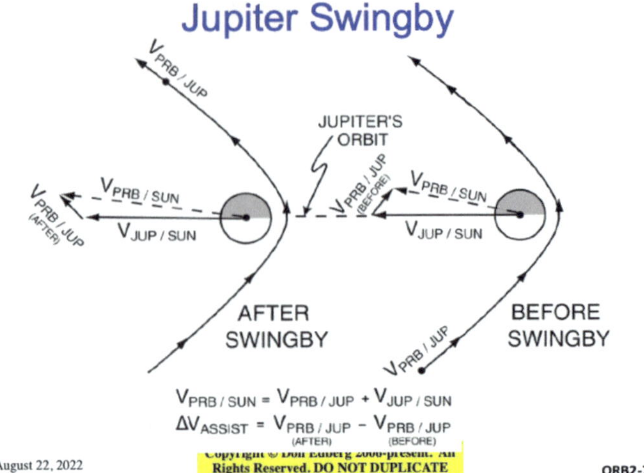

Gravity Assists (cont.)

- Leads to the following relationship for the given case: $|v_0| > |v_i|$
- Relative to the sun, the *heliocentric* velocity has (or can be) *increased*.
- Different flyby cases:

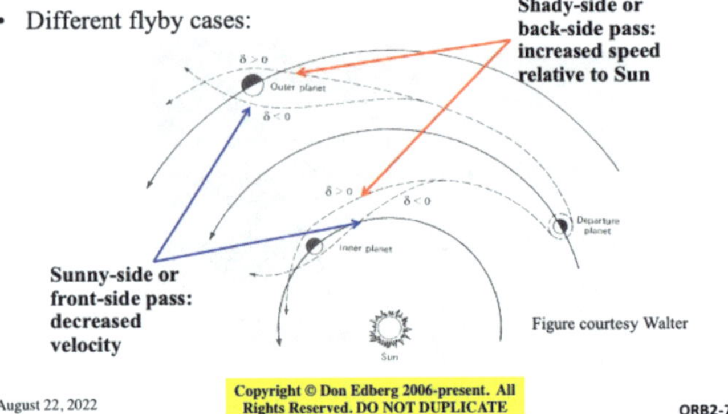

Figure courtesy Walter

Gravity Assists (cont.)

- Different flyby cases:

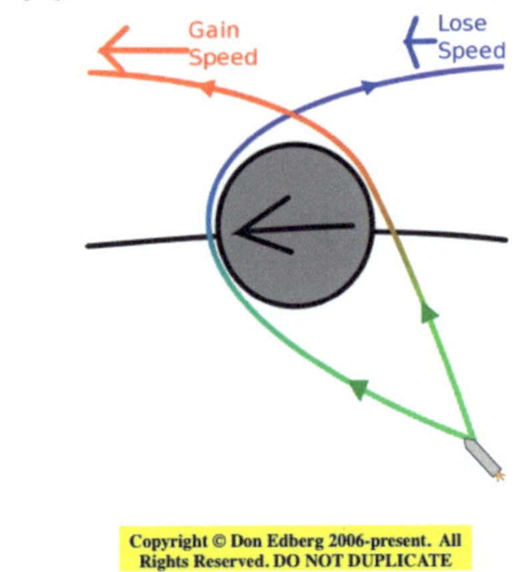

Gravity Assist Maneuvers

- Very energetic maneuvers
- Do not consume propellant
- Momentum exchanged with planet (planet slows, SC speeds up relative to sun)
- May be mission-enabling (*Pioneer*, *Galileo*, *Ulysses*, *Cassini*, *MESSENGER*)
- Velocity vector turns path $180° - 2\delta$

Voyager 1 Gravity Assist out of Solar System

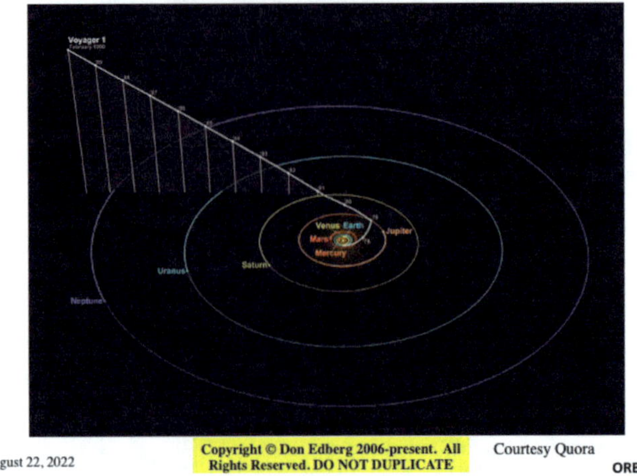

Courtesy Quora

Galileo VEEGA Trajectory

FLYBY (1) DEC 90
FLYBY (2) DEC 92
LAUNCH OCT/NOV '89
EARTH
VENUS
FLYBY FEB 90
ASTEROID BELT
IDA
GASPRA
PROBE RELEASE
JUPITER
ARRIVAL DEC 7, 1995
TIME TICKS:
S/C = 30 days
EARTH = 30 days
VENUS = 30 days
JUPITER = 100 days
*LAUNCH PERIOD OPENING

August 22, 2022

Cassini Trajectory if no Movie

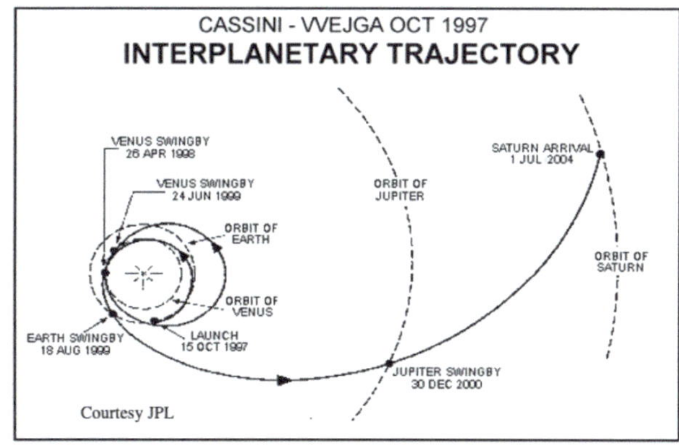

August 22, 2022

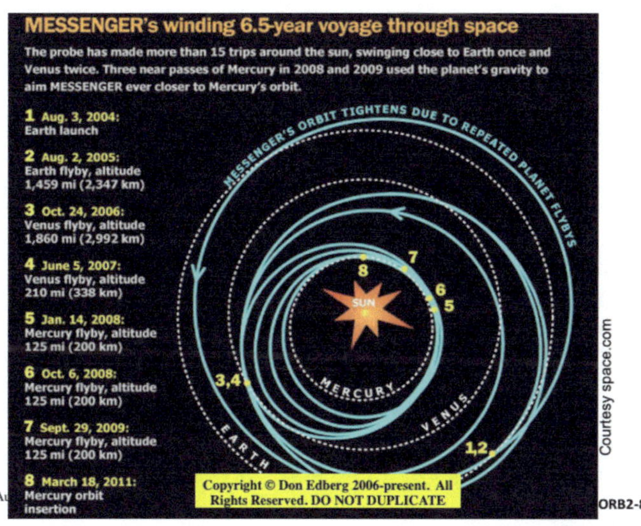

Orbit Plotting & Visualization Software

- Planetary Transfer Calculator (July 2022): `https://transfercalculator.com`
- Celestia: free space simulation lets you explore our universe in three dimensions. Celestia runs on Windows, Linux, and Mac OS X (Jul. 2022) `https://celestia.space`
- Trajectory Optimization Tool 2.1 (July 2022): `www.orbithangar.com/searchid.php?ID=5418`
- NASA Ames Mission Design Ctr. Traj. Browser (ok July 2022): `http://trajbrowser.arc.nasa.gov/traj_browser.php`

August 22, 2022

Low-Energy Transfers

- Routes in space using very little propellant
- Work in the Earth-Moon, Jupiter-moons, and solar system
- Drawback: they take longer to complete than higher-energy (more-propellant) transfers such as Hohmann transfers
- Low-energy transfers also known as "weak stability boundary trajectories" or "ballistic capture trajectories" `http://en.wikipedia.org/wiki/Low_energy_transfer`

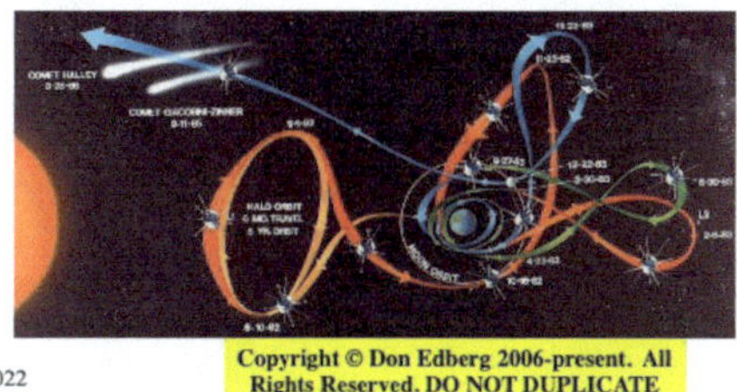

August 22, 2022

The Interplanetary Superhighway

Videos by Shane Ross, Caltech (watched July 2022):

- The Interplanetary Superhighway: Space Transportation Architecture for the 21st Century: `https://youtu.be/c00f1DiHygI`
- The Interplanetary Transport Network: mapping chaotic motion through the solar system `https://youtu.be/FwHDEB1VS_0`

August 22, 2022

Orbital Maneuvering & Rendezvous

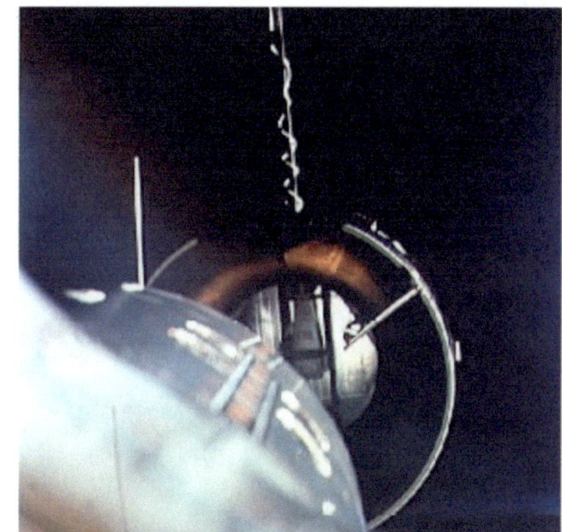

Courtesy NASA

Topics:

- Change of position within given orbit
- Orbit phasing
- Rendezvous maneuvering
- Atmospheric drag effects
- Historical comments about Apollo & Shuttle
- Take-aways

Course Layout

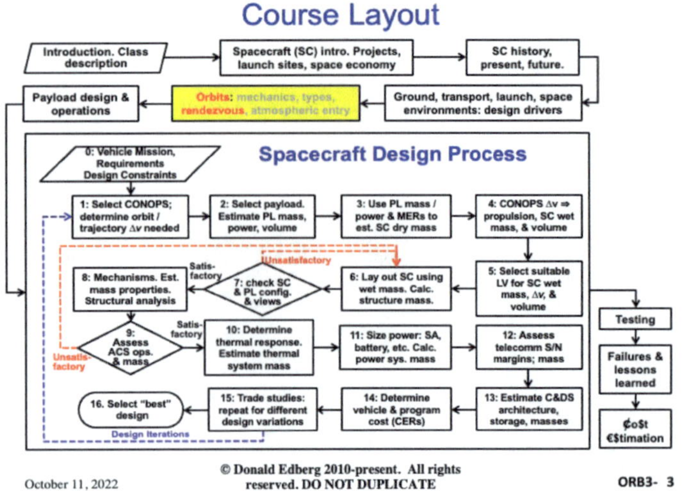

Introduction. Class description
Spacecraft (SC) intro. Projects, launch sites, space economy
SC history, present, future.
Payload design & operations
Orbits: mechanics, types, rendezvous, atmospheric entry
Ground, transport, launch, space environments: design drivers
Spacecraft Design Process
0: Vehicle Mission, Requirements Design Constraints
1: Select CONOPS; determine orbit / trajectory Δv needed
2: Select payload. Estimate PL mass, power, volume
3: Use PL mass / power & MERs to est. SC dry mass
4: CONOPS Δv ⇒ propulsion, SC wet mass, & volume
Unsatisfactory
8: Mechanisms. Est. mass properties. Structural analysis
Satis-factory
7: check SC & PL config. & views
6: Lay out SC using wet mass. Calc. structure mass.
5: Select suitable LV for SC wet mass, Δv, & volume
9: Assess ACS ops. & mass
Satis-factory
Unsatis-factory
10: Determine thermal response. Estimate thermal system mass
11: Size power: SA, battery, etc. Calc. power sys. mass
12: Assess telecomm S/N margins; mass
16. Select "best" design
15: Trade studies: repeat for different design variations
14: Determine vehicle & program cost (CERs)
13: Estimate C&DS architecture, storage, masses
Design iterations
Testing
Failures & lessons learned
¢o$t €$timation

Topics:

- **Change of position within given orbit**
- Orbit phasing
- Rendezvous maneuvering
- Atmospheric drag effects
- Historical comments from Apollo & Shuttle
- Take-aways

Move Location Ahead in *Same* Orbit (used to reposition GEO SC)

- Two burns (Δvs) used: a "mini-Hohmann" transfer
- 1st burn puts SC into slightly lower or higher "drift" orbit depending on desired repositioning:
  - Burn for lower orbit is retrograde, SC travels slightly faster & moves "forward" in orbit towards East, or CCW
  - Burn for higher orbit is prograde, SC travels slightly slower & moves "behind" in orbit towards West, or CW
- 2nd burn is opposite direction of 1st, returning SC to original orbit in new position

Ref: Fortescue

To Move Location East (CCW) in *Same* Orbit

- First burn is *retrograde*: slightly decrease velocity to enter lower drift orbit (lower & faster to drift *ahead* of current orbit)
- Second burn is *prograde* and at top of orbit, when spacecraft arrives at desired longitude

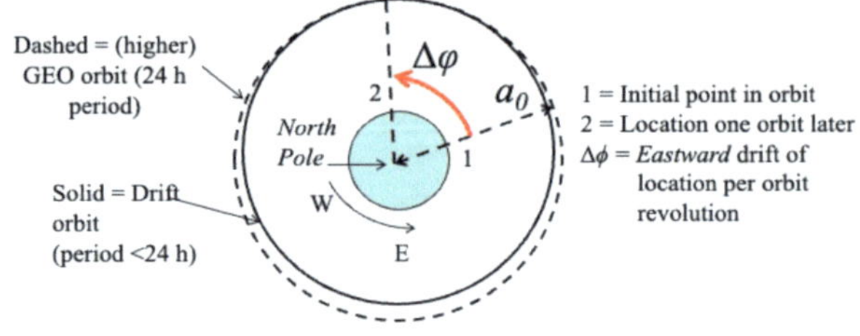

Dashed = (higher) GEO orbit (24 h period)
Solid = Drift orbit (period <24 h)
Δφ
a0
2
1
North Pole
W
E
1 = Initial point in orbit
2 = Location one orbit later
Δϕ = Eastward drift of location per orbit revolution

Ref: Fortescue

To Move Location West (CW) in Same Orbit

- First burn is *prograde*: slightly increase velocity to get into higher and slower orbit to drift *behind* current orbit
- Second burn is *retrograde* and at top of orbit, when spacecraft arrives at desired longitude (out of drift)

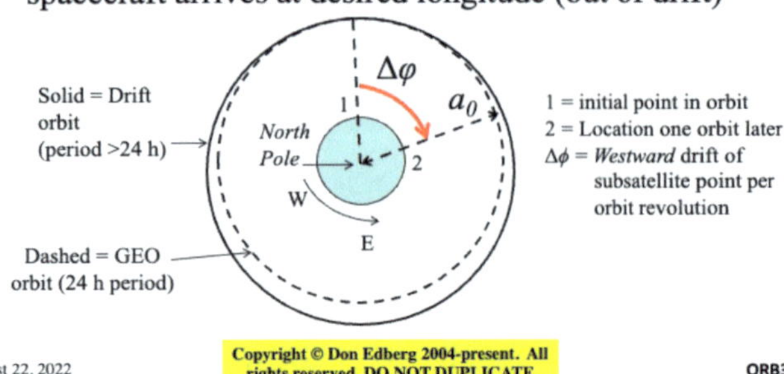

Solid = Drift orbit (period >24 h)
Dashed = GEO orbit (24 h period)
Δφ
a0
1
2
North Pole
W
E
1 = initial point in orbit
2 = Location one orbit later
Δϕ = Westward drift of subsatellite point per orbit revolution

Δv Required For Position Change

- Required $\Delta v = -\frac{1}{3} a_0 \frac{\Delta\varphi}{\Delta t}$ for each burn
  - two burns required
  - a_0 = radius
  - $\Delta\varphi$ = angle change in *radians* (+ = east, – = west)
  - Δt = time for the change to occur
- NOTE: *Longer transfer time* = less Δv required

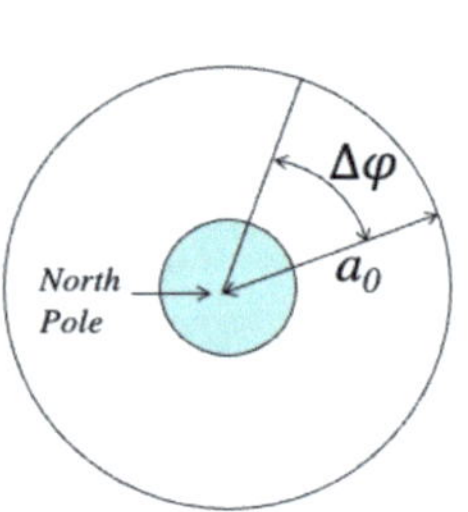

Δφ
North Pole
a0

Longitude Change Example

- Required: shift GEO SC 12° west in three revolutions = three days:
  $a_0 = 42{,}164$ km
  $\Delta\varphi = 12° = -\pi/15$ rad
  $\Delta t \sim 72$ h (3 days or orbits)

$$\Delta v = -\frac{a_0}{3}\frac{\Delta\phi}{\Delta t} = -\left(\frac{42164 \text{ km}}{3}\right)\left(\frac{-\pi/15}{72 \text{ h}}\right)\left(\frac{1000 \text{ m/km}}{3600 \text{ s/h}}\right)$$
$$= 0.473 \text{ m/s}$$

- A second burn in opposite direction occurs at arrival

August 22, 2022 ORB3- 9

Topics:

- Change of position within given orbit
- Orbit phasing
- Rendezvous maneuvering
- Atmospheric drag effects
- Historical comments from Apollo & Shuttle
- Take-aways

August 22, 2022 ORB3-10

"Phasing" Orbits

Used to "catch up" to an existing orbit for rendezvous

- Similar to change of longitude given earlier

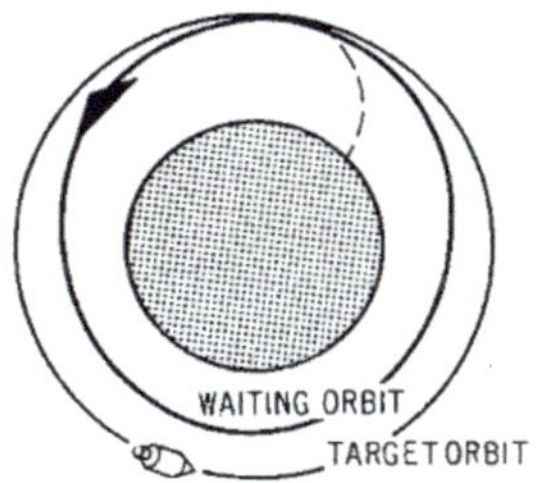

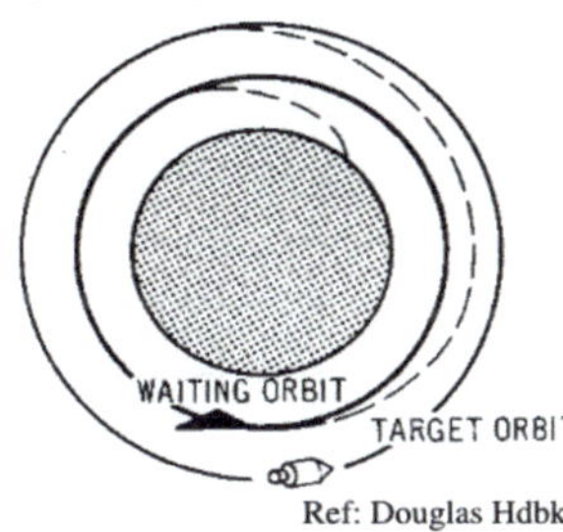

Ref: Douglas Hdbk

August 22, 2022 ORB3-11

Topics:

- Change of position within given orbit
- Orbit phasing
- Rendezvous maneuvering
- Atmospheric drag effects
- Historical comments from Apollo & Shuttle
- Take-aways

August 22, 2022 ORB3-12

The Rendezvous Problem
"RVD" = Rendezvous & Docking

Gemini 6&7 rendezvous.mp4

August 22, 2022 ORB3-13

What Happens If You Approach Rendezvous As A Race Driver

"… They were going to catch that … [Titan stage] target vehicle, and these astronauts were used to those Corvettes, and when you want to pass somebody you push the foot to the floor and zoom around them. Well, they line up behind this target, they're going to catch it, and they shove the throttle forward, and the more they burned their rockets, the further behind they got and the further behind they got, and they couldn't understand why for a while."
~ Henry Pohl, NASA, chief, Propulsion and Power Division, Project Gemini

August 22, 2022 ORB3-14

Rendezvous Problem

- There are two satellites in same orbit
- Second is known range behind the first
- *What direction would you throw a baseball to make it travel from one to the other?* (Hint: not intuitive!)
- Required baseball velocities are same as those needed for rendezvous problem
- Answer depends on range & *time of flight* (ToF)
- Shorter ToF requires larger Δv

Rendezvous Geometry

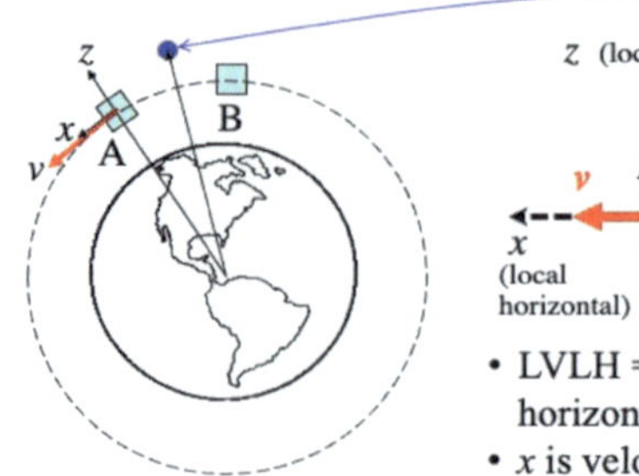

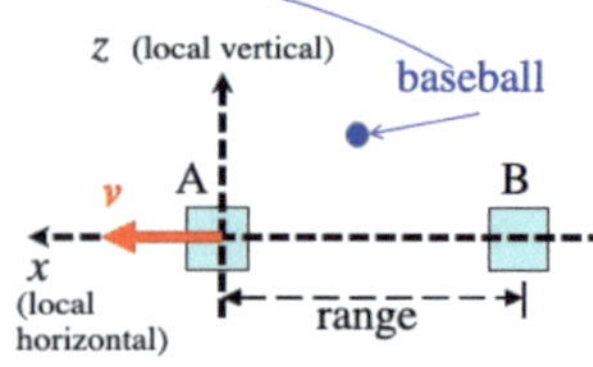

- LVLH = local vertical, local horizontal frame
- x is velocity direction
- y is sideways, towards you, out of screen
- z is local vertical

Geometry (cont.)

- $\mathbf{R}_A$ = distance from Earth center to platform A
- $\mathbf{R}$ = distance from Earth center to baseball
- $\mathbf{r}$ = baseball position relative to platform
- m_A = mass of platform A
- μ = gravitational constant

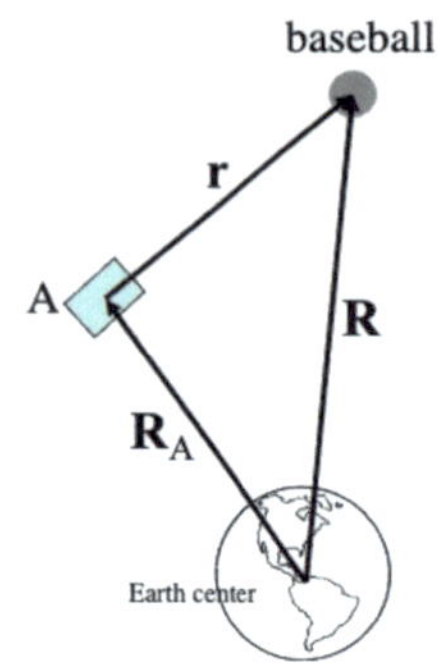

Equations of Motion

- After much algebra, the result is: $\frac{d^2\mathbf{r}}{dt^2} = -\frac{\mu}{R_A^3}\cdot\frac{\mathbf{r}}{R_A}\left\{\frac{\mathbf{r}}{r} - 3\left[\frac{\mathbf{r}}{r}\cdot\frac{\mathbf{R_A}}{R_A}\right]\frac{\mathbf{R_A}}{R_A} + \dots\right\}$
- Convert to xyz coordinates: $\mathbf{r} = x\hat{\mathbf{i}} + y\hat{\mathbf{j}} + z\hat{\mathbf{k}}$
- The orbital rate is defined as n: $\vec{n} = n\hat{\mathbf{j}}$ Note: $n = \sqrt{\frac{\mu}{R_A^3}} = \frac{2\pi}{T_{orbit}}$
- The position of platform A: $\mathbf{R_A} = R_A\hat{\mathbf{k}}$

Equations of Motion

- After much algebra, the result is:

$$\frac{d^2\mathbf{r}}{dt^2} = -\frac{\mu}{R_A^3}\cdot\frac{\mathbf{r}}{R_A}\left\{\frac{\mathbf{r}}{r} - 3\left[\frac{\mathbf{r}}{r}\cdot\frac{\mathbf{R_A}}{R_A}\right]\frac{\mathbf{R_A}}{R_A} + \cdots\right\}$$

- Convert to xyz coordinates: $\mathbf{r} = x\hat{\mathbf{i}} + y\hat{\mathbf{j}} + z\hat{\mathbf{k}}$
- The orbital rate is defined as $\mathbf{n}$: $\mathbf{n} = n\hat{\mathbf{j}}$, where $n = \sqrt{\frac{\mu}{R_A^3}} = \frac{2\pi}{T_{orbit}}$
- The position of platform A: $\mathbf{R_A} = R_A\ \mathbf{k}$

Equations Of Motion (cont.)

- Splitting the vector equation into its three scalar components yields (assume circular orbits with n = constant): $\ddot{x} + 2n\dot{z} = \frac{F_x}{m}$

$$\ddot{z} - 3n^2 z - 2n\dot{x} = \frac{F_z}{m}$$

$$\ddot{y} + n^2 y = \frac{F_y}{m}$$

- If thrusters F_i $(i = x, y, z)$ fire, these equations are called *Hill's* equations (F_i are thruster forces)

Clohessy-Wiltshire (CW) Eqs. Of Motion

- When thruster forces $F_i = 0$, there is *no* continuous thrusting or applied accelerations: the equations are known as the *Clohessy-Wiltshire* (CW) equations
- Can solve for relative motion due to impulsive burns with initial speeds & positions stated

$$\ddot{x} + 2n\dot{z} = 0$$
$$\ddot{z} - 3n^2 z - 2n\dot{x} = 0$$
$$\ddot{y} + n^2 y = 0$$

- Recall x ~ velocity direction, z ~ vertical, y ~ sideways
- Motion in orbital plane (x- and z-directions) is *coupled*
- y motion (perpendicular to orbit plane) is decoupled from (*independent* of) x-z plane (= orbital plane) motion

Hill Equation Solutions

- Closed-form solutions available to coupled linear DEs
- Initial positions: x_0, y_0, z_0; initial velocities: $\dot{x}_0, \dot{y}_0, \dot{z}_0$ applied accelerations: $a_x = \frac{F_x}{m}$, $a_y = \frac{F_y}{m}$, $a_z = \frac{F_z}{m}$

$$x(t) = x_o - \left(\frac{2\dot{z}_o}{n}\right)(1-\cos nt) + \left(\frac{4\dot{x}_o}{n} + 6z_o\right)\sin nt - \left(6z_o + \frac{3\dot{x}_o}{n}\right)nt$$
$$+\frac{2a_z}{n^2}(nt - \sin nt) + a_x\left[\frac{4}{n^2}(1-\cos nt) - \frac{3t^2}{2}\right]$$

$$z(t) = \left(4z_o + \frac{2\dot{x}_o}{n}\right) - \left(\frac{2\dot{x}_o}{n} + 3z_o\right)\cos nt + \left(\frac{\dot{z}_o}{n}\right)\sin nt$$
$$+\frac{2a_x}{n^2}(\sin nt - nt) + \frac{a_z}{n^2}(1-\cos nt)$$

Equations courtesy Ley et al

Hill Equation Solutions (cont.)

$$y(t) = y_0 \cos nt + \left(\frac{\dot{y}_0}{n}\sin nt\right) + \frac{a_y}{n^2}(1 - \cos nt)$$

- Motions contain sin *nt* and cos *nt*, so are *periodic* (they "come back" regularly)
- Terms containing a_x, a_y, a_z refer to steady accelerations (thrusters or atmospheric drag)

Equations courtesy Ley et al

Lateral/Out-of-Plane Motion: *y*-direction

$$y(t) = y_o \cos nt + \left(\frac{\dot{y}_o}{n}\sin nt\right)$$

- No forcing: y equation describes a simple sinusoid
- If astronaut throws ball to the left, it travels to the left for ¼ orbital period, then it reverses direction and begins coming back at him as if on a spring!
- It returns after ½ orbital period with the same speed but in the opposite direction
- If he ducks, it flies to the right and comes back; after one complete orbit it hits him going the same direction as initially thrown
- Basically a mini-plane change

Vertical & Forward/Aft Motion: *x*- & *z*-directions

- *Example*: in orbit, a baseball pitcher throws the ball *downwards* at 100 ft/s. What does it do?
- Here $x_0 = y_0 = z_0 = \dot{x}_0 = \dot{y}_0 = 0$, and

$$\dot{z}_0 = -100\ \frac{\text{ft}}{\text{s}}$$

The no-force motion equations become:

$$x(t) = \left(-\frac{2\dot{z}_o}{n}\right)(1+\cos nt) = \frac{200\text{ ft/s}}{n}(1+\cos nt)$$
$$z(t) = \frac{\dot{z}_o}{n}(\sin nt) = -\frac{100\text{ ft/s}}{n}(\sin nt)$$

Example *x-z* Motion: –100 ft/s Downwards

Example Motion (*x* & *z*) (cont.)

1. The ball goes straight down initially… It curves forward as it moves down below
2. After ¼ an orbital period, the ball is down 26 km and forward at a distance of 52 km!
3. After ½ orbital period, the ball passes directly in front, moving upwards, at a distance of 103 km!
4. After ¾ orbital period, the ball is 26 km up, 52 km ahead, and coming towards you ...
5. After one orbital period ($1/n$), it arrives coming straight down at 100 ft/s! (Get out of the way!)

The equations show the ratio between horizontal and vertical maximum motions is always *2:1*

How To Transfer From B forwards to A?

- Thrust ***backward*****, away from** the receiver A, at just the right velocity so that object arrives at A, approaching from the *front*, in one orbital period.
- By thrusting aft, you drop into lower & faster orbit, overtake target, climb back up, then equalize v with target
- Required velocity changes (2) found by solving CW equations

Transfer Trajectory for 1-Orbit Transfer

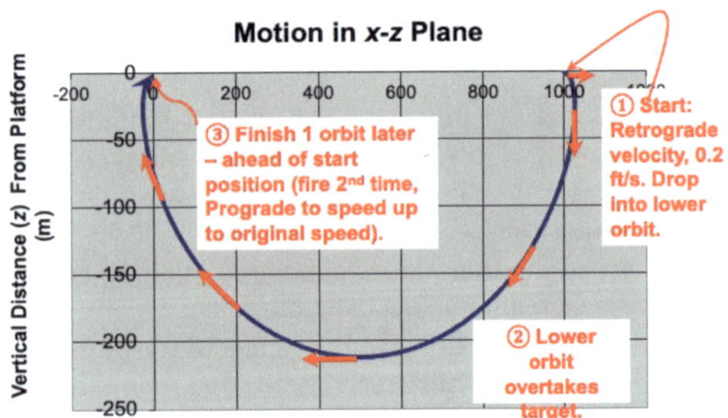

General Form of CW Equations

$$\{\delta\mathbf{r}(t)\} = \mathbf{\Phi}_{rr}(t)\delta\mathbf{r}_0 + \mathbf{\Phi}_{rv}(t)\delta\mathbf{v}_0$$
$$\{\delta\mathbf{v}(t)\} = \mathbf{\Phi}_{vr}(t)\delta\mathbf{r}_0 + \mathbf{\Phi}_{vv}(t)\delta\mathbf{v}_0$$

where:

$$\Phi_{rr}(t) = \begin{bmatrix} 4-3\cos nt & 0 & 0 \\ 6(\sin nt - nt) & 1 & 0 \\ 0 & 0 & \cos nt \end{bmatrix}, \quad \Phi_{rv}(t) = \begin{bmatrix} \frac{1}{n}\sin nt & \frac{2}{n}(1-\cos nt) & 0 \\ \frac{2}{n}(\cos nt - 1) & \frac{1}{n}(4\sin nt - 3nt) & 0 \\ 0 & 0 & \frac{1}{n}\sin nt \end{bmatrix}$$

$$\Phi_{vr}(t) = \begin{bmatrix} 3n\sin nt & 0 & 0 \\ 6n(\cos nt - 1) & 0 & 0 \\ 0 & 0 & -n\sin nt \end{bmatrix}, \quad \Phi_{vv}(t) = \begin{bmatrix} \cos nt & 2\sin nt & 0 \\ -2\sin nt & 4\cos nt - 3 & 0 \\ 0 & 0 & \cos nt \end{bmatrix}$$

Use CW Eqs. To Find Δ*v*s Needed for Two-Impulse Rendezvous

The initial burn velocity change is

$$\Delta\mathbf{v}_0 = \delta\mathbf{v}_0^+ = -\mathbf{\Phi}_{RV}^{-1}\mathbf{\Phi}_{rr}\delta\mathbf{r}_0$$

The final burn velocity change is

$$\Delta\mathbf{v}_f = -\delta\mathbf{v}_f^- = -\mathbf{\Phi}_{vr}^{-1}\delta\mathbf{r}_0 + \mathbf{\Phi}_{vv}\delta\mathbf{v}_0^+$$

How To Ensure NO Return?

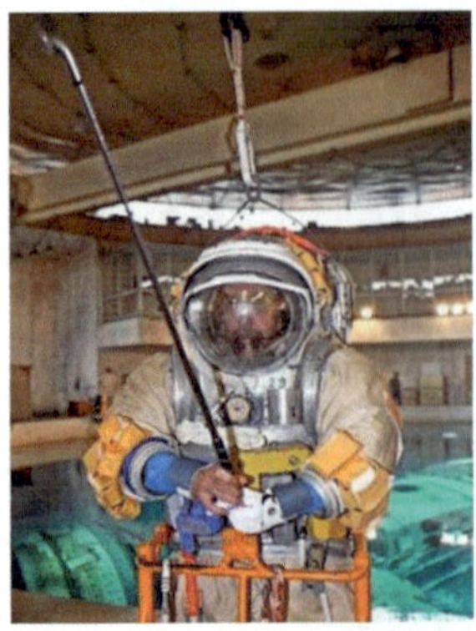

- 2006 Publicity stunt: hit golf ball from ISS
- Problem is to make sure NO return
- Problem of permanently jettisoning trash is similar
  - Photo of Mikhael Tiourine courtesy Nicolas Pillet

Observations

- Only way to "get rid" of something (i.e. trash on ISS) is to use initial velocity in x direction
- Range oscillates but also grows linearly with time

$$x(t) = \frac{4\dot{x}_o}{n}\sin nt - 3\dot{x}_o t$$

August 22, 2022 ORB3-33

Russian Nanosat Launch

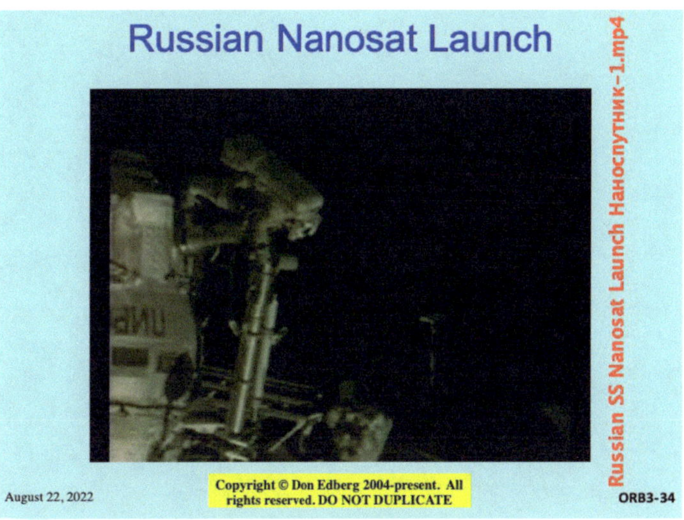

August 22, 2022 ORB3-34

Topics:

- Change of position within given orbit
- Orbit phasing
- Rendezvous maneuvering
- **Atmospheric drag effects**
- Historical comments from Apollo & Shuttle
- Take-aways

August 22, 2022 ORB3-35

Effect of Atmospheric Drag

- To 1st order, air drag is approximately constant over one orbit: insert the *difference* in air drag between the two objects as constant deceleration in Hill equations
- Results shown for ATV ejected from ISS on next chart
- Initial negative drift *away* from ISS reversed after three orbits to a positive drift (*towards* ISS), a *potential safety issue*

August 22, 2022 ORB3-36

Free Drift Trajectory with Δ Aero Drag (blue) & No Δ Drag (red)

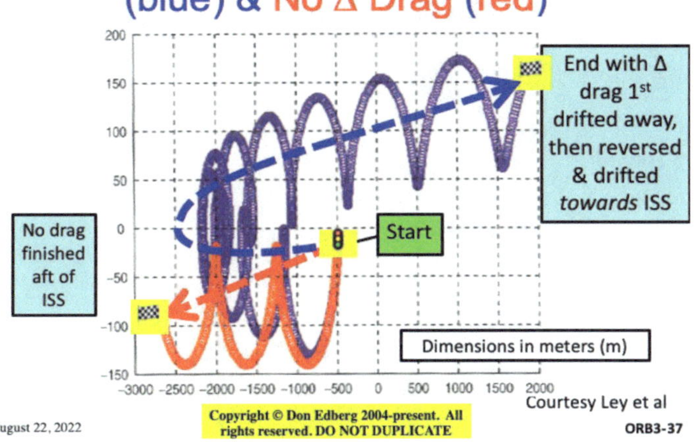

Courtesy Ley et al

August 22, 2022 ORB3-37

Topics:

- Change of position within given orbit
- Orbit phasing
- Rendezvous maneuvering
- Atmospheric drag effects
- **Historical comments from Apollo & Shuttle**
- Take-aways

August 22, 2022 ORB3-38

Historical Notes

- During 1960s, *Gemini* & *Apollo* program astronauts had to learn how to rendezvous without information from ground (in case radio communications with ground lost)
- Relative SC positions and velocities obtained from on-board L-band radar
- Position data difficult to accurately measure from earth, and were not needed!
- Those earth-orbital missions resembled the material assembled here

Apollo IX Rendezvous Diagram

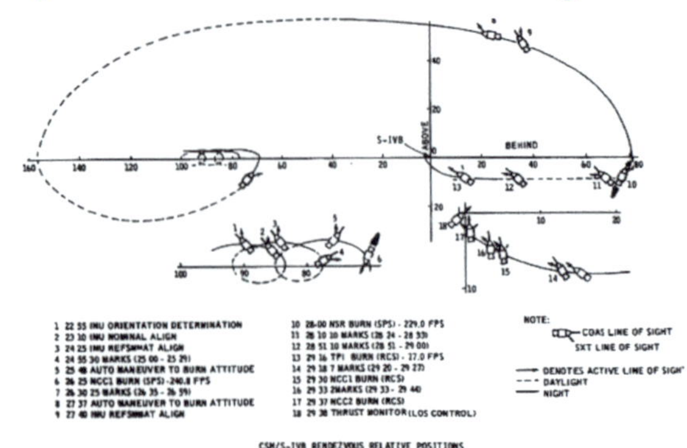

Shuttle-ISS Rendezvous Types

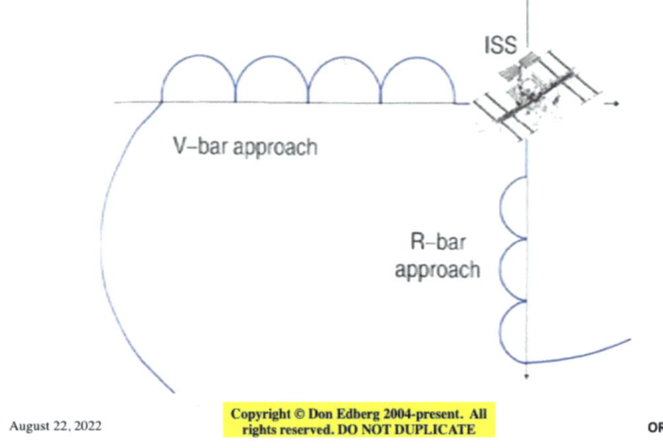

"V-Bar" Approach

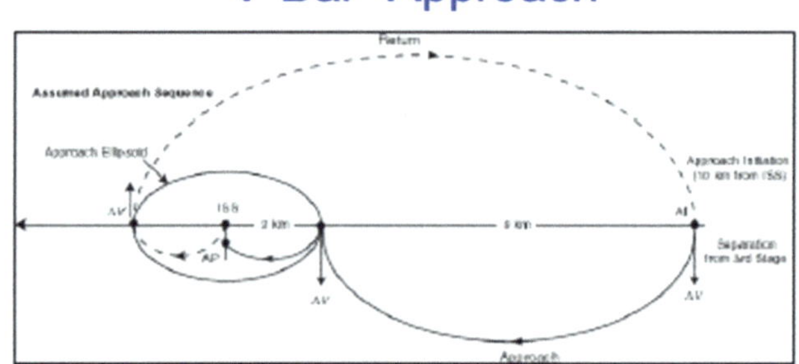

Ref: Collins, Meissinger, & Bell, "Small Orbit Transfer Vehicle (OTV) for On-Orbit Satellite Servicing and Resupply," 15th USU Small Satellite Conference, 2001

Shuttle "R-Bar" Approach

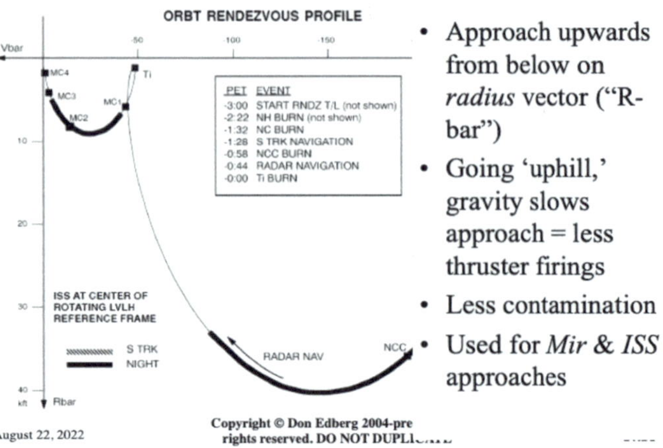

- Approach upwards from below on *radius* vector ("R-bar")
- Going 'uphill,' gravity slows approach = less thruster firings
- Less contamination
- Used for *Mir* & *ISS* approaches

Why R-Bar? Exhaust plumes!

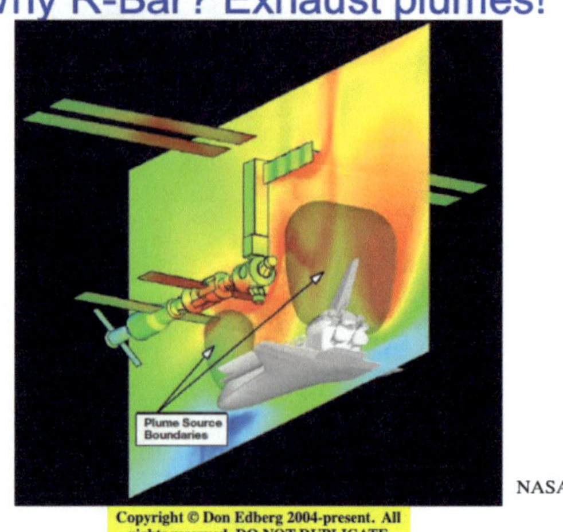

NASA

Topics:

- Change of position within given orbit
- Orbit phasing
- Rendezvous maneuvering
- Atmospheric drag effects
- Historical comments from Apollo & Shuttle
- **Take-aways**

August 22, 2022 ORB3-45

Maneuvering & RDV Take-Aways

- Orbit pursuits & phasing done via 'mini Hohmann' transfers
- Longitude changes take little Δv if time available
- RDV maneuvering (C-W equations) *not intuitive*!!
  - Forward speed towards target ahead moves one further away!
  - Downwards speed creates elliptical path ahead that returns to initial position!
  - Aft speed moves forward towards target! Two impulses needed to stop at target
  - Sideways motion repeats like harmonic oscillator
  - Drag can negatively affect separations
- Shuttle 'R-bar' approach minimized plume effects

August 22, 2022 ORB3-46

Blank slide

August 22, 2022 ORB3-47

Low-Thrust Maneuvering, Orbital Decay, & Atmospheric Entry

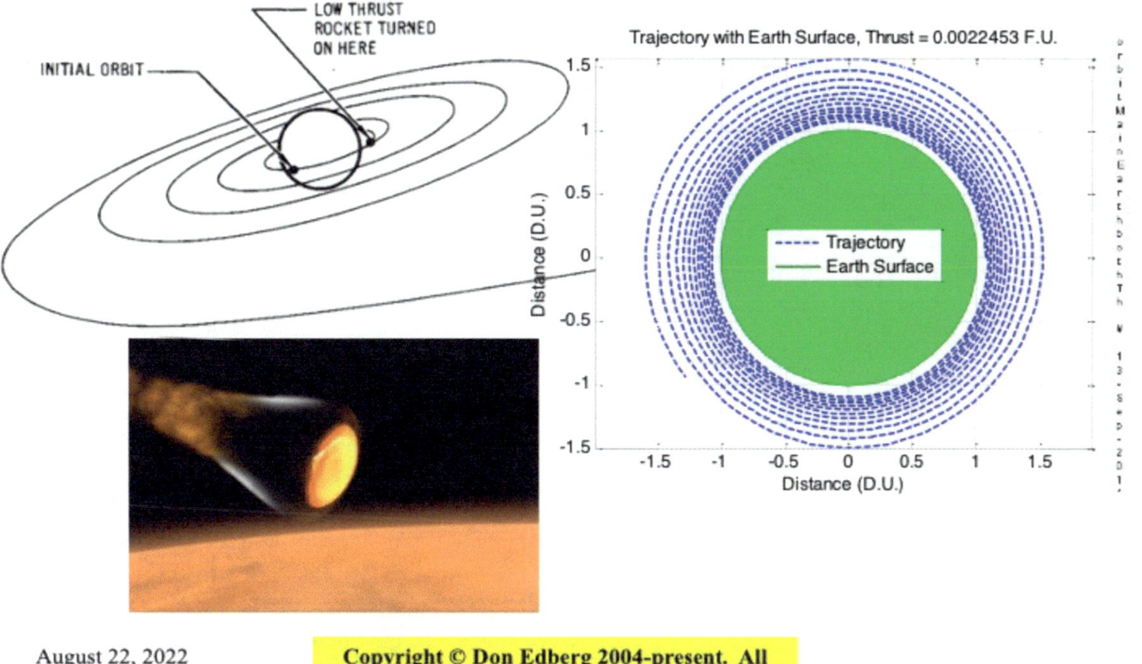

August 22, 2022

Topics

- Low-thrust orbital mechanics
  - Equations of motion
  - Efficiency compared to impulsive systems
- Atmospheric entry: energy to dissipate
  - Critical parameters
  - Orbital decay
  - Entry deceleration
  - Entry heating
  - Thermal protection systems
- Aerobraking & aerocapture

August 22, 2022

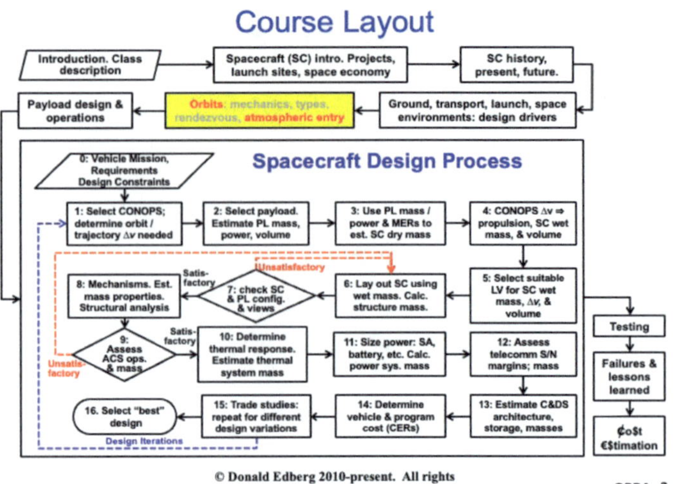
Course Layout
Introduction. Class description
Spacecraft (SC) intro. Projects, launch sites, space economy
SC history, present, future.
Payload design & operations
Orbits: mechanics, types, rendezvous, atmospheric entry
Ground, transport, launch, space environments: design drivers
0: Vehicle Mission, Requirements Design Constraints
Spacecraft Design Process
1: Select CONOPS; determine orbit / trajectory Δv needed
2: Select payload. Estimate PL mass, power, volume
3: Use PL mass / power & MERs to est. SC dry mass
4: CONOPS Δv ⇒ propulsion, SC wet mass, & volume
Unsatisfactory
Satis-factory
8: Mechanisms. Est. mass properties. Structural analysis
7: check SC & PL config. & views
6: Lay out SC using wet mass. Calc. structure mass.
5: Select suitable LV for SC wet mass, Δv, & volume
Testing
9: Assess ACS ops. & mass
10: Determine thermal response. Estimate thermal system mass
11: Size power: SA, battery, etc. Calc. power sys. mass
12: Assess telecomm S/N margins; mass
Failures & lessons learned
Unsatis-factory
16. Select "best" design
15: Trade studies: repeat for different design variations
14: Determine vehicle & program cost (CERs)
13: Estimate C&DS architecture, storage, masses
¢o$t €$timation
Design Iterations

Topics

- Low-thrust orbital mechanics
  - Equations of motion
  - Efficiency compared to impulsive systems
- Atmospheric entry: energy to dissipate
  - Critical parameters
  - Orbital decay
  - Entry deceleration
  - Entry heating
  - Thermal protection systems
- Aerobraking & aerocapture

Low Thrust (Spiral) Transfers

- Utilize low thrust engines (ion, etc.)
- Take a LONG time
- Energy change/time = **T**hrust·**V**elocity/mass
- Max energy change: force = thrust *parallel* to velocity
- Acceleration $a = T/m = thrust \div mass$, m varies. Orbital radius is $r(t)$.

$$\text{work done per unit mass} = \frac{\vec{T}\cdot\vec{v}}{m} = \dot{\varepsilon} = \vec{a}\cdot\vec{v}$$

Low Thrust (cont.)

- Assume orbit is quasi-circular radius $r(t)$, speed is that of circle of current radius
- Recall definition of energy
- Energy change with semimajor axis change

$$v = \sqrt{\frac{\mu}{r}} \text{ for a circular orbit}$$

$$\varepsilon = -\frac{\mu}{2r} \text{ so } \frac{d\varepsilon}{dt} = \frac{\mu}{2} r^{-2} \frac{dr}{dt}$$

$$\text{But } \left|\frac{d\varepsilon}{dt}\right| = a \cdot v = a\sqrt{\frac{\mu}{r}}, \text{ solve for}$$

$$\frac{dr}{dt} = \frac{2r^{3/2}a}{\mu^{1/2}}$$

Two cases: ***constant*** *&* ***varying*** *acceleration*

Low Thrust, *Constant* Acceleration

- Propellant mass << SC mass, so $a \approx$ constant
- Rate of radius change with respect to energy change
- Separate and integrate directly
- Result is time needed as a function of start and finish radius (or orbital speeds v_c)
- Required $\Delta v = a \cdot t_{burn}$

$$\int_{r_0}^{r_f} \frac{dr}{r^{3/2}} = \frac{2a}{\sqrt{\mu}} \int_0^{t_{burn}} d\tau$$

$$t_{burn} = \frac{\sqrt{\mu}}{a}\left(\frac{1}{\sqrt{r_0}} - \frac{1}{\sqrt{r_f}}\right)$$

$$= \frac{v_{c0} - v_{cf}}{a} = \frac{\Delta v}{a}$$

$$\Delta v_{total} = a \cdot t_{burn}$$

(Based on **Spaceflight Dynamics**, W. Wiesel, with mods.)

Topics

- Low-thrust orbital mechanics
  - Equations of motion
  - Efficiency compared to impulsive systems
- Atmospheric entry: energy to dissipate
  - Critical parameters
  - Orbital decay
  - Entry deceleration
  - Entry heating
  - Thermal protection systems
- Aerobraking & aerocapture

Low Thrust Const. Accel. Example

- Transfer from 200 km parking orbit to GEO at 42,164 km (r_{earth} = 6,378 km)
- Acceleration = 0.001 m/s² (~ 1/10,000 of g_0)

$$t = \frac{\sqrt{3.986\times10^5 \text{km}^3/\text{s}^2}}{0.001 \text{ m/s}^2}\left(\frac{1}{\sqrt{6578 \text{ km}}} - \frac{1}{\sqrt{42164 \text{ km}}}\right) = 54.5 \text{ days}$$

- Total $\Delta v = at_{burn}$ = (0.001 m/s²)(54.5 days) = 4.71 km/s for low thrust
- Hohmann Δv = 3.9 km/s for same radii
- 21% more fuel required … if same I_{sp}

Low-Thrust Comments

- Difference of generated Δvs (higher for low-thrust) due to *kinematic inefficiency*
  - Normally want impulsive speed changes for maximum efficiency
- Trade higher Δv for far less fuel mass (ion engine)
- Long duration of burn not an issue for ion engines
- Long duration through van Allen radiation belts CAN BE an issue

Topics

- Low-thrust orbital mechanics
  - Equations of motion
  - Efficiency compared to impulsive systems
- Atmospheric entry: energy to dissipate
  - Critical parameters
  - Orbital decay
  - Entry deceleration
  - Entry heating
  - Thermal protection systems
- Aerobraking & aerocapture

Entry/Reentry: Coming Into An Atmosphere

- Ballistic entry
  - Suborbital Re-entry Vehicles
  - Orbital (Mercury and Gemini)

This class will only deal with ballistic entry

- Skip Entry
  - Apollo
- Gliding Entry
  - Shuttle

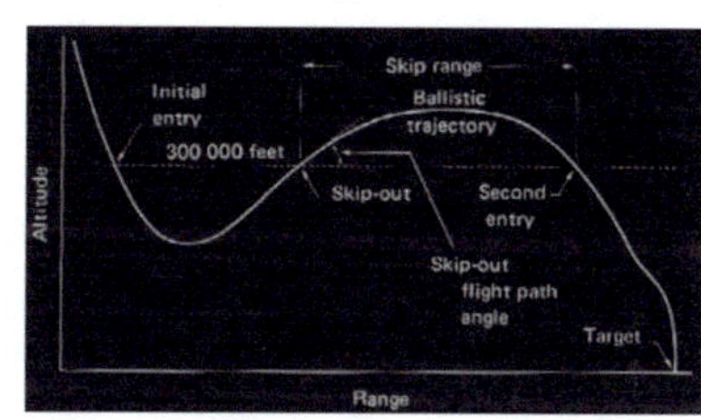

Objects in Orbit Have (Lots of) Energy!

- For 1 kg mass in a 500 km earth orbit:
- Kinetic Energy = 2.9×10^7 J
- Potential Energy = 4.9×10^6 J
- If deceleration occurs over ~10 minutes, average heating rate is 282 kW/kg

Topics

- Low-thrust orbital mechanics
  - Equations of motion
  - Efficiency compared to impulsive systems
- Atmospheric entry: energy to dissipate
  - Critical parameters
  - Orbital decay
  - Entry deceleration
  - Entry heating
  - Thermal protection systems
- Aerobraking & aerocapture

The Major Issues

- **Deceleration**
- **Thermal heating**
- **Landing accuracy**

The values are determined by:

- Entry speed
- Entry angle
- Mass/drag ratio

Entry Coordinates

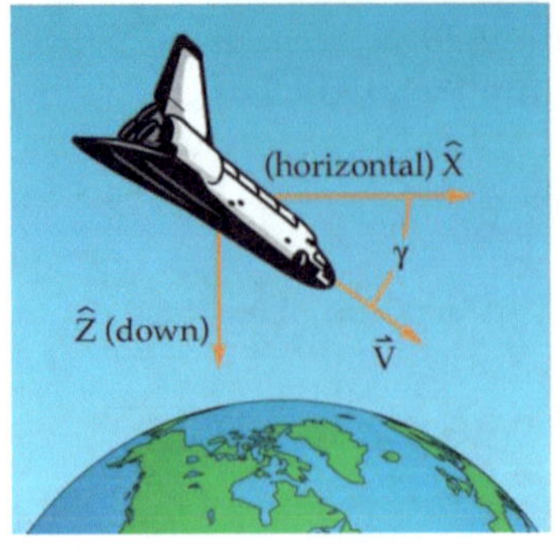

- The flight angle γ is measured from local horizontal
- Local horizontal is perpendicular to the radius vector from Earth center

Courtesy Sellers

Mass/Drag Area ratio = "Ballistic Coefficient"

- Ballistic Coefficient = BC = $\frac{m}{C_D A} = \beta$
- A HIGH ballistic coefficient represents vehicle with low drag (i.e. a "bullet")
- LOW ballistic coefficient represents a vehicle with relatively high drag (a "ping-pong ball")
- Sometimes defined by weight instead of mass
- Higher β = longer time to decelerate

Equations of Motion

- Write down Newton's equation with external forces in horizontal and vertical directions:
  - Gravity $W = mg$ (assume gravity ≈ constant)
  - Drag $D = C_D A \ ½\rho v_{rel}^2$
  - Define *exponential* atmospheric density ρ at height h using the relation $\rho(h) = \rho_0 \, e^{-h/h_0}$ (e = base of natural logs, not eccentricity!)
  - ρ_0 is density at zero altitude (1.225 kg/m^3)
  - h_0 is the "*scale height*," for which the density drops by $1/e$ ($h_0 \approx$ 7,194 m for earth)
  - R_e is earth radius (= 6,378 km)
- Integrate equations to obtain motion profiles

Topics

- Low-thrust orbital mechanics
  - Equations of motion
  - Efficiency compared to impulsive systems
- Atmospheric entry: energy to dissipate
  - Critical parameters
  - Orbital decay
  - Entry deceleration
  - Entry heating
  - Thermal protection systems
- Aerobraking & aerocapture

Orbital Decay (1 of 4)

- Air drag robs orbit of energy E; radius r shrinks with time. With $\dot{r} = \frac{dr}{dt}$,

$$\frac{dE}{dt} = \dot{E} = \frac{\mu}{2r^2}\dot{r}t = v \cdot \text{deceleration } a$$

- Deceleration a = drag ÷ mass (from $\mathbf{F} = m\mathbf{a}$):

$$a = \frac{D}{m} = \frac{C_D A \frac{1}{2}\rho v^2}{m} = \beta q$$

$$\beta = \frac{m}{C_D A} = \text{ballistic coefficient}$$

q = dynamic pressure

Orbital Decay (2 of 4)

- Rewritten, $\dot{r} = \frac{\rho_0\sqrt{\mu r}}{\beta} e^{-\frac{r-R_e}{h_0}}$
- After separating variables and assuming the speed is approximately orbital, altitude $H(t)$ is:

$$H(t) = h_0 \ln\left[\exp\left(\frac{H_0}{h_0}\right) - \frac{\rho_0\sqrt{\mu R_e}}{\beta h_0} t\right]$$

- LEO altitude $H_0 = 20 \sim 40$ scale heights h_0
- 1st term (e^{H_0/h_0}constant) in equation dominates
- Eventually, as t increases, 2nd term becomes comparable and reduces $H(t)$ rapidly

August 22, 2022 ORB4-25

Decay (3 of 4)

- As 2nd term increases, altitude drops rapidly
- $h = 0$ when ln[ ] = 1
- Time to decay to zero, t_d:

$$t_d = \frac{h_0\beta}{\rho_0\sqrt{\mu R_e}}\left(e^{H_0/h_0} - 1\right)$$

- For 25 scale heights ≈ 180 km initial height, decay time ~100 days (depends on β)

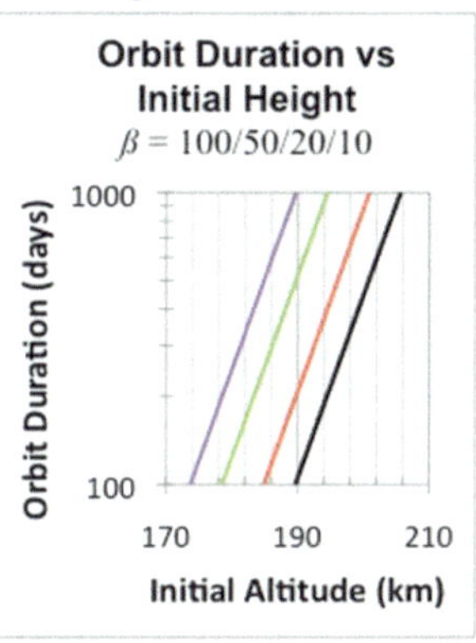

August 22, 2022 ORB4-26

Decay (4 of 4)

- Satellite lifetime proportional to β & H_0
- Lifetime grows exponentially with initial altitude
- Easier to increase altitude than reduce drag (can't change density)
- Increasing H_0 by *one* scale height (6 ~ 8 km) extends lifetime ~ factor of *three*
- Very difficult to predict decay times: varies with scale height & β, which depend on sun & vehicle's drag (attitude)

"NASA cannot confirm the exact trajectory or time of the UARS satellite's plunge, which depend on solar weather, variations in Earth's gravitational field, and the orientation of the satellite. However, as UARS' re-entry draws near, NASA should be able to offer more precise predictions." 9/14/2011

August 22, 2022 ORB4-27

U.S. Spy Satellite Decay 2007-08

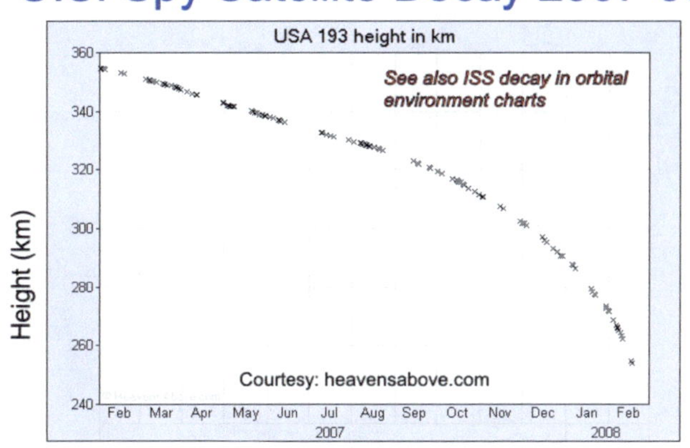

August 22, 2022 ORB4-28

Deliberate Deorbiting

- Orbital decay difficult to predict
- De-orbit forced by retrorocket
- Gives a non-horizontal entry
- Entry & touchdown easier to control

Δv to Deorbit to 50 km Perigee	
400 km	100 m/s
500 km	130 m/s
600 km	160 m/s
700 km	185 m/s

The Russian Space Station *Mir* burns up as it enters Earth's atmosphere over Nadi, Fiji, in March 2001. Photograph: Rob Griffith/AP

 ORB4-29

Orbital Decay References

- *Satellite orbital decay prediction*: calculates decay rates & orbital lifetimes of satellites in ~circular orbits < 500 km altitude: http://www.lizard-tail.com/isana/lab/orbital_decay/
- *Satview forecast for reentry of space junk*: http://www.satview.org/spacejunk.php
- **Satellite Orbital Decay Calculations,** Australian Gov't. Bureau of Meteorology: http://www.sws.bom.gov.au/Category/Educational/Space%20Weather/Space%20Weather%20Effects/SatelliteOrbitalDecayCalculations.pdf

Accessed Dec. 2021

August 22, 2022 ORB4-30

Topics

- Low-thrust orbital mechanics
  - Equations of motion
  - Efficiency compared to impulsive systems
- Atmospheric entry: energy to dissipate
  - Critical parameters
  - Orbital decay
  - **Entry deceleration**
  - Entry heating
  - Thermal protection systems
- Aerobraking & aerocapture

August 22, 2022 ORB4-31

Entry Equations Integrated

- Once atmosphere entered, integration yields:

$$a_{\max} = \frac{v_{entry}^2 \sin\gamma}{2h_0 e} \quad \text{and} \quad h_{a_{\max}} = h_0 \ln\left(\frac{\rho_0 h_0}{\beta \sin\gamma}\right)$$

- a_{max} = max acceleration
- $h_{a\ max}$ = height of max acceleration
- v_{entry} = entry speed
- γ = flight path angle
- ρ_0 = density at sea level = 1.225 kg/m^3

Source: Sellers

August 22, 2022 ORB4-32

Deceleration vs. Altitude for various Ballistic Coefficients (250 kft alt., 22,500 ft/s, −12° entry)

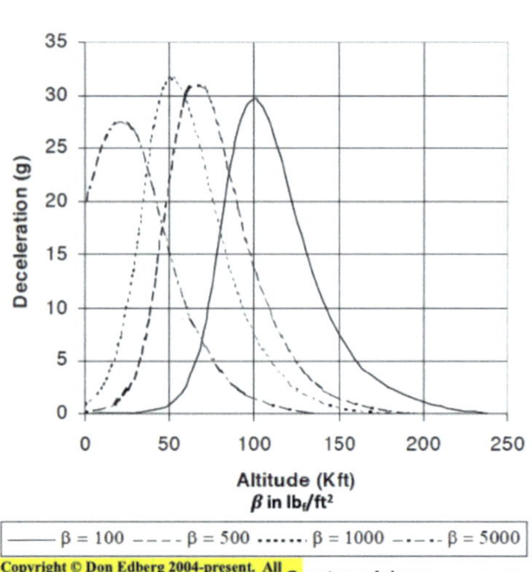

Courtesy: Adams

August 22, 2022 ORB4-33

"Friendship 7" *Mercury* Capsule Entry

Entry: 76.2 km, 1.5° angle, 23,000 ft/s speed — over 8 g peak!

(1,208 kg, $A = 2.812$ m^2, $C_D = 1.6$, $\beta = 268.5$ kg/m^2)

a. Velocity
b. Entry Time (sec)
c. Range (miles)
d. Deceleration (g)

August 22, 2022 ORB4-34

Decel. in g, Entry Angle γ = 20°

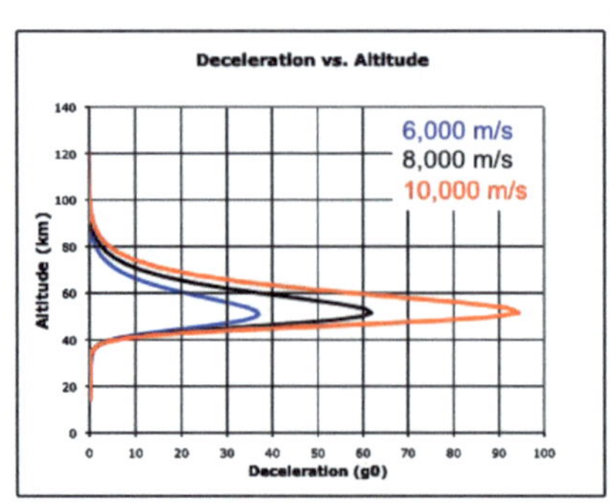

- For a FIXED entry angle, the larger the entry speed, the greater the maximum deceleration
- Note:
  m = 1,000 kg
  A = 50.3 m^2
  C_D = 1.0
  β = 19.9 kg/m^2

August 22, 2022 ORB4-35

Deceleration: v_{entry} = 8 km/s

Deceleration vs. Altitude
γ = -5.0°
γ = -10.0°
γ = -15.0°
Altitude (km)
Deceleration (g0)

For constant entry speed, the steeper the entry angle, the greater the maximum deceleration

August 22, 2022 ORB4-36

Decel. vs. Ballistic Coeff. β, v_{entry} = 8 km/s, γ = –15°

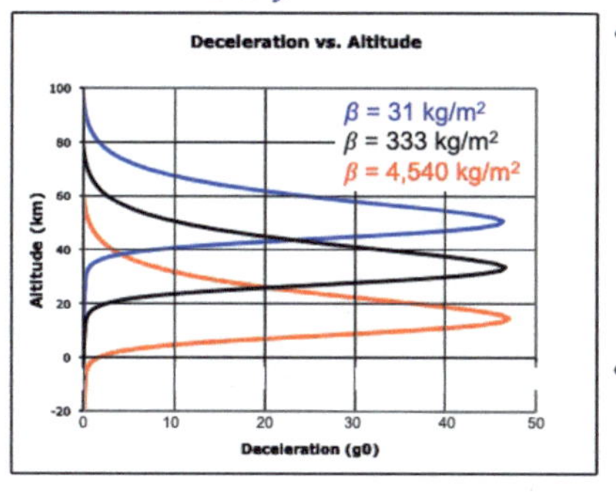

- Regardless of the shape, ***all vehicles experience the same maximum deceleration*** (but at differing altitudes)
- Bodies with more drag decelerate at higher altitudes

Deceleration vs. Entry Angle & L/D

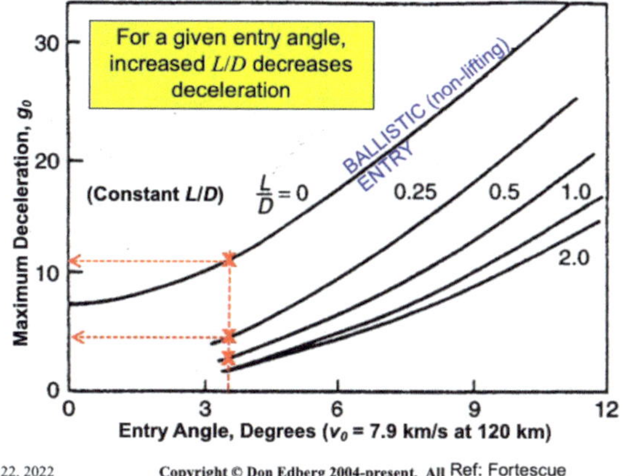

Ref: Fortescue

Velocity vs. Altitude for various Ballistic Coefficients β (250 kft alt., 22,500 ft/s, −12° entry)

Velocity (ft/sec)

Altitude (Kft)
β in lb_f/ft^2

—— β = 100 – – – – β = 500 …… β = 1000 – . – . – β = 5000

Courtesy: Adams

Topics

- Low-thrust orbital mechanics
  - Equations of motion
  - Efficiency compared to impulsive systems
- Atmospheric entry: energy to dissipate
  - Critical parameters
  - Orbital decay
  - Entry deceleration
  - Entry heating
  - Thermal protection systems
- Aerobraking & aerocapture

Heating

- All the kinetic & potential energy has to go somewhere
- Conservation of energy makes it turn into heat
- Shock waves, ionization, etc. occur
- Define $dq/dt = \dot{q}$ or "q-dot" as heating rate (energy/time, in cal/s or W or BTU/s)
- You can change the RATE at which heating occurs by changing the trajectory, but the TOTAL heat generated is fixed

Heating Rates

Can calculate empirical heating rate:

$\dot{q} \cong 1.83\times10^{-4} v^3 \sqrt{\frac{\rho}{r_{nose}}}$, where

$\dot{q}$ = vehicle heating rate $\left(\frac{\mathrm{W}}{\mathrm{m}^2}\right)$,

v = vehicle velocity $\left(\frac{\mathrm{m}}{\mathrm{s}}\right)$,

ρ = air density $\left(\frac{\mathrm{kg}}{\mathrm{m}^3}\right)$,

r_{nose} = vehicle "*nose*" radius (m)

Source: Sellers

Heating vs. Shape: Blunt vs. Pointy

- Blunt means r_{nose} is *large*
- *Detached* shock wave
- Heating spread over *large* area

- Pointy means r_{nose} is *small*
- *Attached* shock wave
- Heating *concentrated* at nose

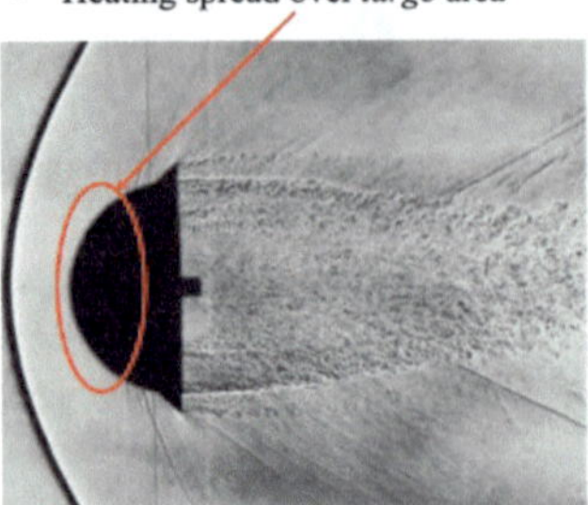

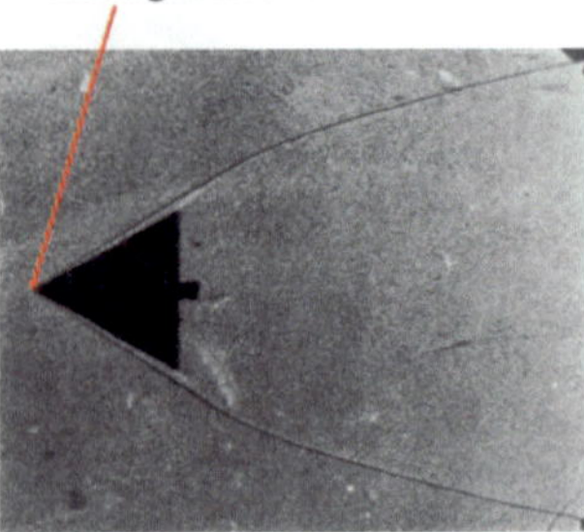

Figure 1: Atmosphere Entry Simulator Schlieren photographs illustrating the detached bow shock wave generate by a blunt reentry body compared to the attached shock wave with a pointed reentry body. The detached bow wave dissipates heat well away from reentry body (D. D. Baals and W.R. Corliss, Wind Tunnels of NASA,(Washington, D.C., 1981), SP-440, 76).

Courtesy: NASA

Pointy Nose *Melts* in Wind Tunnel

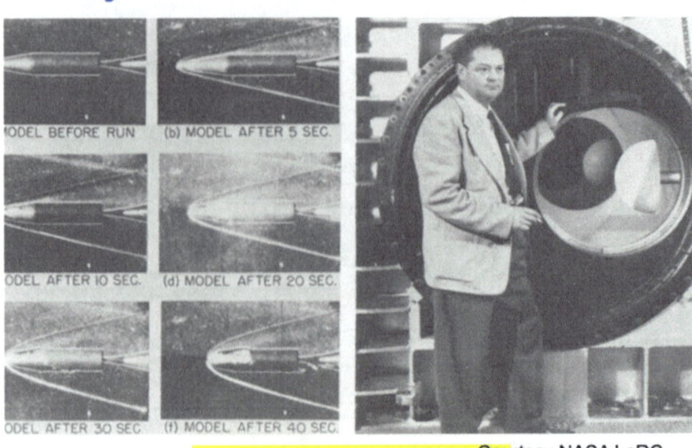

Courtesy NASA LaRC

Altitude and Speed at Max Heating

- Relations for speed and altitude:

$$h_{\dot{q}_{max}} = h_0 \ln\left(\frac{\rho_0 h_0}{3\beta \sin\gamma}\right)$$

$$v_{\dot{q}_{max}} = 0.846\, v_{entry}$$

Source: Sellers

Example: Orion Spacecraft

- Landing weight 16,174 lb_f
- Diameter 198 in, $C_D = 1.0$
- $\beta = m/(C_D A) = 2.3491$ slug/ft^2 = 369 kg/m^2
- Nose radius 237.6 in = 5.03 m
- Entry speed 8 km/s, angle −8°
- Max heat rate of 2,454 kW/m^2 occurs at altitude 88.6 km
- That's ~2.5 MEGAWATTS/m^2 = 30.47 BTU/ft^2/s! (in other words, lots of heating)

Data courtesy AvWeek

Heating Rate vs. Velocity

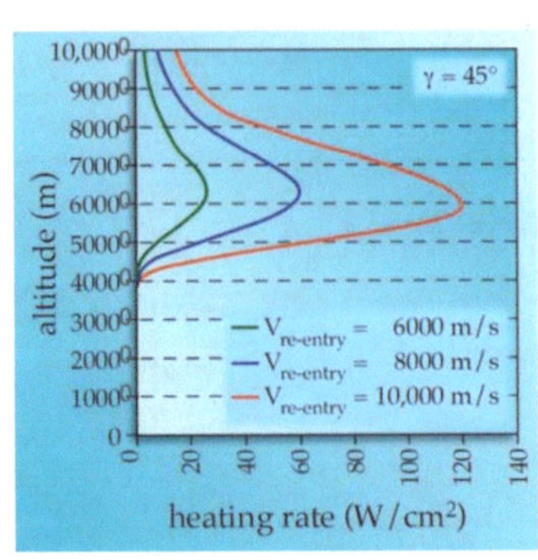

- As you would expect, the heating rate increases with increased entry speed

Source: Sellers

Heating Rate vs. Entry Angle

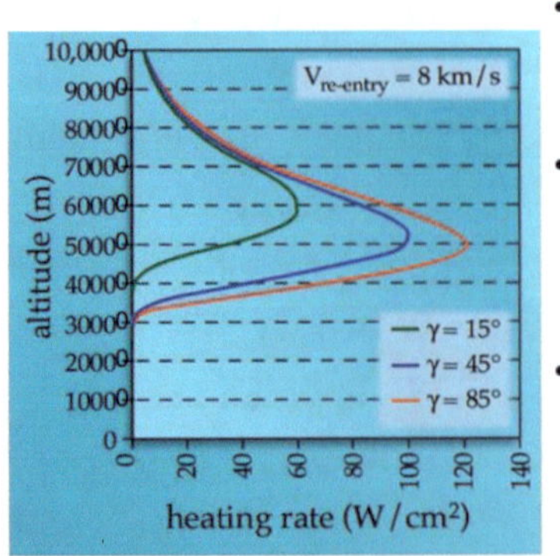

- The steeper the entry angle, the higher the peak heating rate
- Steep entry angles cause high max heating rates but for a *short* duration
- Shallow entry angles cause low max heating rates but for a *long* time

Source: Sellers

Heat Rates vs. Ballistic Coeff. (BC or β)

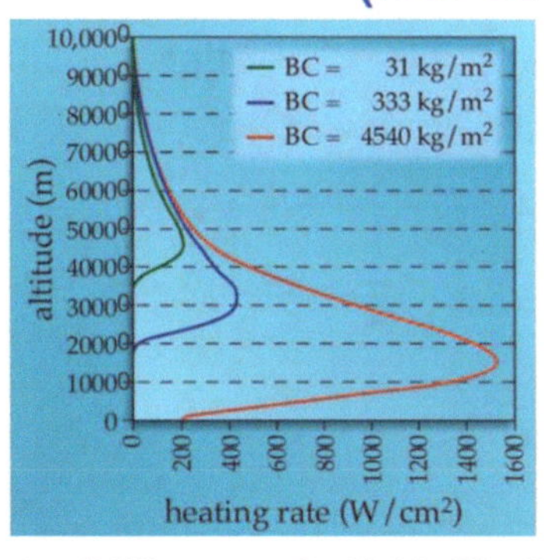

- High BC vehicles (pointy ones) have higher heat rates lower in the atmosphere than low BC vehicles (blunt ones)
- Heat concentrated at tips of sharp objects

Source: Sellers

Topics

- Low-thrust orbital mechanics
  - Equations of motion
  - Efficiency compared to impulsive systems
- Atmospheric entry: energy to dissipate
  - Critical parameters
  - Orbital decay
  - Entry deceleration
  - Entry heating
  - Thermal protection systems
- Aerobraking & aerocapture

Ways To Deal With Heat: Thermal Protection System (TPS)

- Heat Sink
- Ablation
- Transpiration / film-cooling
- Insulation & Radiation

Heat Sink

- A slab of material absorbs heat
- Practical for suborbital reentries (warheads, X-15s)
- For orbital reentries, too much weight, not practical.

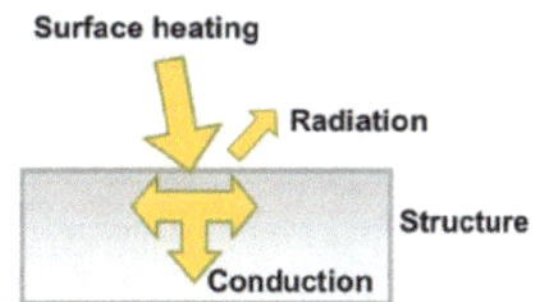

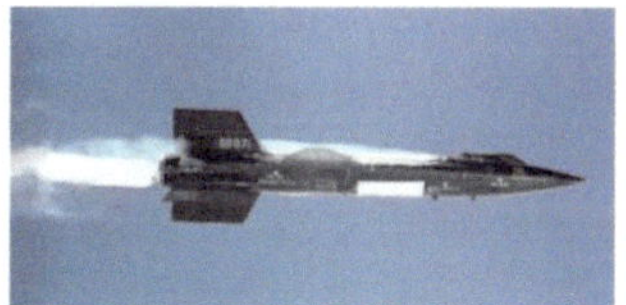

Figure 5. Schematic and photograph (X-15) of a heat sink structure.

Courtesy D. Glass, LaRC

RVs Make Up MIRV Warhead (nose ablates)

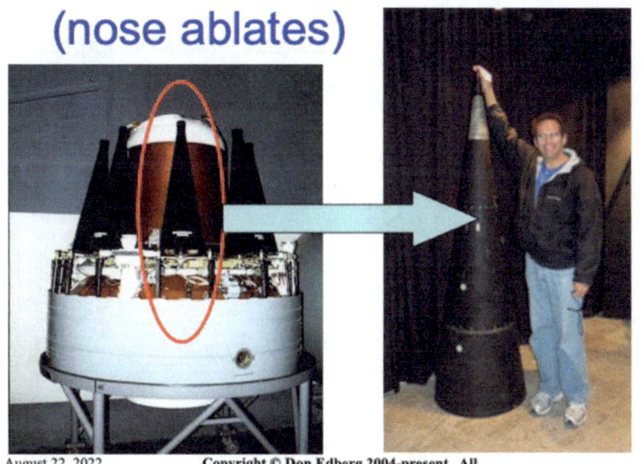

Metallic (Tungsten?) Nose Portion Burns Off

Ablative (sample)

- Material absorbs heat, turns into gas, makes a relatively cool layer to reduce heat absorbed
- "Charring ablative" makes a layer of char that radiates some heat
- Ablative materials not reusable
- Used in *Mercury*, *Gemini*, *Apollo*
- Types: AVCOAT phenolic, L-M SLA-561V, PICA, SpaceX PICA-X, SIRCA

Courtesy D. Glass, LaRC

August 22, 2022 ORB4-55

Transpiration / Film-Cooling

- Similar to ablation
- Use a liquid like water, instead of using a solid
- Limited by quantity of liquid carried

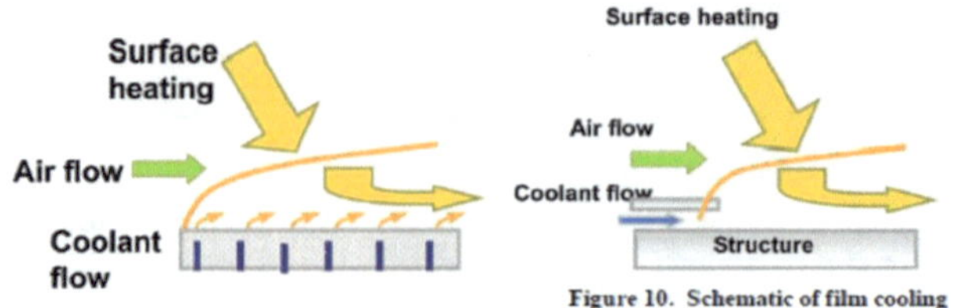

Figure 10. Schematic of film cooling

Courtesy D. Glass, LaRC

August 22, 2022 ORB4-56

Insulation & Radiation

- Entry path affects maximum acceleration, peak heat rate, and total heat
- A steeper reentry means higher peak heat rate but the lower total heat
- Space shuttle uses black tiles to radiate heat effectively
- Tiles insulate & radiate at same time

Surface heating
Radiation
Conduction
Insulation
Structure

Courtesy D. Glass, LaRC

August 22, 2022 ORB4-57

Space Shuttle TPS (sample)

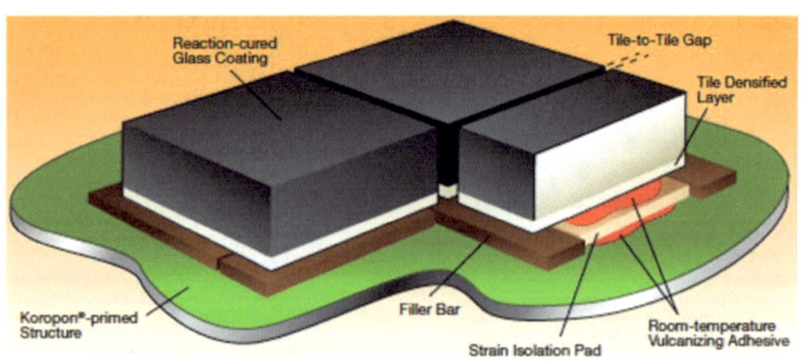

Source: NASA

August 22, 2022 ORB4-58

LOWER SURFACE
UPPER SURFACE

Reinforced Carbon-Carbon (RCC)
High-Temperature, Reusable Surface Insulation (HRSI) —
Low-Temperature, Reusable Surface Insulation (LRSI)
Coated Nomex Felt Reusable Surface Insulation (FRSI)
Metal or Glass

Element*	Area, sq m (sq ft)	Weight, kg (lb)
FRSI	332.7 (3581)	532.1 (1173)
LRSI	254.6 (2741)	1014.2 (2236)
HRSI	479.7 (5164)	4412.6 (9728)
RCC	38.0 (409)	1697.3 (3742)
Miscellaneous		918.5 (2025)
Total	1105.0 (11895)	8574.7 (18,904)

*Includes bulk insulation, thermal barriers, and closeouts

Coloring
HRSI - Black
LRSI - Off White
FRSI - White
RCC - Light Gray

Thermal Protection System, Orbiter 102

Shuttle TPS Data

Material	Areal Density kg/m²	Color
Nomex Felt Reusable Surface Insulation (FRSI)	1.6	white
Low-temperature reusable surface insulation (LRSI) Tiles	3.98	off-white
High-temperature reusable surface insulation (HRSI) tiles	9.2	black
Reinforced Carbon-Carbon (RCC)	44.7	gray

August 22, 2022 ORB4-60

Hot Structure

- Alternative to thermal insulation: make structure or surfaces out of materials tolerant of high temperatures
- Envisioned for *X-20 DynaSoar*, *X-33* vehicles

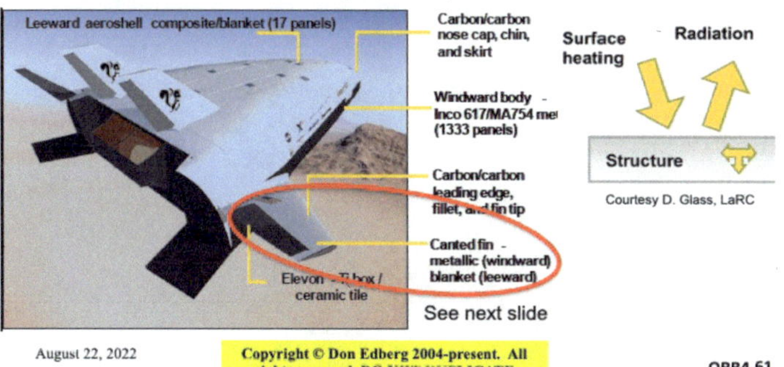

Courtesy D. Glass, LaRC

See next slide

X-33 Hardware

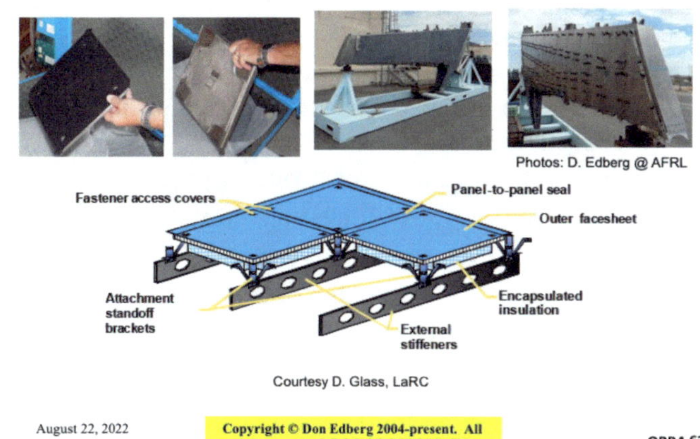

Courtesy D. Glass, LaRC

Thermostructure Design Summary

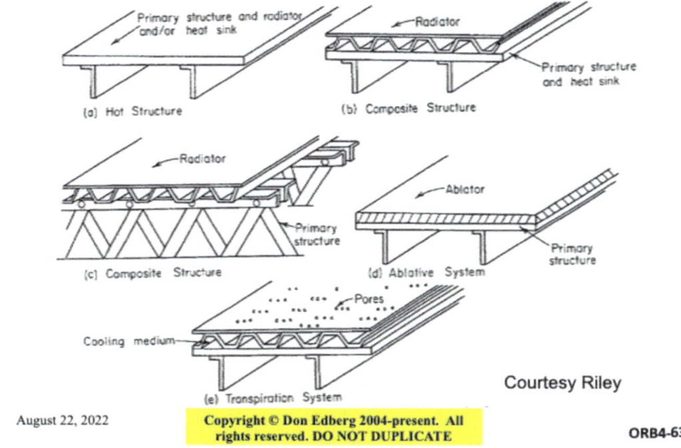

Courtesy Riley

Summary of Key Entry Info

- For given angle γ, deceleration ~ entry speed
- For a given entry speed: the steeper the entry, the greater the deceleration and heating rate (but shorter heat duration). Shallower entries have smaller deceleration & heating rates, but longer exposure
- The total energy to be dissipated is *constant*

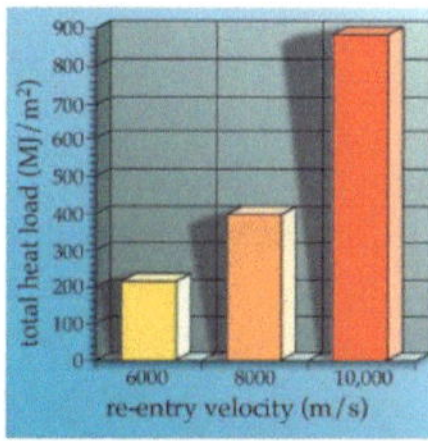

Courtesy: Sellers

Energy to be Dissipated

Condition	Kinetic Energy per Unit Mass, K (kJ)	Potential Energy per Unit Mass, U (kJ)	Total Energy per Unit Mass (kJ)	Total Energy Ratio, Orbit/Staging (Staging = 1)
Staging: v = 4 km/s, h = 100 km	8,000	966	8,966	1.0
Orbit: v = 8 km/s, h = 200 km	32,000	1,931	33,931	3.78

Heat Flux for Different Flight Paths: Area Under Curve (= Energy) is Same

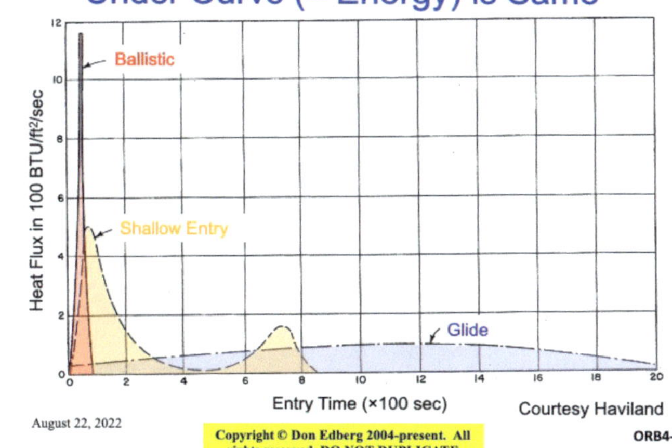

Courtesy Haviland

Key Entry Information (cont.)

- For a given entry speed
  - ⋆ Maximum deceleration same
  - ⋆ Altitude of max deceleration depends on ballistic coefficient β
- Blunt vehicles have lower heating rates
- Pointy vehicles provide better target accuracy
- *Radio blackout* occurs during peak heating… do not count on telemetry during this time (though shuttle could communicate to antennas above)

Topics

- Low-thrust orbital mechanics
  - Equations of motion
  - Efficiency compared to impulsive systems
- Atmospheric entry: energy to dissipate
  - Critical parameters
  - Orbital decay
  - Entry deceleration
  - Entry heating
  - Thermal protection systems
- **Aerobraking & aerocapture**

Aerobraking & Aerocapture (1 of 2)

- Use to decelerate and enter orbit around planet without using propellant for speed change at periapsis
- After capture & passing out of atmosphere, *need to add speed at apoapsis* to raise periapsis out of atmosphere

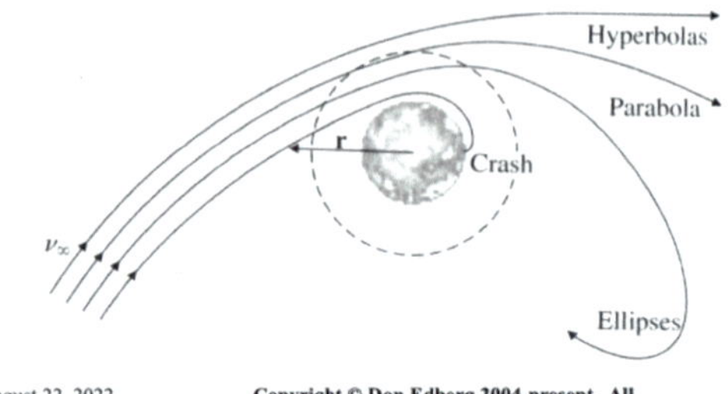

Figure: Wiesel

Aerobraking & Aerocapture (2 of 2)

Instead of using a rocket, dip into the atmosphere

- Lower existing orbit: *aerobraking*
- Brake into orbit: *aerocapture*

Demonstrated by *Magellan*–Venus, *Mars Global Surveyor*, *Space Shuttle*, *Mercury*, *Gemini*, *Apollo* return from moon

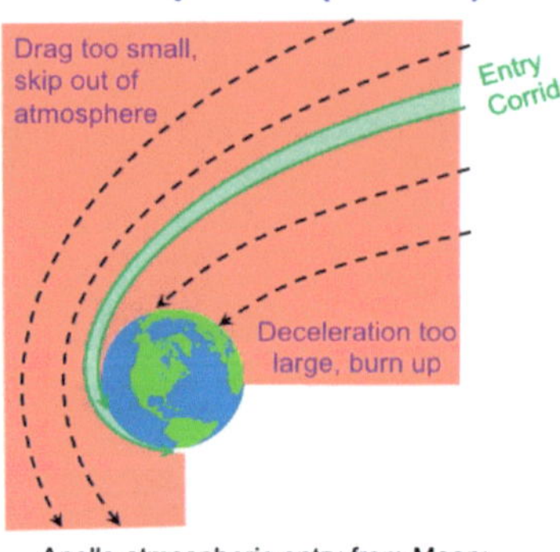

Apollo atmospheric entry from Moon: "Threading the Needle"
Source: Walter

Aerobraking (ii)

Note: after aerobraking (nos. 2-3-4-5-6), propellant needed for two Δvs: 1st at #7 (apoapsis) to raise periapsis #4 out of atmosphere; 2nd at #8 to circularize!

Hyperbolic Approach Trajectory
Atmospheric Entry Interface
Begin Bank Angle Modulation
Periapsis
Circularization Maneuver
Raise Periapsis Burn
End Bank Angle Modulation
Science Orbit
Atmosphere Exit
1 2 3 4 5 6 7 8

Courtesy **Journal of Cosmology**

MRO Aerobraking Movie

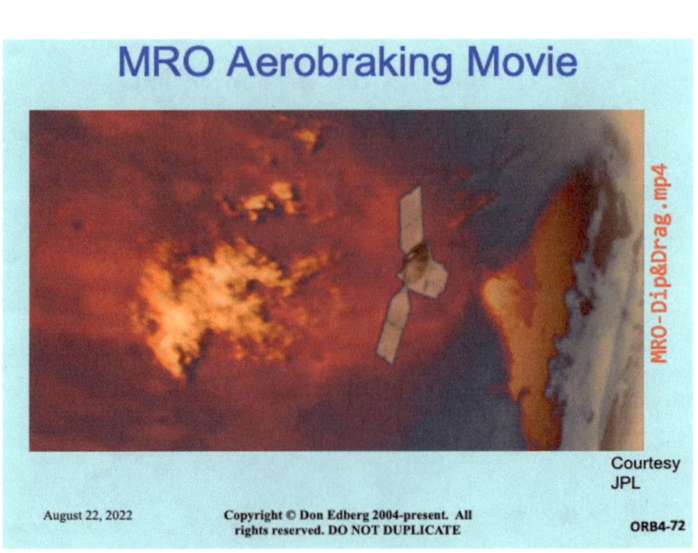

Courtesy JPL

Spacecraft Mission & Payload Development

Topics

- Space mission design cycle timeline
- Phases of design, & planning
- Types of spacecraft, missions, & payloads
- Direct-sensing payloads
- Remote-sensing payloads
- Active sensing instruments
- Payload design & selection, practical considerations

Course Layout

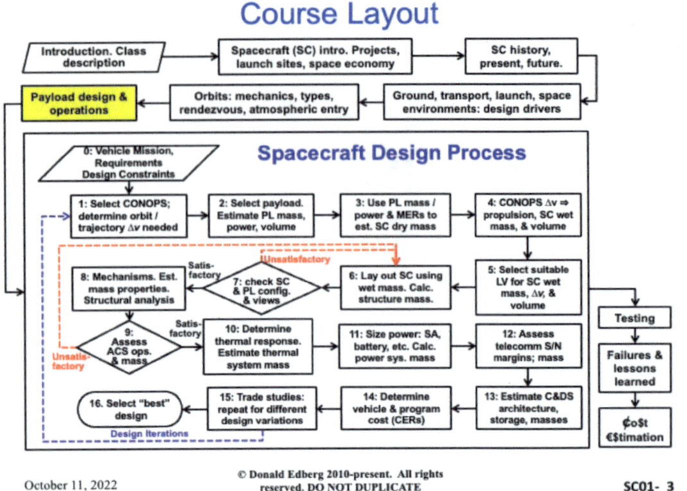

Introduction. Class description
Spacecraft (SC) intro. Projects, launch sites, space economy
SC history, present, future.
Payload design & operations
Orbits: mechanics, types, rendezvous, atmospheric entry
Ground, transport, launch, space environments: design drivers
0: Vehicle Mission, Requirements Design Constraints
Spacecraft Design Process
1: Select CONOPS; determine orbit / trajectory Δv needed
2: Select payload. Estimate PL mass, power, volume
3: Use PL mass / power & MERs to est. SC dry mass
4: CONOPS Δv ⇒ propulsion, SC wet mass, & volume
Unsatisfactory
Satis-factory
8: Mechanisms. Est. mass properties. Structural analysis
7: check SC & PL config. & views
6: Lay out SC using wet mass. Calc. structure mass.
5: Select suitable LV for SC wet mass, Δv, & volume
Testing
9: Assess ACS ops. & mass
Unsatis-factory
10: Determine thermal response. Estimate thermal system mass
11: Size power: SA, battery, etc. Calc. power sys. mass
12: Assess telecomm S/N margins; mass
Failures & lessons learned
16. Select "best" design
15: Trade studies: repeat for different design variations
14: Determine vehicle & program cost (CERs)
13: Estimate C&DS architecture, storage, masses
Design Iterations
¢o$t €$timation

Topics

- Space mission design cycle timeline
- Phases of design, & planning
- Types of spacecraft, missions, & payloads
- Direct-sensing payloads
- Remote-sensing payloads
- Active sensing instruments
- Payload design & selection, practical considerations

Spacecraft/Systems Engineering Design Cycle

- Starts with an idea or mission requirement:
- The **Mission** defines *what* the system must do
- Mission is function-oriented, e.g.:
  - Communicate *X* data within *Y* hours
  - Get pictures of site *X*
  - Return samples from *Y*
  - "Land a man on the moon and return him safely to Earth"

Mission = What Will The Spacecraft Do?

- Typical flight profile
- Data rate, resolution, payload, etc.
- Limitations & environment
- Special requirements (sample return, etc.)

Example: Mars Orbiter (MO) System Requirements Document (1 of 6)

1.0 **Program Overview**

NASA is developing an overall strategy for the exploration of Mars of the next decade involving both orbiters and landers. This document describes the requirements for a Mars orbiter of this new series

2.0 **Payload**

A wide variety of instruments is planned for the Mars Orbiter series of launches. For this study, the instruments chosen for the payload are:

Mars balloon relay	MBR
Mars orbiter camera space	MOC
Thermal emission spectrometer	TES
Mars observer laser altimeter	MOLA

These instruments are described in Appendix A

The payload is supplied by the customer. Payload mass reserve and power reserve are managed by the customer. Mass and power requirements listed in Appendix A are maximums.

Example: Mars Orbiter System Requirements Document (2 of 6)

3.0 Mission Design

The mission for this study will be a Type I trajectory to Mars with a launch from ETR on March 22, 2001 and arrival at Mars on October 8, 2001. The spacecraft cell establish a circular mars orbit with an altitude of 378 km.

4.0 Launch Vehicle

The maximum allowable spacecraft launch mass is 3,230 kg (launch vehicle limit). Spacecraft launch weight includes all propellants and gases as well as the spacecraft launch vehicle adapter. The physical interface with the launch vehicle is 18 equally-spaced bolts on a 97.3-inch diameter bolt circle. The maximum dynamic envelope inside of the payload shroud is 110 inches x 26 feet as shown in Figure 1.

5.0 Mission Life

The spacecraft shall be designed for a three-year life in orbit. The design shall include redundancy such that no single, non-structural, malfunction can cause loss of the mission.

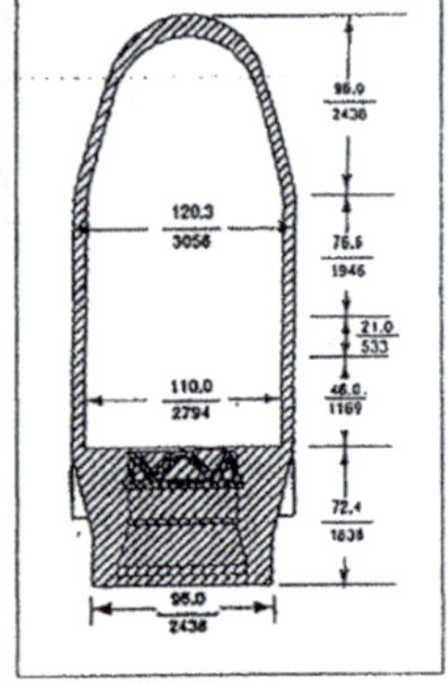

Fig. 1 Payload Fairing

Example: Mars Orbiter System Requirements Document (3 of 6)

6.0 Propulsion

The spacecraft shall be capable of providing a Mars orbit insertion velocity change of 2.80 km/s.

7.0 Attitude Control

The vehicle shall be nadir-pointing, and the attitude control system shall provide the pointing accuracy required by the science instruments. The maximum spacecraft moments of inertia shall be

About x-x axis = 1,700 slug-ft^2

About y-y axis = 1,100 slug-ft^2

About z-z axis = 900 slug-ft^2

It must be possible to make a 180° rotational maneuver about any axis in 30 seconds. The solar arrays shall be pointed towards the sun within a 5° maximum array pointing error.

Example: Mars Orbiter System Requirements Document (4 of 6)

8.0 Power

The power system shall be a direct energy transfer-type with two rigid fold-out solar panels with 2 x 4 cm GaAs cells. The panel shall have 1 degree-of-freedom rotation about an axis perpendicular to nadir. The system shall use batteries and shall be capable of conducting the mission with one battery out. The nominal bus voltage shall be 27.5 V.

9.0 Thermal Control

The thermal control system shall maintain spacecraft bus equipment temperatures 5° higher than the lower temperature limit in Table 1, and 5° below the upper temperature limit listed in table 1 (5° thermal margin).

Table 1 – Temperature Limits

Component	Lower temp. limit, °C	Upper temp. limit, °C
Electronics	0	+40
Batteries	-10	+20
Solar arrays	-100	+100
Propellant, Hydrazine	+7	+35
Structures	-45	+65

Example: Mars Orbiter System Requirements Document (5 of 6)

10.0 Command & Data Handling

The C&DH shall provide for an engineering data rate of 1,200 bps. The science and engineering data shall be Reed Solomon (255, 223) coded. Overhead on science and engineering data shall not exceed 10%.

The command data rate shall be 62.5 bps maximum and shall be BCH (63, 56) coded (CCSDS standard).

The science data will be recorded 24 hours per day, 22 hours at minimum rates and two hours at maximum rates. Engineering data shall be recorded 24 hours/day. See Appendix A for instrument data rates. Redundant data storage shall be provided.

11.0 Telecommunication

The assigned frequencies are downlink 8.4177 GHz; uplink: 7.1646 GHz

The Earth-Mars range is maximum 375E6 km; minimum 20E6 km

For science and engineering data, the maximum bit error rate is 10^{-3}: the minimum Eb/N0 equals 4.5 dB. For commands, the maximum bit Era rate is 10^{-5} and the minimum Eb/N0 = 5.5 dB. The ground stations will be DSM 34-m HEF, see Table 2. DSN time is limited to 12 hours per day during normal operations.

Example: Mars Orbiter System Requirements Document (6 of 6)

Table 2 – 34-m High Efficiency Station Characteristics-MO Frequencies

Receiving Parameters	
Nominal Frequency	8.4177 GHz
Antenna Gain	68.1 dBi
System Noise Temp., Zenith	30°K
Transmitting Parameters	
Nominal Frequency	7.1646 GHz
Transmitter Power	0.4 to 20 kW
Antenna Gain	67 dBi

1) Only the 34-m stations have uplink capability
2) 34-m parameters are for diplexed operation
Data courtesy Tyler et al and J. Geophys. Research, V97, p. 7775

Example: Appendix A – Mars Orbiter Instrument Payload I/F Reqts. (1 of 10)

Payload Element	**Mars Observer Laser Altimeter (MOLA)**
Electrical power, 28 VDC	
Cruise	21 W Repl. Heater, Thermostatically controlled
Orbital Average	23.1 W warm environment
Orbital Peak	33.1 W Normal ops. (Includes 10 W supplemental)
Power regulation	28 V ± 2 V
Data Rate Range	618 bps
Direct Commands	8
No. TLM Channels	2
Mass	25.894 kg
Sun avoidance angle	28° cone about Telescope axis
Temp. Limits, Op & non-Op	-20 to +30 C Op, -30 to +40 C non-Op.
Thermal I/F	Conductively & radiatively isolated from SC
Active thermal control	Thermostatic heaters
Purge	10 SCFH dry Nitrogen
Contamination control	Optics

Example: Appendix A – Mars Orbiter Instrument Payload I/F Reqts. (2 of 10)

Mars Observer Laser Altimeter (MOLA)

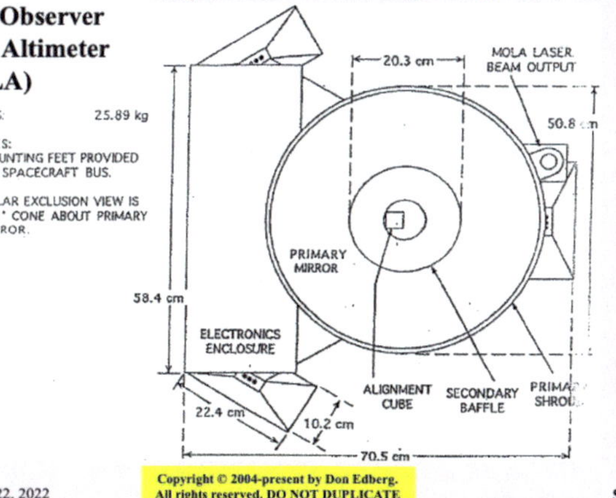

Example: Appendix A – Mars Orbiter Instrument Payload I/F Reqts. (3 of 10)

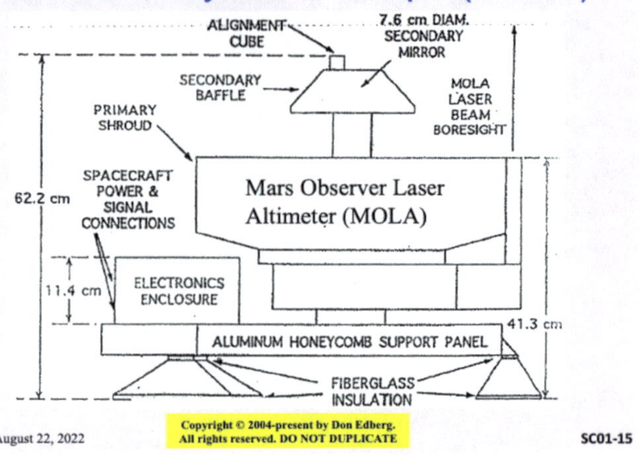

Example: Appendix A – Mars Orbiter Instrument Payload I/F Reqts. (4 of 10)

Payload Element **Mars Observer Camera (MOC)**
Electrical power, 28 VDC
- Cruise: 11 W Repl. Heater, 52.4 W bakeout heater
- Orbital Average: 6 W
- Orbital Peak: 12.5 W

Power regulation: 28 V ± 6 V
Data Rate Range: 700 to 80,000 bps
Direct Commands: 10
No. TLM Channels: 4
Mass: 20.955 kg
Sun avoidance angle: 28° cone about Telescope axis
Temp. Limits, Op & non-Op: -20 to +30 C Op, -30 to +40 C non-Op.
Thermal I/F: Conductively & radiatively isolated from SC, radiates to space
Active thermal control: Heaters
Purge: 3.0 to 4.5 SCFH dry Nitrogen
Contamination control: Optics
Pointing, Nadir dir. (SC): 10 mrad control, 3 mrad knowledge
Magnetic & Radiation environments: N/A

Example: Appendix A – Mars Orbiter Instrument Payload I/F Reqts. (5 of 10)

Mars Observer Camera (MOC)

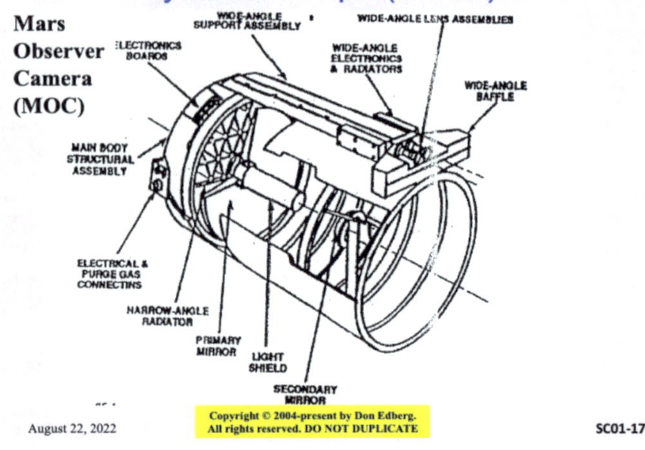

Example: Appendix A – Mars Orbiter Instrument Payload I/F Reqts. (6 of 10)

Mars Observer Camera (MOC)

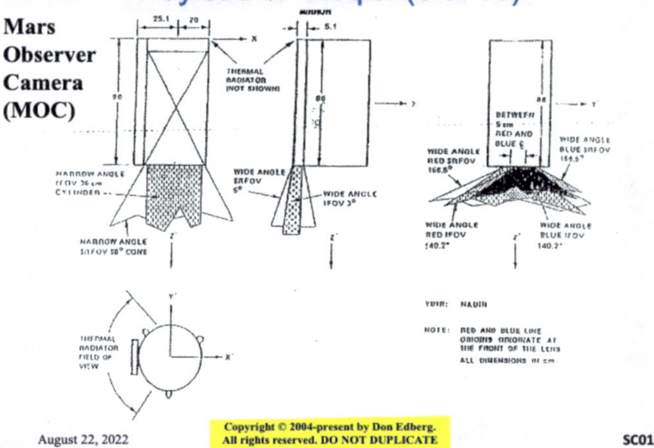

Example: Appendix A – Mars Orbiter Instrument Payload I/F Reqts. (7 of 10)

Payload Element **Mars Balloon Relay (MBR)**
Electrical power, 28 VDC
- Cruise: 9.3 W
- Orbital Average: 9.3 W
- Orbital Peak: 9.3 W

Power regulation: 28 V ± 0.56 V
Data Rate Range: 700 to 40,042 bps
Direct Commands: 5
No. TLM Channels: 2
Mass: 8.829 kg
Temp. Limits: Electronics, -10 to +40 C Op, -30 to +40 C non-Op. Antenna, -80 to +80 C Op, -100 to +100 C non-Op.
Thermal I/F: Antenna & electronics radiatively coupled to SC
Active thermal control: N/A
Purge: N/A
Contamination control: N/A
Magnetic & Radiation environments: N/A

Example: Appendix A – Mars Orbiter Instrument Payload I/F Reqts. (8 of 10)

Mars Balloon Relay (MBR)

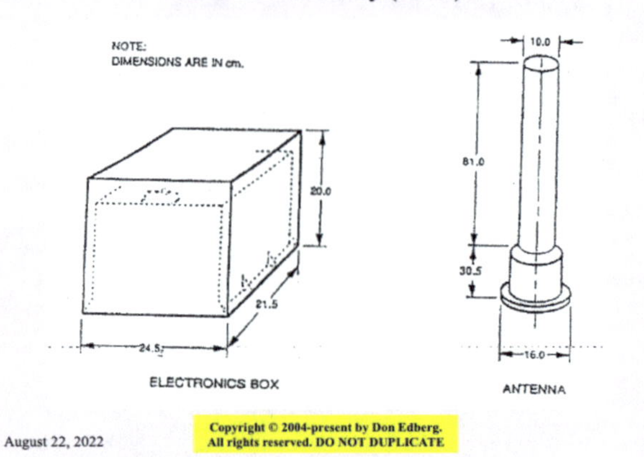

Example: Appendix A – Mars Orbiter Instrument Payload I/F Reqts. (9 of 10)

Payload Element	**Thermal Emission Spectrometer (TES)**
Electrical power, 28 VDC	
Cruise	5.2 W Repl., thermostatically controlled. 8.4 W safe state
Orbital Average	15.2 W
Orbital Peak	17.6 W
Power regulation	28 V +4, -2 V
Data Rate Range	150 to 4,992 bps (programmable)
Direct Commands	4
No. TLM Channels	2
Mass	14.139 kg
Temp. Limits	-20 to +30 C Op, -20 to +50 C non-Op.
Thermal I/F	Conductively isolated, radiatively coupled to SC
Active thermal control	Heaters
Purge through Launch	27-29 SCFH Dry nitrogen. Note: Hygroscopic optics. Purge must be continuous.
Contamination control	Optics
Magnetic & Radiation environments	N/A

Example: Appendix A – Mars Orbiter Instrument Payload I/F Reqts. (10 of 10)

Thermal Emission Spectrometer (TES)

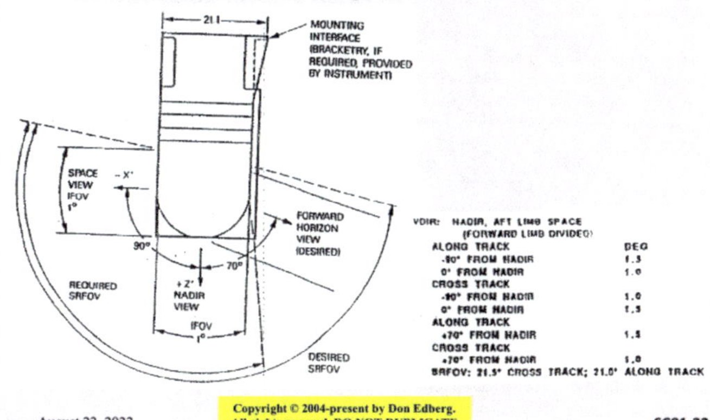

Mission Type Affects Design

- Missions break down into a few basic types.
- Each type makes different demands
  - Different sensitivities
  - Needs map to vehicle concepts and features
- Examining the mission gives designer guidance about:
  - Design drivers
    - Things to emphasize
    - Things to avoid
  - Critical sensitivities
  - General concept approaches

Mission Types & Figures of Merit

"**Orbiter**": Do something useful in Earth orbit

- Data transfer and/or
- Sensor resolution (angular or ground objects)
- Payload mass, power, size
- Operational lifetime/reliability
- Cost

Planetary Mission: Escape from earth, go somewhere

- Science data/sensor resolution
- Data transmission rates
- Payload mass, power, size
- Operational lifetime/reliability
- Cost

Spacecraft Design Features Determined by Mission Type

- Differing importance of mass and power changes resolution, data rates, etc.
- *Launch vehicle availability and selection strongly affects mission capabilities*

Principal Reqts. & Constraints for SC Design

Requirements and Constraints		***Information Needed***
Mission:	Operations concept	Type, mission approach
	Spacecraft life & reliability	Mission duration, success criteria
	Communications	Command, control, comms approach
	Security	Level, requirements
	Programmatic constraints	Cost & schedule
Payload:	Physical parameters	Size, weight, shape, power
	Operations	Duty cycle, data rates, fields of view
	Pointing	Reference, accuracy, stability
	Slewing	Magnitude, frequency
	Environment	Max/min temperatures, cleanliness
Orbit:	Defining parameters	Altitude, inclination, eccentricity
	Eclipses	Maximum duration, frequency
	Lighting conditions	Sun angle and viewing conditions
	Maneuvers	Size, frequency

Requirements & Constraints (cont.)

Requirements and Constraints	*Information Needed*
Environment:	
Radiation dosage	Average, peak
Particles, micrometeoroids, Space debris	Size, density, probability of impact
Hostile environment	Type, level of threat
Launch:	
Launch strategy	Single; shared, upper kick stage
Boosted mass	Launch capabilities
Envelope	Size, shape
Environments	Acceleration, vibe & acoustics, temp.
Launch site	Location, allowed azimuths
Ground System Interface:	
Degree of autonomy	Required autonomous operations
Ground stations	Number, locations, performance
Space links	Space-to-space link, performance

Space Mission Design Process

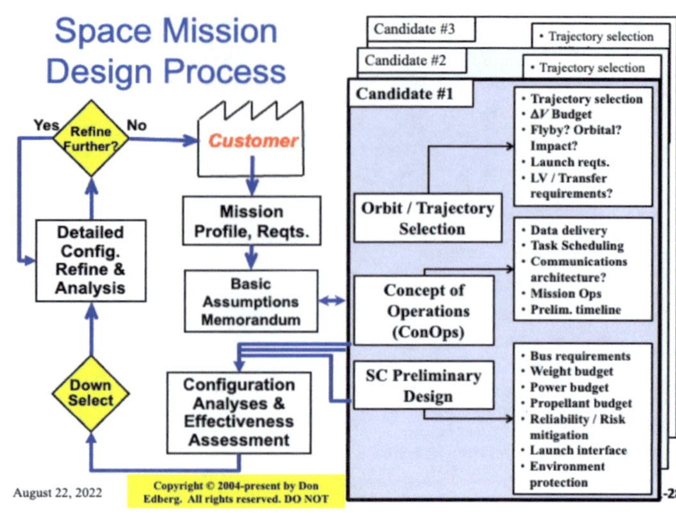

Build A Spacecraft Quickly

How2Build a SC Pt1.mp4 (courtesy NASA GSFC)

Topics

- Space mission design cycle timeline
- Phases of design, & planning
- Types of spacecraft, missions, & payloads
- Direct-sensing payloads
- Remote-sensing payloads
- Active sensing instruments
- Payload design & selection, practical considerations

Phases of Design

- Conceptual
  - Requirements definition
  - Initial configuration concepts defined & evaluated
  - Basic design trades
  - Initial weight, cost, performance estimates
- Preliminary
  - Configuration freeze
  - Testing and Analysis
  - Major component & software layout & design
  - "Real" cost
- Detail
  - Design parts to be built
  - Make or buy decisions (subcontracting)
  - Design tooling & software
  - Major component & systems testing

Stages of Mission Development

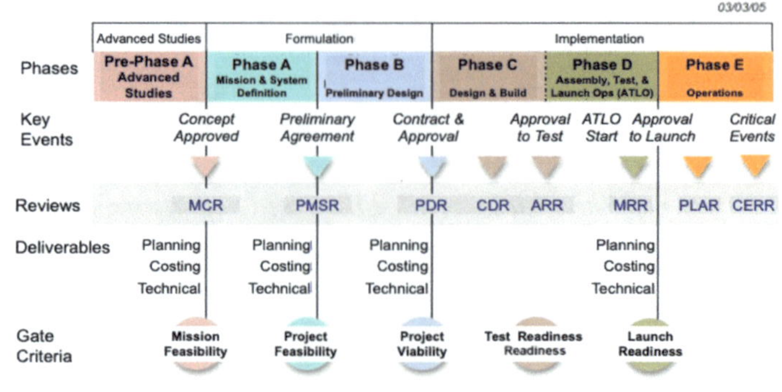

Courtesy: NASA

Purposes of NASA Reviews

Review	Title	Purpose
P/SRR	Program Requirement Review	The P/SRR is used to ensure that the program requirements are properly formulated and correlated with the Agency and mission directorate strategic objectives
P/SDR	Program Definition Review, or System Definition Review	The P/SDR ensures the readiness of the program for making a program commitment agreement to approve project formulation startups during program Implementation phase.
MCR	Mission Concept Review	The MCR affirms the mission need and examines the proposed mission's objectives and the concept for meeting those objectives
SRR	System Requirement Review	The SRR examines the functional and performance requirements defined for the system and the preliminary program or project plan and ensures that the requirements and the selected concept will satisfy the mission
MDR	Mission Definition Review	The MDR examines the proposed requirements, the mission architecture, and the flow down to all functional elements of the mission to ensure that the overall concept is complete, feasible, and consistent with available resources
SDR	System Definition Review	The SDR examines the proposed system architecture and design and the flow down to all functional elements of the system.
PDR	Preliminary Design Review	The PDR demonstrates that the preliminary design meets all system requirements with acceptable risk and within the cost and schedule constraints and establishes the basis for proceeding with detailed design. It will show that the correct design options have been selected, interfaces have been identified, and verification methods have been described
CDR	Critical Design review	The CDR demonstrates that the maturity of the design is appropriate to support proceeding with full-scale fabrication, assembly, integration, and test. CDR determines that the technical effort is on track to complete the flight and ground system development and mission operations, meeting mission performance requirements within the identified cost and schedule constraints.
PRR	Production Readiness Review	A PRR is held for FS&GS projects developing or acquiring multiple or similar systems greater than three or as determined by the project. The PRR determines the readiness of the system developers to efficiently produce the required number of systems. It ensures that the production plans; fabrication, assembly, and integration enabling products; and personnel are in place and ready to begin production.

August 22, 2022 SC01-33

Program & Project Life Cycles (per NASA)

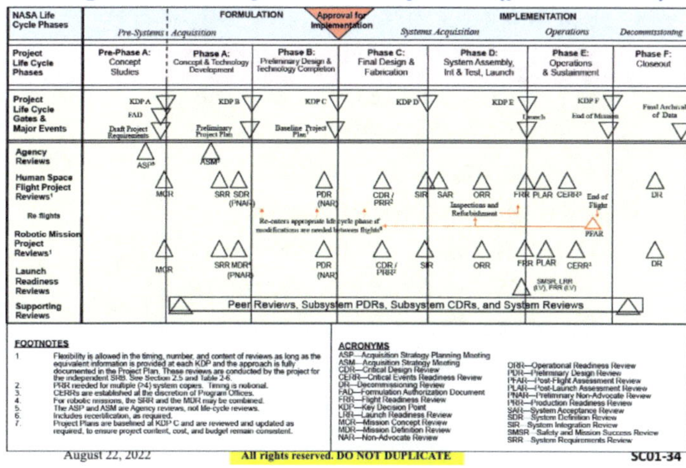

Definitions of All the Reviews You'll Need!

ARR - Assembly, Test & Launch Ops Readiness Review
ASM – Acquisition Strategy Meeting
ASP – Acquisition Strategy Planning Meeting
CDR - Critical Design Review
CERR - Critical Events Readiness Review
DADR - Data Archive Delivery Review
DR - Decommissioning Review
EIRR - External Independent Readiness Review
ETRR - Environmental Test Readiness Review
FAD – Formulation Authorization Document
FIRDR - Final Interface Requirements & Design Review
FPCDR - Fault Protection Critical Design Review
FPPDR - Fault Protection Preliminary Design Review
FPR - Flight Parameters Review
FRR - Flight Readiness Review
FSCDR - Flight System Critical Design Review
FSPDR - Flight System Preliminary Design Review
HRCR - Hardware Review & Certification Record Review
IA - Independent Assessment
IAR - Independent Annual Review
IMAR - Independent Mission Assurance Review
KDP- Key Decision Point
LRR - Launch Readiness Review
MCR - Mission Concept Review
MDR - Mission Definition Review
MESB - Mission Executive Summary Briefing
MOS/GDS PDR /CDR - Mission Ops System/Ground Data System Preliminary/Critical Design Review
Msn/NavS PDR/CDR - Mission/Navigation System Preliminary/Critical Design Review
MRR - Mission Readiness Review
NAR – Non-Advocate Review
ORR - Operational Readiness Review
PDR - Preliminary Design Review
PETR - Post Environmental Test Review
PFAR – Post-Flight Assessment Review
PIRDR – Prelim. Interface Requirements & Design Review
PLAR – Post-Launch Assessment Review
PNAR – Preliminary Non-Advocate Review
PMSR - Project Mission System Review
PRR – Production Readiness Review
PSR - Pre Ship Review
SAR - System Acceptance Review
SDR - System Definition Review
SIR – System Integration Review
SMSR – Safety And Mission Success Review
SRCR - Software Review & Certification Record Review
SRR - System Requirements Reviews
TRR - Test Readiness Review

August 22, 2022 SC01-35

Designer Judgment is Important

- Little or no data available at the outset
- All design teams have essentially the same data and technology base to work with
- Early decisions have large downstream effects
- Choice of approach and success of the design are rooted in designers' knowledge and experience, and available methodology

August 22, 2022 SC01-36

Good Early Estimates are Important

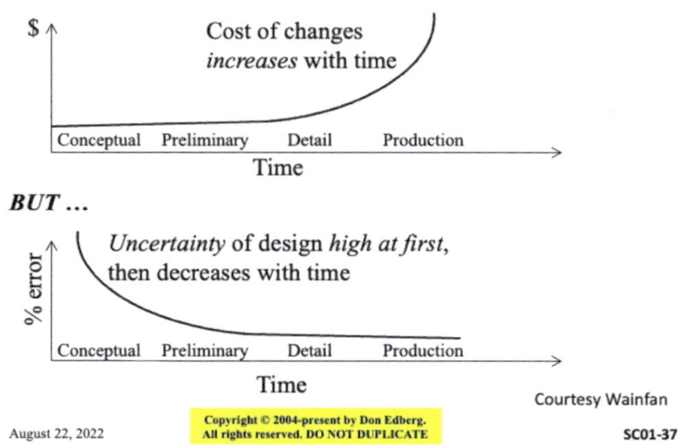

August 22, 2022 SC01-37

Typical Program Cost Profile

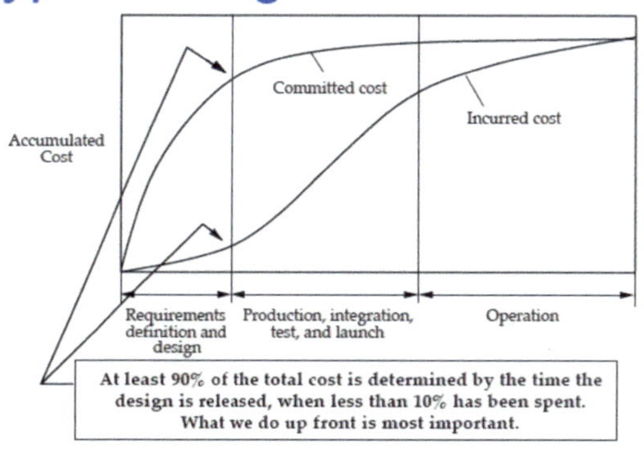

Cost Profile Showing Committed Cost for a Typical Space Program. The *committed* cost is the amount locked in at any given time in the program.

Courtesy Sarafin

August 22, 2022 SC01-38

Design Decision-Making

- Mission & Merit Guide Decisions
  - Features should be evaluated in mission environment/ operations
- Mission first, then configuration
  - No single magic configuration
- Don't fall in love with a concept too early:
  - Concept can end up driving whole design
- Lives can depend on making the right decisions (inhabited missions)
- Beware of "Requirements Creep"

August 22, 2022 SC01-40

Early Space Shuttle Concepts

22-Aug-22 SC01-41

Proposal vs. Actual!

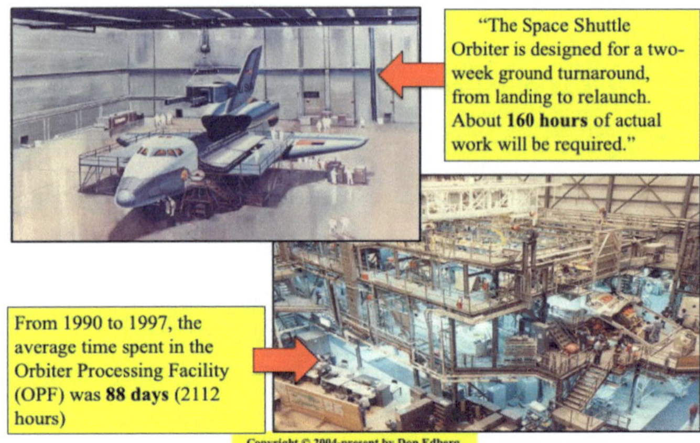

"The Space Shuttle Orbiter is designed for a two-week ground turnaround, from landing to relaunch. About **160 hours** of actual work will be required."

From 1990 to 1997, the average time spent in the Orbiter Processing Facility (OPF) was **88 days** (2112 hours)

Courtesy NASA

August 22, 2022 SC01-42

What is Good?

- Space survival & structural integrity are always necessary
- More of something not needed is useless
- Compromises and tradeoffs are inevitable
- Design Requirements:
  - Must meet constraints
  - Goals: Highly desirable
  - Figures of merit: (more is better)

August 22, 2022 SC01-43

Phase A: Conceptual Design

- Can we come up with a reasonable spacecraft to do the job?
- Any special risks? (technical, schedule, cost)
- How much $?
- How long?
- What trade studies will be done?
- Task planning, make/buy decisions
- May conclude with **conceptual design review** (**CoDR**)

August 22, 2022 SC01-44

Phase B: Preliminary Design

- Money usually awarded to contractor(s) to find the BEST design
- Conduct trade studies for designs/missions.
- Come up with answers to what is the best spacecraft for the mission, and why?
  - What are the risks?
  - How will the spacecraft be built?
  - What is the cost? How long will it take?
- Ends with **preliminary design review**, **PDR**, where customer reviews design.

August 22, 2022 SC01-45

Phase C: Detail Design

- Winner of Phase B gets money to continue
  - but it always changes, funding can take many years.
- Define requirements and performance so that detailed design drawings & flowcharts can be made
- Emphasis on functional performance, requirements definition, and interface definition
- Ends with **Critical Design Review, CDR**. Customer reviews design to see if it meets requirements.

August 22, 2022 SC01-46

Phase D: Development

- Detailed design process
  - Build drawings are made
  - Software is coded
- "Make-Buy" decisions made
- Subcontracts issued for items to be bought or made elsewhere

August 22, 2022 SC01-47

Assembly, Test, & Launch Operations (ATLO)

- Parts and **assemblies** are available for test
- System level **testing** includes:
  - Functional testing
  - Thermal-vacuum (thermal-vac) } "Shake & Bake"
  - VS&A (vibration, shock, and acoustics) } "Shake & Bake"
  - Communications, RFC/RFI
  - Software
  - Mission simulations
- **Pre-ship readiness review (PRR)** which evaluates whether spacecraft is acceptable
- Spacecraft shipped to launch site

August 22, 2022 SC01-48

Pre-Launch, Launch, & Ops Phases

- Reassemble spacecraft at launch site
- Retest
- **Flight readiness review = FRR**
- Successful FRR leads to ***launch***

Mission Ops

- Control the spacecraft
- Make it do its thing
  - Teams can be 10s to 100s of people
  - Ops can be a large fraction of project cost

August 22, 2022 SC01-49

Task Estimation

- Each box in the design & flow diagram is split up into different tasks
  - There can be *thousands* of subtasks
- The better the work can be divided up early in the program, the better the cost/schedule estimate
- WBS = Work Breakdown Structure
  - It's called a "Work Breakdown Structure" because the *Work* remaining will grow until you have a *Breakdown*, unless you enforce some *Structure* on it [*accountant humor* ☺]
- Add schedule margin for unanticipated delays: "shifting to the right" for strikes, weather, …

August 22, 2022 SC01-50

Topics

- Space mission design cycle timeline
- Phases of design, & planning
- Types of spacecraft, missions, & payloads
- Direct-sensing payloads
- Remote-sensing payloads
- Active sensing instruments
- Payload design & selection, practical considerations

August 22, 2022 SC01-51

Types of Spacecraft

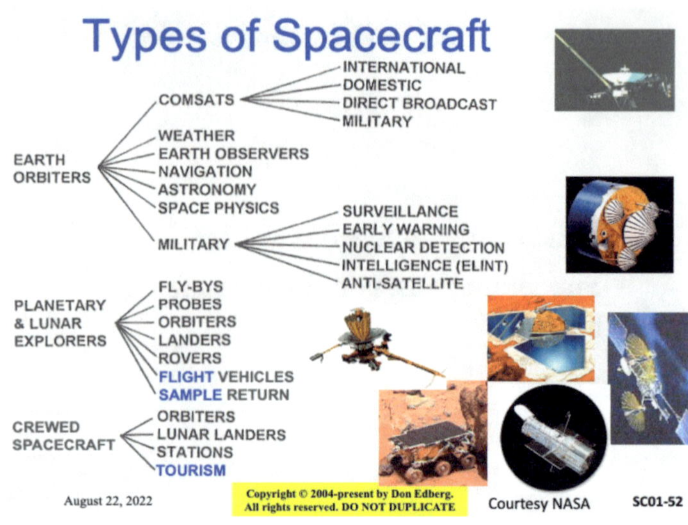

Courtesy NASA

Spacecraft Payload Types

… depend on spacecraft's mission!

- Applications: navigation, communications, data relay, entertainment
- Observation: science investigations with sensors
- Demonstrations & "special" payloads

Types

- Direct-sensing
- Remote-sensing
- Probe or Sample return

Applications Payloads

- Telecommunications
  - Fixed & mobile
- Relay & distribution
  - Entertainment, broadcasting
- Observation: weather, resources
- Navigation

Examples: *INTELSAT*, *TDRSS*, DirecTV, XM radio, GPS, *LANDSAT*, *TIROS*, *GOES*

Relay & Distribution Example

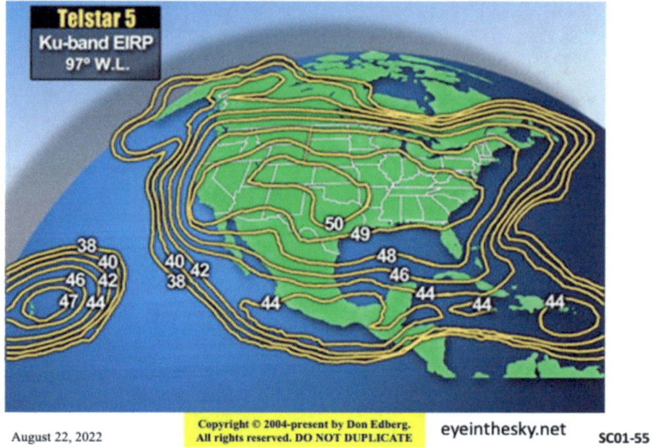

eyeinthesky.net

Science Payloads

- Earth-looking (resources, weather, reconnaissance, oceanography, etc.)
- Remote-sensing (astronomical observations, planetary characteristics)
- Remote probes & sample returns
- Examples: HST (*Hubble Space Telescope*), *Spitzer*, *Mariners*, *Pioneers*, *Voyagers*, *Surveyor*, *Viking*, *MER*, *Huygens*, *Galileo*, *Cassini*, *Venera*, *Magellan*, *Stardust*, etc.

Science Payloads: *Pioneer*

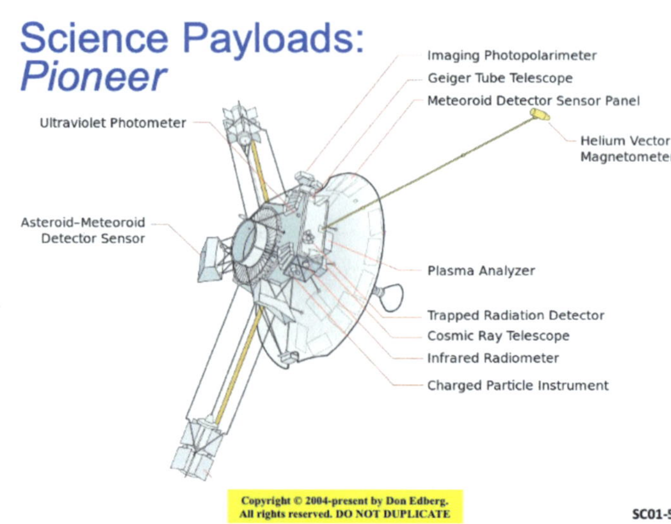

Science Payloads: ESA *Solar Orbiter*

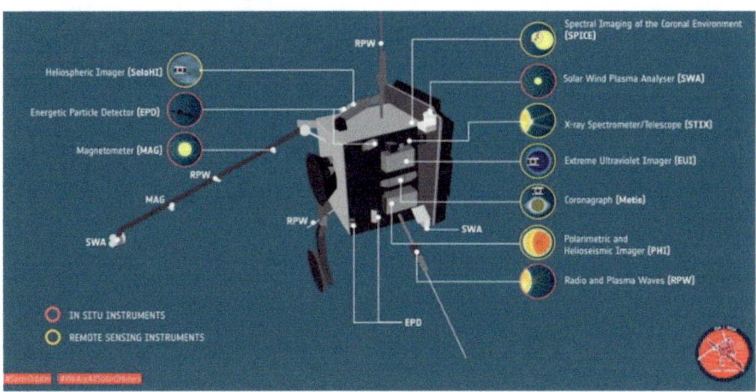

Courtesy: ESA

Special Payloads: Usually Military/Government

- Observation: nuclear burst, boost-phase tracking, etc.
- Test beds, anti-satellite weapons
- Intelligence, jamming
- Surveillance, tracking
- Orbital refueling, Space Tugs
- Examples: *Lacrosse*, *DMSP*, *DSP*, secret
- "National Technical Means" = spy satellites

Topics

- Space mission design cycle timeline
- Phases of design, & planning
- Types of spacecraft, missions, & payloads
- **Direct-sensing payloads**
- Remote-sensing payloads
- Active sensing instruments
- Payload design & selection, practical considerations

DIRECT-SENSING PAYLOADS

Payloads that sense some physical parameter in the space that they are IN

Direct-Sensing Payloads

- **High-energy particle detector**: measures energy of trapped energetic electrons, & atomic nuclei energy & composition
- **Low-energy charged particle detector**: characterizes composition & energies of charged particles
- **Plasma detector**: measures density, composition, temperature, velocity and 3-D distribution of plasmas
- **Dust detector**: measures velocity, mass, charge, flight direction and number of dust particles striking instrument

GLL Photopolarimeter Radiometer

Cassini Cosmic Dust Detector

Galileo Heavy Ion Counter (HIC)

Courtesy JPL

Direct-Sensing Instruments

- **Magnetometer**: measures strength and direction of magnetic (mag) fields, typically in three planes
- **Plasma wave detector**: measures 3-D electrostatic and electromagnetic components of local plasma waves
- **Heavy Ion Counter**: registers the characteristics of ions in the spacecraft's vicinity that actually enter the instrument
- **Mass Spectrometer**: reports the species of atoms or molecules that enter the instrument

Voyager & MAG Boom

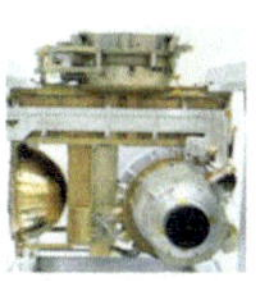

Cassini CAPS (Cassini Plasma Spectrometer Subsystem)

Sojourner APXS (Alpha-Particle X-ray Spectrometer)

Courtesy JPL

Topics

- Space mission design cycle timeline
- Phases of design, & planning
- Types of spacecraft, missions, & payloads
- Direct-sensing payloads
- Remote-sensing payloads
- Active sensing instruments
- Payload design & selection, practical considerations

REMOTE-SENSING PAYLOADS

Payloads that sense some physical parameter at a distance – observation

Remote Sensing Payload Types

- **Imaging systems: cameras, telescopes, radar dishes**
- **Infrared radiometer**: measures the intensity of infrared (thermal) energy radiated by target
- **Polarimeter**: measures direction & extent of polarization of light reflected from target
- **Photometer**: measures source's light intensity
- **Spectrometer**: splits light received from an object into its component wavelengths with a diffraction grating & measures amplitudes of individual wavelengths
- **Bolometer**: measures EM radiation intensity
- **LIDAR**: LIght Detection And Ranging

Remote-Sensing Payloads

- Utilize electromagnetic spectrum
- Angular resolution depends on wavelength and aperture diameter
- High resolution favors *optical* over RF or radar
- *Swath width*, spacecraft speed both depend on orbital height
- Amount of data determines power, transmission, storage, etc.
- All are interrelated!

Common feature: ***Camera*** *or* ***Telescope*** *focuses an image onto detector*

Remote Sensing: View from Spacecraft Chosen for Swath Width

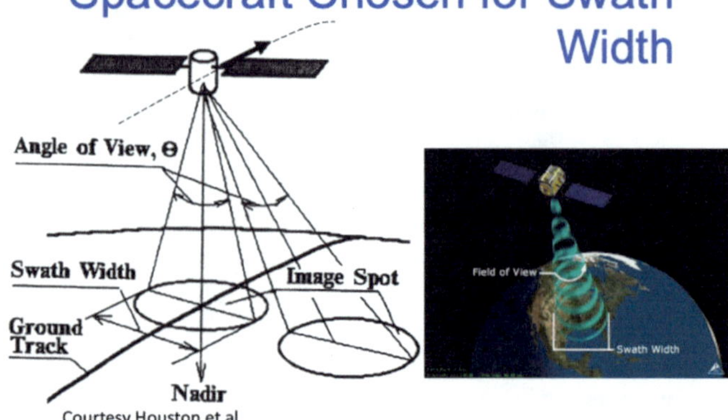

Courtesy Houston et al

Nadir & Side-Looking Sensors

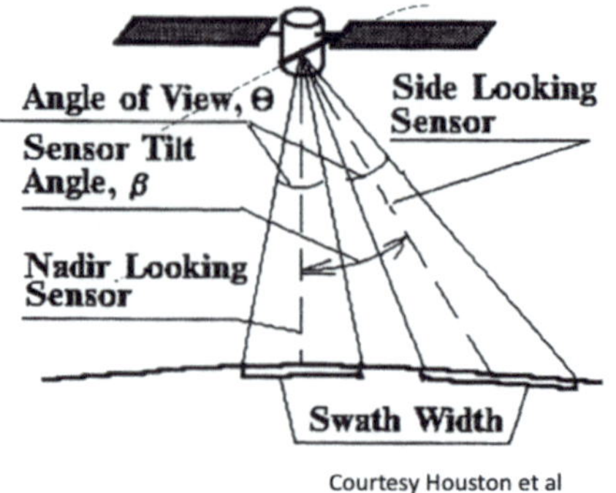

Courtesy Houston et al

Satellite Coverage Geometry

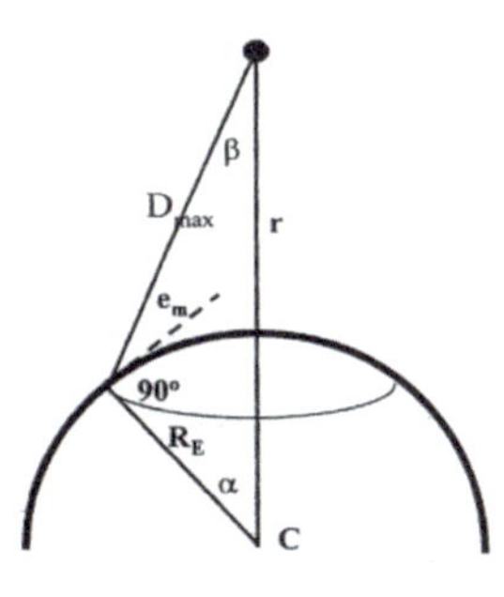

- Largest coverage angle is to horizon (less better)
- Half-angle β: $\beta = \sin^{-1}[(R_E/r)\cos e_m]$ where e_m is minimum elevation angle (5~10°)
- The field of view 2β or half-cone angle β decreases as r increases

Courtesy: Pisacane

Solar-Scatter Geometry

Courtesy Aerospace *Crosslink*

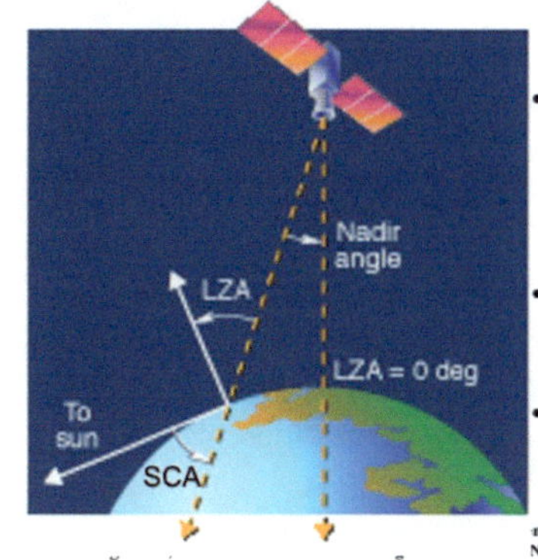

- As sensor LOS shifts from nadir to limb:
  - Both range to target & path length through atmosphere increase
  - The minimum possible solar-scatter angle (SCA) decreases.
- When viewing low-altitude targets at the limb (LZA = 90°), the sun can be directly in the sensor line of sight (i.e., SCA can be 0°)
- Unfortunately, much surface area covered by space sensor lies at the larger LZAs
- Mitigate by overlapping coverage from many sensors

… Orbits Often Chosen For Ground Track

- High-inclination orbit covers most of earth

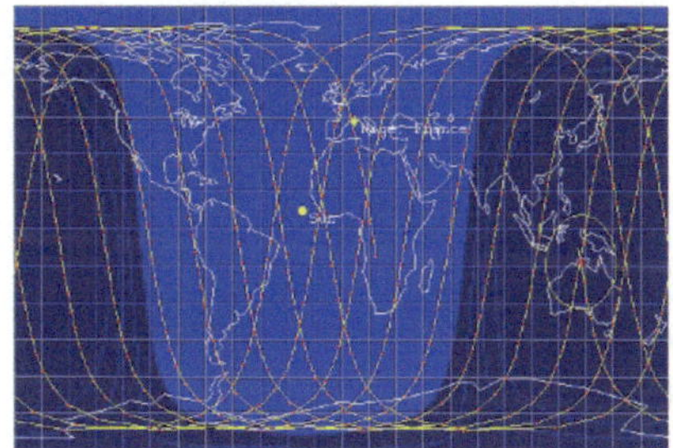

A Basic Camera (or Telescope) System

Image Plane
Pixel of Resolution in the Image Plane
The system's "*f*" number = *f*/*D*
f
Lens Aperture, D
Angular Field of Veiw
H
Angular Resolution, Θ_R
Field of View
Ground Resolution

Courtesy Houston et al

Remote-Sensing Payloads – Imaging

- At the image plane, a **detector** converts image to electrical signals

Two families of detectors:

1. **Charged coupled device** or CCD (right): an imaging device consisting of a large-scale integrated circuit with a 2-D array of charge-isolated "wells." Each well is a *pixel.*
2. **Vidicon**: vacuum tube imaging device

CHARGE-SPREADING BARRIERS
CHARGE ACCUMULATION AREAS (WELLS OR PIXELS)
ELECTRODES FOR CONTROL AND READOUT
MINUS SIGNS REPRESENT ACCUMULATION OF NEGATIVE CHARGE

Figure courtesy JPL

 # pixels/image plane area determines resolution.

Effect of Resolution on Images

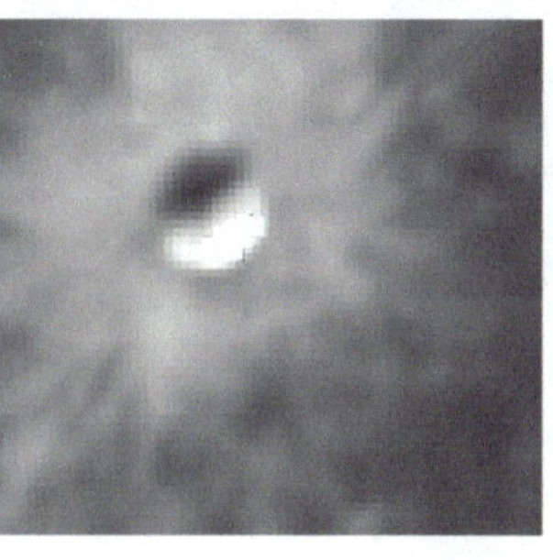

Clementine UV-VIS camera (100 m resolution)

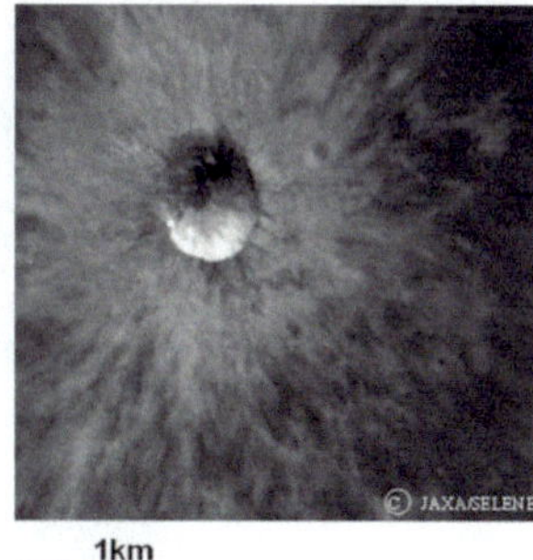

1km

JAXA Kayuga (Selene) Multi-band Imager (20 m resolution)

Effect of Resolution on Images

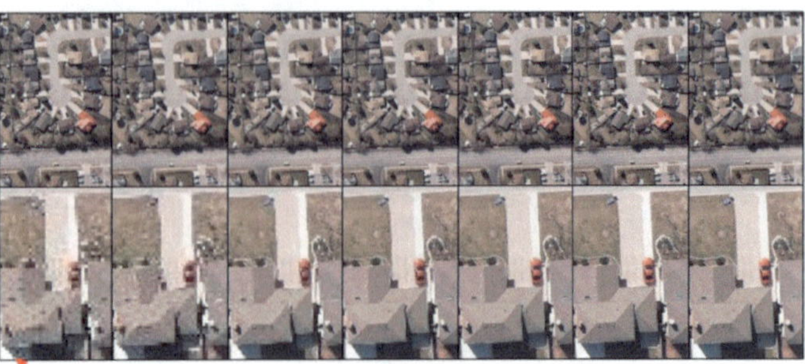

Image "pixel-ated" (sawtooth) at larger resolution values (smaller lengths = better)

Courtesy: Quora

Actual: Orbiting Telescope Images Iran Launch Site from 382 km

- Trump tweeted photo taken apparently by USA 224, an optical reconnaissance SC, 0944 UTC on 2020 Aug. 29, 382 km altitude
- SC has 2.4 m mirror comparable to Hubble's
- What is resolution required for USA 224 to be able to read an auto's license plate from orbit?

Courtesy Forbes

What *IS* USA 224 Resolution?

- Minimum 5 vertical by 3 horizontal pixels to distinguish letters ("H" shown)
- CA plate letters are ~7.5 cm high (3 in)
- Required resolution element needs to be 7.5 cm ÷ 5 pixels = 1.5 cm at telescope's focal surface.
- 1.5 cm from 382 km is angle $\theta_{req} = \frac{h_{req}}{r} = \frac{0.015 \text{ m}}{382{,}000 \text{ m}} = 3.93\times10^{-8}\text{rad}$
- Telescope resolution is $\theta = \frac{1.22\lambda}{d}$
- Assume orange-red light, $\lambda = 0.6$ µm (micrometers) to improve visibility through the atmosphere
- So $\theta_{USA224} = \frac{1.22\lambda}{d} = \frac{1.22(0.6\times10^{-6})\text{m}}{2.4 \text{ m}} = 30.5\times10^{-8}$ m, ≈ *7.7× less than required*
- Can't distinguish license plates (also *slant range* > 382 km)

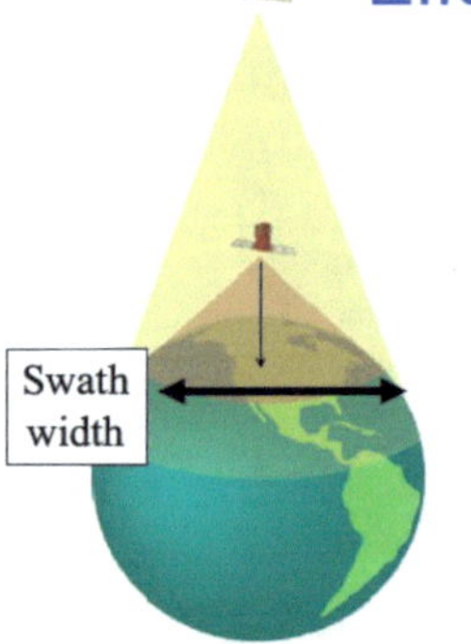

Modern photoreconnaissance satellite features a high-resolution digital camera, on-board fuel supplies and a data-relay antenna to downlink imagery.

Spacecraft Earth Coverage: Effects of Orbits

Swath width

- Area of coverage *increases* with altitude
- Resolution of electro-optical (EO) sensor *decreases* with altitude
- High-resolution EO systems = lower altitudes

Courtesy Houston et al

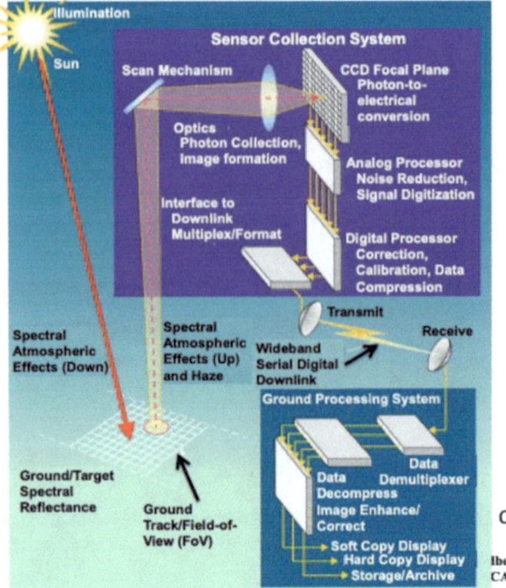

Electro-Optical Sensor Operation

Courtesy Aerospace *Crosslink*

Topics

- Space mission design cycle timeline
- Phases of design, & planning
- Types of spacecraft, missions, & payloads
- Direct-sensing payloads
- Remote-sensing payloads
- Active sensing instruments
- Payload design & selection, practical considerations

Active Sensing Instruments

Power is transmitted away and monitored

- Synthetic aperture radar (SAR): A radar imaging instrument which illuminates with Radio Frequency (RF)
  - Capable of imaging surfaces covered by clouds and haze.
- Altimeter: A device that measures altitude above the surface of a planet or moon
  - Time the round trip of RF signals bounced off the surface

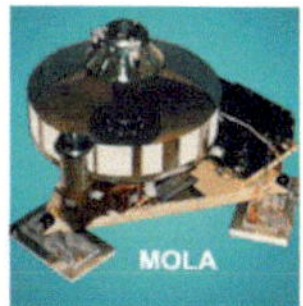

Mars Orbiter Laser Altimeter Courtesy JPL

Radar vs. Synthetic Aperture Radar (SAR)

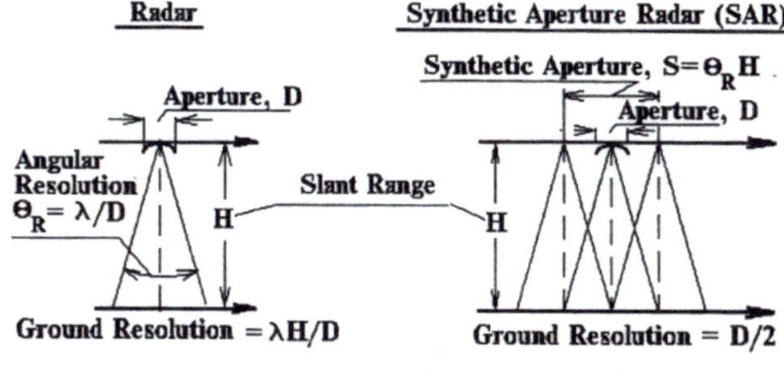

Courtesy Houston et al

SAR System Operation

For a SAR system to develop optical equivalent resolution:

- SC's position & velocity must be known with great precision
- SC attitude must be controlled tightly
- SC navigation data must be updated frequently
- SAR images are constructed of a matrix where lines of constant distance or range intersect with lines of constant Doppler shift

SPACECRAFT MOTION
LINES OF CONSTANT DOPPLER
LINES OF CONSTANT RANGE (DELAY)
SAR FOOTPRINT ON PLANET SURFACE

Figure Courtesy JPL

Topics

- Space mission design cycle timeline
- Phases of design, & planning
- Types of spacecraft, missions, & payloads
- Direct-sensing payloads
- Remote-sensing payloads
- Active sensing instruments
- **Payload design & selection, practical considerations**

Payload Selection & Design

1. Identify requirements, constraints	What are mission objectives & goals? What are restrictions (time, cost, cooperation)? Refer to SRD = System Requirements Document
2. Select & size payload instruments	Consider sensitivity, pointing accuracy, coverage, resolution. Are multiple instruments ok? Will they work well together?
3. Generate concept of operation	Tasking, scheduling? Data processing (on-board vs. on-ground). How to send data to users?
4. Select design features of payload	Specify details of mechanical configuration, frequency bands, details of operation, etc.
5. Define support (bus) requirements	Consider payload dimensions, mass, power, pointing/field-of-view, stability, data rates
6. Iterate to improve	Until the entire system makes sense & meets requirements

Payload Field-of-View (FoV)

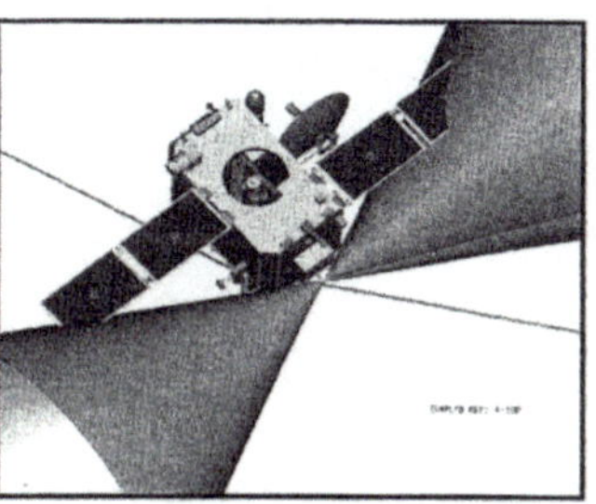

(a) Instrument

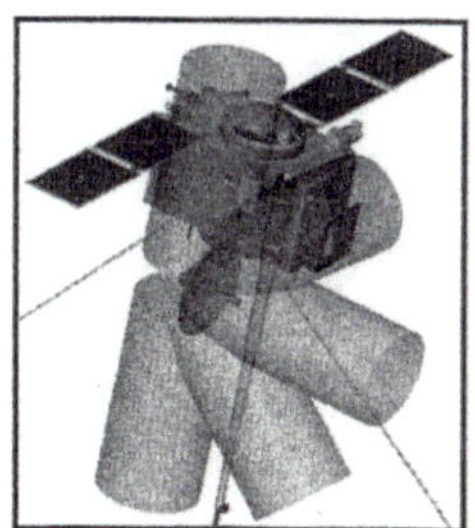

(b) High gain antenna

Courtesy: Pisacane

Payload Field-of-View Important

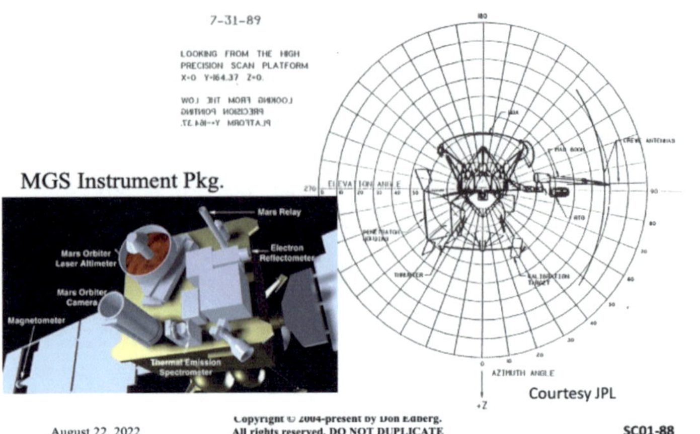

Courtesy JPL

Scan Platform

- An articulated, powered appendage to the spacecraft bus
- Points in commanded directions
- Allows observations to be taken independent of the spacecraft's attitude
- Example: Galileo

D. Edberg photo

Space-Based Sensor System Design

Driving Mission Requirements
Minimum detectable target
Report time
Area-of-interest size
Tactical-parameter accuracy

→ **Constellation Architecture**
Altitude
Inclination
Number of satellites
Communications approach

→ **Driving Sensor Requirements**
Noise equivalent target
Ground sample distance
Coverage area(s)
Revisit time
Spectral band(s)

→ **Sensor Concept Definition**
Example:
Theater IR scanner
Scan pattern
Line-of-sight (LOS) agility, stability defined

→ **Driving Mission Requirements**
Derive field of view (FOV), scan rate, instantaneous FOV
Find aperture & focal-plane array temperature

→ **Optical Subsystem**
Optical design
LOS scan mechanism

→ **Focal-Plane/Processing Subsystems**
Focal-plane size, topology
Pixel size
Pixel count
Sensitivity
Line rate
Analog rates
Digital rates
Power dissipation
Processing functions

→ **Thermal Subsystem**
Size thermal radiator
Select heat transport hardware

→ **Payload Description**
Mass
Power
Size
Performance

→ **Spacecraft/ Launch Vehicle Synthesis**
Spacecraft design/sizing
LV selection

Payload Description → Mission performance analysis

→ **System & performance acceptable?**
No → (back to Driving Mission Requirements)
Yes → Proceed with detailed system design

Dashed lines indicate potential sensor/payload/concept iterations

Courtesy Aerospace *Crosslink*

Approx. Densities for Instruments (kg/m^3)

Spectrometers ~	250	
Mass spectrometers ~	800	
Synthetic aperture radar ~	32	
Rain radars ~	150	thickness / diameter ~ 0.2
Cameras ~	500	Small telescopes w/ camera ~ 325
Small instruments ~	1000	

Scaling Laws: If a smaller instrument exists as a model, then if SF is the linear dimension scale factor…

Area proportional to	SF^2	Mass proportional to	SF^3
Area inertia proportional to	SF^4	Mass inertia proportional to	SF^5
Frequency proportional to	1/ SF	Stress proportional to	SF

BEWARE THE SQUARE-CUBE LAW! STRESS WILL INCREASE WITH SF!

Courtesy R. Farley

Spacecraft Systems Design

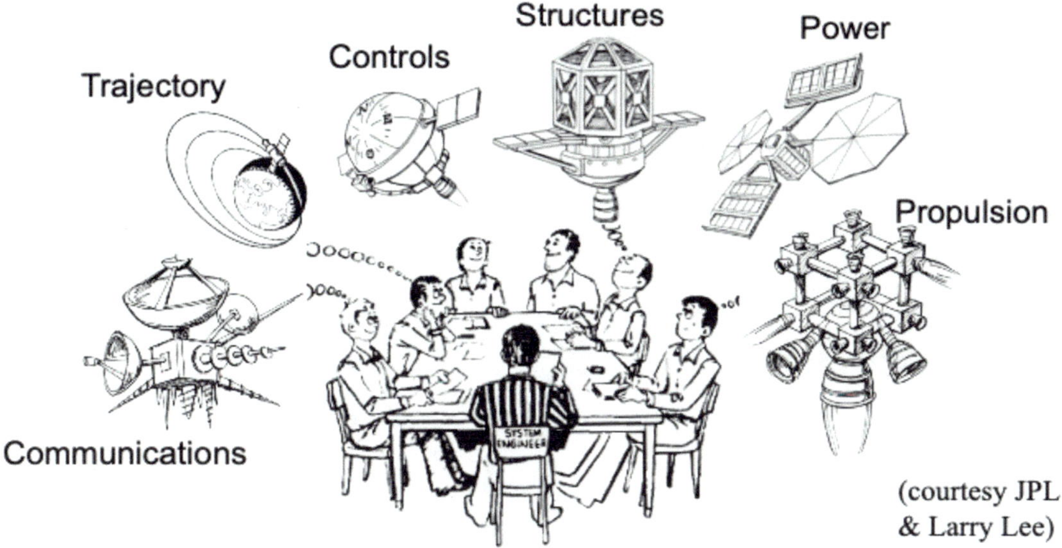

(courtesy JPL & Larry Lee)

Figure 4-1. Before launch, a spacecraft design team did a lot of brainstorming to hammer out the dozens of major considerations (and thousands of smaller details) needed to design and build the amazing Voyager robots.

Spacecraft Systems Topics

- Spacecraft project elements
- Spacecraft payload needs & requirements
- The spacecraft bus, examples
- Spacecraft bus components
- Non-spacecraft mission elements
- Major spacecraft design drivers
- Initial mass estimation
- Mass growth & contingencies; subsystem allocations
- Initial power estimation; subsystem allocations; margins
- Spacecraft volume considerations

Spacecraft Design Process

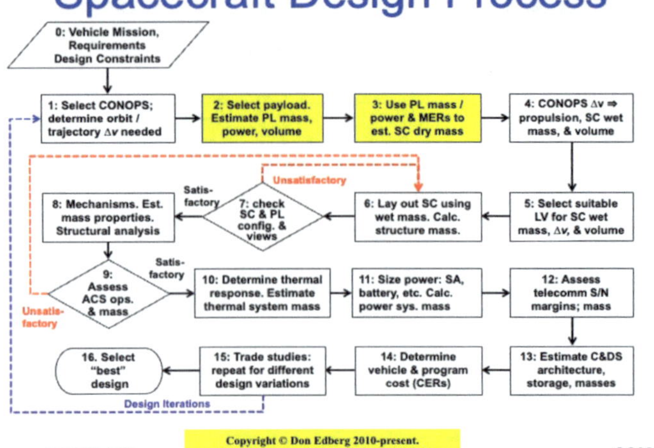

0: Vehicle Mission, Requirements Design Constraints
1: Select CONOPS; determine orbit / trajectory Δv needed
2: Select payload. Estimate PL mass, power, volume
3: Use PL mass / power & MERs to est. SC dry mass
4: CONOPS Δv ⇒ propulsion, SC wet mass, & volume
5: Select suitable LV for SC wet mass, Δv, & volume
6: Lay out SC using wet mass. Calc. structure mass.
7: check SC & PL config. & views
Unsatisfactory
Satis-factory
8: Mechanisms. Est. mass properties. Structural analysis
9: Assess ACS ops. & mass
Satis-factory
Unsatis-factory
10: Determine thermal response. Estimate thermal system mass
11: Size power: SA, battery, etc. Calc. power sys. mass
12: Assess telecomm S/N margins; mass
13: Estimate C&DS architecture, storage, masses
14: Determine vehicle & program cost (CERs)
15: Trade studies: repeat for different design variations
16. Select "best" design
Design Iterations

Spacecraft Systems Topics

- **Spacecraft project elements**
- Spacecraft payload needs & requirements
- The spacecraft bus, examples
- Spacecraft bus components
- Non-spacecraft mission elements
- Major spacecraft design drivers
- Initial mass estimation
- Mass growth & contingencies; subsystem allocations
- Initial power estimation; subsystem allocations; margins
- Spacecraft volume considerations

Spacecraft Project Elements

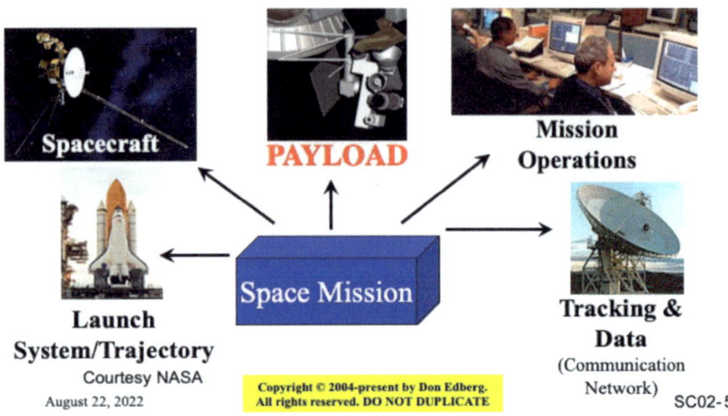

Spacecraft
PAYLOAD
Mission Operations
Space Mission
Launch System/Trajectory
Tracking & Data
(Communication Network)

Courtesy NASA

Spacecraft System

- Mission Planning
  - Launch, Trajectory, Orbit, Maneuvering, Decay, Recovery / Disposal
- Payload
  - Payload Requirements (Power, Data, Volume, Temp.), Payload Design
- Spacecraft Design
  - Bus Design: Structures, Power, Control, Thermal, C&DS, Comm., Propulsion, Manufacturing, Testing, etc.
- Launch & Mission Ops
  - Transportation, Reassemble & Test, Payload Encapsulation, LV Integration, Testing, Launch, Operations, Tracking, Data Transmission

Spacecraft Systems Topics

- Spacecraft project elements
- **Spacecraft payload needs & requirements**
- The spacecraft bus, examples
- Spacecraft bus components
- Non-spacecraft mission elements
- Major spacecraft design drivers
- Initial mass estimation
- Mass growth & contingencies; subsystem allocations
- Initial power estimation; subsystem allocations; margins
- Spacecraft volume considerations

Payload Needs

- Attitude control (point correct direction)
- Power (support payload ops)
- Data management, communicate to ground
- Command (operate payload)
- Thermal (keep payload at proper temp)
- Mechanical support/deployment
- Propulsion (maintain orbit/trajectory)
- Etc.

All needs supplied by spacecraft bus!

How does mission design affect SC?

- Launch site, launch providers
- Payload mass & size
- Sensor visibility, eclipse times
- SC propulsion, orbit maintenance
- Thermal issues
- Radiation amounts
- Power generation
- Navigation
- Ground tracking, visibility
- Link budget & data transfer

August 22, 2022

SC02-9

Spacecraft Systems Topics

- Spacecraft project elements
- Spacecraft payload needs & requirements
- **The spacecraft bus, examples**
- Spacecraft bus components
- Non-spacecraft mission elements
- Major spacecraft design drivers
- Initial mass estimation
- Mass growth & contingencies; subsystem allocations
- Initial power estimation; subsystem allocations; margins
- Spacecraft volume considerations

August 22, 2022

SC0210

Spacecraft Bus

Carries the payload & delivers all of the payload's needs

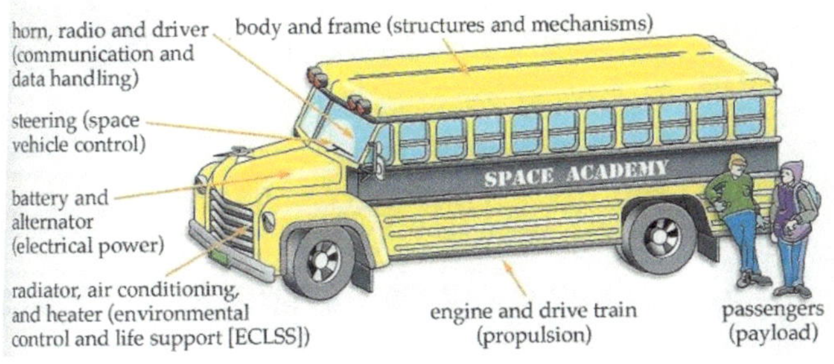

Courtesy Sellers et al

August 22, 2022

SC0211

Bus Example: *Pioneer*

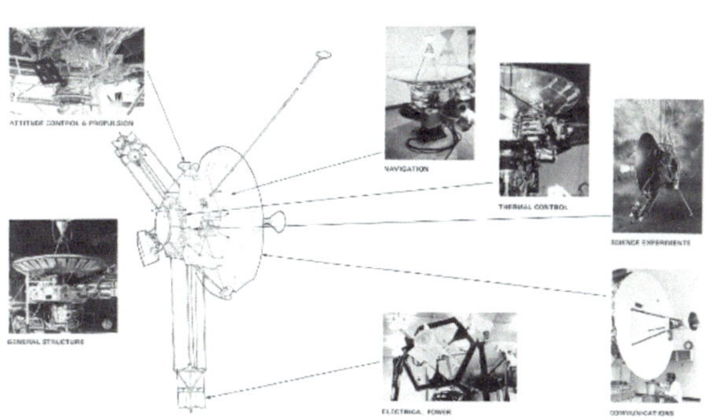

Courtesy JPL

August 22, 2022

SC0212

Bus Example: *Cassini*

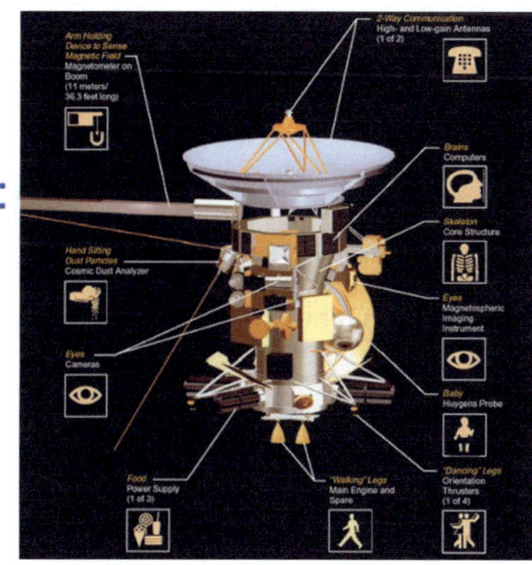

Courtesy JPL

August 22, 2022

Spacecraft Systems Topics

- Spacecraft project elements
- Spacecraft payload needs & requirements
- The spacecraft bus, examples
- **Spacecraft bus components**
- Non-spacecraft mission elements
- Major spacecraft design drivers
- Initial mass estimation
- Mass growth & contingencies; subsystem allocations
- Initial power estimation; subsystem allocations; margins
- Spacecraft volume considerations

August 22, 2022

SC0214

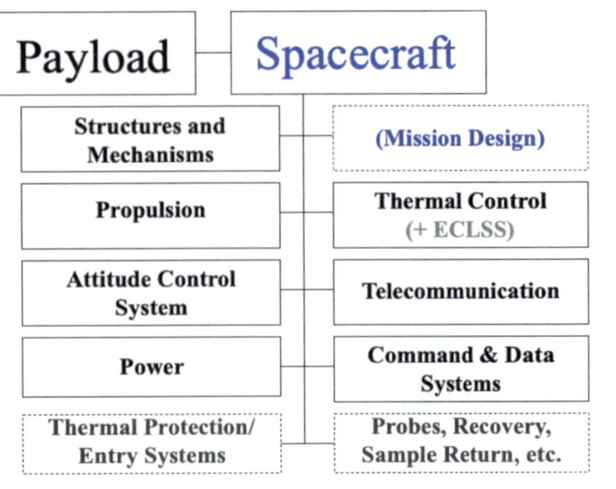

Structure Subsystem

- The "bus structure"
- Maintains general arrangement for launch and for during the mission
- Carries the payload and keeps everything oriented
- Provides all mechanisms to convert between different mission modes
- Mates spacecraft to the launch vehicle (LV) with a payload attach fitting (PAF) or launch vehicle adapter (LVA)
- Provides actuators or mechanisms to articulate antennas, solar panels, cameras, telescopes, etc.

Propulsion System

- Provides Δv needed to establish and maintain required orbit(s) and/or trajectory
- Must control thrust direction and duration
- Can also provide control of vehicle rotation about all axes

Attitude Control System (ACS)

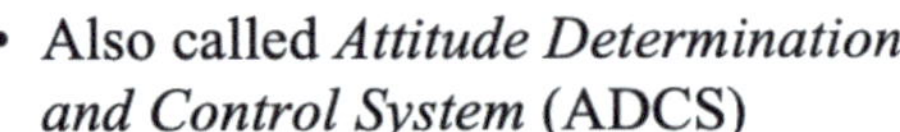
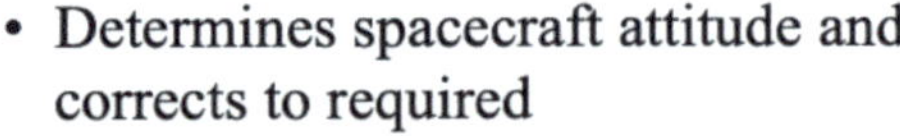

- Also called *Attitude Determination and Control System* (ADCS)
- Determines spacecraft attitude and corrects to required
- Implements and controls changes in velocity or attitude
- Controls articulation of appendages

Power System

- Manages electrical power for all equipment in all modes
  - Generates
  - Distributes
  - Regulates
  - Stores
- Switches equipment on and off as required by the command system
- Protects against faults

Thermal Control System

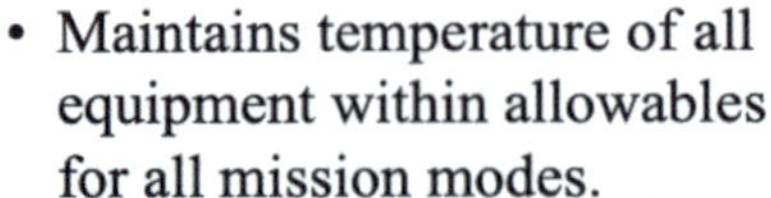

- Maintains temperature of all equipment within allowables for all mission modes.
- For inhabited spacecraft, this includes **ECLSS** (**E**nvironmental **C**ontrol and **L**ife **S**upport **S**ystem). Adds huge complications.

Command & Data Systems (CDS)

- Receives commands from the ground via telecomm system
- Decodes, stores, distributes, & initiates commands
- Collects, processes, formats, stores, & delivers data to the telecomm system

Telecommunications

- Receives commands from ground (uplink) & provides to the C&DS system
- Transmits science, payload, & engineering data to the ground (downlink)
- Receives & retransmits navigation and tracking signals

Mission Design

- Not really a subsystem — a critical part of the spacecraft design process done on ground.
- Develops mission trajectory & Δvs required
- Determines the launch vehicle and spacecraft propulsion system impulses required

Other *Possible* Subsystems

- Probe (separates to perform separate mission; really an independent spacecraft)
- Rover
- Thermal protection, entry systems
  - Heatshields, etc.
- Recovery system
  - Parachutes, retro rockets, airbags, etc.
- Sample return

All of these "custom" and very specific

Typical Bus Component 'Parts'

ACS
- Reaction wheels
- Torquer bars
- Nutation damper
- Star trackers
- Inertial reference unit
- Earth scanner
- Digital sun sensor
- Coarse sun sensor
- Magnetometer
- ACE electrical box

Communication
- S-band omni antenna
- S-band transponder
- X-band omni antenna
- X-band transmitter
- Parabolic dish reflector
- 2-axis gimbal
- Gimbal electronics
- Diplexers, RF switches
- Band reject filters
- Coaxial cable

Power
- Batteries
- Solar array panels
- Articulation mechanisms
- Articulation electronics
- Array diode box
- Shunt dissipaters
- Power Supply Elec.
- Battery a/c ducting

Mechanical
- Primary structure
- Deployment mechanisms
- Fittings, brackets, struts, equipment decks, cowling, hardware
- Payload Adapter Fitting

Thermal
- Radiators
- Louvers
- Heat pipes
- Blankets
- Heaters
- Heat straps
- Sun shield
- Cryogenic pumps
- Cryostats

Propulsion
- Propulsion tanks
- Pressurant tanks
- Thrusters
- Pressure sensors
- Filters
- Fill / drain valve
- Isolator valves
- Tubing

Electrical
- C&DH box
- Wire harness
- Instrument electronics
- Instrument harness

Courtesy R. Farley

Spacecraft Systems Topics

- Spacecraft project elements
- Spacecraft payload needs & requirements
- The spacecraft bus, examples
- Spacecraft bus components
- **Non-spacecraft mission elements**
- Major spacecraft design drivers
- Initial mass estimation
- Mass growth & contingencies; subsystem allocations
- Initial power estimation; subsystem allocations; margins
- Spacecraft volume considerations

Launch System

- Launch vehicle (LV) puts spacecraft into orbit and/or on escape trajectory
- LV dictates launch loads, mass, and geometry
- Added safety reviews for crewed vehicles
- Planning for integration with LV may begin 3-4 YEARS in advance.
- Best LV reference from manufacturer: *Payload Planner Guide* or *Payload User Guide* or *Mission Planner's Guide* or *Payload User's Guide*

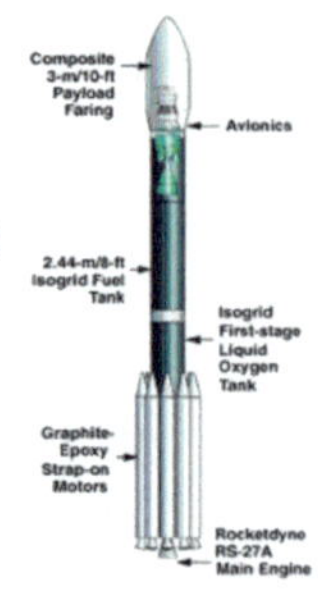

August 22, 2022

SC0227

Tracking & Data Systems

- Communications link to SC after launch
- Maintained by tracking and data stations
- Receive & relay SC downlink to mission ops
- Uplink commands from mission ops to SC
- *Tracking conflicts*: SC using the Deep Space Network (DSN) or other networks may be forced to "compete" for time

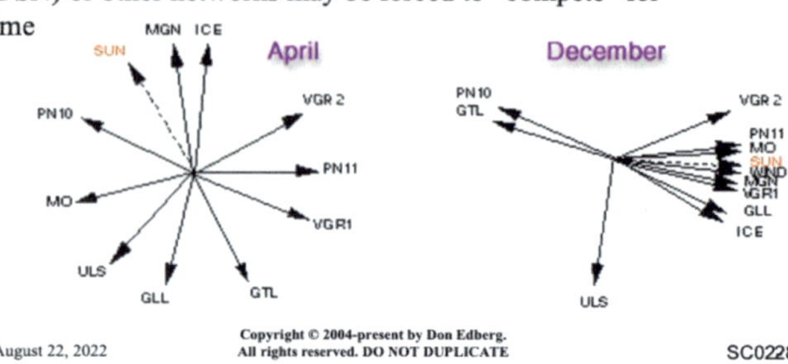

August 22, 2022

SC0228

Mission Operations

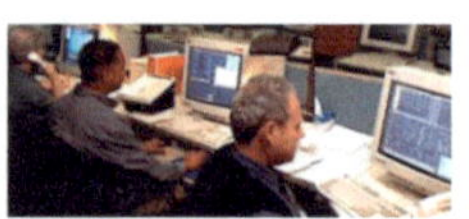

- Spacecraft control handed to the mission ops team after launch
- Monitors & analyzes spacecraft performance based on downlink
- Provides commands for uplink
- Deals with any anomalous behavior
  - i.e. Mars rover that had to be rebooted
- Hours or days of boredom punctuated by moments of sheer terror!

August 22, 2022

SC0229

Mission Ops Diagram

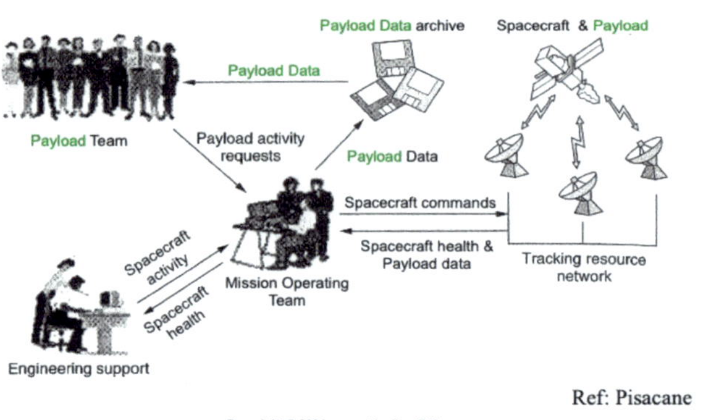

Ref: Pisacane

August 22, 2022

SC0230

Mission Ops (cont.)

- Critical to meet mission objective
- Run simulation ("sims") with intentional problems to test crew & reactions ("what if …?")
- Must also check and verify all commands to SC: use SC simulator to check the commands before issuing
- MUST NOT send bad commands, as there is no repairman to fix things. Many spacecraft have been lost by incorrect commands being sent
- Can consume up to 60% of the life-cycle cost!

August 22, 2022

SC0231

Spacecraft Systems Topics

- Spacecraft project elements
- Spacecraft payload needs & requirements
- The spacecraft bus, examples
- Spacecraft bus components
- Non-spacecraft mission elements
- **Major spacecraft design drivers**
- Initial mass estimation
- Mass growth & contingencies; subsystem allocations
- Initial power estimation; subsystem allocations; margins
- Spacecraft volume considerations

August 22, 2022

SC0232

The Three Major Spacecraft Design Drivers

- These three things pretty much dictate the spacecraft design
- All are determined by the launch vehicle

1. ***Mass***
2. ***Size***
3. ***Power***

Design Driver #1 is MASS

- Limited by the launch vehicle (LV)
- Cannot exceed the mass that the launch vehicle can send to the desired trajectory or available C3
- Must be carefully monitored
- Must keep current record as design proceeds
- ALWAYS increases during a project
- Design will include margins so that it can increase without overrunning the allowable mass

Mass (cont.)

- Keep track of center of mass and moments of inertia of each component
- Mass & inertia properties very important for stability and control of spacecraft
  - oblate vs. prolate spin characteristics
  - thrust line offset from CM location, etc.
- Keep a "mass properties" spreadsheet
  - Track item, mass, CM location, and $I = mr^2$ for each item in all three axes
  - Data used to provide mass properties to ACS

Mass Definitions

- Science or payload mass = mass of the science instruments plus any support equipment: mounting brackets, wiring, instrumentation, thermal control heaters, blankets, and radiators
  - **The spacecraft calls this the "payload."**
  - **DO NOT CONFUSE with the payload of the launch vehicle (the ENTIRE spacecraft)**
- Bus or platform mass = DRY (no propellants or gasses) mass of the spacecraft subsystems
  - Includes structure, propulsion, thermal, data, power, ACS, C&DS, telecom.

Mass Definitions (cont.)

- **On-Orbit Dry mass** = mass of the unfueled SC with payload, after separation from LV
- **Launch mass** = mass of the WET (fueled and gassed) spacecraft *including* its attachment hardware
- **Cruise mass** = Wet spacecraft **without** the LV attachment hardware
- **Mass margin or contingency** = difference between current launch mass estimate and the launch vehicle capability

Mass Definitions

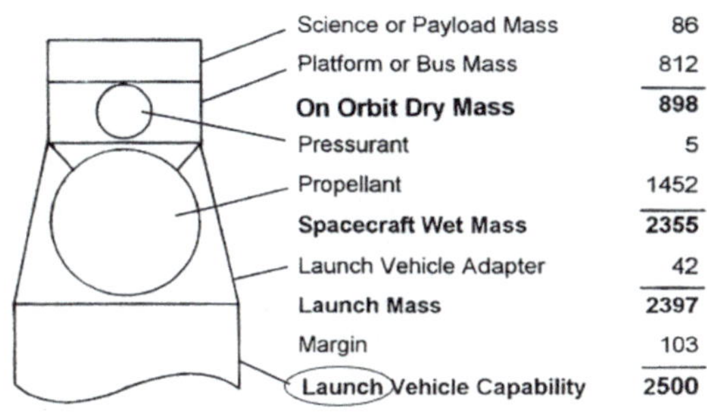

Spacecraft mass definitions.

Courtesy: Brown

SC Attachment to LV

- Name = "Payload Attach Fitting" (*PAF*) or "Launch Vehicle Adapter" (*LVA*)
- Supplied by LV company, but … mass allocated to *spacecraft* for launch
- Stays attached to LV's upper stage; reduced spacecraft mass after separation.
- Dimensional & mass data from LV manufacturer: *Payload Planner Guide* or *Payload User Guide* or *Mission Planner's Guide* or *Payload User's Guide* [preferred]
- ROM estimate using SC launch mass [worse]:

LVA mass (kg) = 0.0755 × SC *launch mass* + 50 kg (Brown eq. 2.3)

August 22, 2022 SC0239

Payload Attach Fitting

PAF or LVA mass = mass of the structure, separation devices, cabling, and any other equipment needed to mount the spacecraft to LV

August 22, 2022 SC0240

Delta IV PAFs (courtesy Boeing)

August 22, 2022 SC0241

Comments about SC Layouts

- On-orbit dry mass is used for orbiting SC. If the SC is going to other destinations, we should call it SC dry mass
- If the SC separates during a mission, EACH PART counts as a distinct spacecraft
- The separating part(s) may be considered 'payload' of the full SC, but this payload is an entire spacecraft, and should not be considered scientific or commercial payload (although it may carry such a payload in addition to its own bus)

August 22, 2022 SC0242

Spacecraft Systems Topics

- Spacecraft project elements
- Spacecraft payload needs & requirements
- The spacecraft bus, examples
- Spacecraft bus components
- Non-spacecraft mission elements
- Major spacecraft design drivers
- **Initial mass estimation**
- Mass growth & contingencies; subsystem allocations
- Initial power estimation; subsystem allocations; margins
- Spacecraft volume considerations

August 22, 2022 SC0243

Estimating SC Mass

- Preliminary estimates based on statistical analysis of existing spacecraft
- Relation is between m_{PL}, the payload mass, and OODM, the SC dry mass
- *Different types of spacecraft have different trends*
- Considerable scatter in curve fits, so 2-3 significant digits are all that make sense
- May not show improvements with technology and miniaturization (based on *existing* spacecraft)
- Estimates for preliminary design only — replace with detailed, "bottoms-up" estimates ASAP

August 22, 2022 SC0244

Comm. & Earth Orbit SC Mass

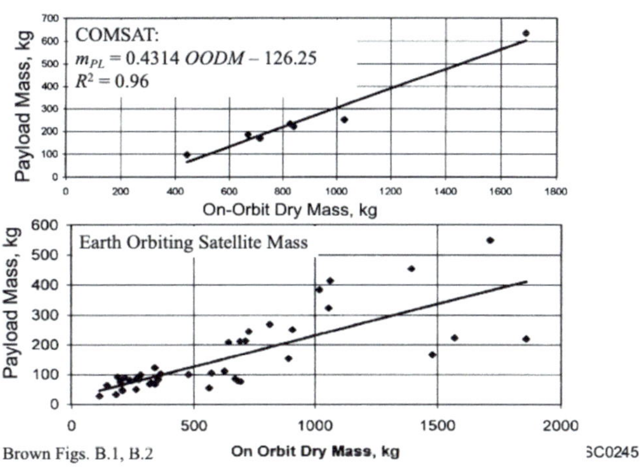

Brown Figs. B.1, B.2 SC0245

Payload vs Total Mass Planetary Spacecraft

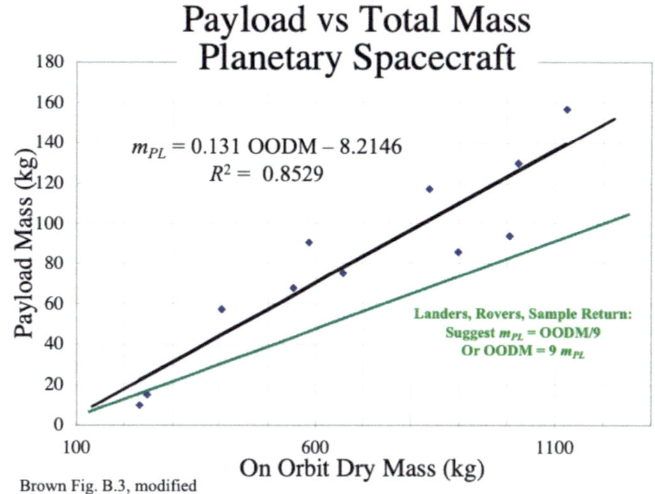

Brown Fig. B.3, modified

Mars Orbiter Mass Estimation Example: *LV Choice Unrestricted*

Goal: calculate maximum mass of *Mars Orbiter* SC

- SC has four instruments, total mass = 69.83 kg
- From plot, on-orbit dry mass OODM = 600 kg
- Assume propellant = 1,788 kg, pressurant = 8 kg
- Wet SC mass = 600 + 1,788 + 8 kg = 2,396 kg
- LV adapter = 0.0755 × 2,396 + 50 kg = 231 kg
- Launch mass = 2,396 + 231 kg = **2,627 kg** (this mass & required C3 determines possible LVs)
- Bus mass = OODM – PL = 600 – 70 = 530 kg

August 22, 2022 SC0247

Spacecraft Systems Topics

- Spacecraft project elements
- Spacecraft payload needs & requirements
- The spacecraft bus, examples
- Spacecraft bus components
- Non-spacecraft mission elements
- Major spacecraft design drivers
- Initial mass estimation
- **Mass growth & contingencies; subsystem allocations**
- Initial power estimation; subsystem allocations; margins
- Spacecraft volume considerations

August 22, 2022 SC0248

Mass Growth

- **MASS ALWAYS GROWS!** Always need to include margin
- Average growth rate of several programs from **ATP (authority to proceed)** to launch was **27%**!

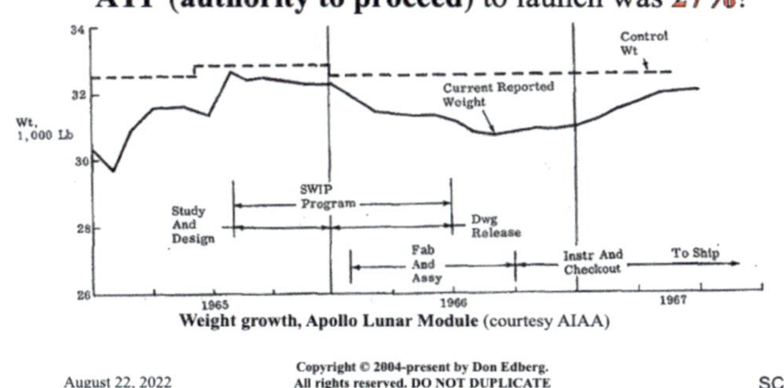

Weight growth, Apollo Lunar Module (courtesy AIAA)

August 22, 2022 SC0249

Space Vehicle Mass Growth: 39 Programs (Fall 2006 Aerospace *Crosslink*)

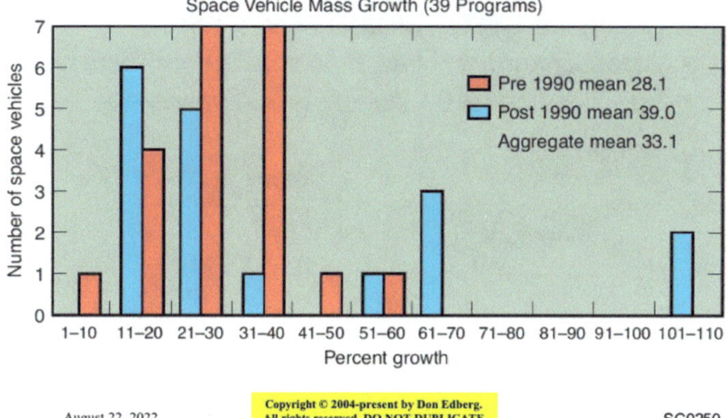

August 22, 2022 SC0250

Nine Reasons for Mass Change (1 of 2)

1. *Better design definition*: as design progresses, criteria & requirements are better defined. Optimistic proposal assumptions cannot be justified. Changes generally early in the program but prior to drawing release.
2. *Out-of-scope changes*: mass affected when customer adds/changes reqts beyond scope of original proposal.
3. *Redesign*: original design criteria may need change because of repackaging, component failure during testing, effects of other subsystem changes, etc.
4. *Maturing component design*: mass estimates are affected by drawing updates after original release resulting in improved mass analysis

Nine Reasons for Mass Change (2 of 2)

5. *Error in estimates*: errors in calculations for a previous estimate, or for added components that were overlooked.
6. *Uncontrolled vendor changes*: mass reported by a hardware supplier differs from contractual requirements
7. *Mass reduction activity*: mass changes due to official mass-reduction efforts
8. *Measured versus calculated:* mass estimates can reflect the differences between actual measured mass of components & latest calculated values
9. *Cost reduction*: mass changes occur to save money, e.g., by swapping out costly materials or by eliminating expensive machined parts

Mass Estimation Process

1. Determine max spacecraft launch mass (determined by selected LV)
2. Determine PAF mass, subtract from max launch mass
3. Determine required propellants and pressurants for the mission (usually found using required Δvs and rocket equation)
4. Determine total mission or on-orbit dry mass

(continues)

Mass Estimation (cont.)

5. Establish total allowable payload mass (may be specified by the customer)
6. Determine the recommended margin to be set aside, using the AIAA guidelines
7. Allocate mass budgets to each of the spacecraft subsystems
8. (If joined spacecraft are used, size the separated pieces first, then join them using support hardware and find propellant mass. See following slides.)

Dealing with SC that Separate

- Spacecraft C = an orbiter (O) + lander (L)
- During cruise, O transports L to target
- After separation, O & L operate independently
- So, O & L must be treated as *independent* SC

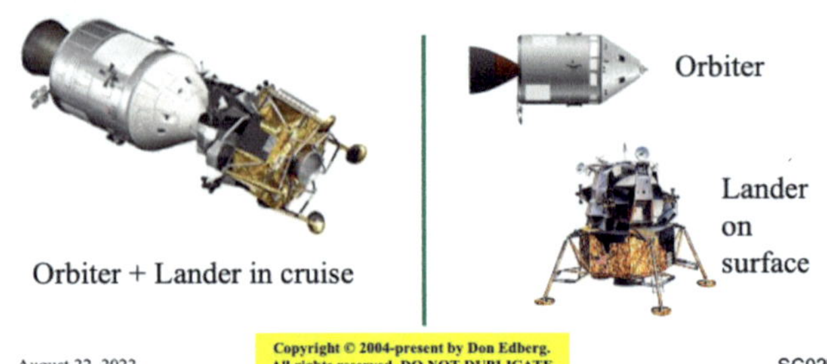
Orbiter + Lander in cruise

Orbiter

Lander on surface

Separating SC: Mass Estimating

1. Size Lander L: get L bus mass from MER; get L's propellant mass from L's Δv requirements
2. Size Orbiter O: O bus mass from MER; add $1.15 \times m_L$ to O dry mass; O's propellant mass from Δv requirements of [dry $O + 1.15 \times$ (wet m_L)]
3. Combined SC C = wet O + wet L

Spacecraft	Mass buildup	Notes
C Combined	$m_C = m_O + m_L$	O+L together after separating from LV
O (Orbiter)	$m_O = m_{PL\_O} + m_{dry\_O} + m_{prop\_O} + 0.15\, m_L$	m_{dry_O} from m_{PL_O}, $m_{propellant\_O}$ from Δv reqts. Add 15% of 'passenger' L's mass to attach L
L (Lander)	$m_L = m_{PL\_L} + m_{dry\_L} + m_{prop\_L}$	m_{dry_L} from m_{PL_L}, m_{prop_L} from Δv reqts.

How To Deal With Mass Growth

- *Remove* a margin, or "fudge factor" from mass estimates to allow for mass growth while still staying under estimates
- Estimate mass margins using AIAA recommendations
- Note: recommended margins categorized by
  - Spacecraft mass
  - Development stage / maturity

August 22, 2022 SC0257

Mass Contingencies, % (AIAA Recs.)

1 = New, one of a kind spacecraft
2 = Next-generation based on previously developed family
3 = Multi-unit development based on existing design

Minimum standard weight contingencies, %

Description/ categories	Proposal stage: Bid Class 1	2	3	Design development stage: CoDR Class 1	2	3	PDR Class 1	2	3	CDR Class 1	2	3
Category AW, 0–50 kg 0–110 lb	50	30	4	35	25	3	25	20	2	15	12	1
Category BW, 50–500 kg 110–1102 lb	35	25	4	30	20	3	20	15	2	10	10	1
Category CW, 500–2500 kg 1102–5511 lb	30	20	2	25	15	1	20	10	0.8	10	5	0.5
Category DW, 2500 kg and up	28	18	1	22	12	0.8	15	10	0.6	10	5	0.5

[a]Data copyright AIAA, reproduced with permission; Ref. 11.

SC0258

Subsystem Mass Allocation – Depends on SC Type (Brown)

Table 2.6 Subsystem on-orbit dry mass allocation guide

GFE = Gov't Furnished Eqpt.

Subsystem	Comsats[a] with P/L[c]	Comsats GFE P/L	Metsats[b] with P/L	Metsats GFE P/L	Planetary with P/L	Planetary GFE P/L	Other with P/L	Other GFE P/L
Structure, %	21	29	20	29	26	29	21	30
Thermal, %	4	6	3	4	3	3	3	4
ACS, %	7	10	9	13	9	10	8	11
Power, %	26	35	16	23	19	21	21	29
Cabling, %	3	4	8	12	7	8	5	7
Propulsion, %	7	10	5	7	13	15	5	7
Telecom, %	—	—	4	6	6	7	4	6
CDS, %	4	6	4	6	6	7	4	6
Payload, %	28	—	31	—	11	—	29	—

[a]Comsat = communication satellite. [b]Metsat = meteorology or weather satellite. [c]P/L = payload.

Mass Margin: *Mars Orbiter* Example

- Mass contingency table: Cat. CW, Bid Class 1 = 30%
- Allocation for bus = bus mass ÷ (1 + contingency) = 1070 ÷ 1.3 = 823 kg
- Mass margin = bus mass – allocation = 247 kg
- **823 kg** of bus mass allocated per Brown Table 2.6, column 6
- **247 kg** *reserved* (left unassigned) for future mass growth

Subsystem	Budget %	Budget kg
Structure	29%	239
Thermal	3%	25
ACS	10%	82
Power	21%	173
Cabling	8%	66
Propulsion	15%	123
Telecom	7%	58
CDS	7%	58
Payload	0%	0
Total		823

August 22, 2022 SC0260

Project's Mass Recordkeeping

- "*Mass Maturity*" keeps track of the origin of the recorded value – how the mass is calculated
- Three categories & their accuracy:
  - Estimated = worst accuracy
    - Based on historical statistics, or similar vehicles
    - "Better than nothing," but replace ASAP
  - Calculated = better accuracy
    - Uses drawing dimensions & engineering materials
  - Actual = best accuracy
    - Obtained from measurement of actual hardware

August 22, 2022 SC0261

Spacecraft Systems Topics

- Spacecraft project elements
- Spacecraft payload needs & requirements
- The spacecraft bus, examples
- Spacecraft bus components
- Non-spacecraft mission elements
- Major spacecraft design drivers
- Initial mass estimation
- Mass growth & contingencies; subsystem allocations
- **Initial power estimation; subsystem allocations; margins**
- Spacecraft volume considerations

August 22, 2022 SC0262

Second Payload Driver: Power

- Start with the payload requirements
- Total spacecraft power is a function of the payload power
- Use statistical methods to estimate power based on mission type
- Add contingencies to allow for growth
- Replace early estimates with better ones found as the design matures.

Estimating Power Needed

- Consider power utilization during mission
- Example: "replacement heaters" used to protect turned-off instrument. Turned OFF if instrument ON
  - Don't add replacement power to the maximum power, or you're double-booking
- "Bake-out" heaters bake out volatiles so they don't land on optics/solar arrays and reduce performance
  - These are typically on and later shut off long before the instruments are turned on
- Neither affect total power

Power Estimating Relationships

- SV total power statistically a function of *payload* power in watts.

Spacecraft Mission	Total Power Estimate (W)
Communications	$P_t = 1.1568\ P_{pl} + 55.497$
Meteorology	$P_t = 602.18 \ln(P_{pl}) - 2{,}761.4$
Planetary	$P_t = 332.93 \ln(P_{pl}) - 1{,}046.6$
Other missions	$P_t = 1.3\ P_{pl} + 210$

Ref: Brown T2.9

Power Allocation Process

Recommended power allocations from statistical studies of many conventional spacecraft

Subsystem	Comsats	Metsats	Planetary	Other
Thermal Control	30%	48%	28%	33%
Attitude Control	28	19	20	11
Power	16	5	10	2
Command & Data	19	13	17	15
Communication	0 (incl. in C&DS)	15	23	30
Propulsion	7	0	1	4
Mechanisms	0	0	1	5

(From Brown)

Power Margins

- Margin = total capability – **current best estimate (CBE)**

$$\% \text{ Margin} = \frac{\text{Margin}}{\text{Total Capability}} \times 100\%$$

"Total Capability" = max system power output

- Can vary depending on distance from the sun (solar-powered missions)
- Categorized by
  - Magnitude of the power
  - Development stage

  (similar to the mass contingency table earlier)

Recommended Power Margins

Table 2.11 AIAA recommended power contingencies[a] (From Brown)

1 = New, one of a kind spacecraft
2 = Next-generation based on previously developed family
3 = Multi-unit development based on existing design

	Minimum standard power contingencies, %											
	Proposal stage			Design development stage								
	Bid Class			*CoDR* Class			*PDR* Class			*CDR* Class		
Description/ categories	1	2	3	1	2	3	1	2	3	1	2	3
Category AP, 0–500 W	90	40	13	75	25	12	45	20	9	20	15	7
Category BP, 500–1500 W	80	35	13	65	22	12	40	15	9	15	10	7
Category CP, 1500–5000 W	70	30	13	60	20	12	30	15	9	15	10	7
Category DP, 5000 W and up	40	25	13	35	20	11	20	15	9	10	7	7

[a]Copyright AIAA, data reproduced with permission; Ref. 11.

Example Power Calculation

Goal: calculate power for *Mars Orbiter* SC

- SC has four instruments as given in SRD.
- Total power = 9.3 + 12.5 + 17.6 + 33.1 W = 73 W
- Table: $P_t = 332.93 \ln(P_{pl}) - 1{,}046.6 = 380$ W
- Subsystem allocation = 380 – 73 = 307 W
- Proposal, class 1, new: 90% contingency on subsystems = 307 × 0.90 = 276 W
- Total power = 73 + 307 + 276 W = **656 W**

August 22, 2022 SC0269

Spacecraft Systems Topics

- Spacecraft project elements
- Spacecraft payload needs & requirements
- The spacecraft bus, examples
- Spacecraft bus components
- Non-spacecraft mission elements
- Major spacecraft design drivers
- Initial mass estimation
- Mass growth & contingencies; subsystem allocations
- Initial power estimation; subsystem allocations; margins
- **Spacecraft volume considerations**

August 22, 2022 SC0270

3rd Major Driver: Payload Volume

- Spacecraft must fit inside the **payload fairing (PLF)** of selected launch vehicle
- PLF provides clean environment & protects SC from heating during ascent
- PLF dimensional information may be obtained in the ***Payload Planner's Guide*** or equivalent for the launch vehicle you choose, or from Isakowitz book
- Must reduce the envelope based on dynamic loads and vibrations ("rattle space")

August 22, 2022 SC0271

3-View Layout, Launch Config.

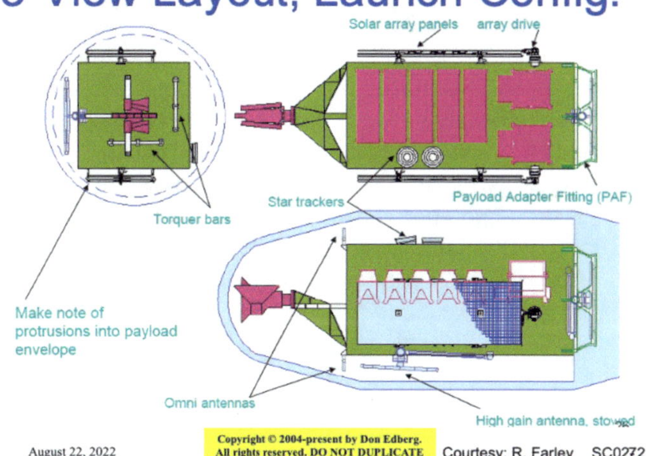

August 22, 2022 Courtesy: R. Farley SC0272

Spacecraft In Payload Fairing (PLF)

Left: *Mars Global Surveyor* spacecraft being encapsulated into PLF (courtesy NASA). Below: *Glory* spacecraft in Taurus PLF (courtesy OSC).

August 22, 2022 SC0273

Payload Fairing Dimensions

Hashed area = "keep-out zone"
(courtesy Boeing)

August 22, 2022 SC0274

Payload Fairing Internal Dimensions from Payload Planners' Guides

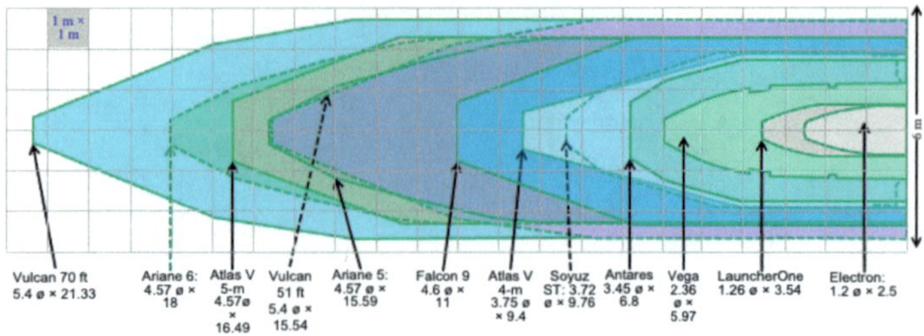

All dimensions in meters unless specified

8/22/22 SC02-75

Dual-Manifest Spacecraft

- Two SC, one LV
- Must have compatible missions
- SC #2 depends on SC #1 deployment

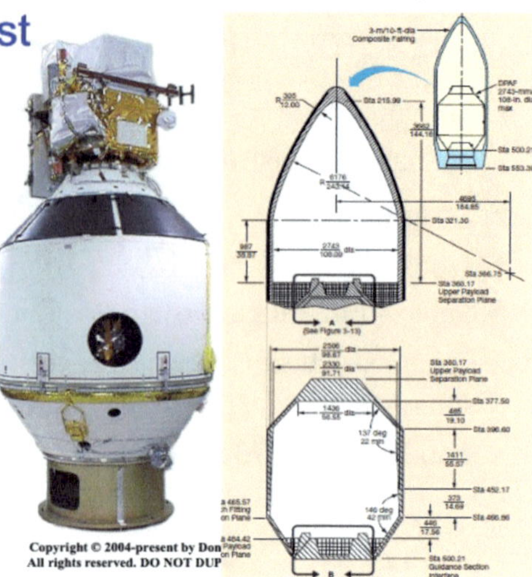

Courtesy Boeing/Delta II PPG

August 22, 2022

Arianespace Faces Issues With Stacking Heavier Payloads On *Ariane 5*

"Europe's heavy-lift *Ariane 5* rocket will be unable to launch any commercial missions for the next six months because of worsening payload compatibility issues." Designers hope the vehicle's planned upgrade will address the issues, but the *Ariane 5* Midlife Evolution … "will not make its first flight before 2016 at the earliest …" This means Arianespace "will have to cope with the gradually increasing weight of the average telecommunications satellite by matching payloads as best it can while staying within *Ariane 5*'s current limits.
Space News (9/24/2011)

August 22, 2022 SC0277

Typical Component Densities (kg/m^3)

- Once masses are known, the density values below may be used to estimate each component or system's volume
- Once volumes are known, approximate shapes can be assumed, and a rough layout can be made
- More layout considerations may be found in Structures section

Item	Density (kg/m^3)
"Average" electronic box	320
Cameras	500
Small instruments	1,000
Spectrometers	250
Mass spectrometers	800
Synthetic Aperture Radar	32
Rain radars	150

Courtesy Farley

August 22, 2022 SC0278

Spacecraft Propulsion Systems

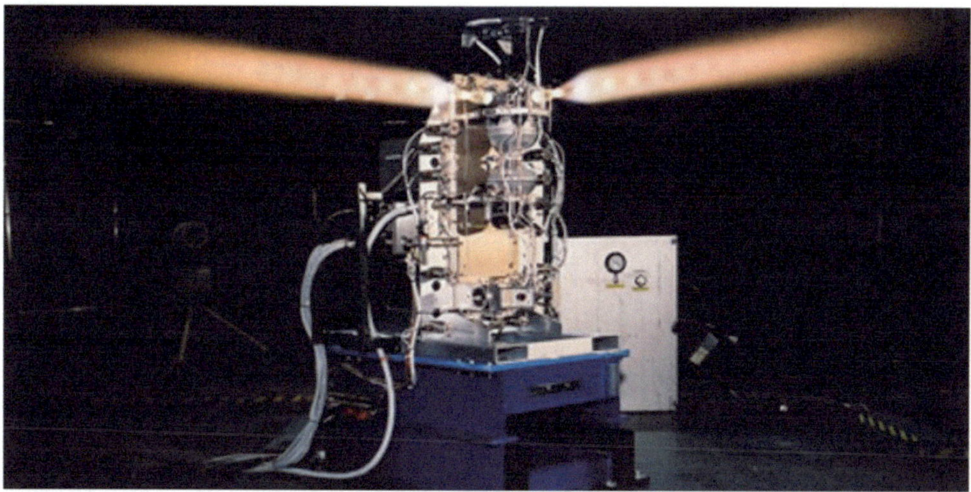

Source: AFRL

22 August 2022

(89)

Propulsion Topics

- Need for propulsion & definitions
- Types of space propulsion systems
  - Solid propellants
  - Liquid: mono- vs. bi-propellant, dual-mode systems
  - Liquid engines & systems
  - Liquid propellant management
  - Electric & nuclear propulsion
- Propulsion system design considerations
- Solid system mass calculations
- Liquid system mass calculations

22 August 2022

Spacecraft Design Process

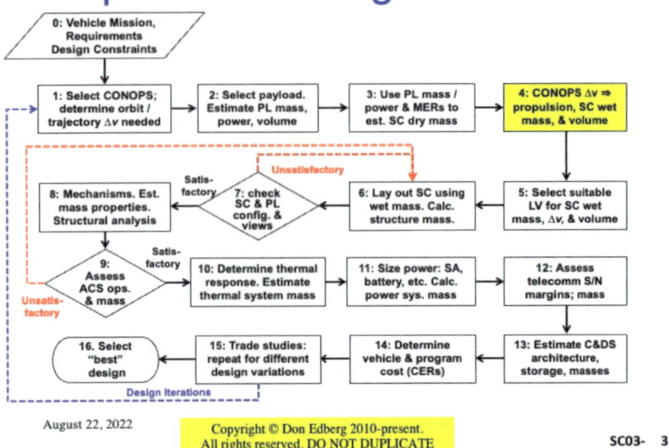
0: Vehicle Mission, Requirements Design Constraints
1: Select CONOPS; determine orbit / trajectory Δv needed
2: Select payload. Estimate PL mass, power, volume
3: Use PL mass / power & MERs to est. SC dry mass
4: CONOPS Δv ⇒ propulsion, SC wet mass, & volume
5: Select suitable LV for SC wet mass, Δv, & volume
6: Lay out SC using wet mass. Calc. structure mass.
7: check SC & PL config. & views
Unsatisfactory
Satis-factory
8: Mechanisms. Est. mass properties. Structural analysis
9: Assess ACS ops. & mass
Satis-factory
Unsatis-factory
10: Determine thermal response. Estimate thermal system mass
11: Size power: SA, battery, etc. Calc. power sys. mass
12: Assess telecomm S/N margins; mass
13: Estimate C&DS architecture, storage, masses
14: Determine vehicle & program cost (CERs)
15: Trade studies: repeat for different design variations
16. Select "best" design
Design Iterations

Propulsion Topics

- Need for propulsion & definitions
- Types of space propulsion systems
  - Solid propellants
  - Liquid: mono- vs. bi-propellant, dual-mode systems
  - Liquid engines & systems
  - Liquid propellant management
  - Electric & nuclear propulsion
- Propulsion system design considerations
- Solid system mass calculations
- Liquid system mass calculations

Propulsion Needs

- Orbital changes:
  - Shaping
  - Transfer (to GEO or other)
  - Plane changes
  - Escapes
- Operations
  - Spin-up & Spin-down
  - Orbit corrections & stationkeeping
  - Drag management
  - Deliberate de-orbiting
- Attitude control in space
  - Pointing & Stabilization
  - Reaction wheel desaturation

Propulsion Applications

System → / Application	High thrust liquid engine w/ turbopump	Medium-to-low thrust liquid engine	Multiple pulsing liquid thrusters	Cold gas	Large solid rocket motor (segmented)	Medium-to-small solid motor	Electric: ion, Hall, arcjet, plasma
Launch vehicle booster	XX				XX		
Strap-on motor/engine	XX				XX		
Orbit injection & orbit transfers		XX				XX	X
Speed adjustments; flight path corrections; orbit raising		X	XX				X
Orbit / orbital position maintenance; maneuvering			XX	X			XX
Spacecraft docking			XX				
De-orbit		X	X			X	X
Entry, landing, emergency maneuvers		X	X				

Legend: X = in use. XX = preferred for use. Based on: Sutton

Key Terms

- **Thrust**, T: force generated by a rocket
  - Measured in pounds (lb_f) or newtons (N)

  Thrust depends on mass flow rate $\dot{m}$, exit velocity v_e, nozzle exhaust area A_e, exhaust pressure p_e, and ambient pressure p_∞:

$$T = \dot{m} v_e + A_e \left(p_e - p_\infty \right)$$

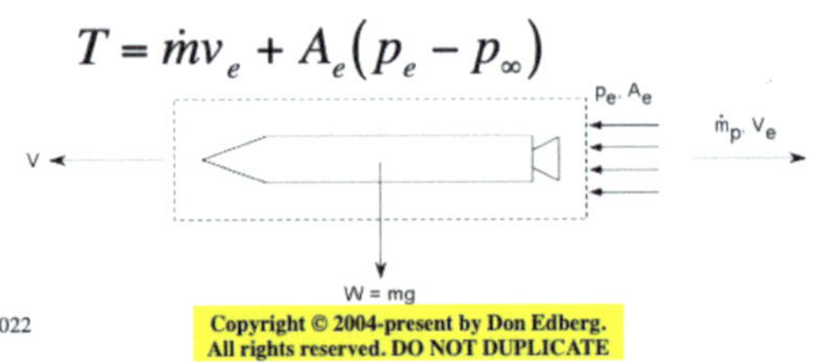
pe, Ae
ṁp, Ve
V
W = mg

More Key Terms

Effective Exhaust Velocity (v_e or c): affected by

- Ratio of specific heats (γ)
- Exhaust pressure (p_e)
- Chamber pressure (p_0)
- Exhaust molecular weight ($\mathcal{M}$: $\mathcal{M}_{H_2O} = 10$)
- Note: *lower molecular weight $\mathcal{M}$* = higher velocity = higher efficiency
- Chamber temperature (T_0)

  $\Re$ = Universal gas constant 8314 J/(mol K)

$$v_e = \sqrt{\frac{2\gamma}{\gamma - 1} \frac{\Re T_0}{\mathcal{M}} \left[1 - \left(\frac{p_e}{p_0} \right)^{\frac{\gamma - 1}{\gamma}} \right]}$$

Definition: Specific Impulse

- Specific Impulse is the *amount of thrust* produced by *one unit* of propellant *per unit time*

I_{sp} = Thrust ÷ (Propellant flow per second)

- Measure of rocket motor "efficiency": ***higher I_{sp} is better***

Units may cause problems:

English	SI ('metric')
Weight units (lb_f/s)	Mass units (kg/s)
$I_{sp\text{-}English} = lb_f/(lb_f/s) = s$	$I_{sp\text{-}SI} = N/(kg/s) = m/s$

22 August 2022

SC03- 9

On Specific Impulse Units

- SI system: $\frac{N}{kg/s} = m/s$

  Note that the specific impulse in SI units is the exit velocity in m/s
- English system: $\frac{lb_f}{lb_f/s} = s$

  If English system used slugs/s for mass flow, the specific impulse would be exit velocity in ft/s
- Be careful in using specific impulses with different systems of units

$$I_{sp-English} = \frac{T}{\dot{w}_p} = \frac{T}{g_0 \dot{m}_p} = \frac{c}{g_0} = \frac{v_e}{g_0} \qquad I_{sp-SI} = \frac{T}{\dot{m}_p} = c = v_e$$

22 August 2022

SC03- 10

Typical Specific Impulse Values

Engine Type	Specific Impulse (s)
Chemical (Liquid, Solid, Hybrid)	150 – 450
Nuclear Thermal	825 – 925
Arcjet (Electrothermal)	800 – 1,200
MPD (Electromagnetic)	2,000 – 5,000
Ion, Hall (Electrostatic)	3,500 – 10,000

22 August 2022

SC03- 11

More Terms

- **Mass Ratio *MR*** = initial mass ÷ final or burnout or inert mass: $MR = \frac{m_{initial}}{m_{final}} = \frac{m_0}{m_f}$
  - *MR* is an indication of rocket performance: less empty mass and more propellant are good
- **Propellant Mass Fraction**: percentage of vehicle (or motor) that is propellant
- **Outage**: amount of one propellant left when other is depleted (use 1% for initial guess)
- **Total Impulse, *I***

  $I = \int_0^{t_f} T(t)dt$ (units: $lb_f \cdot s$ or N·s)

22 August 2022

SC03- 12

Still More Terms

- **Mixture Ratio, *r***
  - Mixture ratio (for bipropellant systems) determines T_c, $\mathcal{M}$, and volume ratios of individual propellants – bulk density)

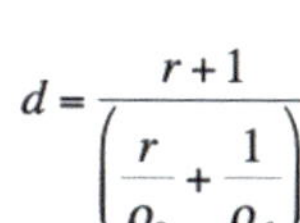

$$r = \frac{\dot{m}_{oxidizer}}{\dot{m}_{fuel}}$$

- **Bulk Density, *d***
  - Measure of the bulk or "average" density of a bipropellant combination taking *r* and individual densities into account

$$d = \frac{r+1}{\left(\frac{r}{\rho_0} + \frac{1}{\rho_f}\right)}$$

- **Ullage**
  - Empty volume on top of tank to allow for pressurant (use ~ 30%)

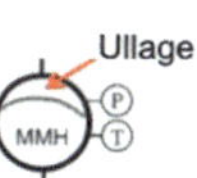

22 August 2022

SC03- 13

Basic Propulsion Calculations

- Propellant(s) required
- Thrust
- Weight (or thrust-to-weight ratio, T/W)
- Exhaust velocity c or c^* or v_e
- Propellant efficiency (specific impulse, I_{sp}). Units are force/(mass/time), not "seconds"
- Higher I_{sp} better

$$I_{sp} = \frac{T}{\dot{w}_p} = \frac{T}{g_0 \dot{m}_p} = \frac{c}{g_0} = \frac{v_e}{g_0},\ g_0 = \text{ sea level standard gravity}$$

22 August 2022

SC03- 14

Propellant Calculations

- Calculate propellant mass from required Δv:
$$m_p = m_i\left(1-e^{-\Delta v/g_0 I_{sp}}\right) = m_f\left(e^{\Delta v/g_0 I_{sp}}-1\right)$$
$$= m_i\left(1-e^{-\Delta v/v_e}\right) = m_f\left(e^{\Delta v/v_e}-1\right)$$
- m_0 = initial mass, m_f = final mass, $m_p = m_0 - m_f$, given I_{sp} or v_e = exit velocity (either determined by propellants selected)
- Can also calculate from required burn time t_b:
$$m_p = \frac{t_b T}{g_0 I_{sp}} = \frac{t_b T}{v_e}$$ where T = thrust.
  - Useful to calculate thruster propellant based on a total number of firings.

22 August 2022

SC03- 15

Propellant Calculations

- Calculate propellant mass m_p from required Δv:
$$m_p = m_f\left(e^{\Delta v/g_0 I_{sp}}-1\right) = m_f\left(e^{\Delta v/v_e}-1\right)$$
$$= m_0\left(1-e^{-\Delta v/g_0 I_{sp}}\right) = m_0\left(1-e^{-\Delta v/v_e}\right)$$

  Here m_0 = initial mass, m_f = final mass, $m_p = m_0 - m_f$, I_{sp} or v_e = exit velocity determined by propellants
- Mass flow rate $\dot{m} = \frac{T}{g_0 I_{sp}}$ where T = thrust
- Calculate $m_p = \frac{t_b T}{g_0 I_{sp}} = \frac{t_b T}{v_e}$ from required burn time t_b

Useful to calculate thruster propellant based on a total number of firings

22 August 2022

SC03- 16

Necessary Propulsion Ingredients

- Rockets operate in space
- Must carry onboard fuel and oxidizer
- Or only fuel if monopropellant
- Or reaction mass if not chemical (electric power needed)

22 August 2022

SC03- 17

Types of Rocket Propulsion

Use for launch vehicles: Hybrid, Solid propellant, Liquid propellant

Use for spacecraft: Solid propellant through Exotic

- Hybrid
- **Solid propellant**
- **Liquid propellant**
  - **Biprop**
  - **Monopropellant**
  - **Dual-mode systems**
- **Cold-gas**
- **Electric Ion /Arc**
- **Nuclear**
- **Exotic**

22 August 2022

SC03- 18

Propulsion Topics

- Need for propulsion & definitions
- **Types of space propulsion systems**
  - **Solid propellants**
  - Liquid: mono- vs. bi-propellant, dual-mode systems
  - Liquid engines & systems
  - Liquid propellant management
  - Electric & nuclear propulsion
- Propulsion system design considerations
- Solid system mass calculations
- Liquid system mass calculations

22 August 2022

SC03- 19

Solid Propellant Motors

- Propellant is a "rubbery" combination of fuel, oxidizer, and binder (i.e. aluminum, ammonium perchlorate, PBAN binder). Gunpowder for model rockets.
- Examples: *IUS*, *PAM-D*, *Star*

☺ Inexpensive
☺ Storable
☺ High T/W_e
☺ Very simple, lightweight
☹ Lower I_{sp} than liquid: I_{sp} ~ 250-300 s
☹ Can't be throttled or shut off easily
☹ Need Thrust Vector Control (TVC) or spin

22 August 2022

SC03- 20

Some Solid Kick Motor Specs

Motor	Total Impulse (Ns)	Loaded Mass (kg)	Prop. Mass Fraction	Average Thrust (N)	Specific Impulse (s)
STAR 13B	1.16E5	47	0.88	7015	285.7
STAR 30BP	1.46E6	543	0.94	26511	292.0
STAR 30C	1.65E6	628	0.95	31760	284.6
STAR 30E	1.78E6	667	0.94	35185	289.3
STAR 37F	3.02E6	1149	0.94	44086	291.0
STAR 48A	6.78E6	2559	0.95	79623	283.9
IUS SRM-2	8.11E6	2995	0.91	80157	303.8
LEASAT PKM	9.26E6	3658	0.91	157356	285.4
IUS SRM-1	2.81E7	10374	0.94	198435	295.5

Courtesy SMAD

22 August 2022

SC03- 21

Propulsion Topics

- Need for propulsion & definitions
- Types of space propulsion systems
  - Solid propellants
  - Liquid: mono- vs. bi-propellant, dual-mode systems
  - Liquid engines & systems
  - Liquid propellant management
  - Electric & nuclear propulsion
- Propulsion system design considerations
- Solid system mass calculations
- Liquid system mass calculations

22 August 2022

SC03- 22

Spacecraft Liquid-Propellant Propulsion Systems

Types:

- Monopropellant
  - Note: *"green" propellant chemistry available*!! HPGP = High performance green propulsion propellant, AF-M315E
- Bipropellant
- Dual-mode systems (combination of mono- and bi-propellant)

22 August 2022

SC03- 23

Monopropellant Systems

- Use a single propellant ("mono"); no oxidizer needed
- Propellant decomposes into hot gases upon contact with a catalyst
- No ignition system needed; simple & reliable
- Lower I_{SP} (~235) than bipropellants (~328)
- Must heat catalyst before burn for best performance
- Example: *Voyager* 445 N engine (**sample**)
- Propellant very poisonous

22 August 2022

SC03- 24

Hypergolic Propellant video

hypergolic demo de Titan II.mp4

22 August 2022

SC03- 25

Hypergolic Propellant video 2

hypergolic.mpg

22 August 2022

SC03- 26

Hypergolics Are NASTY!

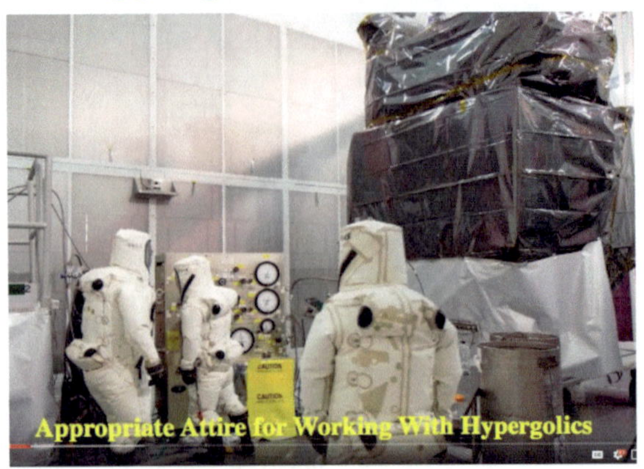

Hypergolics Are NASTY!

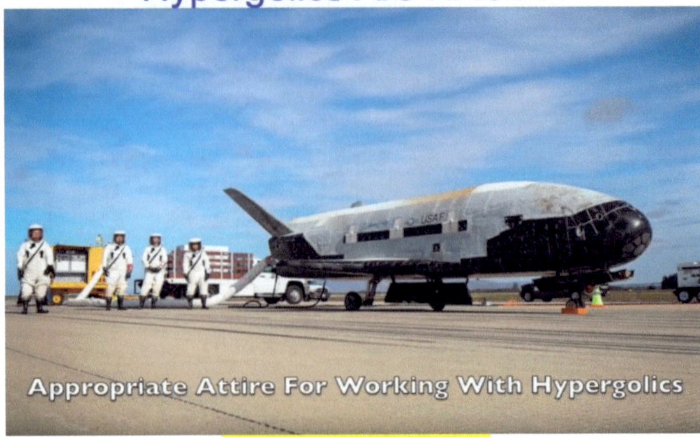

Monoprop Propellant Options

- Hydrazine (or monomethyl hydrazine, MMH)
  $3\ N_2H_4 \rightarrow 4(1-x)NH_3 + (2x+1)N_2 + (6x)H_2 +$ **heat**
- New: LMP-103S reduced toxicity 'green' propellant (ECAPS AB, Swedish Space Corp.)
  $NH_4N(NO_2)_2 + H_2O + CH_3OH + NH_3$
  $\rightarrow H_2O + N_2 + H_2 + CO_2 +$ **heat** (no NH_3 out)
- Hydrogen Peroxide: $2\ H_2O_2 \rightarrow 2\ H_2O + O_2 +$ **heat**
- In all above cases, released heat creates high pressure to exhaust gases at high speed

Comparison of Hydrazine vs. "Green" LMP-103S

Item	Hydrazine	LMP-103S
Air transport approval	No	Yes
Performance	Nominal I_{sp}	6 – 12% better I_{sp}
Sensitive to air & moisture	Yes	No
Toxic, carcinogenic, corrosive	Yes	No
Density	1.01	1.24
Loading: toxic & Propellant waste	470 & 29 kg	3 kg non-toxic & 1 kg
Loading personnel	25	3

Reference: Moog HPGP Presentation

Propulsion Topics

- Need for propulsion & definitions
- Types of space propulsion systems
  - Solid propellants
  - Liquid: mono- vs. bi-propellant, dual-mode systems
  - Liquid engines & systems
  - Liquid propellant management
  - Electric & nuclear propulsion
- Propulsion system design considerations
- Solid system mass calculations
- Liquid system mass calculations

Monopropellant Engine

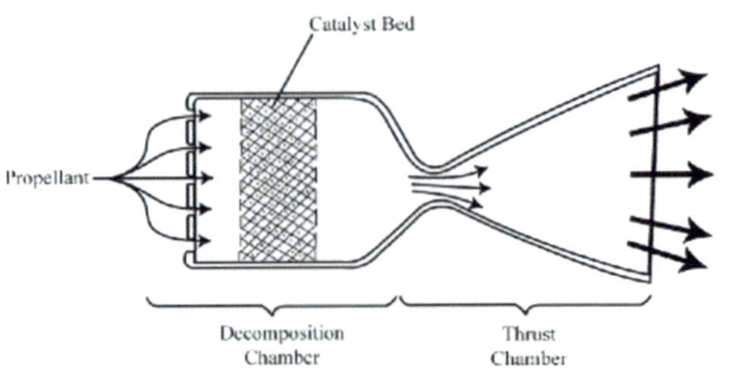

Courtesy Hamilton-Standard

Monopropellant Engine (Internals)

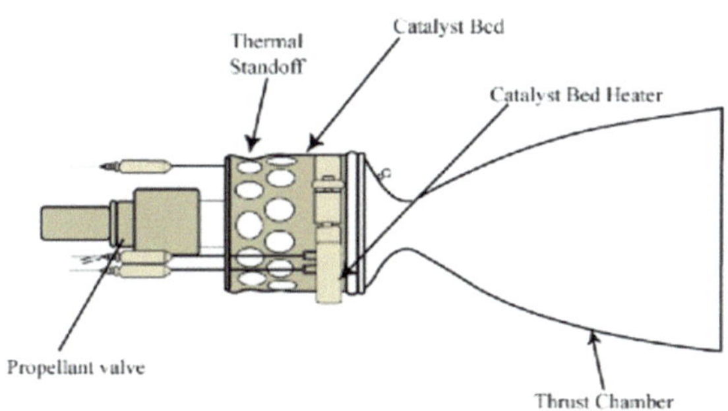

Typical monopropellant thruster. (Courtesy Hamilton Standard.)

Thruster Catalyst Bed

D. Edberg photo

MR-104A/C 440 N (100-lbf) ROCKET ENGINE ASSEMBLY

P/N 28875-305

ICD PS-2598867

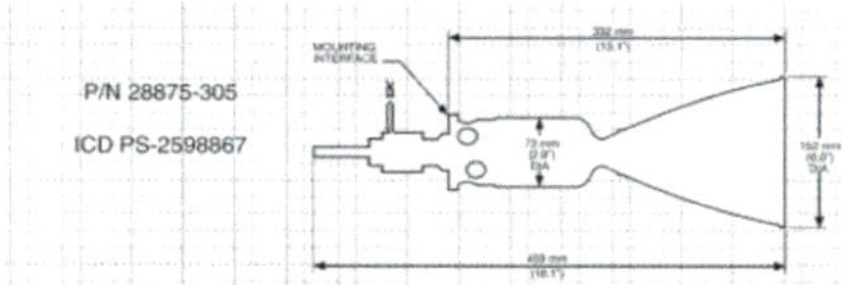

Design Characteristics

- Propellant .... Hydrazine
- Catalyst .... S 405/LCH-202
- Thrust/Steady State .... 572.5 – 204.6 N (128.7 – 46 lbf)
- Feed Pressure .... 28.9 – 6.9 bar (420 – 100 psia)
- Chamber Pressure .... 10.7 – 3.9 bar (– 56 psia)
- Expansion Ratio 53:1
- Flow Rate .... 240.4 – 90.72 g/sec (0.53 – 0.20 lbm/sec)
- Valve .... Single Seat
- Valve Power .... 30 Watts @ 28 Vdc & 21°C
- Cat. Bed Heater Power .... 13.1 Watts @ 28 Vdc & 21°C
- Mass .... 1.86 kg (4.11 lbm)
- Engine .... 1.44 kg (3.17 lbm)
- Valve .... 0.43 kg (0.94 lbm)

Performance

- Specific Impulse .... 239-223 sec (lbf-sec/lbm)
- Total Impulse .... 693,900 N-sec (156,000 lbf-sec)
- Total Pulses 1,742
- Minimum Impulse Bit .... 8.23 N-sec @ 24.13 bar & 22 ms ON .... (1.85 lbf-sec @ 350 psia & 22 ms ON)
- Steady-State Firing .... 2,000 sec – Single Firing .... 2,654 sec – Cumulative

Rev. Date: 4/02/03

11411 139th PL NE • P.O. BOX 97009 • REDMOND, WA 98073-9709
(425) 885-5000 FAX (425) 882-5747

Approved for public release and export

AEROJET

Monopropellant RCS (Reaction Control System) Engine

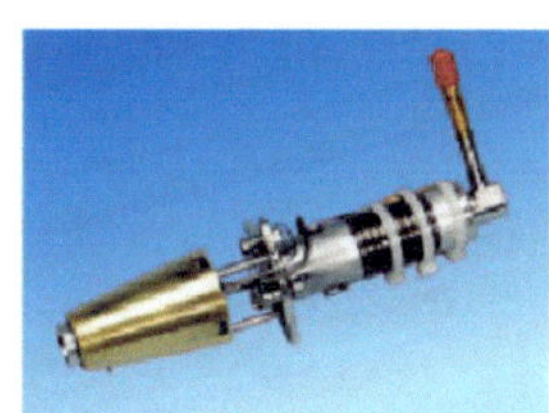

Aerojet MR-111C

- Monopropellant Hydrazine thruster
- 1.3 to 5.3 N thrust
- Minimum impulse bit of 0.08 Ns

Courtesy Aerojet

Mono-propellant System

(*MMH* = Mono-Methyl Hydrazine)

Ullage

High-pressure tank (COPV = Composite Overwrap Pressure Vessel)

Source: Pisacane

Three-Axis Spacecraft Thrusters

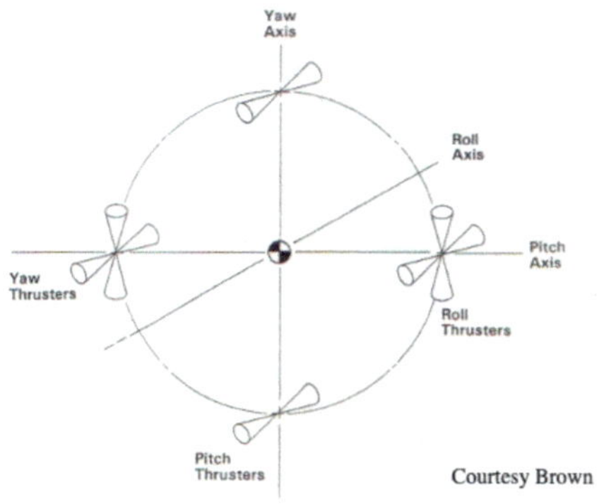

Courtesy Brown

Hydrazine Monoprop Engines

Thrust (N)	Developer/Source	I_{SP} steady	I_{SP} pulse	Mass (kg)
0.09 - 0.22	TRW, Olin RRC, ERNO, Marquardt, HAC	205 - 215	110 - 180	0.1 - 0.2
0.09 - 0.67	TRW, Olin	285 - 320	250 - 290	0.5 - 0.9
2.22	Olin RRC, Marquardt	215 - 230	120 - 200	0.1 - 0.2
4.45	TRW, SEP, HAC, Marquardt, Olin RRC	210 - 230	120 - 210	0.1 - 0.2
13 - 18	TRW, SEP, ERNO	215 - 235	150 - 210	0.2 - 0.3
22 - 36	TRW, Olin RRC, Marquardt, HAC	215 - 240	120 - 210	0.2 - 0.3
45 - 67	Marquardt, TRW	215 - 240	120 - 210	0.3 - 0.5
111	Olin RRC	215 - 240	150 - 220	1.5 - 1.6
133	Marquardt	225 - 242	150 - 225	2 - 3
178 - 222	Olin RRC, TRW	220 - 245	150 - 220	1.4 - 1.8
445 - 689	Olin RRC, Marquardt	225 - 245	150 - 225	1.8 - 2.3
1335	Kidde, Olin RRC	225 - 245	150 - 225	11.3
2669	Olin RRC	225 - 240	N/A	8.2

22 August 2022

Courtesy SMAD

SC03- 46

I_{SP} Transient < I_{SP} Steady

(courtesy Brown)

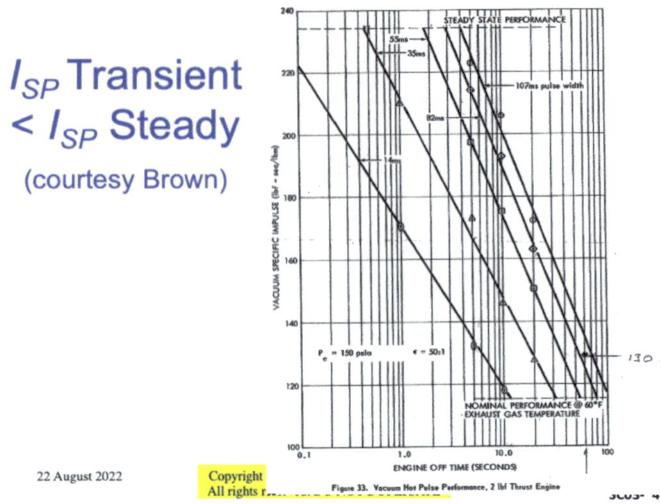

22 August 2022

SC03- 47

Bipropellant Liquid Engines

- Separate liquid oxidizer & liquid fuel
- Higher I_{sp} than monopropellants
- More complex than monopropellants
- Common combinations
  - Storable: MMH/N_2O_4 (monomethyl hydrazine, fuel; dinitrogen tetroxide, oxidizer). "nitrogen tetroxide"
  - ***Hypergolic*** = self-igniting
  - Cryogenic: LOx/LH_2 (high energy)
  - Hydrocarbons: LOx/Kerosene or LOx/RP

22 August 2022

SC03- 48

Bipropellant Engine

- Apollo Service Module biprop engine
- Thrust 445 N

- Aerojet's HiPAT liquid apogee engine
- Bipropellant: N_2O_4/MMH
- Specific impulse of 328 sec
- Thrust 445 N

22 August 2022

SC03- 49

Biprop ACS Schematic

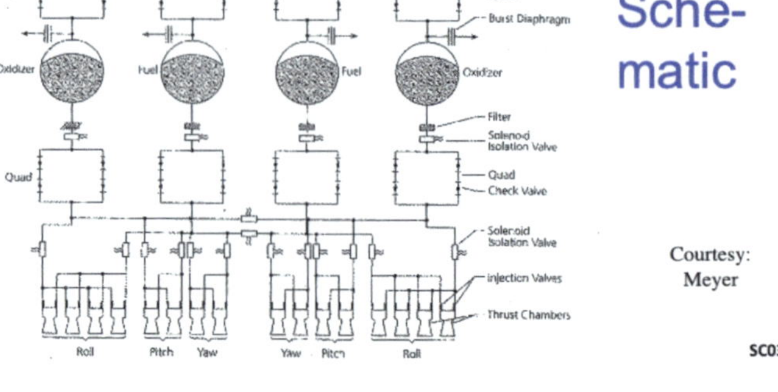

Courtesy: Meyer

SC03- 51

Dual-Mode Propulsion Systems

- Combine bipropellants with monopropellants
- Use more efficient *Bi-* for high impulses (i.e. orbital transfers) to save propellant mass
- Use *Mono-* for fine controls (attitude control) for simplicity with many thrusters
- Commonly used for planetary spacecraft

22 August 2022

SC03- 52

Dual-Mode Propulsion System

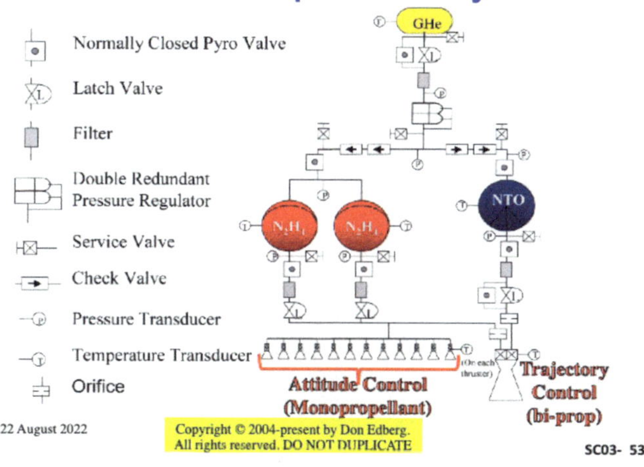

22 August 2022

Propulsion Topics

- Need for propulsion & definitions
- Types of space propulsion systems
  - Solid propellants
  - Liquid: mono- vs. bi-propellant, dual-mode systems
  - Liquid engines & systems
  - **Liquid propellant management**
  - Electric & nuclear propulsion
- Propulsion system design considerations
- Solid system mass calculations
- Liquid system mass calculations

22 August 2022

Propellant Management

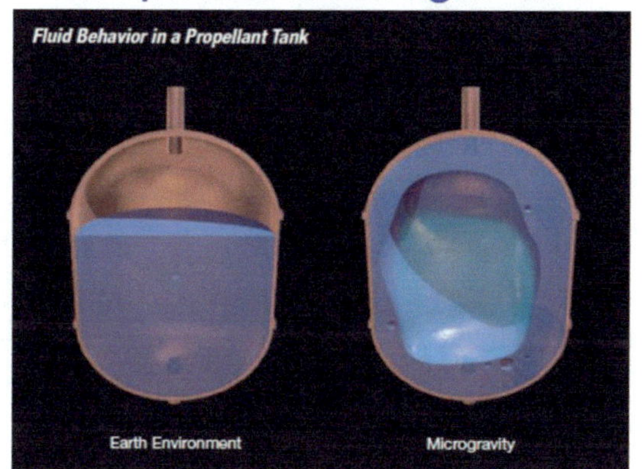

Courtesy: NASA

22 August 2022

Liquid Propellant Delivery

- Regulated systems: provide metered pressure to propellant tanks
- Blowdown systems: pressure to propellant tanks is unregulated
- Microgravity propellant control important (see right)

Pressurant
Propellant
Diaphragm
Capillary
Bladder
Collapsible Cylinder
Trap

Zero-g propellant control devices.
Image adapted from:
SPACECRAFT PROPULSION, by Ch. D. Brown AIAA Education Series, 1996

22 August 2022

Diaphragm Tank

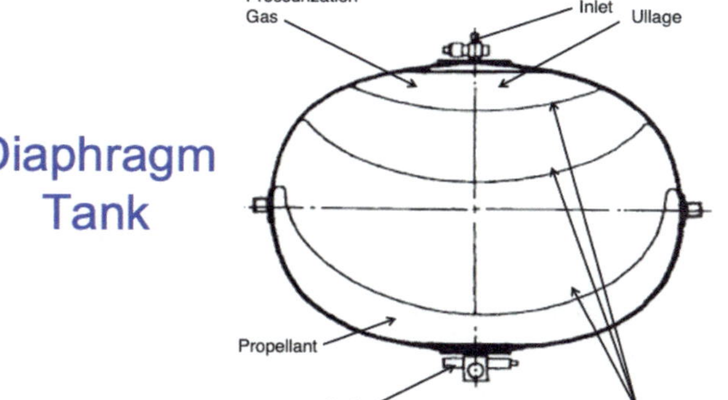

Source: Huzel

22 August 2022

Capillary Propellant Control

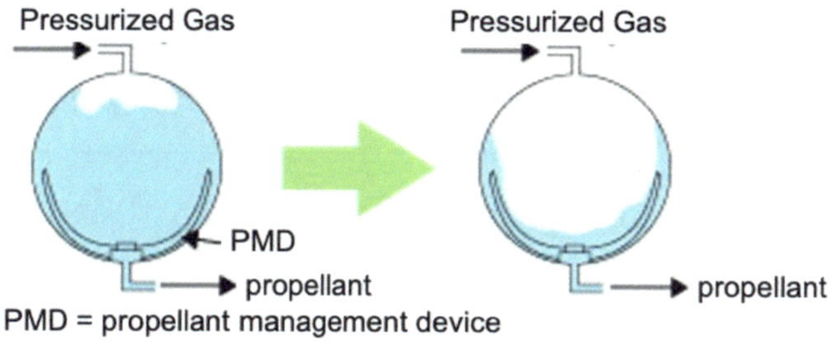

Source: NASDA

22 August 2022

Space Shuttle Propellant Control

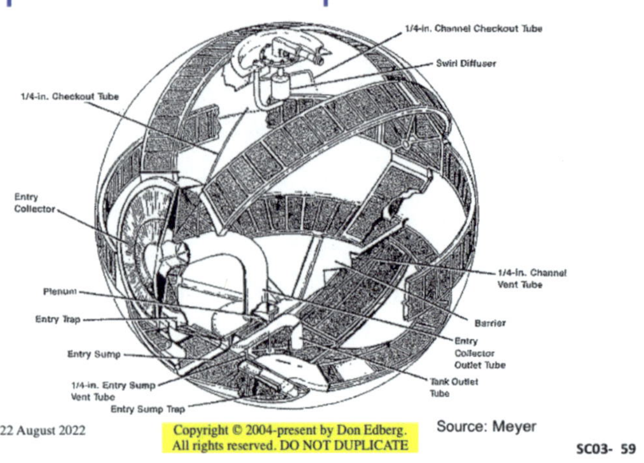

Source: Meyer

Capillary Action Video from ISS

ISS Weightless Tea Party.mp4

Cold Gas Thrusters

- No combustion/contamination
- Pressurized gas expanded through nozzle
- Very simple & inexpensive
- Low specific impulse
- Low thrust levels
- Example: Vacco 8.9 N Thruster (sample: *Vashon* model rocket)

Courtesy: Vacco Industries

Propulsion Topics

- Need for propulsion & definitions
- Types of space propulsion systems
  - Solid propellants
  - Liquid: mono- vs. bi-propellant, dual-mode systems
  - Liquid engines & systems
  - Liquid propellant management
  - Electric & nuclear propulsion
- Propulsion system design considerations
- Solid system mass calculations
- Liquid system mass calculations

Electric Propulsion: Higher I_{sp}, Lower T

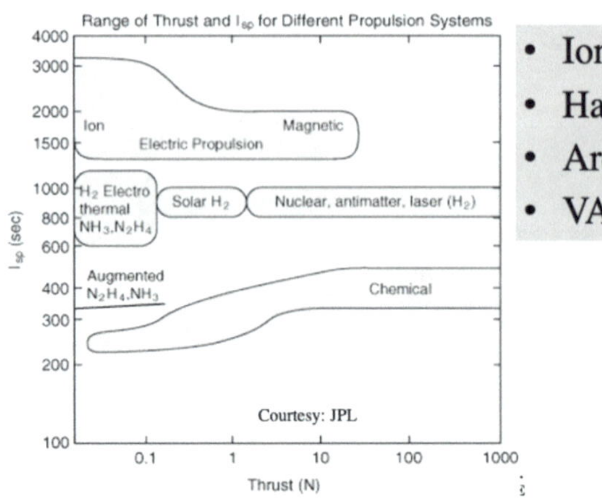

- Ion Engine
- Hall Effect
- Arc Jet
- VASIMR®

Ion Propulsion

Thrust from accelerating charged ions (xenon, cesium, argon)

Energy (electricity) from nuclear or solar generators

Typically used for station-keeping or slow, spiral orbit transfers (Deep Space 1, DAWN)

☺ - Very high I_{sp}: ~ 1,000 – 10,000s

☹ - Very low thrust (mN or μN)

Deep Space 1

Schematic of Ion Engine

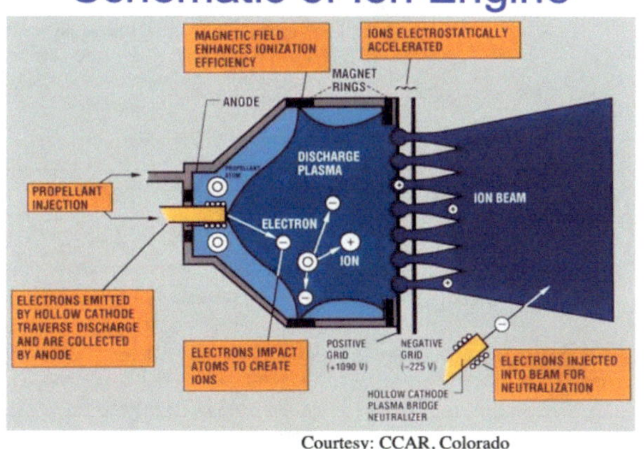

Courtesy: CCAR, Colorado

22 August 2022

SC03- 65

Ion Engine & Support Equipment

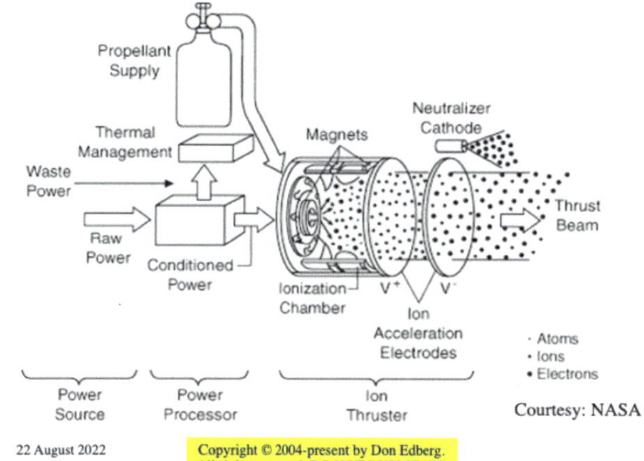

Courtesy: NASA

22 August 2022

SC03- 66

Why Electric Propulsion? Example*

- Consider spacecraft at GEO. \$100M/year BOL revenue drops to \$50M-60M/year at EOL (*numbers fictional, "cartoon" equation from Intelsat*). Use \$60M.
- On 15 year lifetime, ~\$1200M total revenue
- Higher I_{SP} of ion engines significantly reduces required stationkeeping propellant consumption, propellant lasts longer
- Existing propellant mass with efficient ion engines extends lifetime for ~5 more years, produces ~\$1476M revenue over SC lifetime, increase = \$276M
- Or, *MDA Robotic Satellite Refueler* SC could refuel: ~\$50M cost to extend mission similarly: Δ = +\$226M

*Modified from Intelsat cartoon equation, courtesy AvWeek

22 August 2022

SC03- 67

Electric Propulsion: Actual

- In 2005, Boeing offered a Xenon Ion Propulsion System (XIPS) for the *702SP* satellite, 10 × more efficient than liquid propellant systems
- Four 25 cm thrusters provide economical station-keeping, needing only 5 kg propellant per year, "a fraction of what bipropellant or arcjet systems consume"
- XIPS can be used for final orbit insertion to conserve even more mass, as compared to using a traditional on-board liquid apogee engine

22 August 2022

SC03- 68

Hall Thruster: Similar to Ion

- Attractive (–) charge provided by electron plasma at open end of thruster instead of grid
- Higher thrust than ion
- ~ 80 mN thrust (a U.S. quarter weighs 60 mN)
- I_{sp} 1200-1800 s
- 50–70 mN/kW
- Can use many propellants (Bismuth!)
- Used on ESA *Smart 1*

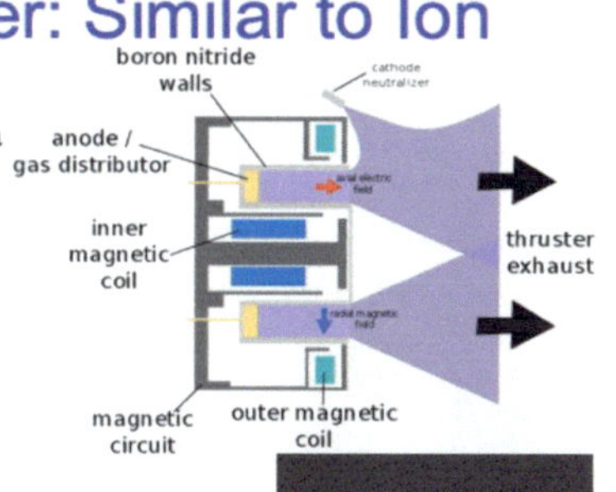

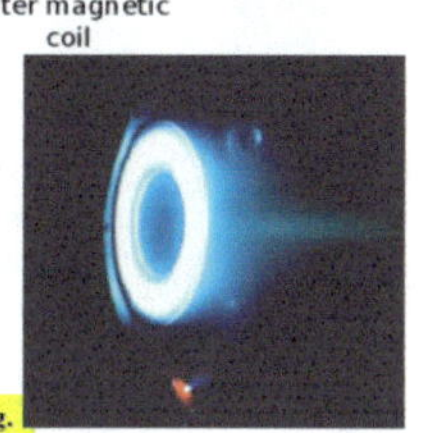

Ref.: upper, Wikipedia; lower CCAR, Colorado

22 August 2022

SC03- 69

Arcjet Engine

- An electrical discharge (arc) is created in a propellant flow, imparts additional energy
- Propellants: hydrazine, ammonia
- High I_{sp}: ~1600 seconds
- Efficiency 40 – 50%

Courtesy CCAR, Colorado

22 August 2022

SC03- 70

VASIMR® (Variable Specific Impulse Magnetoplasma Rocket)

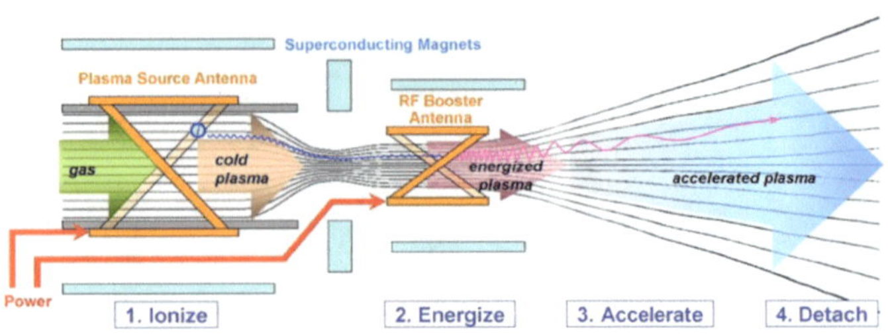

22 August 2022 Courtesy Ad Astra SC03- 71

VASIMR Advantages

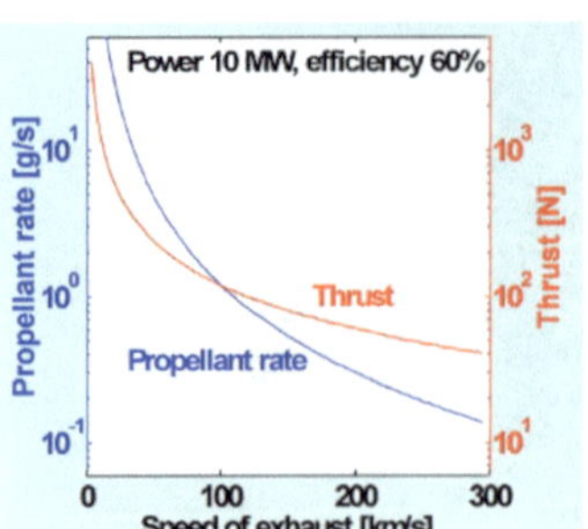

- Variable I_{sp} & thrust at max power to optimally match mission requirements results in lowest trip time with the highest payload for a given propellant load
- Continuous operation
- Driven by RF waves, no physical material electrodes in contact with the hot plasma, resulting in greater reliability & longer life
- Adaptable to slow robotic cargo missions as well as fast human transfers
- Uses less expensive and more abundant propellants: hydrogen, argon, neon
- July 2021: Ad Astra completed 88 hours at a power level of 80 kW (AvWeek)
- Earth-Mars in 7.5 months (vs. 24-36 mos. via chemical)

22 August 2022 Courtesy Ad Astra SC03- 72

Electric Propulsion System Mass Depends on Desired I_{sp} & Power System Mass

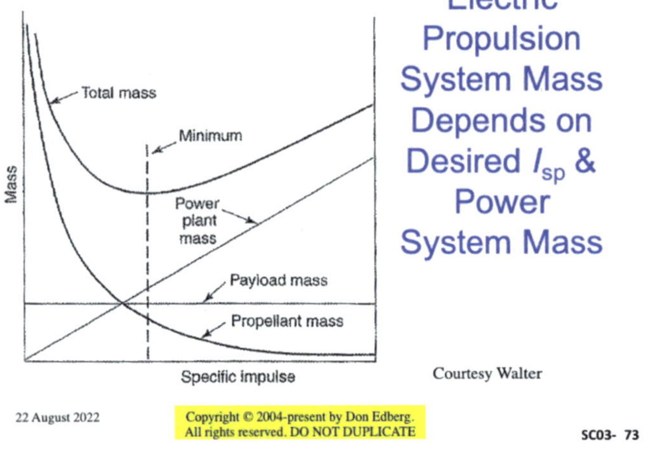

Courtesy Walter

22 August 2022 SC03- 73

Nuclear-Thermal Rocket (NTR)

Nuclear reactor provides heat energy (no chemical reaction)
LH_2 only required working fluid
Examples: *NERVA*, *Topaz*
☺ No need for oxidizer
☺ High T/W_e (> 150)
☺ High I_{sp} ~ 900 s
☹ Political and safety issues
☹ Not acceptable for atmospheric uses (in-space only)

NERVA at MSFC. D. Edberg photo

22 August 2022 SC03- 74

NERVA vs. RL-10 Hydrolox Engine

NERVA/KIWI
- Mass: 2.268 T (5,000 lb_m) includes reactor
- Thrust: 111.1 kN (25,000 lb_f)
- Exhaust speed v_e = 9 km/s
- I_{sp} = 917.7, 2 × RL-10 — ½ fuel needed!

RL-10 (highest-efficiency chemical engine)
- Mass: 301 kg (664 lb_m)
- 110.1 kN (24,800 lb_f) thrust
- Exhaust speed v_e = 4.565 km/s
- I_{sp} = 465.5, ½ of NERVA)

22 August 2022 SC03- 75

NERVA Nuclear Rocket Engine

Courtesy NASA

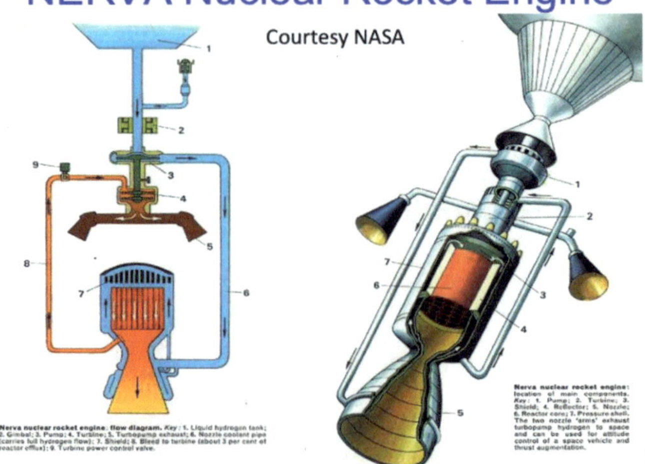

76

Russian Nuclear Rocket Engine

Advantages:

- High I_{sp}: NTR = 971 s, (compare LOx/LH_2 = 455 s)
- Better T/W compared to other high I_{sp} engines

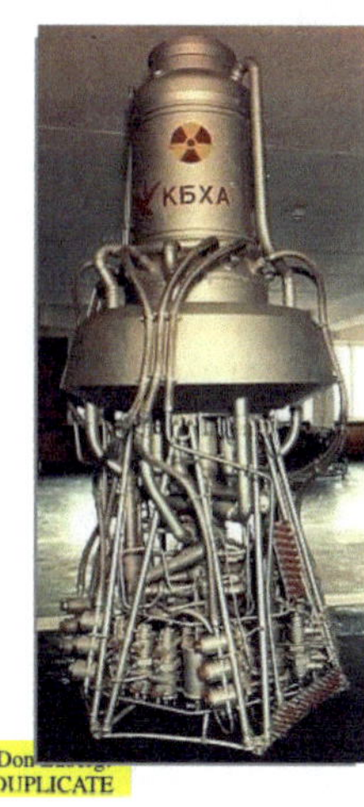

22 August 2022

Propulsion Topics

- Need for propulsion & definitions
- Types of space propulsion systems
  - Solid propellants
  - Liquid: mono- vs. bi-propellant, dual-mode systems
  - Liquid engines & systems
  - Liquid propellant management
  - Electric & nuclear propulsion
- **Propulsion system design considerations**
- Solid system mass calculations
- Liquid system mass calculations

22 August 2022

Propulsion System Design: Many details.

Anyone can design a propulsion system… the devil is in the details!

Liquid Propulsion System Components

- Propellant control: make sure fluids feed to system (liquids only)
- High-pressure reservoir
- Accumulators
- Burst discs
- Engines
- Filters
- Orifices
- Ports
- Regulators
- Tanks
- Transducers
- Valves
- Plumbing/pipes
- Brackets, fittings, etc.

22 August 2022

Symbols on Schematics Indicate: Valves, Regulators, Sensors, & Plumbing!

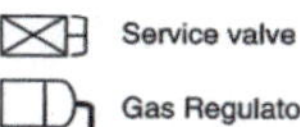

Service valve

Gas Regulator

Filter

Burst Valve

Latch Valve

Access Port

Pyrovalve normally open

Pyrovalve normally closed

Latch Valve

Check Valve, arrow direction of flow

Pressure Sensor

Temperature

Source: Pisacane

22 August 2022

Types of Valves

- Check: allows one-way flow (spring closes)
- Solenoid: can be open & shut electrically
- Latch: can be commanded open & shut and stays until next command
- Pyro: valve either closes, or opens ONE TIME ONLY. Minimal leakage.
- Relief: releases excess pressure
- Service: used for propellant loading, checking

22 August 2022

Pyro Valve (*Astrium*)

Courtesy **Handbook of Space Technology**

22 August 2022

SC03- 83

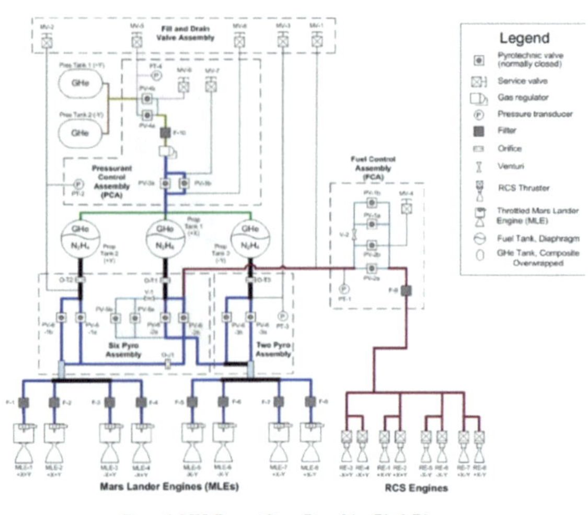

Figure 1. MSL Descent Stage Propulsion Block Diagram.

03- 84

Ulysses RCS

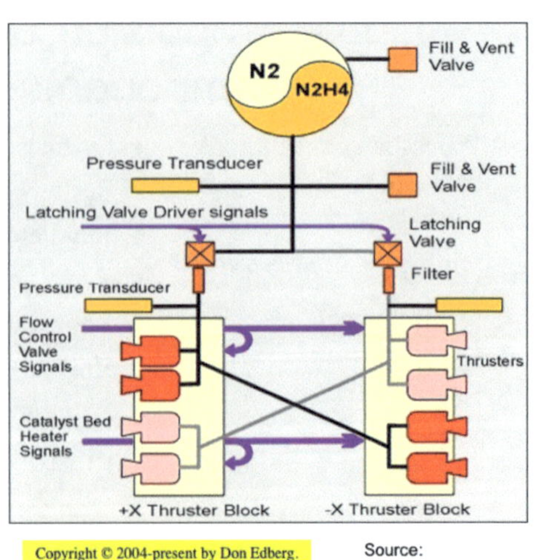

22 August 2022

Source: tudelft.nl

SC03- 85

Ulysses Thruster Block

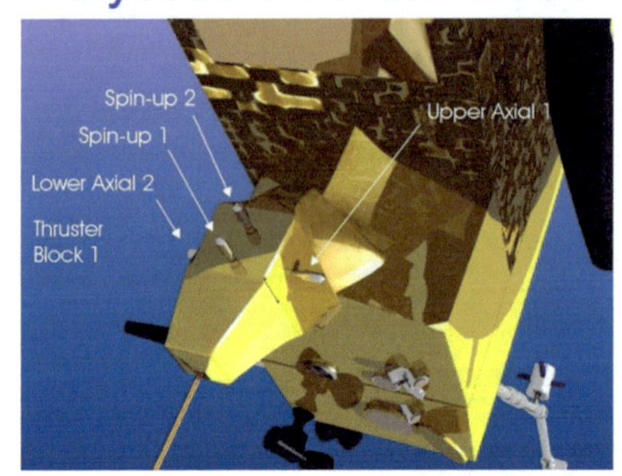

Source: ulysses-ops.jpl.esa.int

22 August 2022

SC03- 86

Modular Propulsion System Example: Shuttle

allows separate loading & testing of propellant system (NASA)

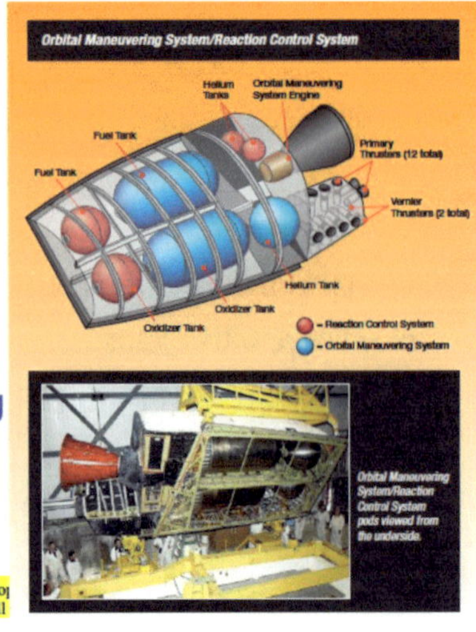

22 August 2022

SC03- 87

Thruster Exhaust Protection

- Exhaust tends to expand to fill vacuum
- Heats sensitive surfaces or contaminates
- Solution: shields or baffles

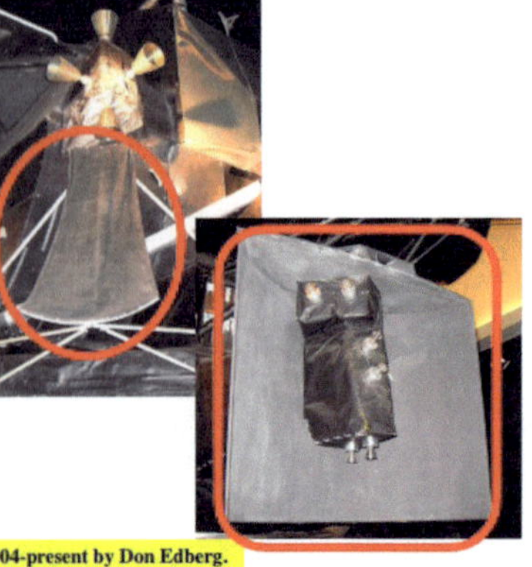

Plume shields on Apollo LM (L) and GLL (R) shown. Credit: Edberg-Cosmosphere & JPL

22 August 2022

SC03- 88

Propulsion Topics

- Need for propulsion & definitions
- Types of space propulsion systems
  - Solid propellants
  - Liquid: mono- vs. bi-propellant, dual-mode systems
  - Liquid engines & systems
  - Liquid propellant management
  - Electric & nuclear propulsion
- Propulsion system design considerations
- **Solid system mass calculations**
- Liquid system mass calculations

22 August 2022 SC03- 89

Solid Propulsion System Mass Calculations

1. Determine final SC mass m_f & Δv requirements
2. Use I_{sp}, m_f, Δv, and propellant mass equation to calculate required propellant mass
3. Size combustion chamber. Include propellant density, pressure requirements, chamber material, allowable stresses, etc. Decide: spherical vs. cylindrical tank geometry; calculate dimensions (Or use a COTS motor)
4. Add extras: mounting structure, TVC system, etc.
5. Update spacecraft mass statement

22 August 2022 SC03- 90

Solid Rocket Motor Example

- Required velocity change = 1,831 m/s, I_{sp} = 290 s
- Burnout mass (including motor) = 1,500 kg
- Use rocket equation to calculate propellant mass:

$$m_p = 1{,}500\text{ kg}\left[e^{\left(\frac{1{,}831\text{ m/s}}{9.8\text{ m/s}^2\times 290\text{s}}\right)} - 1\right] = 1{,}356\text{ kg}$$

- If mass fraction 93%, SRM mass = 1,458 kg
- Total impulse $I = m_p I_{sp} g_0$ = 1,356 kg(290 s)9.8 m/s² = 3.856E6 Ns
- SC bus mass: 1,500 – (1,458 – 1,356) = 1,398 kg
- Propellant density ρ_p = 1,770 kg/m³
- Next: calculate casing volume

22 August 2022 SC03- 91

Solid Motor (cont.)

- Assume 90% loaded (10% empty space for igniter & propellant burning pattern)
- Volume of motor: $VOL = m_p/(\rho_p\ 0.90)$ = 1,356 kg/(1,770 kg/m³ × 0.9) = 0.85 m³
- Motor case: use *mass estimating relation* or *MER* (MERs on later slide):
  $m_{motor\ casing} = 0.135\ m_{propellants}$ = 0.135 × 1,356 kg = 183 kg
- Mounting hardware = 15% of motor mass = 0.15(1,356 + 183)kg = 230 kg
- Total solid system = 1,770 kg

22 August 2022 SC03- 92

Propulsion Topics

- Need for propulsion & definitions
- Types of space propulsion systems
  - Solid propellants
  - Liquid: mono- vs. bi-propellant, dual-mode systems
  - Liquid engines & systems
  - Liquid propellant management
  - Electric & nuclear propulsion
- Propulsion system design considerations
- Solid system mass calculations
- **Liquid system mass calculations**

22 August 2022 SC03- 93

Liquid Propulsion System Mass Calculation Procedure (1 of 2)

1. Determine SC mass & velocity requirements
2. Use I_{sp}, m_f, Δv, and propellant mass equation to calculate required *ideal* propellant mass using
   $$m_p = m_f\left(e^{\Delta v/g_0 I_{sp}} - 1\right)$$
3. A) add engine startup propellant mass:
   for n burns: $m_{p-startup} = t_{startup}\dot{m}$
   Propellant flow $\dot{m} = \frac{T}{g_0 I_{sp}}$ assuming thrust T & time taken to reach full thrust $t_{startup}$
   B) unusable propellant mass: add 2% for trapped fuel & loading error: $m_{p-residual} = 0.02 m_p$

Total propellant $m_{p-total} = m_p + m_{p-residual} + m_{p-startup}$

22 August 2022 SC03- 94

Liquid Propulsion System Mass Calculation Procedure (2 of 2)

4. Separate propellants into fuel & ox. Use propellant densities & ullage volume to size tank volumes
   If possible, use an identical tank for both to save $.
5. Use Mass Estimating Ratios (MERs) to calculate tank masses
6. Size pressurant & its tank using gas equations of state
7. Specify engine, required valves, regulators, filters, transducers, mounting structure (~15% of object mass), check valves, isolation valves, propellant ducting, etc. (item masses below)
8. Determine propulsion subsystem mass, and update spacecraft mass statement

Liquid Engine Example: Hypergolic (1 of 3)

1. Assume: I_{sp} = 320 s; required Δv = 1,831 m/s; 1,500 kg final mass; oxidizer = N_2O_4 = DNTO, fuel = Hydrazine, 445 N thruster
2. Use rocket equation to calculate propellant mass:

$$m_p = 1{,}500\text{ kg}\left[e^{\left(\frac{1{,}831\text{m/s}}{9.8\text{m/s}^2\times 320\text{s}}\right)} - 1\right] = 1{,}189\text{ kg}$$

3. Startup and unusable propellant:
   A) Startup propellant for *10* burns @ 0.2 s startup time each:

$$m_{p-startup} = t_{startup}\dot{m} = \frac{t_{startup}T}{g_0 I_{sp}} = \frac{2\text{ s}\cdot 445\text{ N}}{9.8\frac{\text{m}}{\text{s}^2}\times 320\text{s}} = 0.284\text{ kg}$$

   B) Unusable propellant (2% of total)

$$m_{p-residual} = 0.02 m_p = 23.78\text{ kg}$$

$$m_{p-total} = 1{,}189\text{ kg} + 23.78\text{ kg} + 0.284\text{ kg} = 1{,}213\text{ kg}$$

Liquid Engine Example: Hypergolic (2 of 3)

4. Separate propellants into fuel & ox:

- Hypergolic engines use mixture 1.42:1 ox/fuel by mass

$$m_{DNTO} = \frac{1.42}{2.42}\times 1{,}213\text{ kg} = 711.8\text{ kg},$$

$$m_{Hyd} = \frac{1}{2.42}\times 1{,}213\text{ kg} = 501.3\text{ kg}$$

- Densities: ρ_{DNTO} = 1,450 kg/m³; ρ_{Hyd} = 1,021 kg/m³
  Volume calculations:

$$VOL_{DNTO} = \frac{m_{DNTO}}{\rho_{DNTO}} = \frac{711.8\text{ kg}}{1{,}450\text{ kg/m}^3} = 0.4909\text{ m}^3$$

$$VOL_{Hyd} = \frac{m_{Hyd}}{\rho_{Hyd}} = \frac{501.3\text{ kg}}{1{,}021\text{ kg/m}^3} = 0.4910\text{ m}^3$$

- Tank volume including added 30% volume for ullage

$$VOL_{DNTO\ tank} = 1.3\,VOL_{DNTO} = 0.6382\text{ m}^3$$

$$VOL_{hyd\ tank} = 1.3\,VOL_{Hyd} = 0.6382\text{ m}^3$$

Decide: spherical vs. cylindrical tank geometry; calc. dimensions

Liquid Engine Example: (3 of 3)

5. Use MERs to calculate mass of propellant tanks (& any required insulation if cryogenics)

$$m_{Hyd\text{-}tank}(\text{kg}) = 0.316\,[501]^{0.6} = 13.17\text{ kg}$$

$$m_{DNTO\text{-}tank}(\text{kg}) = 0.316\,[712]^{0.6} = 16.26\text{ kg}$$

   Consider using same (heavier) tank for both

Then, calculate

6. Pressurant mass & volume (see Brown 4.4.3.3); pressurant tank mass (MER, or Brown 4.4.4)
7. Masses of rocket engine(s), valves, transducers, heaters, etc. Mass of support structure/mounting hardware (~15% of loaded tank mass). These all depend on system details: component values are below.
   See structures slides for tank mounting arrangement
8. Use all estimated masses to update spacecraft mass statement

LV Tank Mass-Estimating Ratios (MERs)

- $m_{storables\ tank}(\text{kg}) = 0.316\,[m_{storables}\,(\text{kg})]^{0.6}$
  (storables are RP-1, N_2O_4, hydrazine)
- $m_{solid\ motor\ casing}(\text{kg}) = 0.135(m_{propellants}\,(\text{kg}))$
- $m_{small\ liquid\ tank}(\text{kg}) = 0.1(m_{contents}\,(\text{kg}))$
- $m_{small\ pressurized\ gases\ tank}(\text{kg}) = 2(m_{contents}\,(\text{kg}))$
- $m_{LH2\ insulation}(\text{kg}) = 2.88\text{ kg/m}^2$
- $m_{LOX\ insulation}(\text{kg}) = 1.123\text{ kg/m}^2$
- Remember to calculate propellant load based on the max possible spacecraft mass, worst trajectory, & off-nominal system performance (I_{sp})

MERs courtesy Akins

Estimating Biprop Engine Mass

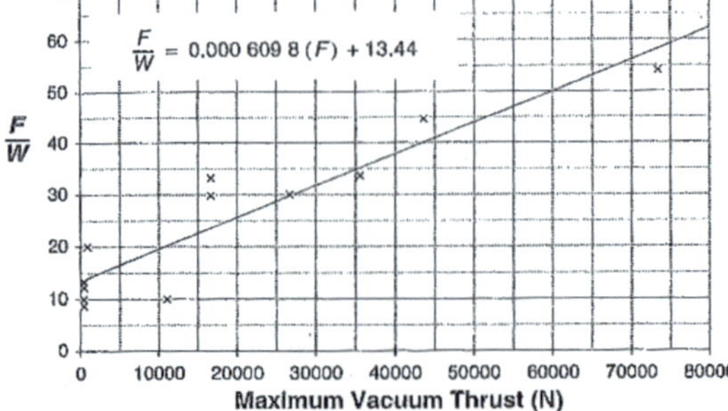

Caution: Use chart to find *weight* of engine (earth newtons) based on engine's vacuum thrust in N (watch units!)

Courtesy Humble

Estimating Monoprop Engine Mass

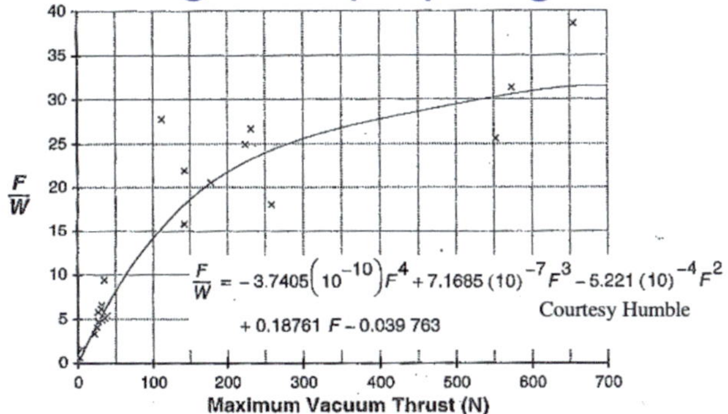

Use chart to find *weight* of engine including its valve (N on earth) based on engine's vacuum thrust in N (watch units!)

22 August 2022

SC03-101

Component Masses

Item	Mass ea. (kg)
Line Heaters	0.90
Squib valve	0.50
Service Valve	0.11
Filter	0.03
Latch Valve	0.50
Pressure Transducer	0.22
High Pressure Transducer	0.50
High Pressure Filter	0.02
High Pressure Service Valve	0.03
Regulator-Series Pair	1.27
Check Valve	0.15

Source: Brown

22 August 2022

SC03-102

Liquid Propulsion System Mass Estimating Rules of Thumb

$m_{\text{biprop system}} = 26.4\text{ kg} + 0.077 \times m_{\text{propellant}}$

$m_{\text{mono primary}}{}^{*} = 9.43\text{ kg} + 0.09 \times m_{\text{propellant}}$

$m_{\text{mono secondary}}{}^{**} = 6.36\text{ kg} + 0.19 \times m_{\text{propellant}}$

$m_{\text{dry blowdown}} = 2.6109\text{ kg} + 0.2337 \times m_{\text{propellant}}$

$m_{\text{mono thruster/valve}} = 0.4\text{ kg} + 0.0033 \times T\text{ (thrust)}$

*Primary = propulsion

**Secondary = station-keeping, attitude control

22 August 2022

SC03-103

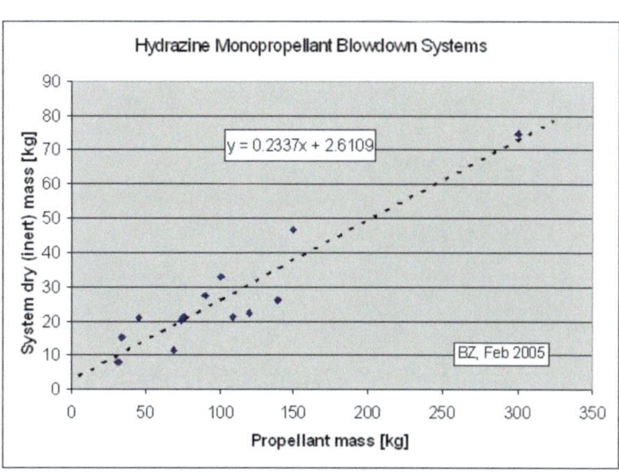

Courtesy Zandbergen@tudelft.nl

22 August 2022

SC03-104

Required Propellant Mass, Initial & Final Masses Known

For a SC with given Δv requirement & I_{sp} or exhaust speed v_{eq}, usable propellant mass m_p may be calculated with

1. Initial mass $m_0 = m_s + m_p + m_{PL}$, or
2. Final mass $m_f = m_s + 0 \times m_p + m_{PL}$ = empty step mass (zero propellant) + payload mass
3. Propellant mass m_p:

$$m_p = m_0\left[1 - e^{\left(-\frac{\Delta v}{g_0 I_{sp}}\right)}\right] = m_0\left[1 - e^{\left(-\frac{\Delta v}{v_{eq}}\right)}\right]$$

$$m_p = m_f\left[e^{\left(+\frac{\Delta v}{g_0 I_{sp}}\right)} - 1\right] = m_f\left[e^{\left(+\frac{\Delta v}{v_{eq}}\right)} - 1\right]$$

8/22/22

SC03-105

Required Propellant Mass For Known Payload, Δv, I_{sp}, & Structural Ratio σ or ε

Given: I_{sp}, required Δv, and structural ratio $\sigma = \frac{m_s}{m_p + m_s}$ or $\varepsilon = \frac{m_s}{m_p}$

Calculate

$$\mu = \frac{m_0}{m_f} = e^{\left(\frac{\Delta v}{g_0 I_{sp}}\right)} = e^{\left(\frac{\Delta v}{v_{eq}}\right)} = e^{\left(\frac{\Delta v}{c}\right)}$$

$$= \exp\left(\frac{\Delta v}{v_{eq}}\right) = \exp\left(\frac{\Delta v}{g_0 I_{sp}}\right) = \exp\left(\frac{\Delta v}{c}\right)$$

We can develop relations between the variables, which are ***very useful for initial sizing purposes***: one can calculate ALL masses if μ, σ (or ε), & ANY mass are given! See table below, from "Design of Rockets & Space Launch Vehicles" by Edberg & Costa

8/22/22

SC03-106

Required Propellant Mass For Known Payload, Δv, I_{sp}, & Structural Ratio σ

Given: I_{sp}, required Δv,
and structural ratio $\sigma = \frac{m_s}{m_p+m_s}$, or $\varepsilon = \frac{m_s}{m_p}$

Calculate $\mu = \frac{m_0}{m_f} = e^{\left(\frac{\Delta v}{g_0 I_{sp}}\right)} = e^{\left(\frac{\Delta v}{v_{eq}}\right)} = e^{(\Delta v/c)}$

$$= \exp\left(\frac{\Delta v}{v_{eq}}\right) = \exp\left(\frac{\Delta v}{g_0 I_{sp}}\right) = \exp\left(\frac{\Delta v}{c}\right)$$

Relations between the variables are ***very useful for initial sizing purposes***: one can calculate ALL masses if μ, σ (or ε), & ANY mass are given! See table below. Note $\pi = \frac{m_{PL}}{m_s+m_p+m_{PL}}$.

9/9/22

5-115

Quantities From Given Δv, I_{sp}, & Structure Ratio σ or ε

Description	In terms of $\sigma = m_s/(m_s + m_p)$	in terms of $\varepsilon = m_s/m_p$
Propellant mass (USEFUL!)	$m_p = m_{PL}\frac{(\mu-1)(1-\sigma)}{1-\mu\sigma}$	$m_p = m_{PL}\frac{(\mu-1)}{1-\varepsilon(\mu-1)}$
Final speed	$\frac{\Delta v}{c} = -\ln[\sigma + (1-\sigma)\pi] = \ln\left(\frac{1}{1-\zeta_p}\right)$	$\frac{\Delta v}{c} = -\ln\left(\frac{\varepsilon+\pi}{1+\varepsilon}\right) = \ln\left(\frac{1+\varepsilon}{\varepsilon+\pi}\right)$
Gross liftoff mass	$m_0 = m_{PL}\frac{\mu(1-\sigma)}{1-\mu\sigma}$	$m_0 = \frac{\mu m_{PL}}{1+\varepsilon(1-\mu)}$
Final or burnout mass	$m_f = m_{PL}\frac{1-\sigma}{1-\mu\sigma}$	$m_f = \frac{m_{PL}}{1+\varepsilon(1-\mu)}$
Payload mass	$m_{PL} = m_p\frac{1-\mu\sigma}{(\mu-1)(1-\sigma)}$	$m_{PL} = m_p\left[\frac{1}{(\mu-1)} - \varepsilon\right]$
Structural mass (no PL)	$m_s = m_p\frac{\sigma}{1-\sigma}$	$m_s = m_p\varepsilon$
Mass ratio	$\mu = \frac{m_p + m_{PL}(1-\sigma)}{\sigma m_p + m_{PL}(1-\sigma)}$	$\mu = \frac{m_{PL}+m_p(1+\varepsilon)}{m_{PL}+\varepsilon m_p}$
Structural fraction	$\sigma = \frac{m_{PL}(\mu-1) - m_p}{m_{PL}(\mu-1) - \mu m_p}$	$\varepsilon = \frac{1}{\mu-1} - \frac{m_{PL}}{m_p}$

5

Spacecraft Structures

Pleiades HR-1 under radiometric testing. Courtesy CNES

Topics

- Functions of structures
- Launch vs. flight configurations
- The General Arrangement; examples of different layouts
- *Spacecraft Configuration Checklist*
- Layout considerations
- Types of structures
- Structural materials (metals & composites) & selection criteria
- Composite fabrication techniques
- Structure mass estimation

Spacecraft Design Process animated

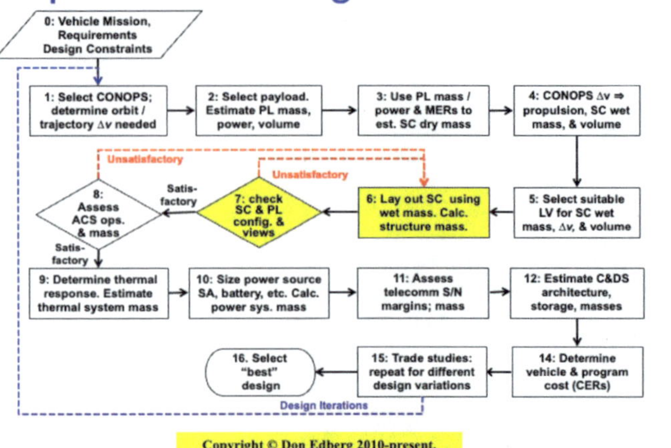
0: Vehicle Mission, Requirements Design Constraints
1: Select CONOPS; determine orbit / trajectory Δv needed
2: Select payload. Estimate PL mass, power, volume
3: Use PL mass / power & MERs to est. SC dry mass
4: CONOPS Δv ⇒ propulsion, SC wet mass, & volume
5: Select suitable LV for SC wet mass, Δv, & volume
6: Lay out SC using wet mass. Calc. structure mass.
7: check SC & PL config. & views
8: Assess ACS ops. & mass
Unsatisfactory
Unsatisfactory
Satis-factory
Satis-factory
9: Determine thermal response. Estimate thermal system mass
10: Size power source SA, battery, etc. Calc. power sys. mass
11: Assess telecomm S/N margins; mass
12: Estimate C&DS architecture, storage, masses
14: Determine vehicle & program cost (CERs)
15: Trade studies: repeat for different design variations
16. Select "best" design
Design Iterations

August 22, 2022 SC04A- 3

Topics

- **Functions of structures**
- Launch vs. flight configurations
- The General Arrangement; examples of different layouts
- *Spacecraft Configuration Checklist*
- Layout considerations
- Types of structures
- Structural materials (metals & composites) & selection criteria
- Composite fabrication techniques
- Structure mass estimation

August 22, 2022 SC04A- 4

Functions of Structures

- Accommodate & protect payload & subsystems ("holds everything together")
- Provide accessibility
- Survive launch vehicle loads (strength, stiffness, & buckling)
- Provide on-orbit environmental protection and alignment
- Provide thermal & electrical paths as needed

August 22, 2022 SC04A- 5

Structure (cont.)

- Should be minimized so mass can be used for other things …
  - More instruments
  - More propellant
  - More transponders
  - Smaller (less expensive?) launch vehicle
  - Etc.

August 22, 2022 SC04A- 6

Structure Types

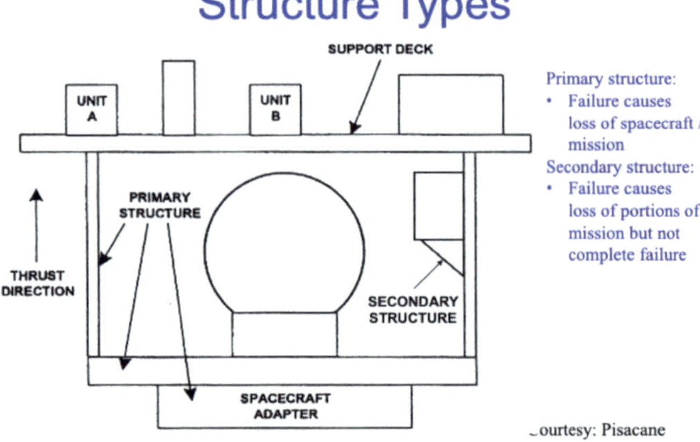
SUPPORT DECK
UNIT A
UNIT B
PRIMARY STRUCTURE
THRUST DIRECTION
SECONDARY STRUCTURE
SPACECRAFT ADAPTER

Primary structure:
- Failure causes loss of spacecraft / mission

Secondary structure:
- Failure causes loss of portions of mission but not complete failure

Courtesy: Pisacane

August 22, 2022 SC04A- 7

ESA Definitions: Primary, Secondary, & Tertiary Structure

(HERSCHEL IR Telescope courtesy ESA)

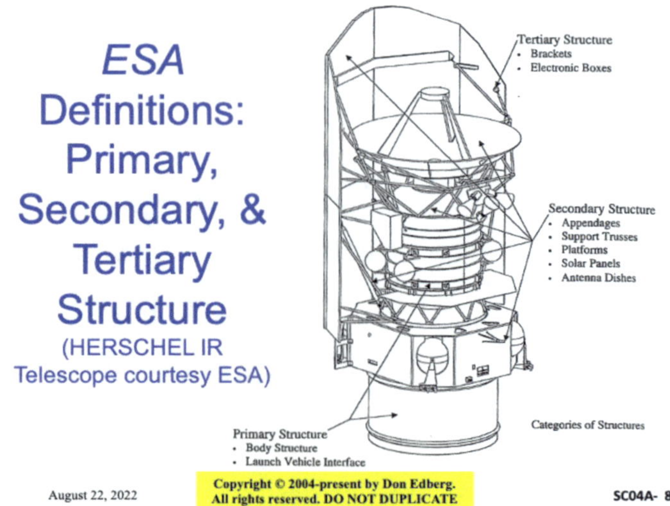
Tertiary Structure
• Brackets
• Electronic Boxes
Secondary Structure
• Appendages
• Support Trusses
• Platforms
• Solar Panels
• Antenna Dishes
Primary Structure
• Body Structure
• Launch Vehicle Interface
Categories of Structures

August 22, 2022 SC04A- 8

Beginning Design

- Make sketches
- Incorporate all requirements, geometric & mission constraints
- Usually there are at least two configurations:
  - Launch configuration
  - Mission configuration(s)

Topics

- Functions of structures
- **Launch vs. flight configurations**
- The General Arrangement; examples of different layouts
- *Spacecraft Configuration Checklist*
- Layout considerations
- Types of structures
- Structural materials (metals & composites) & selection criteria
- Composite fabrication techniques
- Structure mass estimation

MGS (Launch) Configuration …

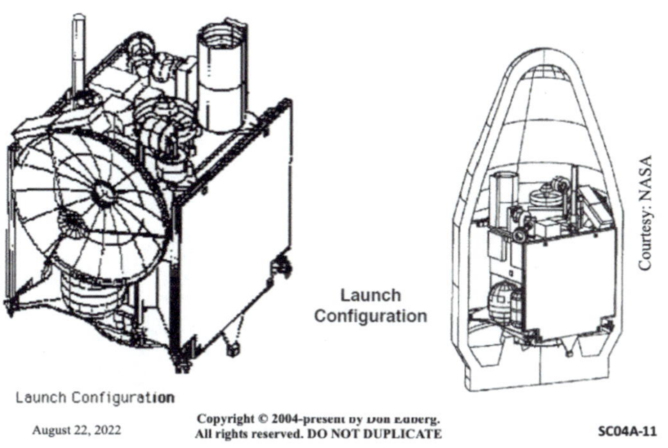

… vs. MGS Flight Configurations

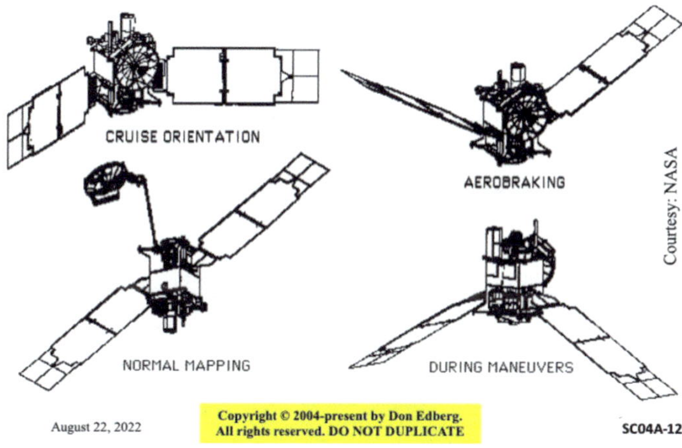

TDRS Stowed & Deployed

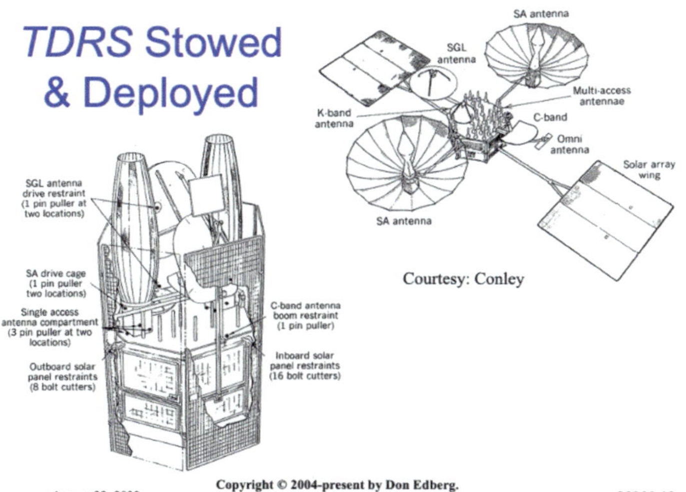

Delta II 6915 Payload Attach Fitting (PAF)

Courtesy Boeing

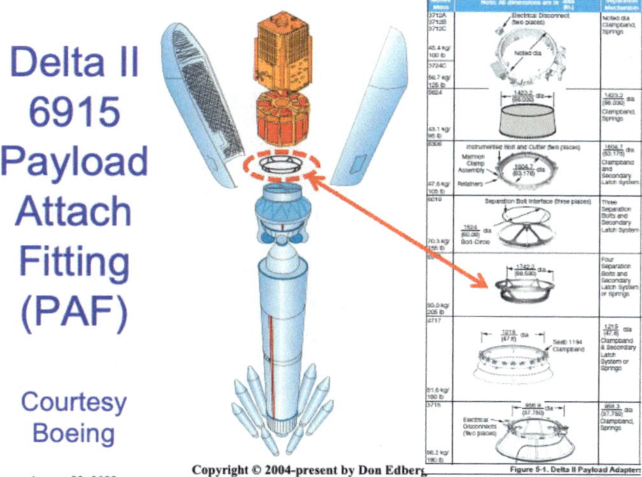

PAF Must Support SC Mass (D2 PAFs)

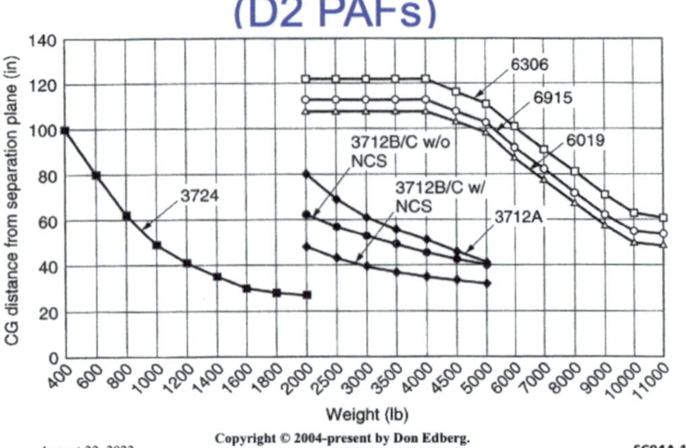

August 22, 2022

SC04A-15

Launch Arrangement

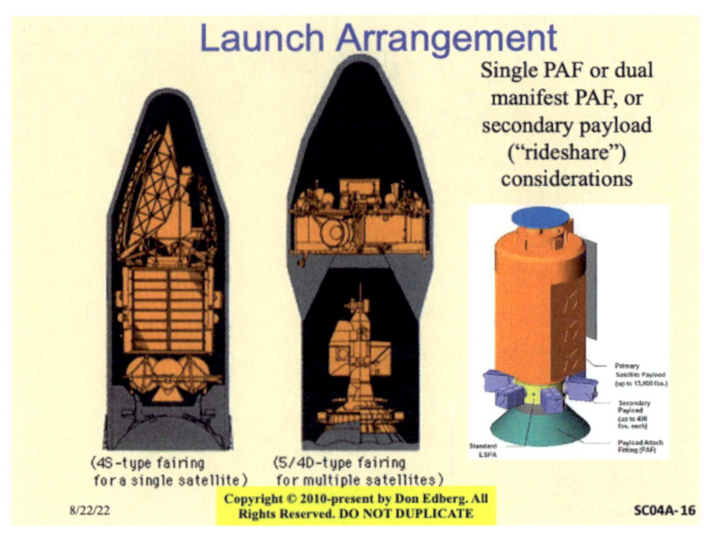

8/22/22

SC04A-16

ULA's "Rideshare" (If Margin Exists!)

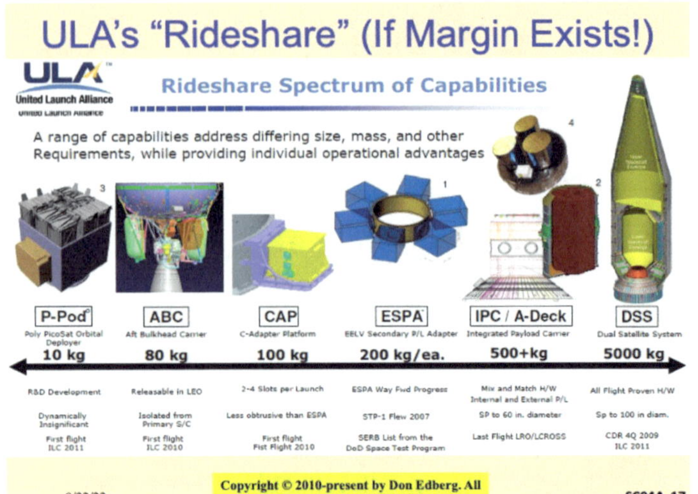

8/22/22

SC04A-17

SpaceX's F9 "Rideshare"

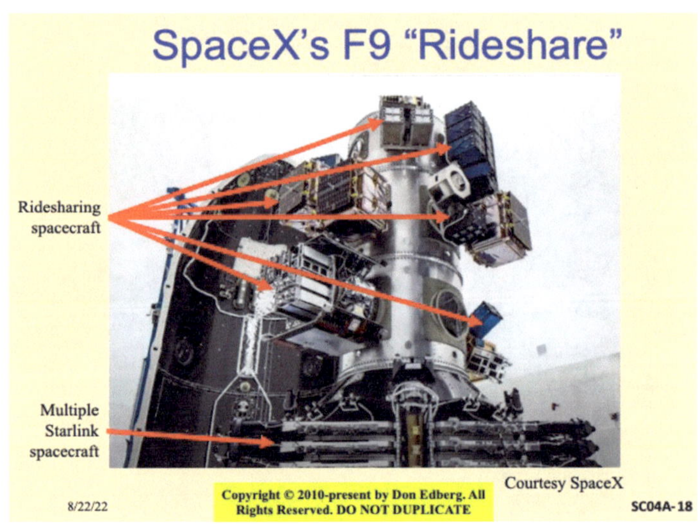

Courtesy SpaceX

8/22/22

SC04A-18

Launch Arrangement (cont.)

- Spacecraft must fit within PLF or shuttle cargo bay
- Galileo (GLL) before unfolding & deployment

Courtesy: NASA

August 22, 2022

SC04A-19

Mission Arrangement: It All Hangs Out!

(GLL)

Engineering
Fields and Particles
Probe
Remote Sensing
LOW-GAIN ANTENNA
SUN SHIELDS
EXTREME ULTRAVIOLET SPECTROMETER
STAR SCANNER
PLASMA-WAVE ANTENNA
MAGNETOMETER SENSORS
ENERGETIC PARTICLES DETECTOR
PLASMA SCIENCE
HEAVY ION COUNTER (BACK)
DUST DETECTOR
RETROPROPULSION MODULE
ABOVE: SPUN SECTION
BELOW: DESPUN SECTION
THRUSTERS (2 places)
RTG
PROBE RELAY ANTENNA
JUPITER ATMOSPHERIC PROBE
SCAN PLATFORM, CONTAINING:
- ULTRAVIOLET SPECTROMETER
- SOLID-STATE IMAGING CAMERA
- NEAR-INFRARED MAPPING SPECTROMETER
- PHOTOPOLARIMETER RADIOMETER
RADIOISOTOPE THERMOELECTRIC GENERATORS (RTG) (2 places)

Courtesy: NASA

20

Topics

- Functions of structures
- Launch vs. flight configurations
- The General Arrangement; examples of different layouts
- *Spacecraft Configuration Checklist*
- Layout considerations
- Types of structures
- Structural materials (metals & composites) & selection criteria
- Composite fabrication techniques
- Structure mass estimation

August 22, 2022 SC04A-21

The General Arrangement

- Once done, potential problems and improvements are used to feed trade studies to develop improved configurations
- Changes become less and less dramatic as time goes on, and the design stabilizes.
- Leads to required structural, thermal, electronics layouts, attitude control layout, propulsion, and mechanisms list

August 22, 2022 SC04A-22

General Arrangement Examples

- *Boeing 601* (comsat)
- *Mars Exploration Rover* (MER)
- *Hubble Space Telescope* (precision platform)
- *Cassini/Huygens*
- *Voyager*
- *LEASAT* (dual spin comsat)
- *Boeing 376*
- *Gravity Probe B*
- *Messenger*

August 22, 2022 SC04A-23

Boeing 601

X-Wing Reflector
Spacecraft Control Process
Inertial Reference Unit
Momentum Wheel
490N Liquid Apogee Motor
22N Thruster
Nickel-Hydrogen Battery Cells
Solar Wing
Sensor Suite
Ku-Band Transponders
L-Band Antenna Array
Ka-Band Beacon Horn
Solar Wing Drive
Mirrored Radiator Panel
L-Band Transponders

Courtesy: Boeing

Boeing 601 Internals: Modular

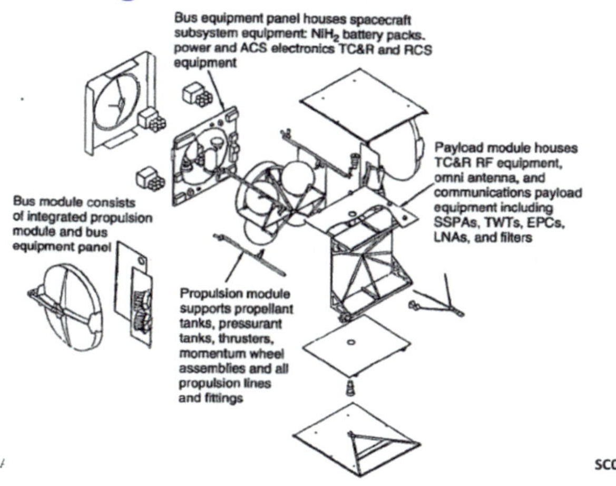
Bus equipment panel houses spacecraft subsystem equipment: NiH2 battery packs, power and ACS electronics TC&R and RCS equipment
Payload module houses TC&R RF equipment, omni antenna, and communications payload equipment including SSPAs, TWTs, EPCs, LNAs, and filters
Bus module consists of integrated propulsion module and bus equipment panel
Propulsion module supports propellant tanks, pressurant tanks, thrusters, momentum wheel assemblies and all propulsion lines and fittings

SC04A-25

MER Spacecraft Configuration

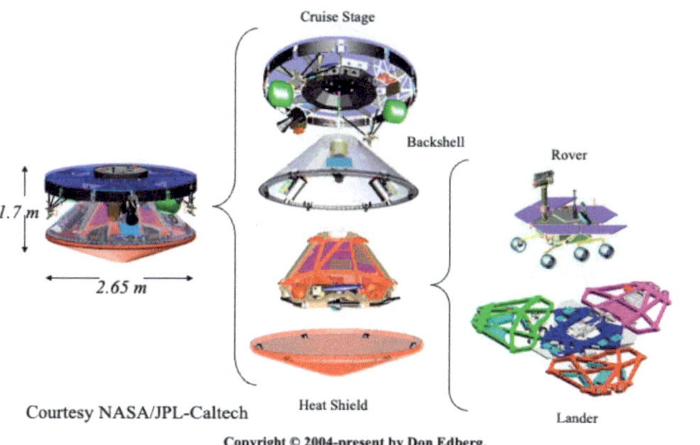
Cruise Stage
Backshell
Rover
1.7 m
2.65 m
Heat Shield
Lander

Courtesy NASA/JPL-Caltech

August 22, 2022 SC04A-26

Hubble Space Telescope

Communications Antenna
Secondary Mirror
Aperture Door
Primary Mirror
Light Shield
Spacecraft Compartment
Telescope Pointing System
Cover
Telescope Compartment
Solar Array
Scientific Instruments
Telescope Pointing System

Courtesy: NASA

27

Hubble, Exploded

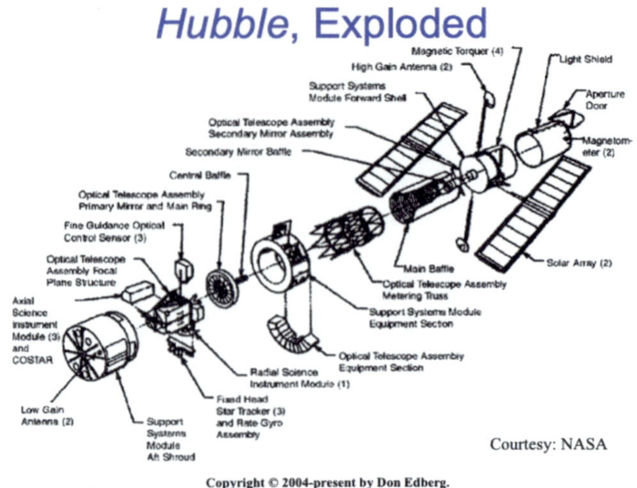

Courtesy: NASA

August 22, 2022

SC04A-28

Cassini Side 1 (Thermal Insulation Not Applied)

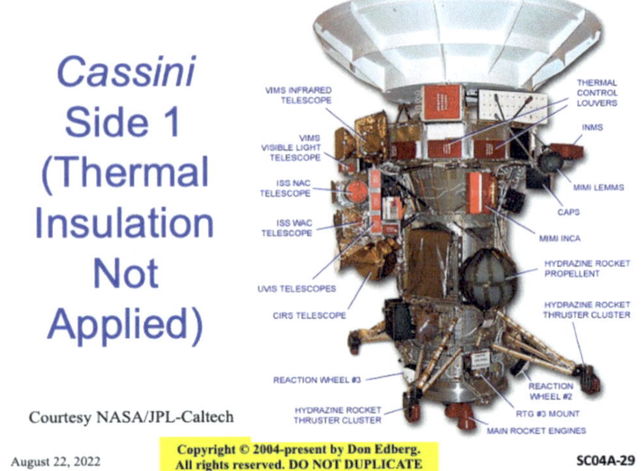

Courtesy NASA/JPL-Caltech

August 22, 2022

SC04A-29

Cassini Side 2 (*Huygens* Probe Not Installed)

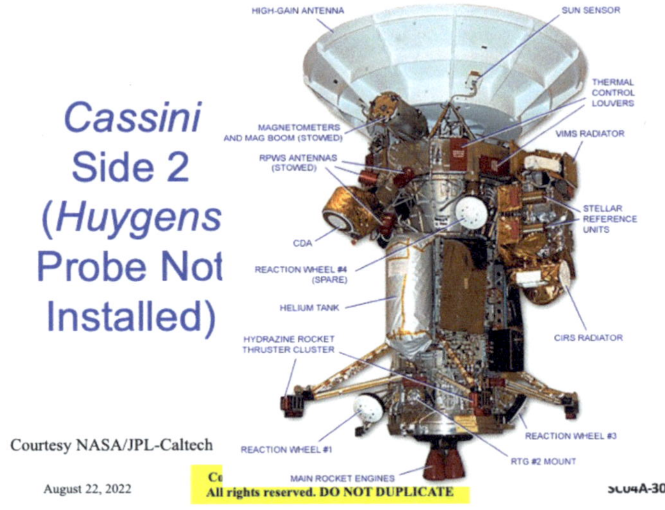

Courtesy NASA/JPL-Caltech

August 22, 2022

SC04A-30

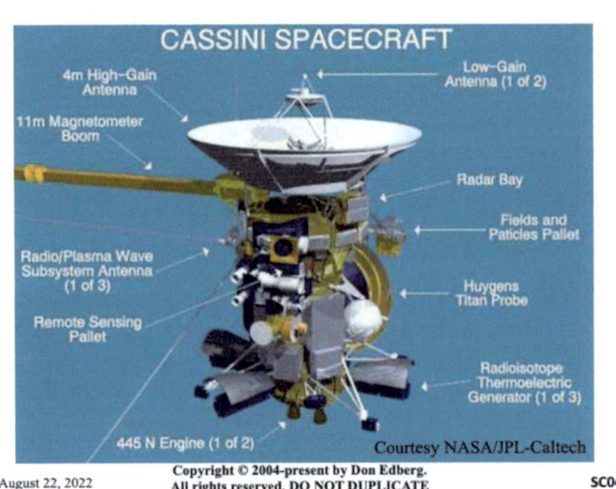

Courtesy NASA/JPL-Caltech

August 22, 2022

SC04A-31

Voyager Structure

Source = **The Voyager Neptune Travel Guide**
Author, NASA

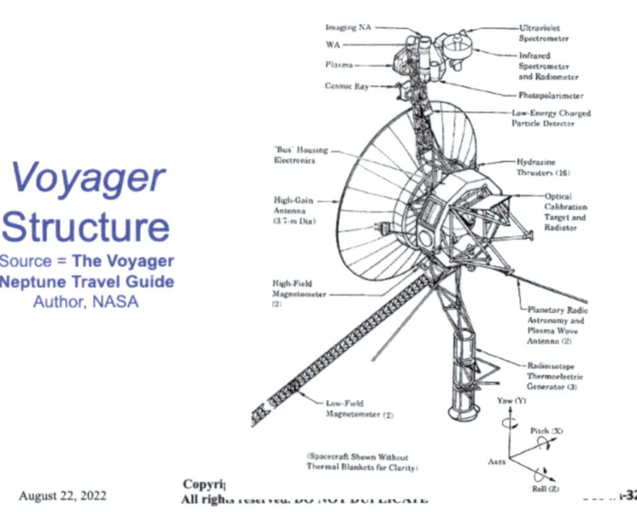

August 22, 2022

-32

LEASAT Dual-Spin SC

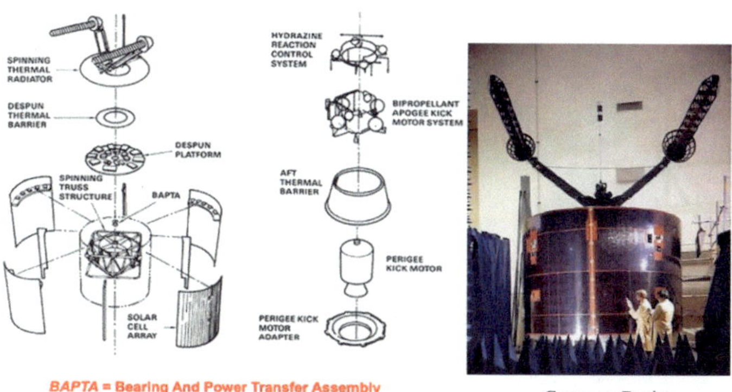

BAPTA = Bearing And Power Transfer Assembly

Courtesy: Boeing

August 22, 2022 SC04A-33

LEASAT Cutaway

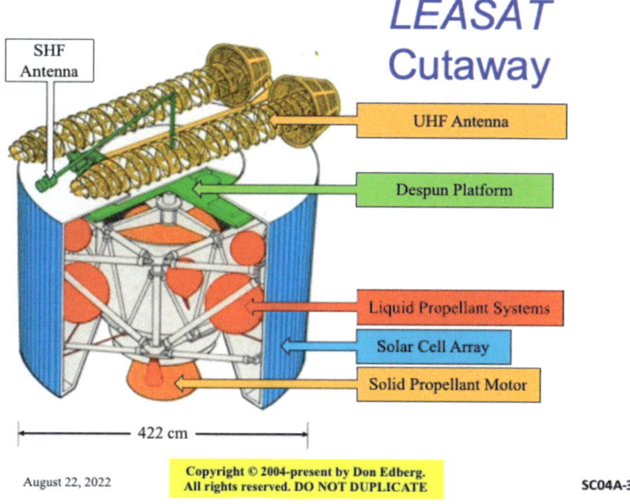

August 22, 2022 SC04A-34

Boeing 376 Dual-Spin Cutaway

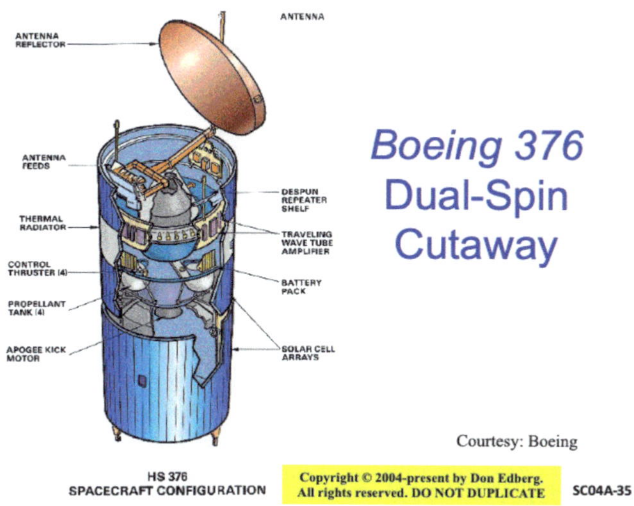

HS 376 SPACECRAFT CONFIGURATION

Courtesy: Boeing

 SC04A-35

Gravity Probe-B (GP-B)

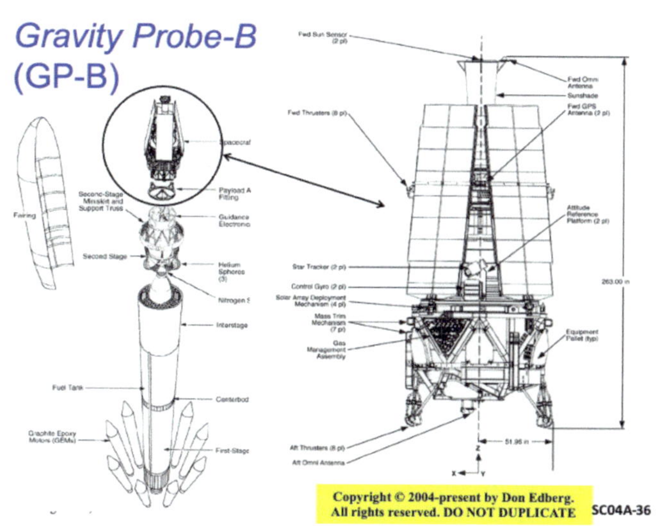

 SC04A-36

GP-B: Flight Configuration

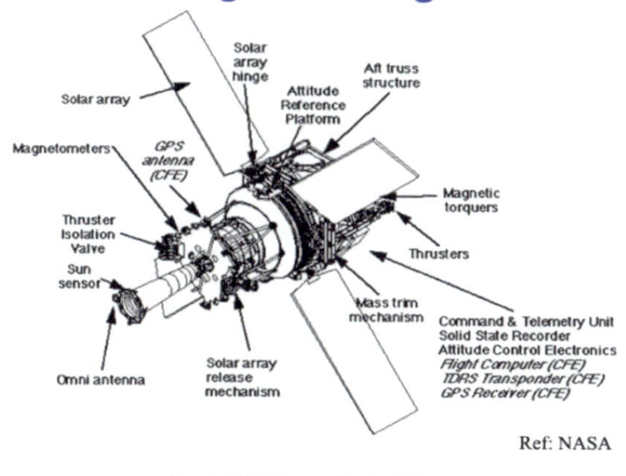

Ref: NASA

August 22, 2022 SC04A-37

Messenger

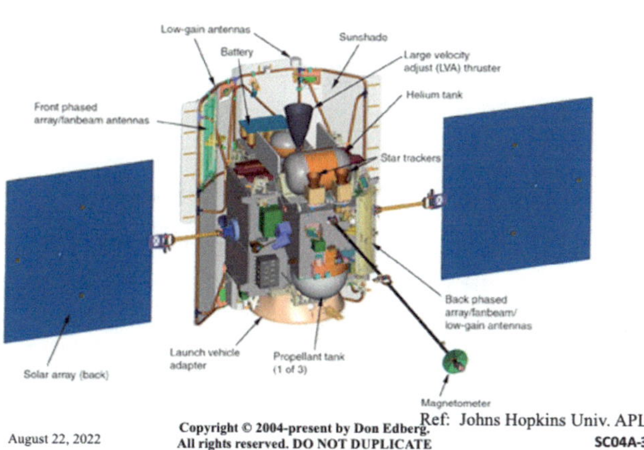

Ref: Johns Hopkins Univ. APL

August 22, 2022 SC04A-38

Design Constraints

- Spacecraft can be any shape fitting inside PLF
- Does not have to be aerodynamic or streamlined (unless it's designed for atmospheric entry, as the shuttle was)
- Shape or Configuration evaluated by: ***Configuration Evaluation Checklist*** (Modified from Brown, *Elements of Spacecraft Design*, p. 521)

August 22, 2022 SC04A-39

Topics

- Functions of structures
- Launch vs. flight configurations
- The General Arrangement; examples of different layouts
- *Spacecraft Configuration Checklist*
- Layout considerations
- Types of structures
- Structural materials (metals & composites) & selection criteria
- Composite fabrication techniques
- Structure mass estimation

August 22, 2022 SC04A-40

Config. Evaluation: FoV Verifications

1. Science instruments field of view (FoV). Can mount on "scan platform" for independent orientation. Nadir pointing if necessary
2. Antenna(e) FoV(s)
3. ACS sensor(s) FoV(s)

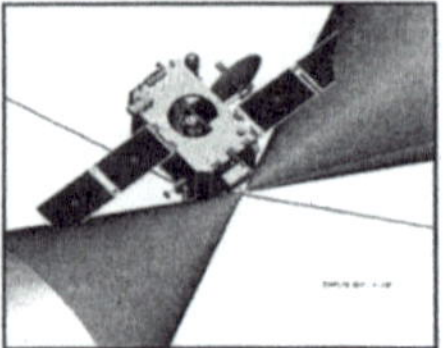

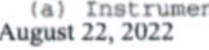

(a) Instrument

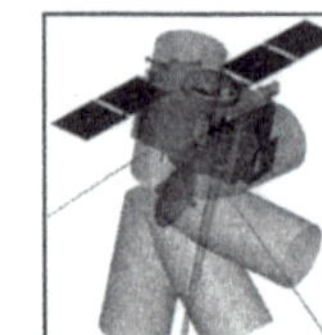

(b) High gain antenna

Ref: NASA, Pisacane

August 22, 2022 SC04A-41

Configuration Evaluation Checklist

4. Articulating structures do not interfere/contact
5. Sun-OK and sun-free faces identified
6. Light-sensitive instruments on sun-free faces
7. Telecomm electronics near antennas

August 22, 2022 SC04A-42

Configuration Evaluation Checklist (cont.)

8. ACS sensors/gyros on common structure
9. ACS thrusters have maximum moment arm, no plume issues
10. Structures identified with reasonable load paths
11. Compatible with payload attach fitting (refer to LV's *Payload Planner's Guide* for dimensional information)

August 22, 2022 SC04A-43

12. Fit in PLF including dynamic envelope? (dimensions from Payload Planners' Guides)

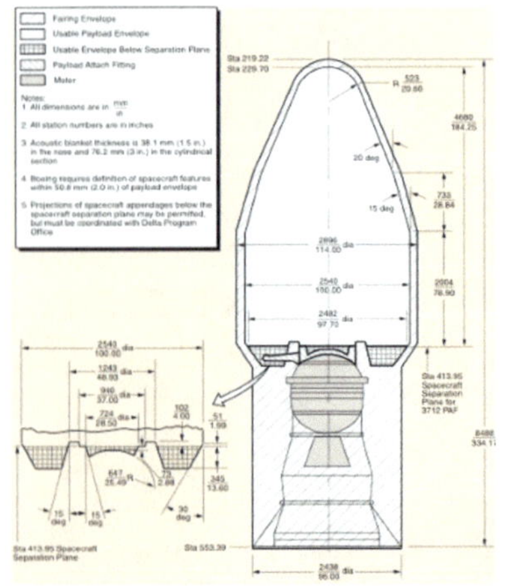

August 22, 2022 SC04A-44

13. Separation planes identified; interference-clear

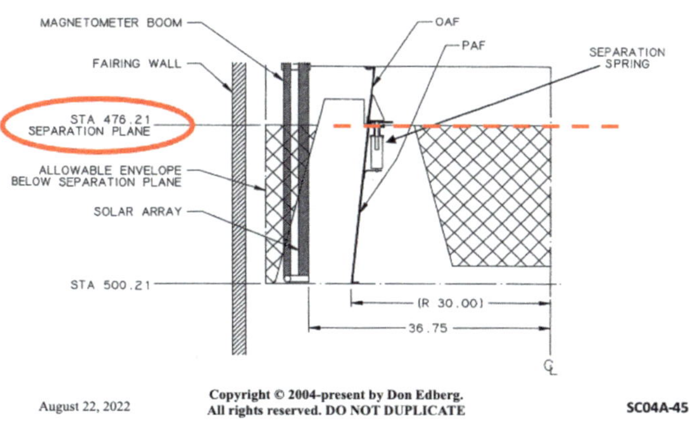

Configuration Evaluation Checklist (cont.)

14. Ordnance and separation systems shock path (keep far from equipment)
15. Propulsion engine plume impingement; clear aft of exit plane
16. Propulsion system modular, equipment w/tanks
17. Propellants centered on CM regardless of load
18. Electronics heat rejection: face views space; no sun

Configuration Evaluation Checklist (cont.)

19. Radiator heat rejection: free view of sun-free space
20. CM near geometric center
21. Acceptable moments of inertia for spinners
22. Solar panels have clear view of sun without shadows
23. Batteries together and near control electronics. C&DH near ACS if common computer used

Configuration Checklist

- If prospective design passes checklist successfully, it's ready to be evaluated analytically
- Analysis process discussed in Structural Analysis section (2nd following)

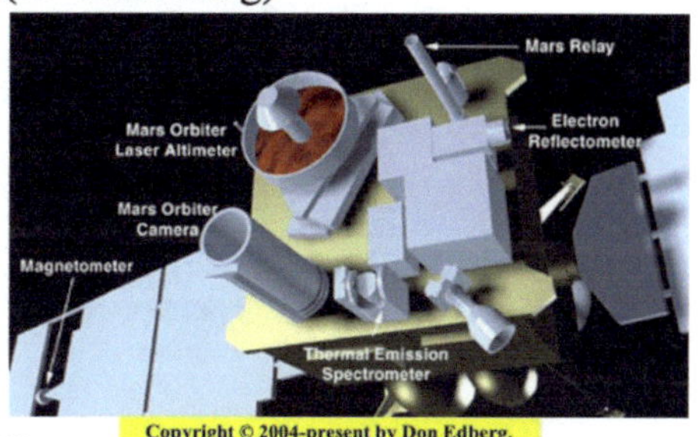

Mars Orbiter courtesy NASA

Topics

- Functions of structures
- Launch vs. flight configurations
- The General Arrangement; examples of different layouts
- *Spacecraft Configuration Checklist*
- **Layout considerations**
- Types of structures
- Structural materials (metals & composites) & selection criteria
- Composite fabrication techniques
- Structure mass estimation

Layout Considerations

- Shape
- Configuration
- Stowage of deployables
- Mounting of tanks: propellant, pressurant, etc.

Possible Solar Array Arrangements

Locate SAs to maximize sun illumination over all orbits and sun angles

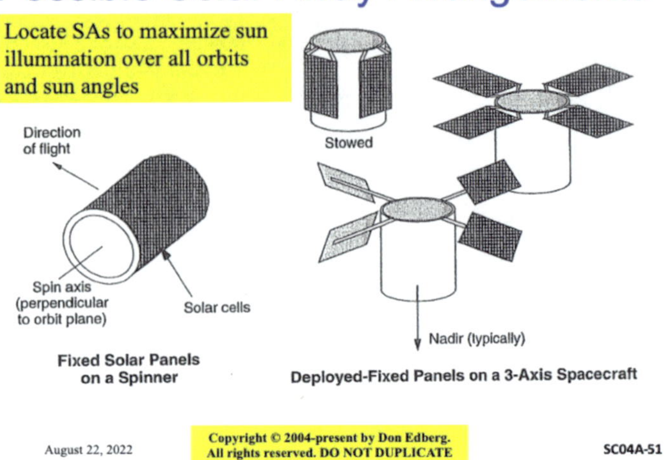

August 22, 2022

Launch Packaging

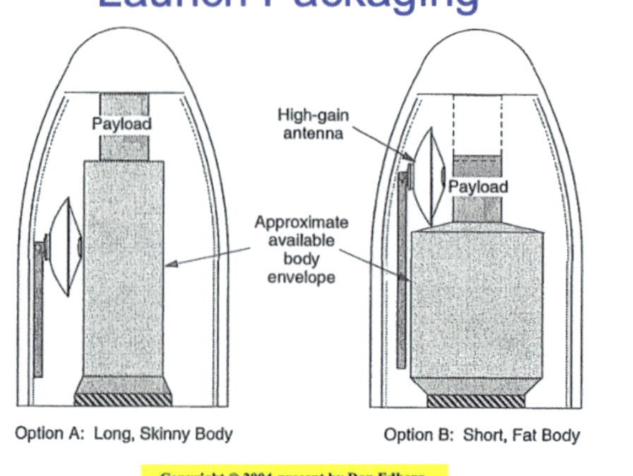

August 22, 2022

One- & Two-DoF Pointing Systems

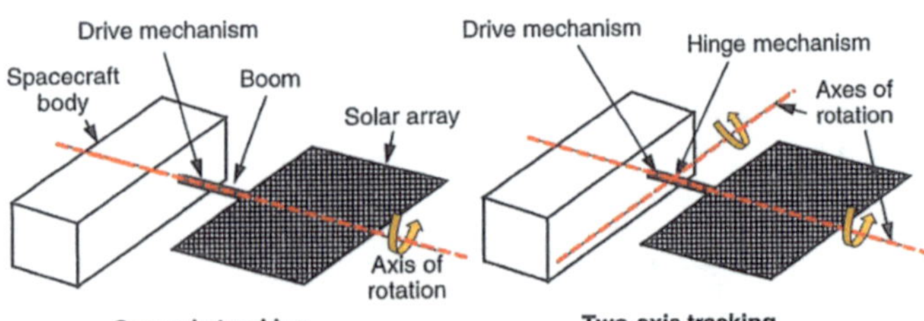

- Rotary joints are sometimes referred to as "alpha" (1st DoF) & "beta" (2nd DoF) joints
- ISS has SARJ = Solar Alpha Rotary Joint

August 22, 2022

Tank Mounting

- Avoid high stresses by introducing loads tangentially
  - Radial loads – "poking" can puncture
- Mount must provide 3-axis support
  - Usually need reinforcement of shell at mounting locations
- Two types of tank mounting
  - Polar
  - Equatorial

August 22, 2022

Polar Tank Mount

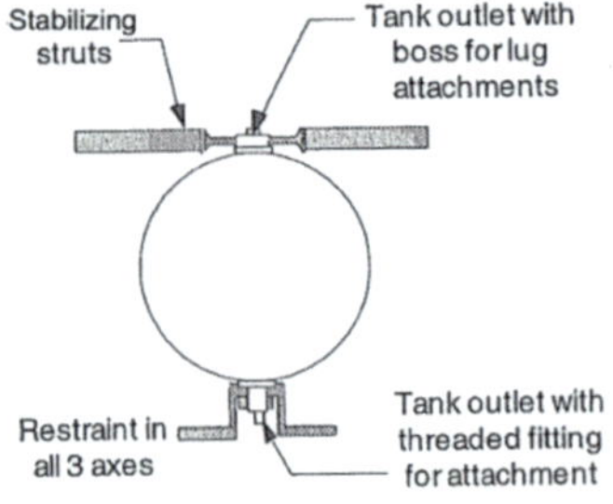

Courtesy Sarafin

- Introduce loads tangentially (NOT radially)
- Shell must be thicker at poles

August 22, 2022

Equatorial Tank Mounts

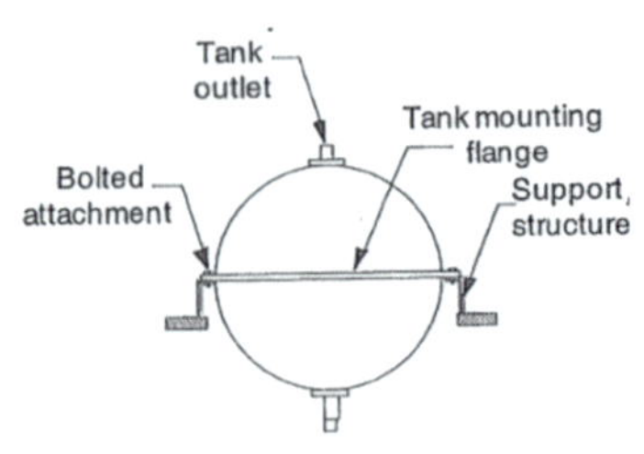

- Avoid high stresses by introducing loads circumferentially (NOT radially)
- Reinforcement required around equator of tank

Courtesy Sarafin

August 22, 2022

Topics

- Functions of structures
- Launch vs. flight configurations
- The General Arrangement; examples of different layouts
- *Spacecraft Configuration Checklist*
- Layout considerations
- Types of structures
- Structural materials (metals & composites) & selection criteria
- Composite fabrication techniques
- Structure mass estimation

August 22, 2022 SC04A-57

Structure Types

- Truss (fixed or deployable, composite or metal)
- Corrugated shell (composite or metal)
- Skin-stringer (composite or metal)
- Honeycomb plates (composite and/or metal)
- Machined waffle or isogrid (usually metal, not always)
- Deployable structures
- Filament wound (usually pressure vessels, tanks)

August 22, 2022 SC04A-58

Trusses

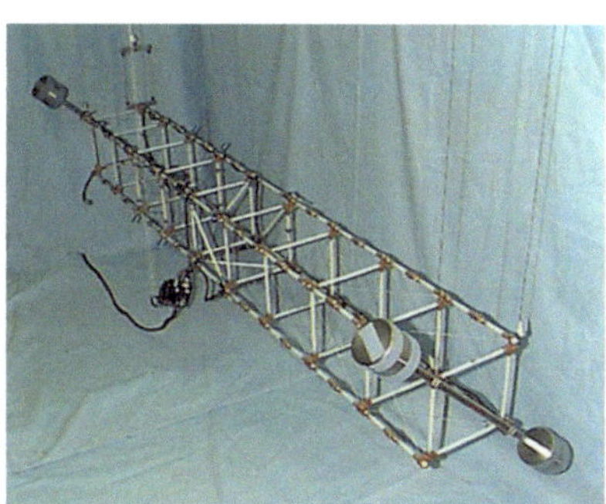

August 22, 2022 SC04A-59

Skin-Stringer Construction

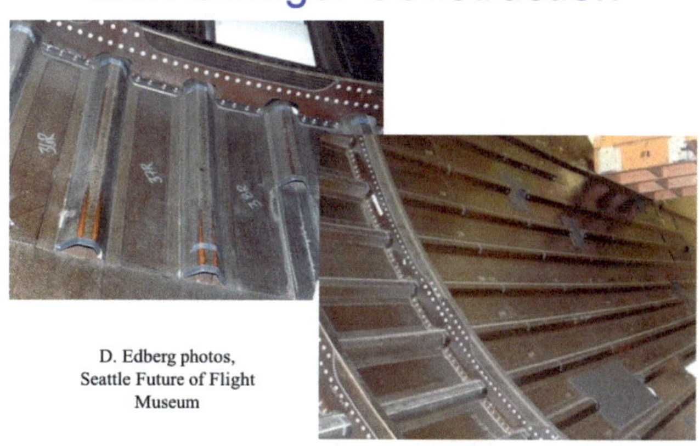

D. Edberg photos, Seattle Future of Flight Museum

August 22, 2022 SC04A-60

"Sandwich" or Honeycomb Construction (sample)

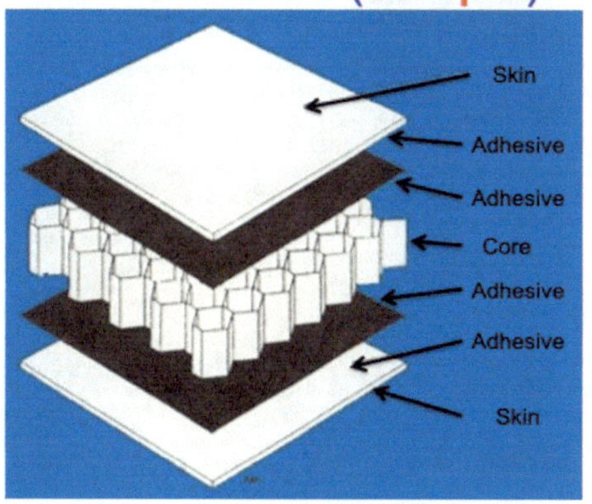

August 22, 2022 SC04A-61

Sandwich Construction Joining & Inserts

- Need to incorporate "hard points" for attachment fasteners and to join panels

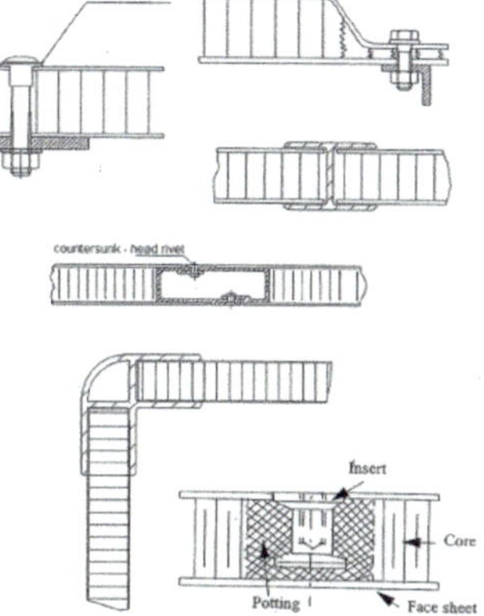

Courtesy **Spacecraft Structures**, J. Wijker

August 22, 2022 SC04A-62

August 22, 2022 SC04A-63

Topics

- Functions of structures
- Launch vs. flight configurations
- The General Arrangement; examples of different layouts
- *Spacecraft Configuration Checklist*
- Layout considerations
- Types of structures
- **Structural materials (metals & composites) & selection criteria**
- Composite fabrication techniques
- Structure mass estimation

August 22, 2022 SC04A-64

Structural Materials

Metallic materials

- Aluminum: easy to use, low strength
- Aluminum-Lithium: ~30% weight savings
- Steel alloys & stainless: high strength & temps
- Titanium: low density, high strength, high temps
- Beryllium: ultra-high stiffness, low mass, carcinogen

Composite materials

- Fiberglass ("glass")
- Aramid (Kevlar®)
- Graphite (Carbon fiber)
  - CFRP = Carbon fiber reinforced plastic
  - HT-CFRP (high strength)
  - HM-CFRP (high modulus)
  - UHM-CFRP (ultra-high modulus)

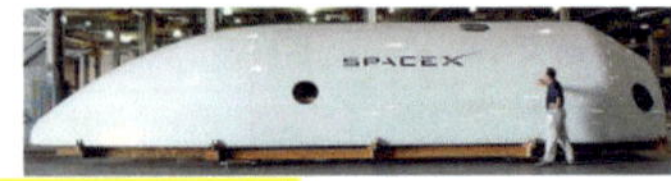

August 22, 2022 SC04A-65

Material Choice Depends On:

- Specific strength, stiffness, buckling *at operating temperature*
- Overall mass
- Machinability & weldability/shape of part to be made (complex shapes favor composites)
- Corrosion resistance/Atomic oxygen
- Outgassing/sublimation/erosion
- Electrical & thermal conductivity

NO miracle materials (*Unobtanium, Balonium, Eludium, Handwavium, Miraculum, Wishalloy, Ohmygodium*)

August 22, 2022 SC04A-66

Material Behavior

(courtesy Casey, UCSC)

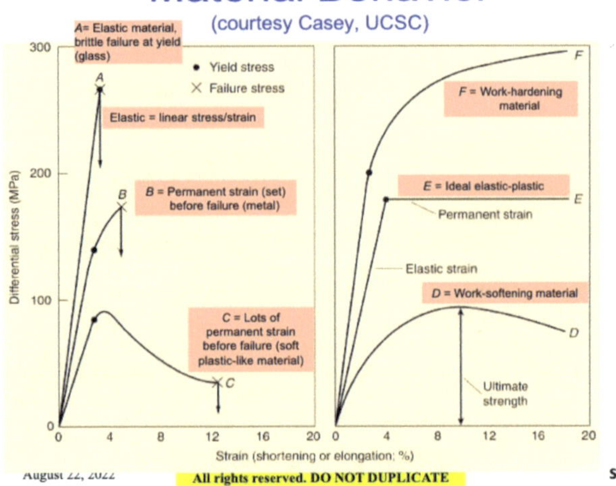

August 22, 2022 SC04A-67

Metal Material Usage Guide

Type	Advantages	Disadvantages	Applications
Aluminum 2219 alloy	High strength, resists stress corrosion cracking.	High CTE, less corrosion resistance, harder to weld.	Tank walls & skins, with coatings
Aluminum 2014 alloy	Good for cryogenics, good machining, welding, low cost	High temperature limitations, high CTE	Cryogenic tank walls
Aluminum 6xxx series	Higher strength, good welding, machining, low cost	High CTE, galling, stress corrosion issues	Truss structure, skins, stringers, brackets
Aluminum 7075 alloy	High strength, ductility, corrosion & fatigue resistance, toughness,	High CTE, susceptible to embrittlement	Highly-stressed structures
Al-Lithium e.g. 2195	Lower density & higher modulus compared to Al	Reduced ductility & fracture toughness in transverse dir.	Skins, stringers, face sheets
Titanium	Low density, high strength, good high temps, low CTE & thermal conductivity	Expensive, difficult to machine	Attach fittings for composites, thermal isolators, flexures
Steels: 4130, D6AC	High stiffness, strength, high temps ok, low cost, weldable	Heavy, magnetic, non-stainless oxidizes. Stainless galls easily	Fasteners, threaded parts, bearings & gears
High-Temp Alloys (i.e. Inconel)	High stiffness, strength at high temperatures, oxidation resistance & non-magnetic	Heavy, difficult to machine	Fasteners, thermal protection system, high temperature parts
Beryllium	ultra-high stiffness, very low mass	SUPER Expensive to machine due to *carcinogen*	Ultra stiff, ultra light structures, mirrors

Courtesy R. Farley, NASA

8/22/22 SC04A-68

Composite Material Usage Guide

Type	Advantages	Disadvantages	Applications
Carbon-Epoxy, high-strength	High strength-to-weight, high modulus-to-weight, low CTE, flight heritage	Outgasses & absorbs water depending on matrix	Monocoque cylinders, face sheets, truss members
Carbon-Epoxy, high-modulus	High strength-to-weight, very high modulus-to-weight, low CTE, high thermal conductivity	Low compressive strength, outgasses & absorbs water depending on matrix, ruptures at low strain	Monocoque cylinders, face sheets, truss members
Glass/Epoxy (continuous fiber)	Low electrical conductivity, RF transparent, well-established manufacturing process	Higher density, lower strength & modulus than Gr/E	Payload fairings, RF antenna covers, printed wire boards
Aramid/Epoxy (Kevlar® or Spectra®)	Impact resistant, lower density than Gr/E, high strength-to-weight	Absorbs water, outgasses, low compressive strength, negative CTE	RF antenna covers, armored structures (around turbopumps, etc.)
Carbon matrix	High modulus vs. weight, high temperature operation (>4,000°F)	Expensive, low strength, poor oxidation resistance	Engine nozzles, re-entry vehicle leading edges
Boron fiber metal matrix	High strength vs. weight, low CTE	Anisotropic, expensive	Truss members, Space Shuttle payload doors

From Sarafin, **Spacecraft Structures and Materials**, Table 15.16

8/22/22 SC04A-69

Topics

- Functions of structures
- Launch vs. flight configurations
- The General Arrangement; examples of different layouts
- *Spacecraft Configuration Checklist*
- Types of structures
- Layout considerations
- Structural materials (metals & composites) & selection criteria
- **Composite fabrication techniques**
- Structure mass estimation

August 22, 2022 SC04A-70

Composite Materials

- Layers or plies of fibers combined with adhesive stacked on top of each other
- Fibers can be carbon/graphite, fiberglass, aramid, others
- Stacked onto a mold, placed into autoclave for curing

Courtesy **Handbook of Space Technology**

August 22, 2022 SC04A-71

Composite Stacking

Composites can be "stacked" for different properties

- Unidirectional = all fibers in a single direction
- Bidirectional = fibers oriented in two directions
- Can be stacked in any orientation to provide desired structure properties: below is $[+\theta/0/-\theta/90]_s$

8/22/22 SC04A-72

Composite Manufacturing Process (1 of 2)

8/22/22 NASA SC04A-73

Composite Manufacturing Process (2 of 2)

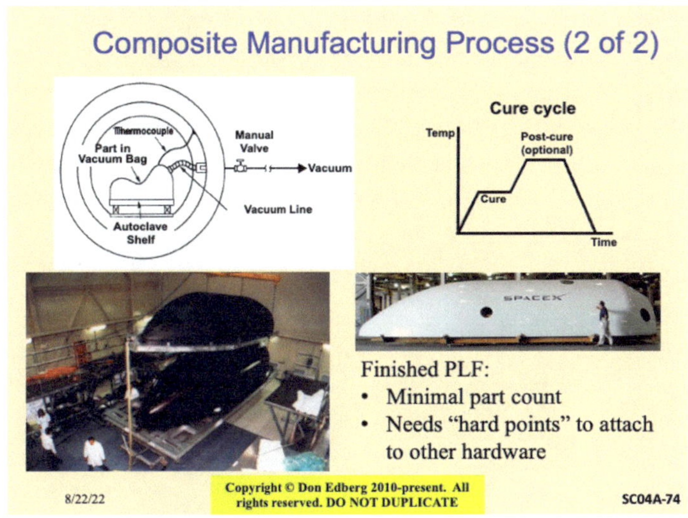

Finished PLF:

- Minimal part count
- Needs "hard points" to attach to other hardware

8/22/22 SC04A-74

Topics

- Functions of structures
- Launch vs. flight configurations
- The General Arrangement; examples of different layouts
- *Spacecraft Configuration Checklist*
- Types of structures
- Layout considerations
- Structural materials (metals & composites) & selection criteria
- Composite fabrication techniques
- Structure mass estimation

August 22, 2022 SC04A-75

Structure Mass Estimation

Structure Type	Metallic	Composite	Notes/Source
Primary structure	5 – 15* kg/m²	4 – 11* kg/m²	Brown §10.2.4.1
Equipment support structure	Use 15% of equipment mass for support structure		Brown §10.2.4.3. For mounting brackets, frames, clips, etc.
PAF or LVA (Payload Attach Fitting or Launch Vehicle Adapter)	Use actuals from LV Payload Planner's Guide. Otherwise, $m_{PAF} = 0.0755\ m_{payload} + 50$ kg		Brown Eq. 2.3, p. 27. *Only use if Payload Planner's Guide not available.*
Solar array, including substrate structure	3.6 – 4.0 kg/m², rigid array. 3.4 kg/m² if body-mounted		Brown § 6.2.4.10.
Lightly-loaded fairing	1 – 4 kg/m²	1 – 3 kg/m²	Brown §10.2.4.2*
Margin	Add 20%: allows for tolerances, joints, thermal coatings, etc.		

*Value depends on materials, factor of safety, loading type (static /dynamic)
1 lb_m/ft^2 = 4.88 kg/m². Info from Brown, **Elements of Spacecraft Design**, with additions.

August 22, 2022 SC04A-76

Simple Rules for Structures

- Empty weight is king: minimize mass, but:
- "When in doubt, make it stout!" (Mike Hand, Designer, Boeing)
- Make short, straight, and continuous load paths
  - Place heavier items lower
  - Minimize cutouts & doors
- Watch out for compression/buckling! To improve or stabilize:
  - Sandwich construction
  - Radial or circular stringers or both

8/22/22 SC04A- 77

Make Short, Straight Load Paths

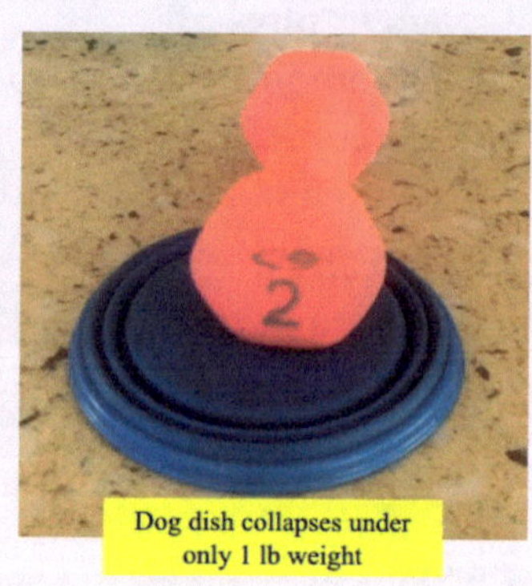

Dog dish collapses under only 1 lb weight

8/22/22 SC04A- 78

Mechanisms

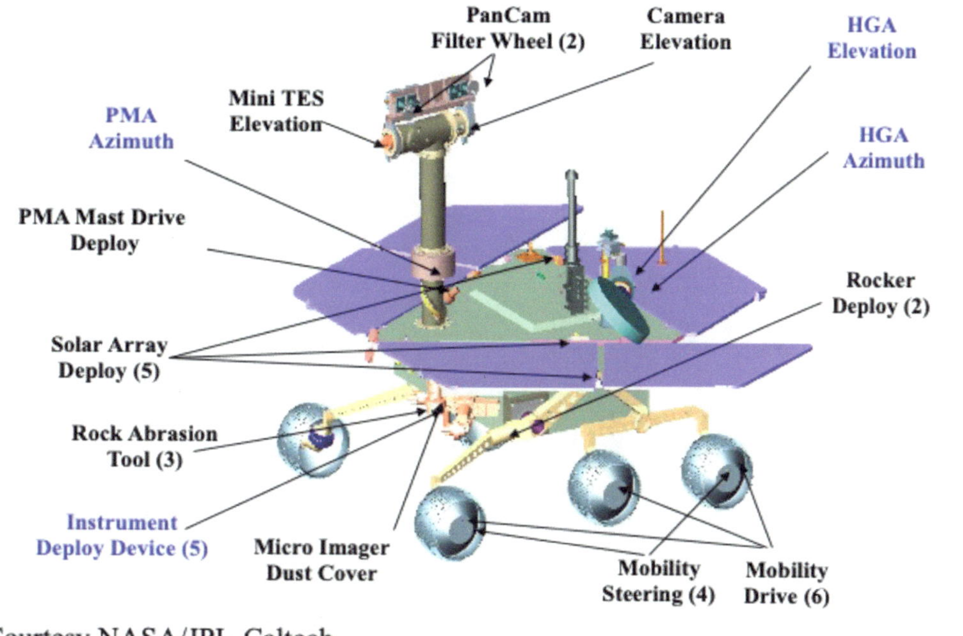

Courtesy NASA/JPL-Caltech

Topics

- Why are mechanisms needed?
- Restraints or launch locks
- Solar panel hinges/deployers
- Cable cutters & guillotines
- Pointing or articulation
- Spin bearings
- Gimbals & scan platforms
- Deployable structures
- Separation mechanisms, ordnance devices

Spacecraft Design Process

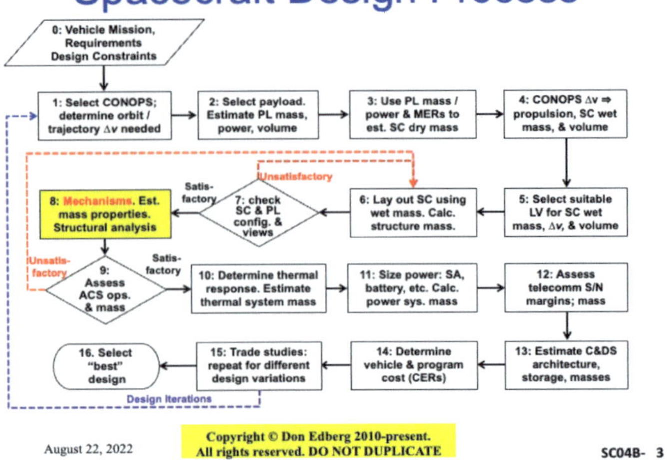
0: Vehicle Mission, Requirements Design Constraints
1: Select CONOPS; determine orbit / trajectory Δv needed
2: Select payload. Estimate PL mass, power, volume
3: Use PL mass / power & MERs to est. SC dry mass
4: CONOPS Δv ⇒ propulsion, SC wet mass, & volume
5: Select suitable LV for SC wet mass, Δv, & volume
6: Lay out SC using wet mass. Calc. structure mass.
7: check SC & PL config. & views
8: Mechanisms. Est. mass properties. Structural analysis
Unsatisfactory
Satis-factory
9: Assess ACS ops. & mass
Unsatis-factory
Satis-factory
10: Determine thermal response. Estimate thermal system mass
11: Size power: SA, battery, etc. Calc. power sys. mass
12: Assess telecomm S/N margins; mass
13: Estimate C&DS architecture, storage, masses
14: Determine vehicle & program cost (CERs)
15: Trade studies: repeat for different design variations
16. Select "best" design
Design Iterations

Topics

- Why are mechanisms needed?
- Restraints or launch locks
- Solar panel hinges/deployers
- Cable cutters & guillotines
- Pointing or articulation
- Spin bearings
- Gimbals & scan platforms
- Deployable structures
- Separation mechanisms, ordnance devices

Mechanisms

- All moving parts require mechanisms to move, extend, hold, or release things
- Mechanisms list includes all the gear needed to deploy all the items from launch to cruise to mission operations
- The number of mechanisms should be minimized — moving parts tend to fail and mess things up
  - MER on title slide had 34 actuators
- There are entire books on mechanisms (i.e. **Space Vehicle Mechanisms**, Peter L. Conley, John Wiley & Sons, 1998: ISBN 047112141X)

Speaking of Mechanisms: JWST

- Deploying & tensioning the sunshield involved
  - ✓ 139 release mechanisms
  - ✓ ~70 hinge assemblies
  - ✓ 8 deployment motors
  - ✓ ~400 pulleys
  - ✓ 90 stainless steel cables totaling 400 m in length
- 1st deployment in zero-g Jan. 2022!

Courtesy AvWeek

Flexures

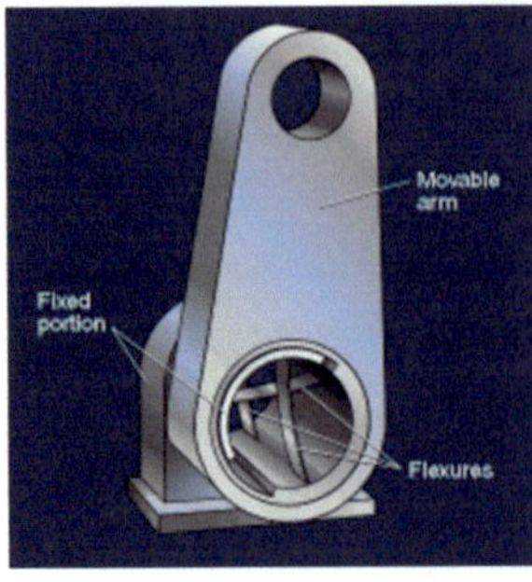
Movable arm
Fixed portion
Flexures

The lower, fixed portion of the hinge is attached to perpendicular blade flexures. The upper, rotating arm is attached to the other end of the blades. As the arm rotates, the blades flex to allow the motion.

(Courtesy Aerospace Corp.)

Topics

- Why are mechanisms needed?
- Restraints or launch locks
- Solar panel hinges/deployers
- Cable cutters & guillotines
- Pointing or articulation
- Spin bearings
- Gimbals & scan platforms
- Deployable structures
- Separation mechanisms, ordnance devices

Restraints, Launch Locks

- Items must be restrained for launch so they don't get damaged by the vibration
- Applies to any item which must be held down for launch but released for mission

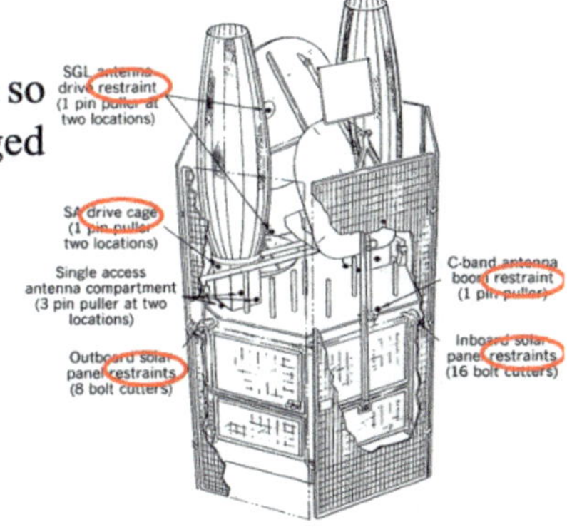

August 22, 2022

SC04B- 9

Solar Panel Launch Locks (Venus Express R, Cup & Cone stack, below)

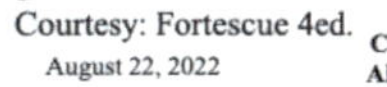

Courtesy: Fortescue 4ed.

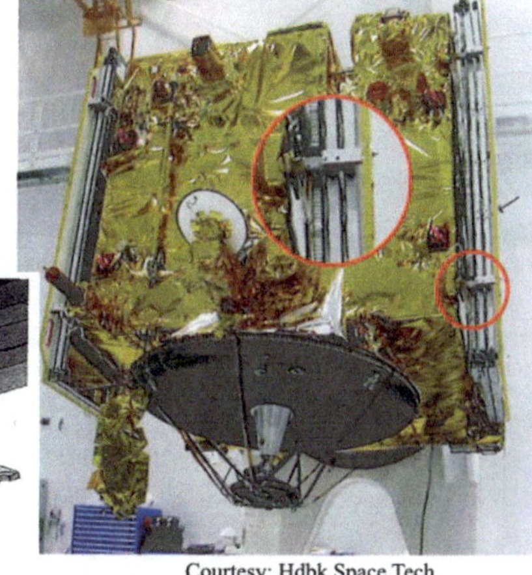

Courtesy: Hdbk Space Tech.

August 22, 2022

SC04B- 10

Topics

- Why are mechanisms needed?
- Restraints or launch locks
- **Solar panel hinges/deployers**
- Cable cutters & guillotines
- Pointing or articulation
- Spin bearings
- Gimbals & scan platforms
- Deployable structures
- Separation mechanisms, ordnance devices

August 22, 2022

SC04B- 11

Solar Panel Deployment

- Something releases panel (pin-puller, cutter, or pyro)
- Spring-loaded hinge(s) or SMA hinge(s) bring to desired shape
- Flexible wires over hinge
- Dampers added to keep motion smooth
- Add latch to keep open

Courtesy Conley

SMA Hinge Courtesy ESA

August 22, 2022

SC04B- 12

Topics

- Why are mechanisms needed?
- Restraints or launch locks
- Solar panel hinges/deployers
- **Cable cutters & guillotines**
- Pointing or articulation
- Spin bearings
- Gimbals & scan platforms
- Deployable structures
- Separation mechanisms, ordnance devices

August 22, 2022

SC04B- 13

Cable Cutter (from Dawn movie)

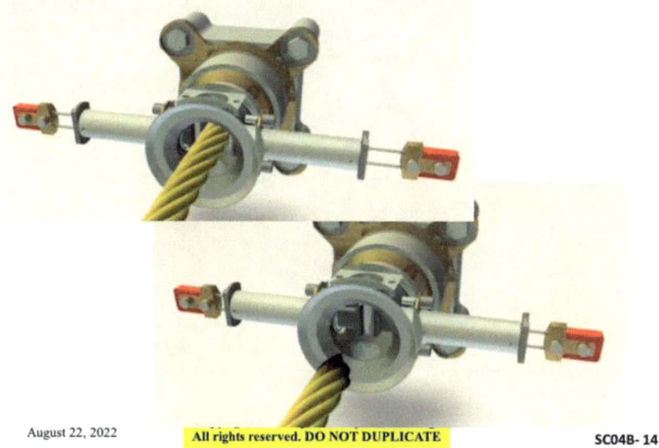

August 22, 2022

SC04B- 14

Dawn Spacecraft Mechanisms Video

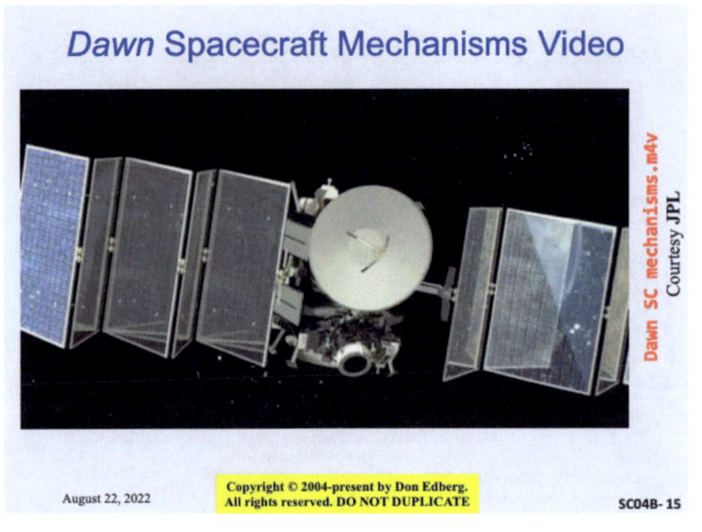

August 22, 2022

SC04B- 15

Lunar Module Guillotine Umbilical Cutter

Initiators and explosive charges are at both ends — cable will be cut as long as one of the two ends fires

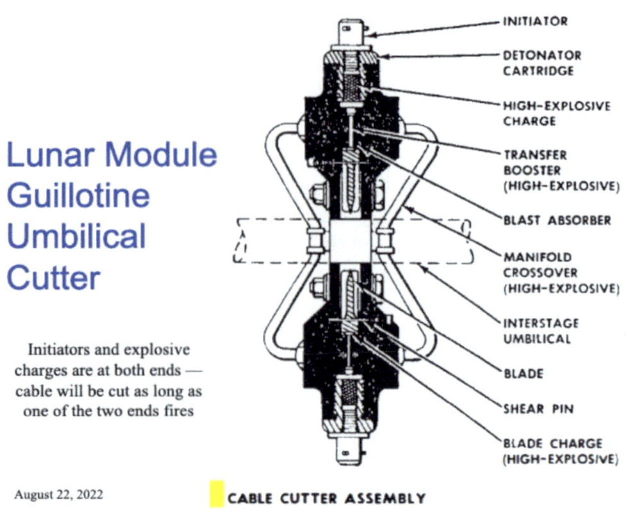

August 22, 2022

CABLE CUTTER ASSEMBLY

Topics

- Why are mechanisms needed?
- Restraints or launch locks
- Solar panel hinges/deployers
- Cable cutters & guillotines
- **Pointing or articulation**
- Spin bearings
- Gimbals & scan platforms
- Deployable structures
- Separation mechanisms, ordnance devices

August 22, 2022

SC04B- 17

Solar Array (SA) Pointing or Articulation

ISS SARJ shown
(from SPG Media)

- SAs continuously need to point at sun & track while payloads look elsewhere
- May operate in one or two axes
- Must pass power across moving joint
  - Most common: slip ring or roll ring (sliding contacts)
  - Less common: cable wrap assembly (like car steering wheel)

August 22, 2022

SC04B- 18

ATV Solar Drive Mechanism

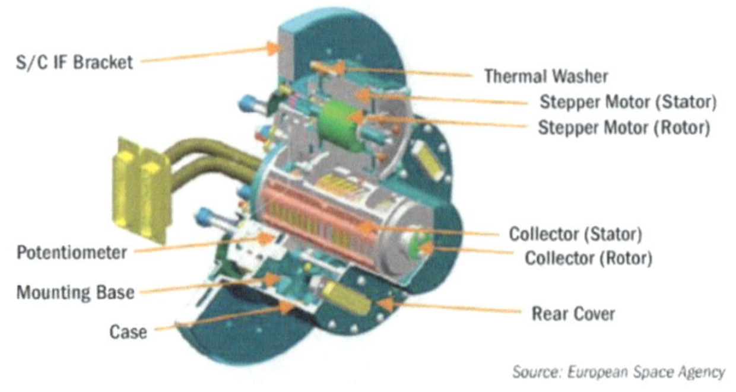

Source: European Space Agency

August 22, 2022

SC04B- 19

Topics

- Why are mechanisms needed?
- Restraints or launch locks
- Solar panel hinges/deployers
- Cable cutters & guillotines
- Pointing or articulation
- **Spin bearings**
- Gimbals & scan platforms
- Deployable structures
- Separation mechanisms, ordnance devices

August 22, 2022

SC04B- 20

Spin Bearings & Slip Rings

- Needed in dual-spin spacecraft
- Transfer mechanical loads and electrical power and signals back and forth
- Expensive to build and difficult to test

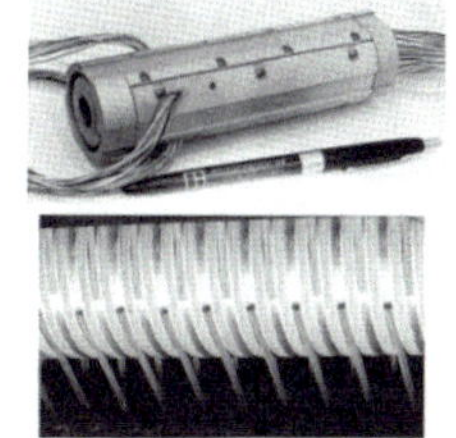

Courtesy: NASA/TRW (Conley)

Topics

- Why are mechanisms needed?
- Restraints or launch locks
- Solar panel hinges/deployers
- Cable cutters & guillotines
- Pointing or articulation
- Spin bearings
- Gimbals & scan platforms
- Deployable structures
- Separation mechanisms, ordnance devices

Gimbals and Scan Platforms

- Control attitude of narrow FOV science instruments or antennae
- Must secure for launch

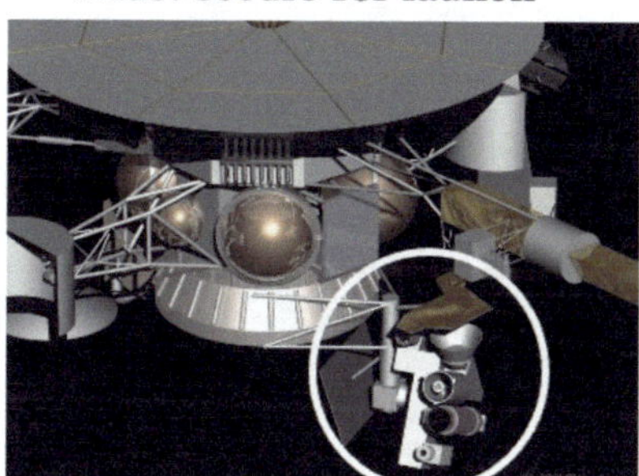

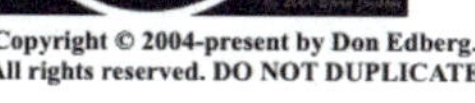

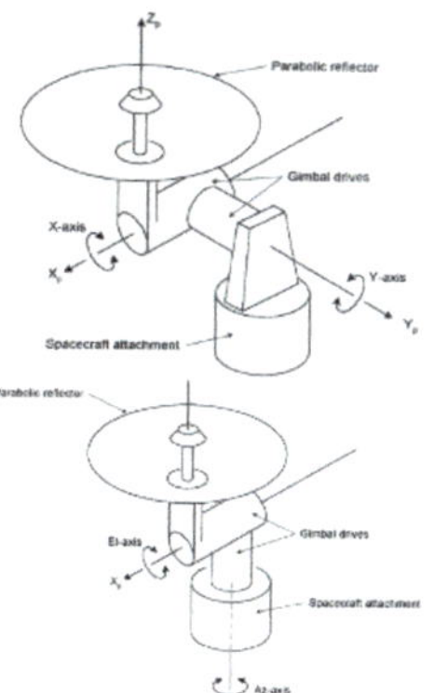

Courtesy: Conley

Topics

- Why are mechanisms needed?
- Restraints or launch locks
- Solar panel hinges/deployers
- Cable cutters & guillotines
- Pointing or articulation
- Spin bearings
- Gimbals & scan platforms
- Deployable structures
- Separation mechanisms, ordnance devices

Deployable Structures

- Booms & trusses used after launch to
  - Separate parts of spacecraft, or
  - Grab samples
- Boom/Truss Types:
  - Tape-Spring joint
  - Prestressed, rolled material
  - Deployable truss, coiled longerons. Very efficient lightweight, stiff structures. (AstroMast, Ablemast)
- Collapsing antennas & solar arrays (SAs)

Tape-Spring Folded Joint

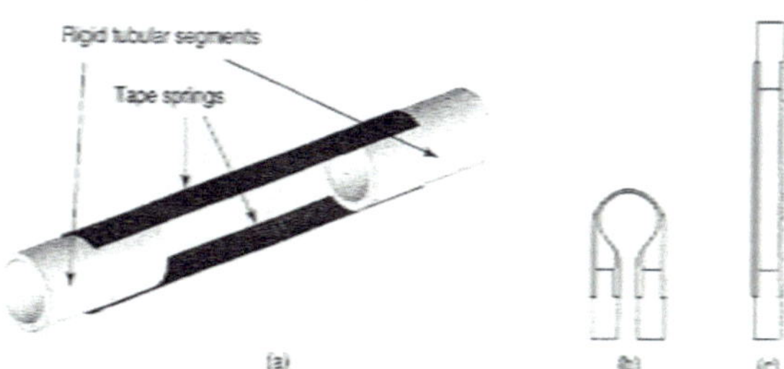

- 3D view of joint (a), folded joint (b), and unfolded joint (c)

Courtesy Fortescue

Spring Hinge at NGC

D. Edberg photo. Looks like part of a tape measure!

Rolled Extendable Booms

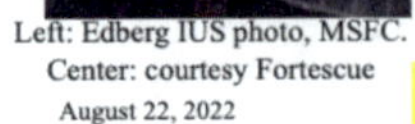

Typical cross sections for deployable tubular booms

(a) Lenticular welded double element

(b) Overlapping single element

(c) Overlapping double element

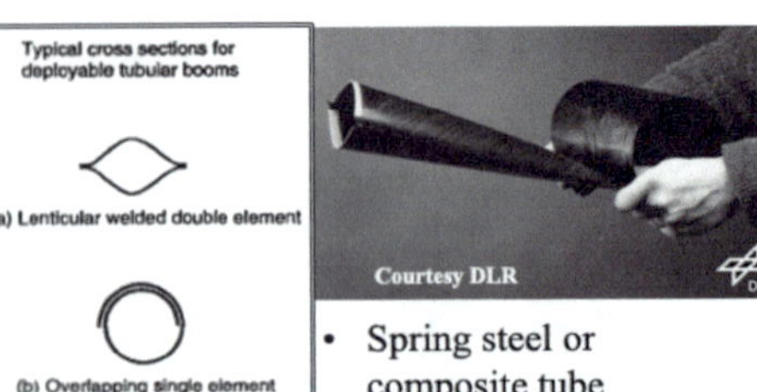

- Spring steel or composite tube
- Like a tape measure!
- Rolls flat but springs outward for bending stiffness

Left: Edberg IUS photo, MSFC.
Center: courtesy Fortescue

Coiled Extendable Booms

ABLE_Boom_deployment_02.avi

Double-click to show movie

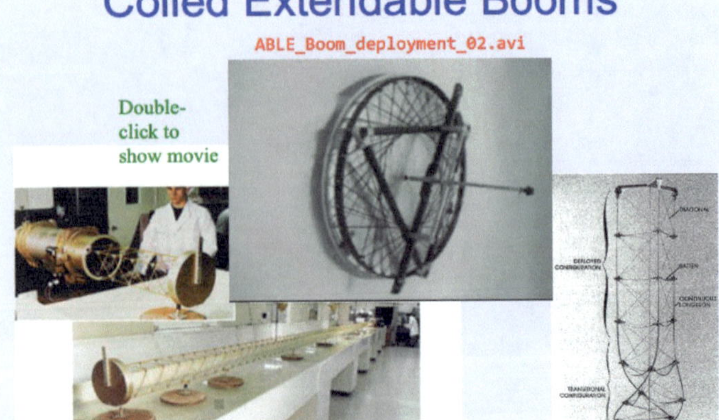

Collapsing Antennas

The 112 kg Alphasat reflector (above) forms a precision 14 x 11 meter RF reflective antenna surface in space. It stows for launch into a compact package (left) Courtesy Northrop Grumman

Mechanisms: James Webb Space Telescope

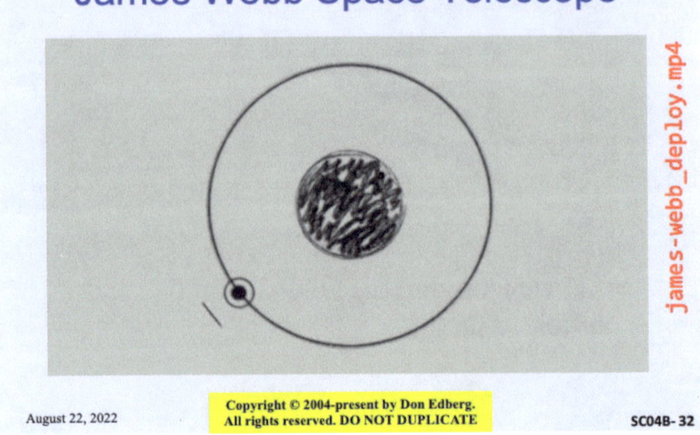

james-webb_deploy.mp4

Mechanism Design Margins

- For actuators commanding motion, add margin to account for increased needs at worst-case conditions:
  - Low temperatures
  - Stiff cabling
  - Vacuum friction (high)
- These mean additional torque or force required
- Margin of at least 100% (doubling) recommended

Topics

- Why are mechanisms needed?
- Restraints or launch locks
- Solar panel hinges/deployers
- Cable cutters & guillotines
- Pointing or articulation
- Spin bearings
- Gimbals & scan platforms
- Deployable structures
- Separation mechanisms, ordnance devices

August 22, 2022

Separating Spacecraft From LV

Pyrotechnic or *pyro* devices utilize explosives & produce shocks:
  - "Frangible" or intentionally-breakable pyrotechnic nuts or bolts (good for point attachments)
  - Clamp band: good for circular shell structures or motor casings where you don't want point loads.
- Non-pyrotechnic release mechanisms: minimize shock:
  - Hold-down & release mechanisms from Ensign-Bickford, others
  - Pin pullers, cable cutters
- More info in Lecture 13 SC03B Mechanisms

August 22, 2022

Ordnance/Pyro Devices

- Squibs or initiators (small charges of explosive a.k.a. "ordnance" or **pyro**technics) used to activate devices
- Ordnance devices are really reliable ways to make things happen ***a single time***
  - Explosive bolts
  - Explosive nuts or pyro nuts
  - Pin pullers
  - Wire cutters/guillotines
  - Valves (in Propulsion section)

August 22, 2022

Frangible Bolt

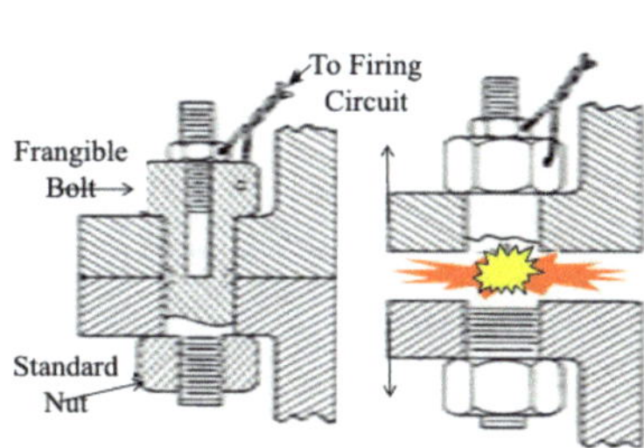

Courtesy Hobbs

Before & After
Courtesy Aerospace *Crosslink*

August 22, 2022

Lunar Module Separation Bolt

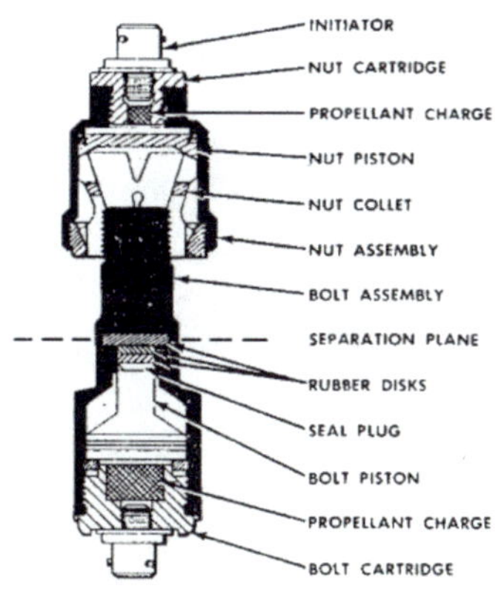

INTERSTAGE STRUCTURAL CONNECTION NUT AND BOLT ASSEMBLY

August 22, 2022

Frangible *Nut*

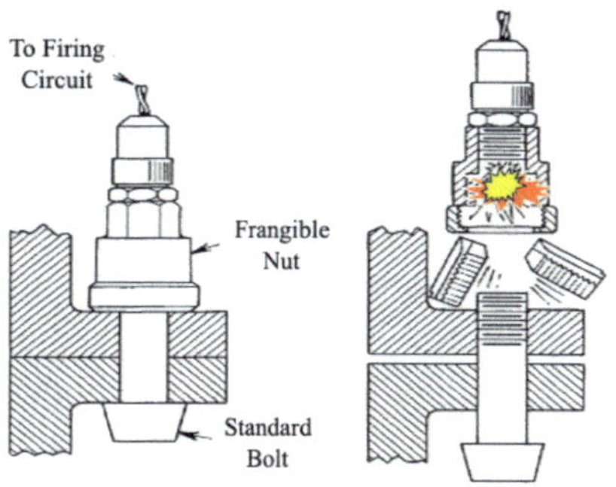

Courtesy Hobbs

August 22, 2022

Clamp Band (Marman Clamp)

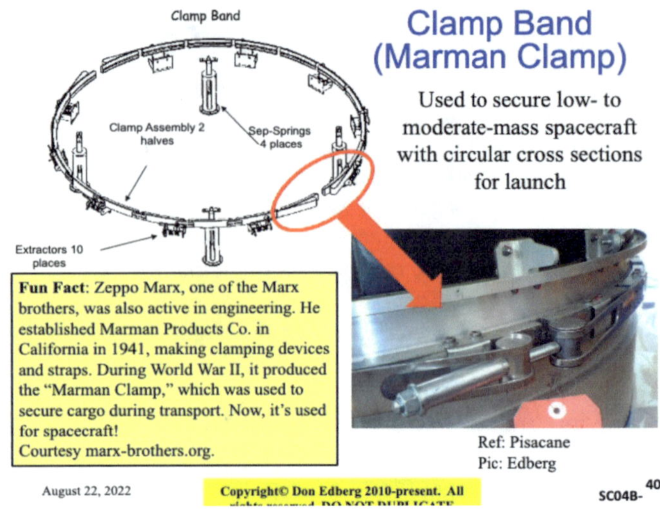

Used to secure low- to moderate-mass spacecraft with circular cross sections for launch

Fun Fact: Zeppo Marx, one of the Marx brothers, was also active in engineering. He established Marman Products Co. in California in 1941, making clamping devices and straps. During World War II, it produced the "Marman Clamp," which was used to secure cargo during transport. Now, it's used for spacecraft!
Courtesy marx-brothers.org.

Ref: Pisacane
Pic: Edberg

Clamp Band Cross-Section

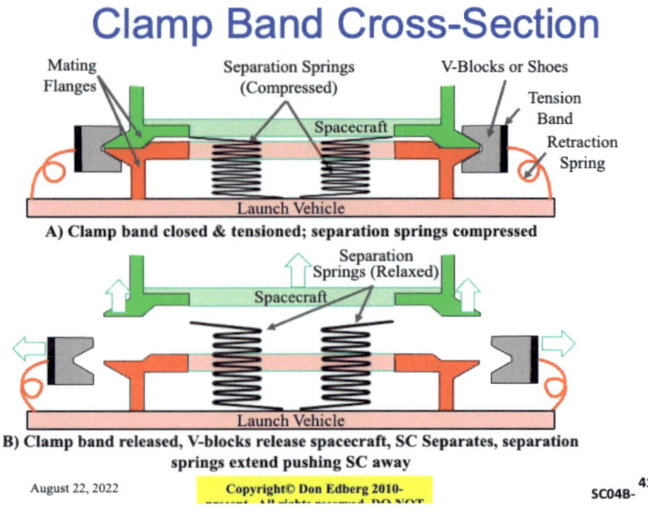

A) Clamp band closed & tensioned; separation springs compressed

B) Clamp band released, V-blocks release spacecraft, SC Separates, separation springs extend pushing SC away

Clamp Band, V-Band, Marmon Clamp Details

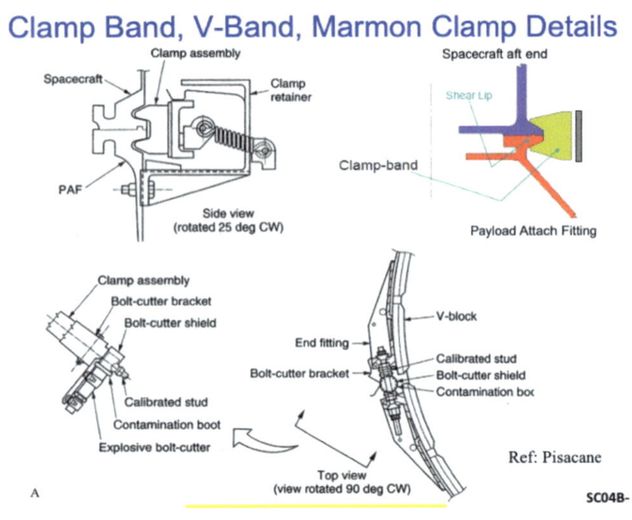

Ref: Pisacane

Lunar Module Guillotine Umbilical Cutter

Initiators and explosive charges are at both ends — cable will be cut as long as one of the two ends fires

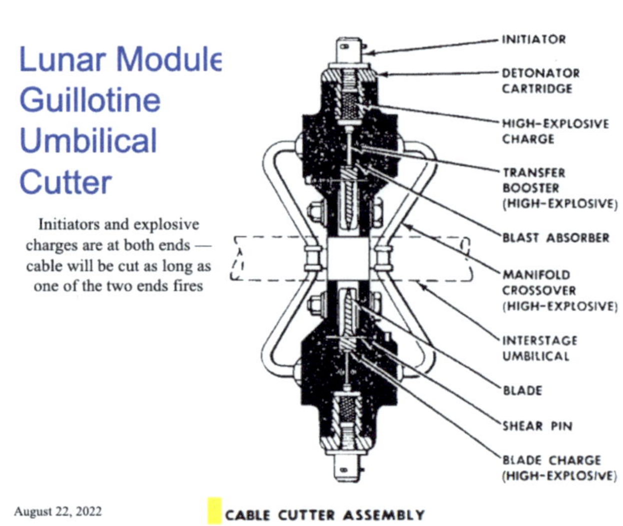

Low-Shock Mechanism Deployment

- Recent low-shock separation mechanisms look promising

Examples:

- Hold Down & Release Mechanism (HDRM) from Sierra Nevada
- HDRM from Ensign-Bickford Aerospace & Defense

Courtesy Sierra Nevada Corp.

Courtesy EBAD

Frangibolt® Low-Shock System

Shape Memory Alloy (Actuator)
Notched Bolt
Separation Plane
Broken Bolt in Tension
Heater and Insulation
Actuator Elongated

Courtesy TiNi Aerospace, Inc.

Pin Pullers

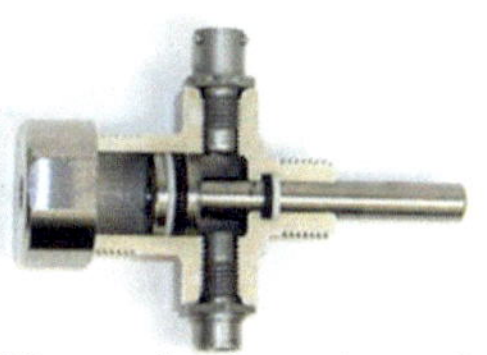

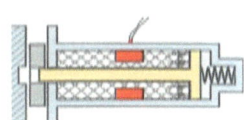

- The mechanism retracts the pin to release an arm or panel that had been held in a compacted or stowed configuration for flight.
- On right: Melting wax pin puller

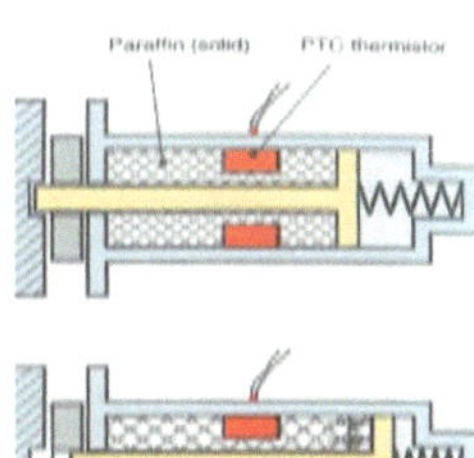

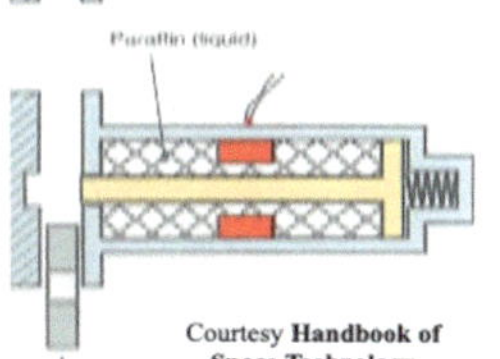

Courtesy **Handbook of Space Technology**

August 22, 2022

SC04B- 46

Non-Pyro: Motorized Lightband (MLB)

- Offered in a range of sizes from 20-96 cm (8-38 in) bolt circle diameter
- Up to 540 kg (1200 lb) payload
- Non-pyrotechnic, low shock
- Extensive flight heritage
- Made of aluminum: lower ring has hinged leaves, springs, and a dual redundant release motor; upper ring has a load-bearing hinge interface
- Some include isolation system

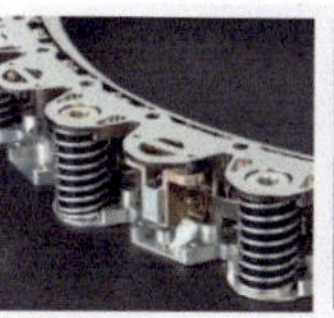

Courtesy: https://planetarysystemscorp.com

August 22, 2022

SC04B- 47

Non-Pyro Payload Adapter: Lightband

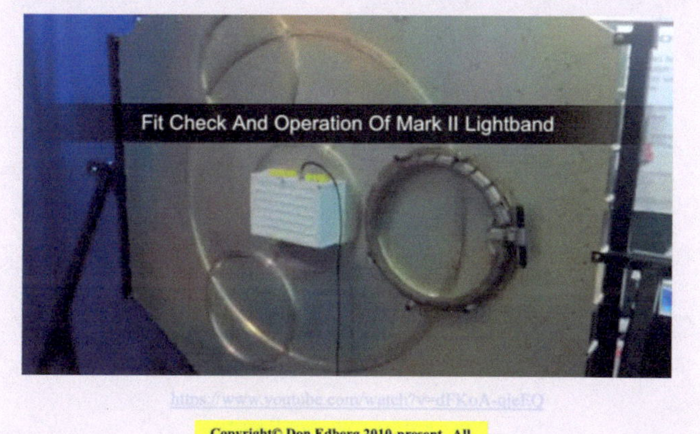

https://www.youtube.com/watch?v=dFKoA-qleEQ

1/28-30/20

SC04B-48

Inflatable Antenna Experiment (L'Garde)

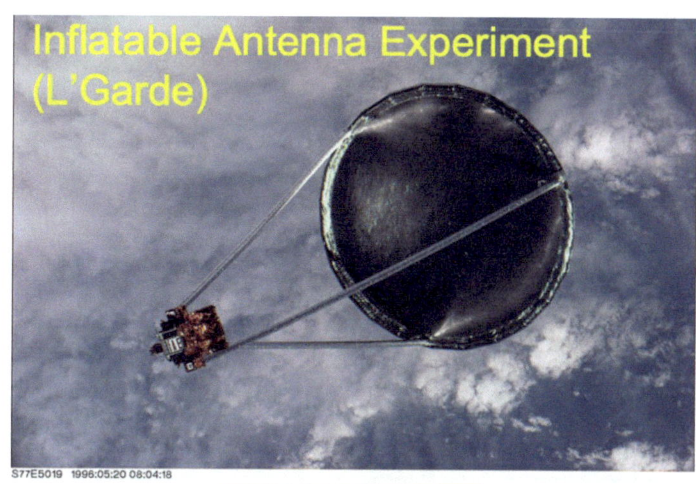

August 22, 2022

SC04B- 49

Analysis of Structures

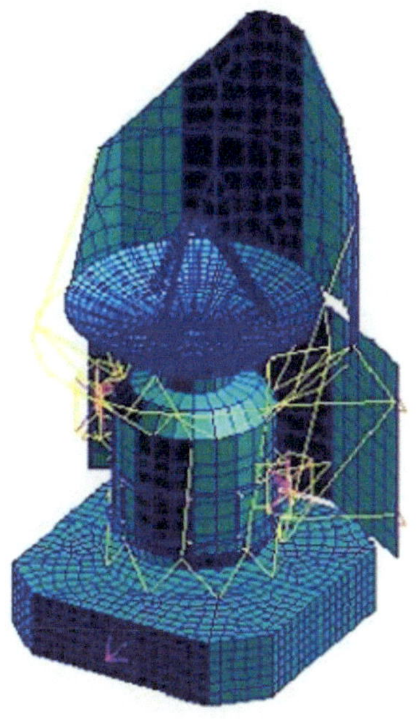

Planck Structural Model, Courtesy ESA

August 22, 2022

Structural Analysis Topics

- Definitions: structure & stress terminology
- Design process
- Static structural analysis
  - Stress/strength calculations
- Dynamic structural analyses
  - Normal modes, forced & random vibration, shock & acoustics
- Finite-Element Modeling (FEM)
  - Coupled-Loads Analysis (CLA)

August 22, 2022

Spacecraft Design Process

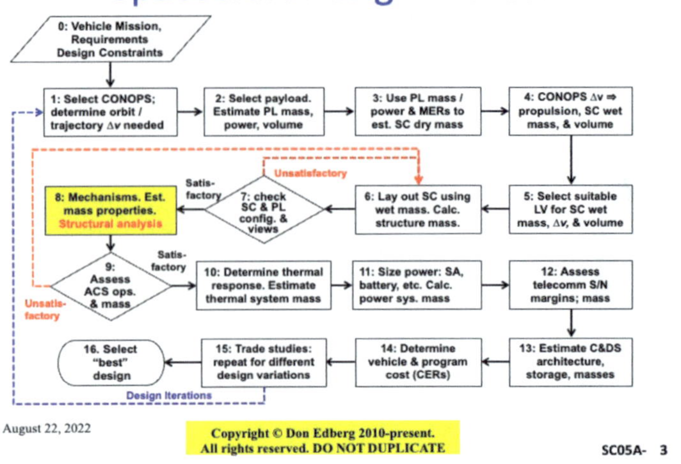

0: Vehicle Mission, Requirements Design Constraints
1: Select CONOPS; determine orbit / trajectory Δv needed
2: Select payload. Estimate PL mass, power, volume
3: Use PL mass / power & MERs to est. SC dry mass
4: CONOPS Δv ⇒ propulsion, SC wet mass, & volume
5: Select suitable LV for SC wet mass, Δv, & volume
6: Lay out SC using wet mass. Calc. structure mass.
7: check SC & PL config. & views
8: Mechanisms. Est. mass properties. Structural analysis
9: Assess ACS ops. & mass
10: Determine thermal response. Estimate thermal system mass
11: Size power: SA, battery, etc. Calc. power sys. mass
12: Assess telecomm S/N margins; mass
13: Estimate C&DS architecture, storage, masses
14: Determine vehicle & program cost (CERs)
15: Trade studies: repeat for different design variations
16. Select "best" design
Unsatisfactory
Satis-factory
Satis-factory
Unsatis-factory
Design Iterations

August 22, 2022

Structural Analysis Topics

- Definitions: structure & stress terminology
- Design process
- Static structural analysis
  - Stress/strength calculations
- Dynamic structural analyses
  - Normal modes, forced & random vibration, shock & acoustics
- Finite-Element Modeling (FEM)
  - Coupled-Loads Analysis (CLA)

August 22, 2022

Structure & Mechanical Terms

- *Stiffness*: deflection under load
- *Strength*: resistance to some failure (cracking, rupture, etc.)
  - Stiffness and strength are two different things: Glass is *stiff* but not *strong* (brittle), Plain plastic is *strong* but not *stiff*
- *Yield*: permanent deformation occurs at sufficient load or stress (paper clip bends)
- *Ultimate*: material begins to fail
- *Failure*: the material physically breaks

August 22, 2022

Tensile Stress

- *Tension*, *Tensile* = stress caused by stretching or 'pulling' loads
- Stress has positive or '+' sign

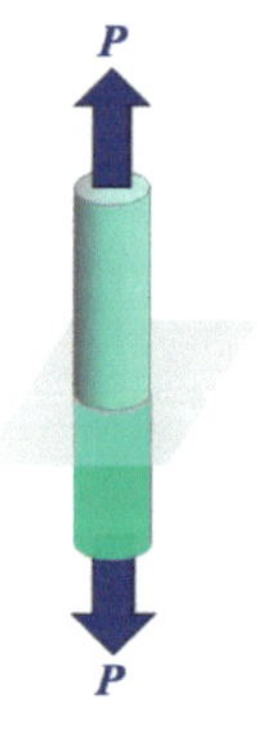

P
P

Tensile Stress

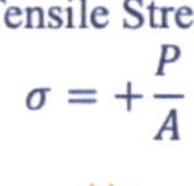

$$\sigma = +\frac{P}{A}$$

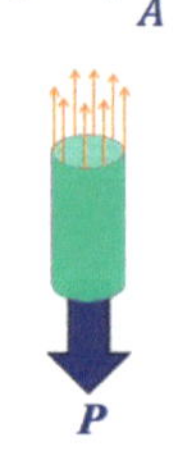

P

8/22/22

Courtesy: efunda.com

Compressive Stresses

- *Compression* or *Compressive* = stress caused by being "squashed" or shortened
- Stress has negative or '–' sign

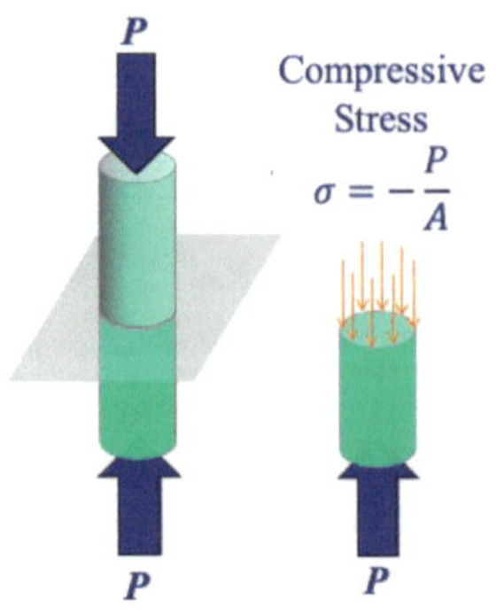

P
P
P

Compressive Stress

$$\sigma = -\frac{P}{A}$$

Courtesy: efunda.com

8/22/22

Bending Stresses/Loads

- *Bending* creates a combination of tension, compression, & shear in different areas of an object

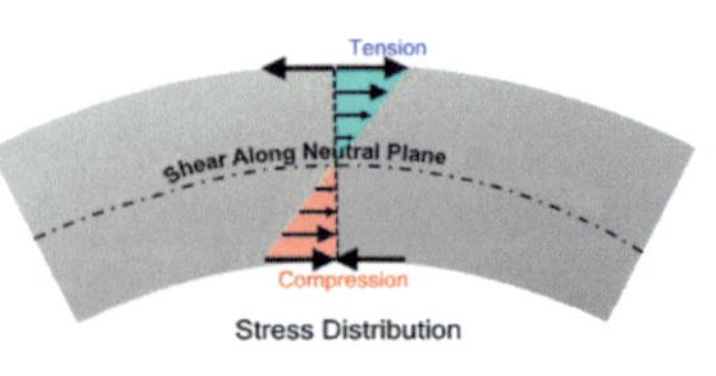

Tension
Shear Along Neutral Plane
Compression
Stress Distribution

Courtesy tutelman.com

8/22/22

Shear Stress From Pure Shear

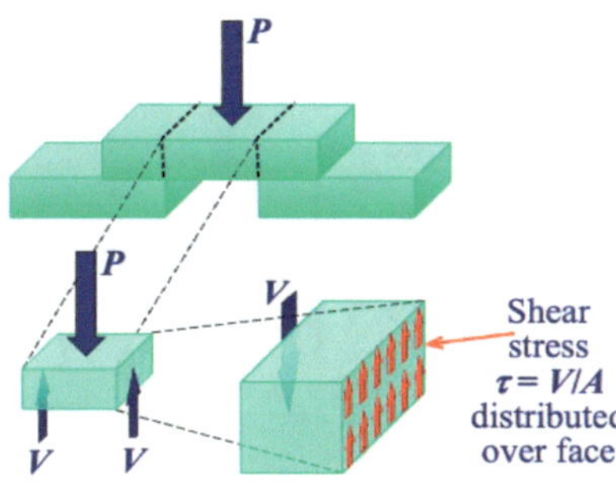

8/22/22

Shear Stress in Fastener

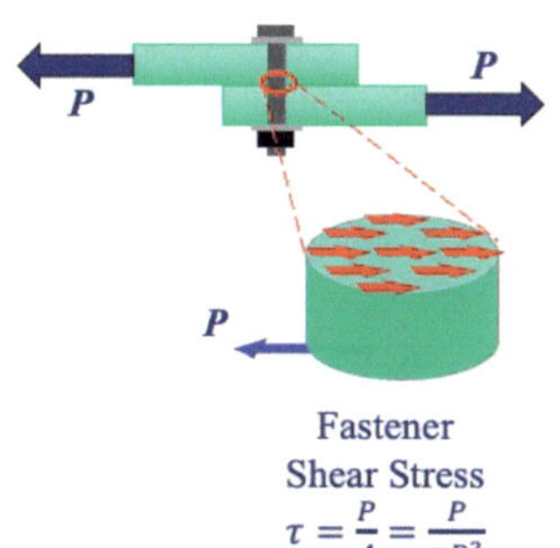

8/22/22

Stress Terms, Subscripts & Material Properties

σ_t = tensile stress (positive, > 0)
σ_c = compressive stress (negative, < 0)
σ_u = ultimate stress
$\sigma_y = Y$ = yield stress
E = modulus of elasticity (Young's modulus)
F.S. = Factor of Safety (multiplier for uncertainty)
G = shear modulus
M.S. = Margin of Safety (a measure of capability of a part after applying load)
ν = Poisson ratio (Greek letter "nu")
α = coefficient of thermal expansion (CTE)

August 22, 2022

Factors & Margins of Safety

- *Factors of safety* (F.S.) tell how much stronger the part has to be than the limit load on the part
  - Typical F.S. values:
    ELVs (expendable launch vehicles) = 1.25
    Shuttle = 1.4
    Pressure vessels = 2.0 ("no test" factory of safety)
- *Design Ultimate Load*: yield condition multiplied by factor of safety
- *Margin of Safety* (M.S.): measures the capability of a part after applying load

$$\text{M.S.} = \frac{\text{allowable stress}}{\text{F.S.} \times \text{actual stress}} - 1$$

M.S. given in percent (%)

August 22, 2022

Example: Margin of Safety

- Protoflight structural member experiences stress of 30 ksi at limit load condition
- Material yield strength is 42 ksi
- Factor of Safety = 1.25 for protoflight
- $\text{M.S.} = \frac{\text{allowable stress}}{\text{F.S.} \times \text{actual stress}} - 1$
  $= \frac{42\text{ ksi}}{1.25 \times 30\text{ ksi}} - 1 = +0.12 = +12\%$
- Always want positive M.S.!

August 22, 2022

Structural Analysis Topics

- Definitions: structure & stress terminology
- Structure design process
- Static structural analysis
  - Stress/strength calculations
- Dynamic structural analyses
  - Normal modes, forced & random vibration, shock & acoustics
- Finite-Element Modeling (FEM)
  - Coupled-Loads Analysis (CLA)

August 22, 2022

Structures Development

- Process must react to changes in design, loads, etc.
- Typically at least three load cycles
  - preliminary design, detailed design, and actual hardware testing
- Cycle prior to PDR will uncover any "big picture" problems the design may encounter.
- Prior to CDR detailed design is done: find and correct ALL problems
- After CDR, materials are ordered and fabrication begins, and there will be many more problems uncovered here.

Structure Design Process

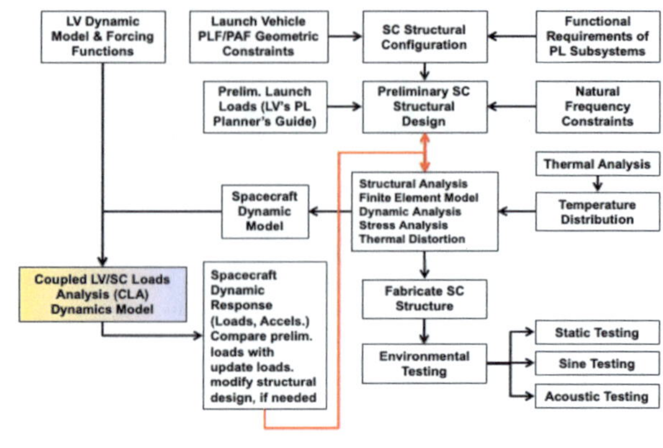

Courtesy Agrawal

Structural Concerns

Three major structure concerns:

- Strength (structure does not fail under quasi-static loads)
- Stiffness (structure exceeds natural frequency requirements and sustains dynamic loads)
- Buckling (structural stability: structure does not collapse under loads due to instability)

Analyze the structure to ensure it satisfies *ALL of these* during *all phases of flight*

Structural Analysis Topics

- Definitions: structure & stress terminology
- Structure design process
- Static structural analysis
  - Stress/strength calculations
- Dynamic structural analyses
  - Normal modes, forced & random vibration, shock & acoustics
  - Coupled-Loads Analysis (CLA)
- Finite-Element Modeling (FEM)

Structural Analysis Requires:

Knowledge of

1. Structural design (layout & geometry)
2. Materials (mass & stiffness properties at appropriate temperatures)
3. Loads & load factors (flight environment: acceleration, vibration, acoustic levels, thermal environment)

1. Strength Design Load Factor

- Use quasi-static load factors based on LV axial & lateral acceleration
- Obtain from the Payload Planners' Guide
- Typical values: 6-8 g axial, 2-3 g lateral
- Resulting loads applied statically to structure

Atlas V Design Load Factors-Graphical

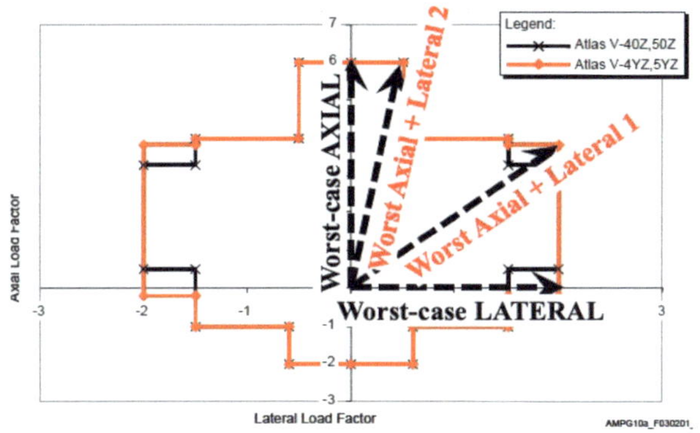

August 22, 2022

Example Design Factors: *Atlas V*

Flight Condition	Lateral (+ = compression,	Axial – = tension)
Liftoff	**±2.0 g**	+1.8 ±2.0 g
Flight Winds	**±0.4 ±1.6 g**	+2.8 ±0.5 g
Strap-On Separation	±0.5 g	+3.3 ±0.5 g
BECO (max axial)	±0.5 g	**+5.5 ±1.0 g**
BECO (max lateral)	±1.5 g	+2.5 ±1.0 g
MECO (max axial)	±0.3 g	+4.5 ±1.0 g
MECO (max lateral)	±0.6 g	±2.0 g

Worst-case loads in Red boxes above.

August 22, 2022

Simple Hand Stress Calculation

- Tensile/Compressive stress calculated as $\sigma_{t\ or\ c} = \pm\frac{P}{A}$ (= load ÷ area)
- Bending stress σ_b calculated as $\sigma_b = \frac{My}{I}$ = Moment × (distance from bending axis) ÷ (area moment of inertia)
- Moment M = force × distance (i.e. apparent weight × CG height)

Total stress = combined bending & tensile/compressive: $\sigma_{total} = \pm\frac{P}{A} \pm \frac{My}{I}$

August 22, 2022

Example: Stress Calculations

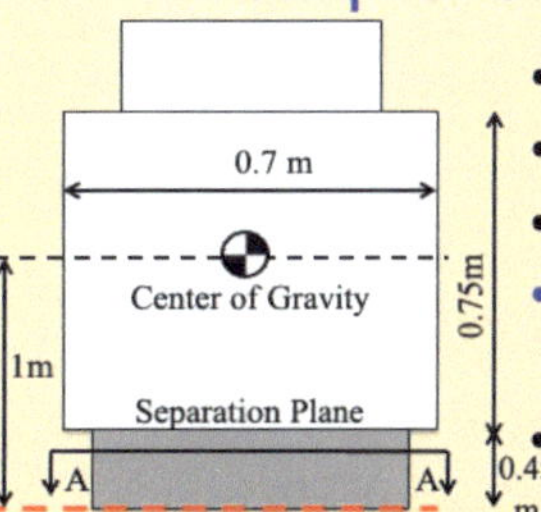

- Assume: LV = *Atlas V*
- SC mass = 500 kg
- Factor of safety = 1.25
- What is max stress in PAF (shaded gray)?
- Consult with table on previous slide. Worst cases are 1-lateral: *liftoff* or *flight winds* @ **2 g**; 2-axial: *BECO max axial* @ +5.5 + 1 = **6.5 g**

August 22, 2022

Stress Calculations Example (cont.)

- Area and Moment of inertia calculations:

$A = 2\pi rt = 2\pi\,(0.225\text{ m})(0.003\text{ m}) = 0.00424\text{ m}^2$

$I = \pi r^3 t = \pi\,(0.225\text{ m})^3(0.003\text{ m}) = 0.000107\text{ m}^4$

P = axial load = $n_a m g_0$
= 6.5(500 kg)9.8 m/s² = 31,850 N for max axial
= 3.3(500 kg)9.8 m/s² = 16,170 N for liftoff/flight wind

F_B = bending *force* (not moment) = $n_b m g_0$
= 0.5(500 kg)9.8 m/s² = 2,450 N (max axial)
= 2.0(500 kg)9.8 m/s² = 9,800 N (liftoff)

We will not consider shear stresses here, but they're important too.

August 22, 2022

Stress Calculations Example (cont.)

- Now, combine loads for two cases, calculate stresses:

$$\sigma = \left(+\frac{P}{A} + \frac{F_B z_{CG} y}{I}\right) FS$$

- For max axial: $\sigma_{max\ axial} = \left(+\frac{31{,}850\text{ N}}{0.00424\text{ m}^2} + \frac{2{,}450\text{ N}(1\text{ m})\frac{1}{2}0.45\text{ m}}{0.000107\text{ m}^4}\right)1.25$
  = [7.51+5.15 MPa]1.25
  = 15.83 MPa (2,296 psi)
- Liftoff/flight wind: $\sigma_{LO\ flt\ wind} = \left(+\frac{16{,}170\text{ N}}{0.00424\text{ m}^2} + \frac{9{,}800\text{ N}(1\text{ m})\frac{1}{2}0.45\text{ m}}{0.000107\text{ m}^4}\right)1.25$
  = [3.81+20.61 MPa]1.25
  = **30.53 MPa** (4,430 psi)

August 22, 2022

Worst-Case Stress

- We find that the Liftoff/flight winds cases (lateral) produce the max stress!

Because spacecraft are usually long and skinny, it is often the bending loads cases produced by lateral loads (max-q) that determine the structural strength, even if axial loads during launch are considerably higher

August 22, 2022 SC05A- 28

Other Stresses to Consider: Tanks

- Liquid propellant SC tanks are usually pressurized to force propellant into engine's combustion chamber (or decomposition if monoprop)
- Internal pressure creates stresses in two directions:
  - Axial stress
  - Hoop stress

August 22, 2022 SC05A- 29

Pressure Effect on Stresses: Axial Stress

- Axial stress $\sigma_{axial} = \frac{pR}{2t}$ (sometimes called "radial" or "longitudinal" stress):
  - Same for cylindrical or spherical tanks under pressure

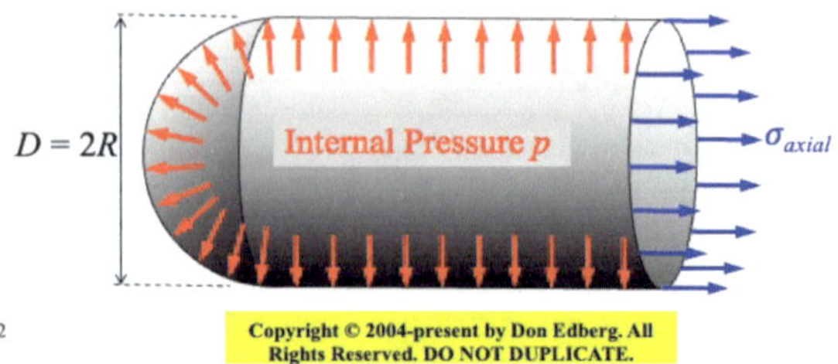

8/22/22 SC05A- 30

Pressure Effect on Stresses: Hoop Stress (Only in Cylindrical Tanks)

- Hoop stress: $\sigma_{hoop} = \frac{pR}{t}$
  - This is perpendicular to the axial pressure stress
  - 2 × axial stress

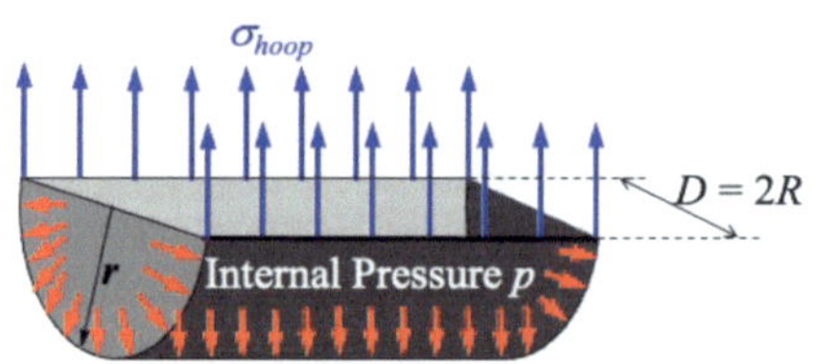

8/22/22 SC05A- 31

Example: Pressure Stresses in Cylinders

- Example: pressure of $p = 344.7$ kPa (50 psi) in tank with wall thickness $t = 1$ mm, radius $R = 1.25$ m
- Axial stress: $\sigma_{axial} = \frac{pR}{2t} = \frac{(344.7\text{ kPa})(1.25\text{ m})}{2(0.001\text{ m})}$ = 215.5 MPa (31,250 psi)
- Hoop stress: $\sigma_{hoop} = \frac{pR}{t} = \frac{(344.7\text{ kPa})(1.25\text{ m})}{0.001\text{ m}}$ = 431 MPa (62,500 psi)

8/22/22 SC05A- 32

What Happens If Allowables Exceeded: *Hoop* Stress Failures

8/22/22 SC05A- 33

2. Stiffness Design Factors

- Each LV has its own frequency requirements, selected to preclude control-structure interaction & instability of LV
- They are *guidelines only* (usually chosen to be very conservative so as to make life easy for the control system on the LV)
- Calculate analytically for SC
- Determine actual loads & frequencies during structures & control system coupled-loads analyses

Example Stiffness Design Factors: *Atlas V*

- Axial direction natural frequency ≥ 15 Hz
- Lateral direction natural frequency ≥ 8 Hz
- Secondary structure/payloads determined by the *non-coupling* or "*octave*" rule: at least double the frequency to preclude coupling and load amplification

Example: Stiffness Reqts

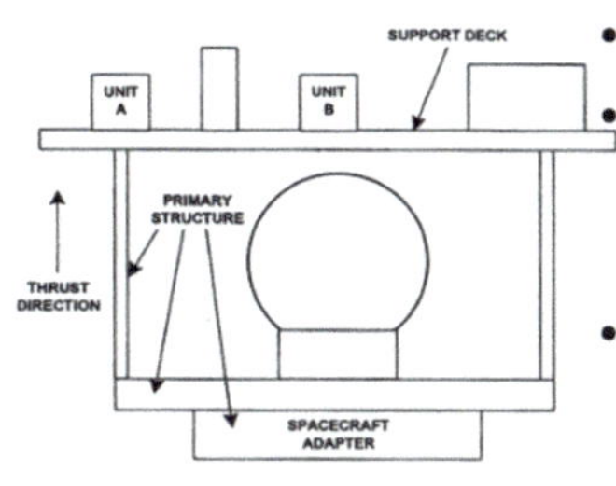

- Launch on *Atlas V*
- Primary structure just meets 15 Hz axial requirement
- What are axial stiffness requirements for unit's brackets A & B, if support deck has 50 Hz natural frequency?

Example: Stiffness Reqts (cont.)

- From table, minimum stiffness is 15 Hz
- *Unit A* mounted directly on primary structure
  - Applying octave rule, *Unit A* brackets should be designed for 2 × 15 Hz = 30 Hz
- *Unit B* behavior dictated by 50 Hz support deck
  - Applying octave rule, *Unit B* support should be designed for 2 × 50 Hz = 100 Hz
  - Note support deck at 50 Hz more than satisfies octave rule (need to be > 2 × 15 Hz = 30 Hz)

Structural Analysis Topics

- Definitions: structure & stress terminology
- Structure design process
- Static structural analysis
  - Stress/strength calculations
- Dynamic structural analyses
  - Normal modes, forced & random vibration, shock & acoustics
- Finite-Element Modeling (FEM)
  - Coupled-Loads Analysis (CLA)

Dynamics Analyses

- Dynamicists like to think of structures as combinations of springs and masses
- Each structure will vibrate at so-called natural frequencies (like a guitar string)
- *Free-Vibration* or "*Normal modes*" FEM analyses used to check FEM & compare with test results

Free Vibration Analysis

- FEM used to calculate spacecraft and equipment normal vibration frequencies f_N, in Hz (cycles/sec), known as "*normal modes*," *characteristic frequencies*, or *eigenvalues*
- Also spits out the *mode shapes* or *eigenvectors*
- Circular or angular frequencies are $\omega_N = 2\pi f_N$
- Period $T = 1/f_N$ (units: s)

Mode Shapes of Beam & Drumhead

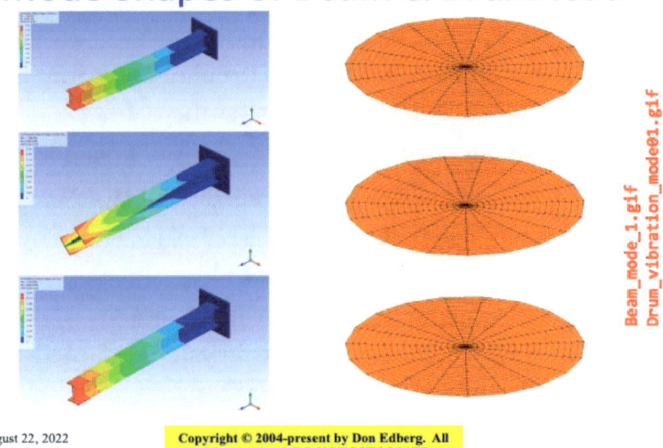

Dynamics Vocabulary

- Stiffness written as k (and is a *matrix* for FEM)
- Mass usually written as m (also a matrix)
- Natural frequency $f_n = \frac{1}{2\pi}\sqrt{\frac{k}{m}}$
- Damping ratio = ζ (Greek letter *zeta*): how fast the system damps out vibration or dissipates energy. Lower ζ, the system "rings" longer
- Quality $Q = \frac{1}{2\zeta}$. Higher quality = less damping (a good, *quality* bell has VERY high Q and rings for a long time)

Forced & Transient Dynamics

- *Forced-Vibration* & *Random* analyses used for shock, random vibrations, transient loads: i.e. transportation, staging, acoustics
- Spacecraft dynamic response highly influenced by its damping
- Damping values assigned to each mode (from test, experience, or previous models)
- Damping ratio ζ varies from 0.001 to 0.1, depends on design, internal connections, etc.
  - Corresponds to Q = 500 to 5
- Damping can also vary with amplitude & temperature (nonlinear)

Random Vibrations

- Some events difficult to characterize analytically
  - Shock events (separations, activations, etc.)
  - Acoustic loads
- Statistical methods used
- Analysis uses forcing functions specified in terms of "power spectral density" or PSD
- PSDs contain given content *spread* over a frequency range
  - Origin of "PSD" from electronics, where $P = \frac{V^2}{R}$

Typical PSD

- RMS acceleration = square root of area under g^2/Hz curve, or $a_{RMS} = \sqrt{\int \frac{g^2}{\text{Hz}} df}$
- Response depends on PSD levels AND damping ratio ζ

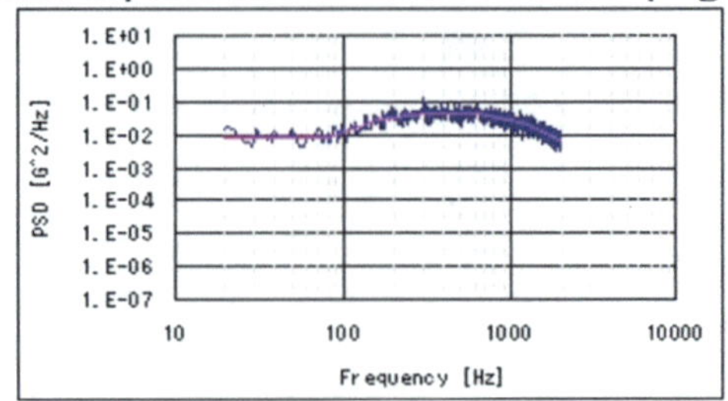

Enveloped "PSD" (actually ASD for Acceleration Spectral Density)

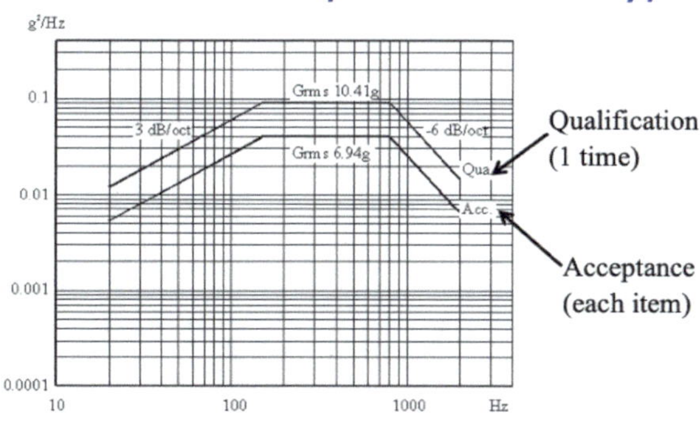

Response to Random Vibration

- The Miles eq.: $a = 3\sqrt{\dfrac{\pi Q s_0 f_0}{2}}$

  a = response in g

  f_0 = system natural frequency (Hz)

  Q = amplification factor or quality

  s_0 = PSD input in g^2/Hz at f_0
- Assumes system can be approximated as single-degree-of-freedom at frequency f_0
- "3" comes from assuming a 3σ statistical value that encompasses 98.99% of all occurrences

Example: Frequency Response

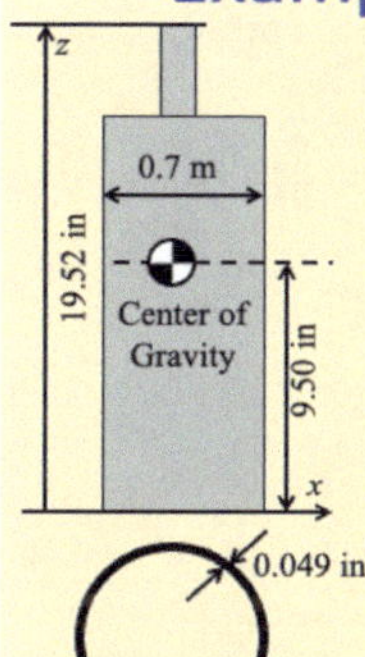

- During launch, 1.21 lb_f antenna assembly subjected to 8.5 *g* in lateral (*x*-dir) and random vibe*, both *x*- & *z*-dir. defined at base. Material: 6061-T4 drawn tube.
- Use amplification factor of 12.
- Yield F.S. = 1.15; Ultimate = 1.25

* Frequency (Hz)	Level (g²/Hz)
10	0.001
10 – 50	+8.6 dB/octave
50 – 150	0.1
150 – 200	–11 dB/octave
200 – 2,000	0.035
Overall g_{RMS} = 8.8	Duration = 60 s

Frequency Response (cont.)

- Area and Moment of inertia calculations:

  $A = 2\pi r t = 2\pi\,(1.076 \text{ in})(0.049 \text{ in}) = 0.331 \text{ in}^2$

  $I = \pi r^3 t = \pi\,(1.076 \text{ in})^3(0.049 \text{ in}) = 0.192 \text{ in}^4$
- Cantilever beam (lateral vibration):

$$k = \frac{3EI}{L^3} = \frac{3(10\times10^6\text{psi})(0.192\text{ in}^4)}{(9.5\text{ in})^3} = 6{,}720\,\frac{\text{lb}_\text{f}}{\text{in}}$$

natural frequency $f_n = \frac{1}{2\pi}\sqrt{\frac{k}{m}} = 230$ Hz.

- Rod (axial vibration):

$$k = \frac{EA}{L} = \frac{(10\times10^6\text{psi})(0.331\text{ in}^2)}{9.5\text{ in}} = 3.48\times10^5\,\frac{\text{lb}_\text{f}}{\text{in}}$$

nat. freq. $f_n = \frac{1}{2\pi}\sqrt{\frac{k}{m}} = 1{,}680$ Hz

Frequency Response Example (cont.)

- Sinusoidal (lateral) Response:

  $a_x = a_y = (8.4\ g)(12) = 100.8\ g$
- Random (axial & lateral): for both frequencies (see table), spectral input = 0.035 $\frac{g^2}{\text{Hz}}$. Use Miles' equation for response.

$$a_x = 3\sqrt{\frac{\pi Q s_0 f}{2}} = 3\sqrt{\frac{\pi(12)(0.035\,g^2/Hz)230\text{ Hz}}{2}} = 36.95g$$

$$a_z = 3\sqrt{\frac{\pi Q s_0 f}{2}} = 3\sqrt{\frac{\pi(12)(0.035\,g^2/Hz)1680\text{ Hz}}{2}} = 99.88g$$

Frequency Response Example (cont.)

- Tube structure stress calculations: bending moments at the base are (including F.S.):

$M_{yield} = 1.15(1.21\text{ lb}_f)(99.88\text{ g})9.50\text{ in} = 1{,}320\text{ in-lb}_f$

$M_{ult} = 1.25(1.21\text{ lb}_f)(99.88\text{ g})9.50\text{ in} = 1{,}435\text{ in-lb}_f$

$$\sigma_{yield} = \frac{M_{yield}c}{I} = \frac{1{,}320\text{ in}\cdot\text{lb}_f(1.125\text{ in})}{0.192\text{ in}^4} = 7{,}735\text{ psi}$$

$$\sigma_{ult} = \frac{M_{ult}c}{I} = \frac{1{,}435\text{ in}\cdot\text{lb}_f(1.125\text{ in})}{0.192\text{ in}^4} = 8{,}410\text{ psi}$$

Material allowables are F_{cy} = 14 ksi, F_{ty} = 16 ksi.

Worst M.S. $= \frac{14\text{ ksi}}{7{,}735\text{ psi}} - 1 = +0.81$

Design is acceptable (positive safety margin)

Acoustic Loads

Acoustics Topics:

- Definitions
- Sound Pressure Level
- Sound Pressure PSD

Sound Pressure

- Sound is an oscillating pressure in air
- Sound pressure level, SPL, defined *logarithmically*:

  $$\text{SPL} = 20 \log_{10} \frac{P}{P_{ref}}$$

  in decibels (dB)
- $P_{\text{ref}} = 2.9 \times 10^{-9}$ psi $= 2 \times 10^{-5}$ Pa = hearing threshold
- NOTE: 3 dB = a factor of 2 in *amplitude*; +3 dB = doubling, –3 dB = halving

Noise Source	Sound Level (dB)
(note: 3 dB ≈ factor of 2)	
Hearing threshold	0
Dishwasher	50-53
Car @ 10 m	60-80
Jack hammer @ 1 m	100
Jet engine @ 100 m	110-140
Jet engine @ 30 m	150
Rocket launch	165
Theoretical limit	194

Acoustic Environments

- Design curves are supplied by LV manufacturers
- Give maximum expected SPLs inside of payload fairing
- Levels vary depending on:
  - LV configuration
  - Payload geometry

Delta II Acoustic Environment

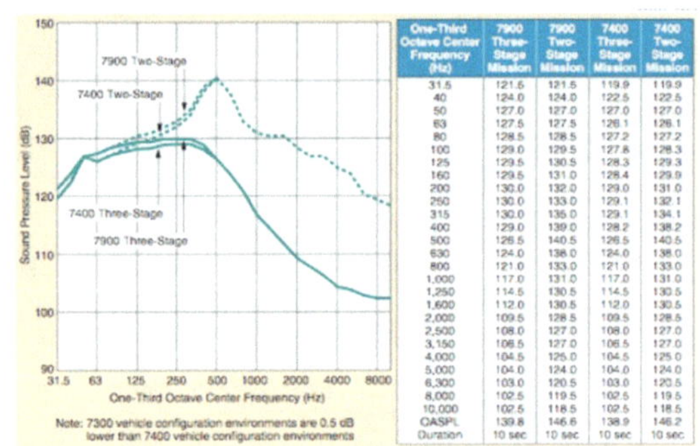

One-Third Octave Center Frequency (Hz)	7900 Three-Stage Mission	7900 Two-Stage Mission	7400 Three-Stage Mission	7400 Two-Stage Mission
31.5	121.6	121.5	119.9	119.9
40	124.0	124.0	122.5	122.5
50	127.0	127.0	127.0	127.0
63	127.5	127.5	126.1	126.1
80	128.5	128.5	127.2	127.2
100	129.0	129.5	127.8	128.3
125	129.5	130.5	128.3	129.3
160	129.5	131.0	128.4	129.9
200	130.0	132.0	129.0	131.0
250	130.0	133.0	129.1	132.1
315	130.0	135.0	129.1	134.1
400	129.0	139.0	128.2	138.2
500	126.5	140.5	126.5	140.5
630	124.0	138.0	124.0	138.0
800	121.0	133.0	121.0	133.0
1,000	117.0	131.0	117.0	131.0
1,250	114.5	130.5	114.5	130.5
1,600	112.0	130.5	112.0	130.5
2,000	109.5	128.5	109.5	128.5
2,500	108.0	127.0	108.0	127.0
3,150	106.5	127.0	106.5	127.0
4,000	104.5	125.0	104.5	125.0
5,000	104.0	124.0	104.0	124.0
6,300	103.0	120.5	103.0	120.5
8,000	102.5	119.5	102.5	119.5
10,000	102.5	118.5	102.5	118.5
OASPL	139.8	146.6	138.9	146.2
Duration	10 sec	10 sec	10 sec	10 sec

Figure 4-20. Predicted Delta II Acoustic Environments for 9.5-ft Fairing Missions

Enveloped Acoustic Environment

- For analysis, environment at liftoff and transonic + max-q phases (separately) enveloped

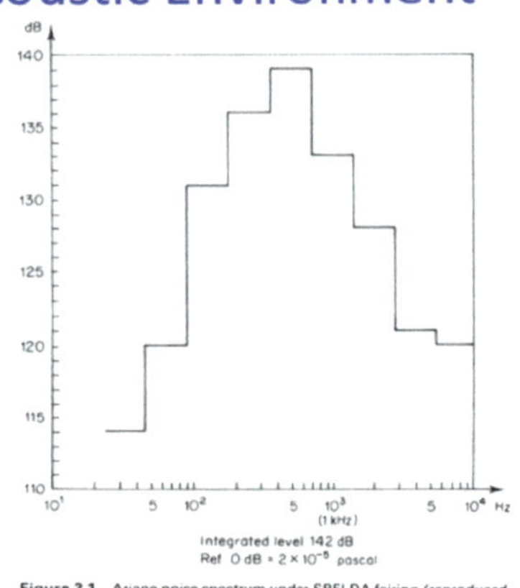

Figure 2.1 Ariane noise spectrum under SPELDA fairing (reproduced by permission of Arianespace [2])

Acoustic Response: Miles' Equation

Equivalent sound pressure spectral density s given by

$$s = \frac{\sqrt{2}\left(P_{ref}\right)^2 10^{K/10}}{f_0}, \text{ where}$$

- s = sound pressure PSD (power spectral density)
- $P_{\text{ref}} = 2 \times 10^{-5}$ Pa = reference pressure level
- K = sound pressure level (SPL) in dB relative to P_{ref}
- f_0 = frequency in Hz

Structural Analysis Topics

- Definitions: structure & stress terminology
- Structure design process
- Static structural analysis
  - Stress/strength calculations
- Dynamic structural analyses
  - Normal modes, forced & random vibration, shock & acoustics
- **Finite-Element Modeling (FEM)**
  - Coupled-Loads Analysis (CLA)

Finite Element Modeling (FEM)

- In reality, these calculations are done by computer on all parts of SC structure
- Math model created using geometric model & FEM software
- Apply loads to structure's finite-element model
- NASTRAN is the industry-standard FEM code
- Two types of FEM runs:
  - **Static** using the load factors (provides deflections, stresses, buckling assessment)
  - **Dynamics** runs (provides natural frequencies, normal modes, transient analysis, peak or dynamic stresses)

Typical Finite Element Model

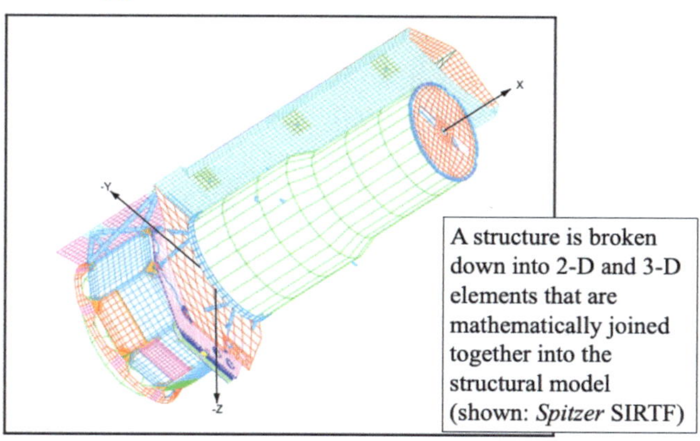

A structure is broken down into 2-D and 3-D elements that are mathematically joined together into the structural model (shown: *Spitzer* SIRTF)

How Does FEM Work?

Stress analysis

- Under FEM all areas of each part must be analyzed to ensure that there is no failure under predicted loads
- "When in doubt, make it stout."
- However making things stout makes them *heavier* — design trades

Simple Natural Frequency Estimation Using *Static* FEM Analysis!

- Do a 1-g XYZ static FEM analysis
- Get CM location of object; find axial (x) deflection δ_x of structure near CM location
- From Hooke's law, $F = k\delta_x = W = mg$, so

$$k = \frac{mg}{\delta_x}, \text{ and}$$

$$f_n = \frac{\omega_n}{2\pi} = \frac{1}{2\pi}\sqrt{\frac{k}{m}} = \frac{1}{2\pi}\sqrt{\frac{g}{\delta_x}}$$

- Approximate axial f_n by knowing mass m and δ_x, static deflection of CM! (get lateral values using y- and z- deflections)

Typical FEM Stress Analysis Results

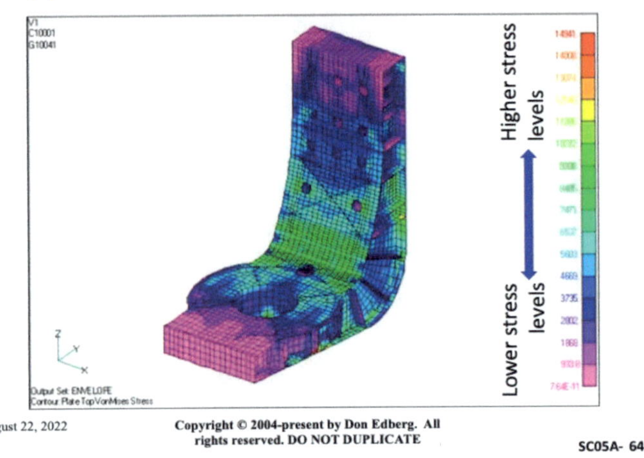

Other Structural Concerns

- Thermal loads
- Buckling and stability
- Stress concentrations, joints/fasteners
- Fatigue, cracks and crack propagation

Structural Analysis Topics

- Definitions: structure & stress terminology
- Structure design process
- Static structural analysis
  - Stress/strength calculations
- Dynamic structural analyses
  - Normal modes, forced & random vibration, shock & acoustics
- Finite-Element Modeling (FEM)
  - Coupled-Loads Analysis (CLA)

FEM Updates

- FEM is updated as design matures
- When hardware exists, it is tested; FEM adjusted to agree with the test results
- Updated model supplied to the LV company for a coupled-loads analysis (CLA)
- CLA simulates various extreme cases during flight (liftoff, maneuvering, max air, staging, etc.)

Coordinate System

- Coordinate system is different from that of LV
- *Must handle carefully*
- There will be a transformation matrix between LV and SC coordinate systems.
- Launch vehicles typically set x-axis origin near bottom of LV & $+x$ is the thrust or launch or LV axial direction.

Some special terms:

- *Nadir* = vertically downwards towards planet center
- *Zenith* = vertically away from a planet = antinadir

Coupled-Loads Analyses: Before

- SC math model supplied to LV company
- LV company mathematically joins ("couples") SC to LV
- LV co. applies flight loads to coupled system
- Multiple analyses determine stresses in all SC components for all critical flight conditions
- Assesses vehicle stability & closed-loop control with actual SC flexibility & natural frequencies

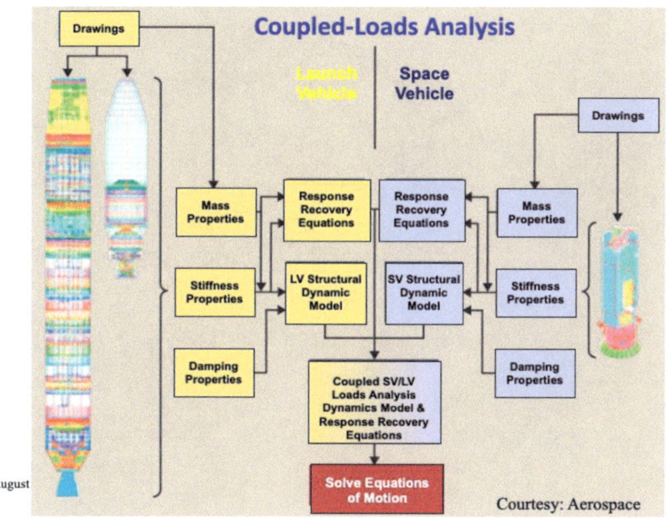

August

Coupled-Loads Analyses: After

- Coupled system results fed back to SC company
- Updated stresses & stability assessment go to SC design group
- SC's structures updated as necessary to correct stress and/or stiffness concerns
- Revised SC re-analyzed
- SC fabricated & tested

After CLA: Revise Design, Build, Test

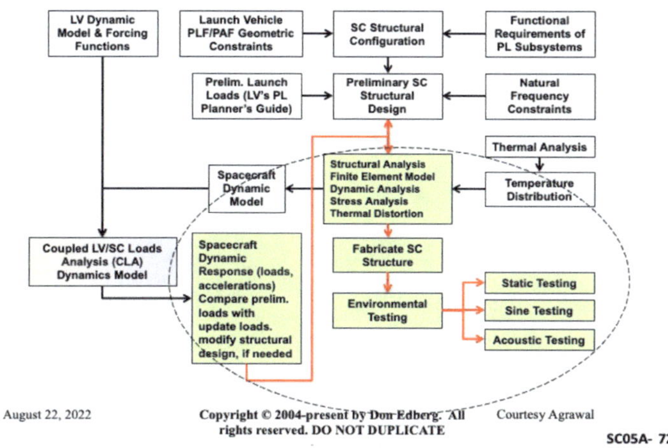

Courtesy Agrawal

Finite-Element Modeling Examples

A number of satellite FEM examples are available at

- http://web.mscsoftware.com/support/online_ex/Scenario/SatelliteStress.cfm (basic)
- http://www.mscsoftware.com/support/online_ex/Scenario/SatelliteStructural.cfm (advanced)

Flight Vehicle Mass Properties

Courtesy: American Aero Services; TBD-Russia; NASA

Mass Properties Topics

- Mass properties & coordinate systems
- Center of mass of assembly of items
- Moment of Inertia of individual items
- Principal axes & planes of symmetry
- Selected MoI values
- Examples of Moment of Inertia Calculation
  - Non-symmetric SC
  - Axisymmetric Bodies (2)
  - Launch vehicle
- Propellants: dealing with liquids
- References

Spacecraft Design Process

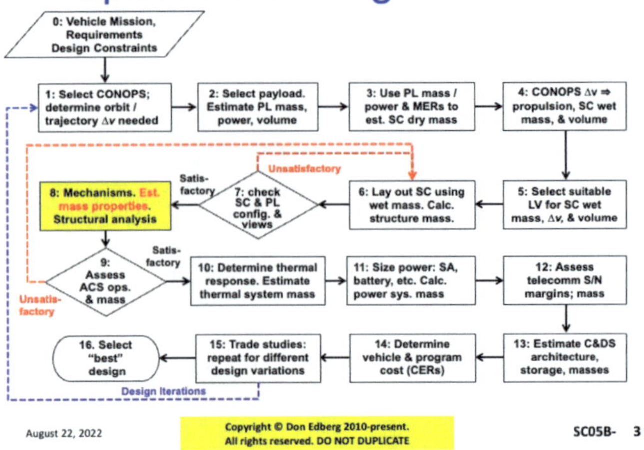

0: Vehicle Mission, Requirements Design Constraints
1: Select CONOPS; determine orbit / trajectory Δv needed
2: Select payload. Estimate PL mass, power, volume
3: Use PL mass / power & MERs to est. SC dry mass
4: CONOPS Δv ⇒ propulsion, SC wet mass, & volume
5: Select suitable LV for SC wet mass, Δv, & volume
6: Lay out SC using wet mass. Calc. structure mass.
Unsatisfactory
7: check SC & PL config. & views
Satis-factory
8: Mechanisms. Est. mass properties. Structural analysis
9: Assess ACS ops. & mass
Satis-factory
Unsatis-factory
10: Determine thermal response. Estimate thermal system mass
11: Size power: SA, battery, etc. Calc. power sys. mass
12: Assess telecomm S/N margins; mass
13: Estimate C&DS architecture, storage, masses
14: Determine vehicle & program cost (CERs)
15: Trade studies: repeat for different design variations
16. Select "best" design
Design Iterations

Mass Properties Topics

- Mass properties & coordinate systems
- Center of mass of assembly of items
- Moment of Inertia of individual items
- Principal axes & planes of symmetry
- Selected MoI values
- Examples of Moment of Inertia Calculation
  - Non-symmetric SC
  - Axisymmetric Bodies (2)
  - Launch vehicle
- Propellants: dealing with liquids
- References

Mass Properties

Statics, stability analyses, & "rigid-body" dynamics of vehicle motion require two types of mass properties:

1. Location of *center of mass* (CM)
2. Values of mass *moments of inertia* (MoIs) in some coordinate system

- Body-centered coordinate system is often used
- If CM changes with time, sometimes a *fixed* reference location is specified, and CM offset is factored in as a function of time
- The *mass properties* group keeps track of all component masses and their locations

Vehicle Coordinate System

+Roll (right)
+x
+Yaw (nose right)
+y
+Pitch (nose-up)
+z

- Often use "airplane" coordinates
- Velocity direction
  = x-axis
  = Roll axis
- Pitch rotation
  = y-axis
- Yaw rotation
  = z-axis

Mass Properties Topics

- Mass properties & coordinate systems
- Center of mass of assembly of items
- Moment of Inertia of individual items
- Principal axes & planes of symmetry
- Selected MoI values
- Examples of Moment of Inertia Calculation
  - Non-symmetric SC
  - Axisymmetric Bodies (2)
  - Launch vehicle
- Propellants: dealing with liquids
- References

Location of Center of Mass

1. Choose a reference location P
   - SC often use bottom center of SC
   - LVs often use location of engine *gimbal block*)
2. Determine the magnitudes & locations of all masses relative to the reference
3. Sum the masses = total mass $M = \sum(m_i)$
4. Sum the "mass moments" = $\sum(m_i l_{ij})$ in each axis j (typically $j = 3$)
5. The location of the CM relative to P is

$$x_{CMj} = \frac{\sum(m_i l_{ij})}{\sum(m_i)} = \frac{\sum(m_i l_{ij})}{M}$$

Mass Properties Topics

- Mass properties & coordinate systems
- Center of mass of assembly of items
- **Moment of Inertia of individual items**
- Principal axes & planes of symmetry
- Selected MoI values
- Examples of Moment of Inertia Calculation
  - Non-symmetric SC
  - Axisymmetric Bodies (2)
  - Launch vehicle
- Propellants: dealing with liquids
- References

Mass Moment of Inertia (MoI)

- An object's *mass moment of inertia* (J) is a measure of an object's resistance to rotation about an axis
  - A point mass has *zero* MoI about *its* location, but *non-zero* MoI if rotation occurs around *another* location
- Two bodies with the same mass may have very different MoIs
- Example: dumbbell with two 1 kg masses at end of a 1 m stick, vs. a 2 kg sphere
  - Sphere is much easier to "spin" compared to twirling the dumbbell around its long axis

Moments of Inertia (*I* matrix)

- Because there are three axes in a coordinate system, a body's MoI may be written as a matrix or tensor:

$$I = \begin{bmatrix} I_{xx} & I_{xy} & I_{xz} \\ I_{yx} & I_{yy} & I_{yz} \\ I_{zx} & I_{zy} & I_{zz} \end{bmatrix}$$

- The inertia matrix/tensor is symmetric about diagonal, so
  $I_{xy} = I_{yx},\ I_{xz} = I_{zx},\ and\ I_{yz} = I_{zy}$

Inertia matrix (cont.)

The six independent terms of the matrix or tensor are calculated by assembling each item i's mass m_i and location coordinates (x_i, y_i, z_i) as follows:

Moments of Inertia (*diagonal* terms)	Products of Inertia (non-*diagonal* terms)
$I_{xx,i} = \sum_i m_i(y_i^2 + z_i^2)$,	$I_{xy,i} = -\sum_i m_i x_i y_i$
$I_{yy,i} = \sum_i m_i(x_i^2 + z_i^2)$,	$I_{xz,i} = -\sum_i m_i x_i z_i$
$I_{zz,i} = \sum_i m_i(x_i^2 + y_i^2)$,	$I_{yz,i} = -\sum_i m_i y_i z_i$

When Masses Change Location:

- If things move, the *parallel axis theorem* is invoked:

$$I_P = I_G + \begin{bmatrix} m\left(y_{G/P}^2 + z_{G/P}^2\right) & -mx_{G/P}y_{G/P} & -mx_{G/P}z_{G/P} \\ -mx_{G/P}y_{G/P} & m\left(x_{G/P}^2 + z_{G/P}^2\right) & -my_{G/P}z_{G/P} \\ -mx_{G/P}z_{G/P} & -my_{G/P}z_{G/P} & m\left(x_{G/P}^2 + y_{G/P}^2\right) \end{bmatrix}$$

- I_P is the (unknown) MoI about point P
- I_G is the (known) MoI about the CM
- $x_{G/P}$, $y_{G/P}$, and $z_{G/P}$ are the x, y, and z distances from CM to point P

Parallel Axis Theorem, Simple Bodies

- Use to calculate MoI of bodies whose CM does not coincide with reference point
- MoI of offset body = MoI of body about own CM + $+ m_{body}$ (body distance from ref. point)^2, or
  $I = I_0 + m_{body} r_{body}^2$, where
  I = MoI of body whose CM is offset from ref. point,
  I_0 = body's MoI about its own CM,
  m_{body} = body's mass, and
  r_{body} = distance from reference point to body CM
- Often $I_0 \ll m_{body} r_{body}^2$ and can be neglected so
  $I \approx m_{body} r_{body}^2$
- **Note: MoI may be written "J" instead of "I" so as not to confuse with area moment of inertia used for beam bending analysis**. This is done in the following notes.

Mass Properties Topics

- Mass properties & coordinate systems
- Center of mass of assembly of items
- Moment of Inertia of individual items
- Principal axes & planes of symmetry
- Selected MoI values
- Examples of Moment of Inertia Calculation
  - Non-symmetric SC
  - Axisymmetric Bodies (2)
  - Launch vehicle
- Propellants: dealing with liquids
- References

8/22/22 SC05B- 15

Inertia Matrix Properties – Principal Directions

- It's always possible to find a set of directions where the inertia matrix is *diagonal* (all off-diagonal terms zero)

$$I = \begin{bmatrix} A & 0 & 0 \\ 0 & B & 0 \\ 0 & 0 & C \end{bmatrix}$$

- These directions are called the *principal* directions

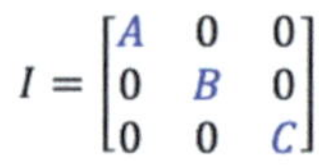

- MoIs A, B, and C are called the *Principal* MoIs

$$I_{xx} < I_{yy} < I_{zz}, \text{ or}$$
$$A < B < C$$

8/22/22 SC05B- 16

Inertia Matrix Properties – Symmetry

- If an object has a plane of symmetry, object is said to be *axisymmetric*
  - Object has y-z symmetry plane
- Two principal moments about the symmetry axis are *equal*

$$I_{yy} = I_{zz} = C,\ I_{xx} = A,\ A < C$$

$$I = \begin{bmatrix} A & 0 & 0 \\ 0 & C & 0 \\ 0 & 0 & C \end{bmatrix}$$

- Many SC & most LVs are axisymmetric

8/22/22 SC05B- 17

Mass Properties Topics

- Mass properties & coordinate systems
- Center of mass of assembly of items
- Moment of Inertia of individual items
- Principal axes & planes of symmetry
- Selected MoI values
- Examples of Moment of Inertia Calculation
  - Non-symmetric SC
  - Axisymmetric Bodies (2)
  - Launch vehicle
- Propellants: dealing with liquids
- References

8/22/22 SC05B- 18

Mass Moments of Inertia of Shells
(Use for Empty Tanks, Fairings, & Skirts)

Shape	$\bar{z}$ CM location	I_x & I_y	I_z	Material Volume
Thin Cylinder	$h/2$	$m(R^2/2 + h^2/12)$	mR^2	$2\pi Rht$
Thin Hemisphere	$R/2$	$5mR^2/12$	$⅔\, mR^2$	$2\pi R^2 t$
Thin Cone	$h/3$	$m(R^2/4 + h^2/18)$	$½\, mR^2$	$\pi Rt\sqrt{R^2+h^2}$

$\bar{z}$ = CM location above base, m = mass, z = axis of symmetry. Note that these coordinates are different from the vehicle coordinates given earlier.

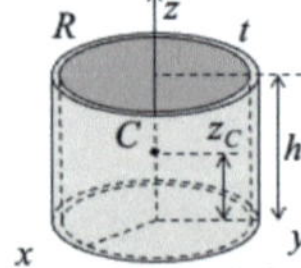
Open Cylindrical Shell (No End Caps)

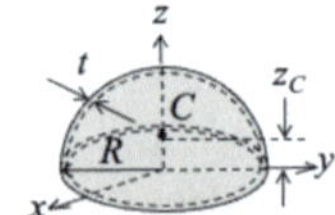
Open Hemispherical Shell (No Base)

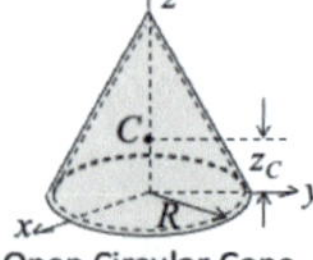
Open Circular Cone Lateral Surface

8/22/22 SC05B- 19

Mass Moments of Inertia of Solid Bodies
(Use for Loaded Tanks & Casings)

Shape	CM location	I_x & I_y	I_z	Int. Volume
Solid Cylinder	½ h	$m(R^2/4 + h^2/12)$	½ mR^2	$\pi R^2 h$
Solid Hemisphere	⅜ R	0.259 mR^2	⅖ mR^2	⅔ πR^3
Solid Sphere	0 (center)	⅖ mR^2	⅖ mR^2	4/3 πR^3

CM location given above base, m = mass, z = axis of symmetry. These coordinates differ from the vehicle coordinates given earlier.

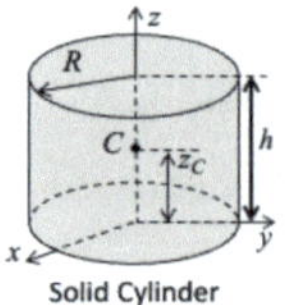
Solid Cylinder

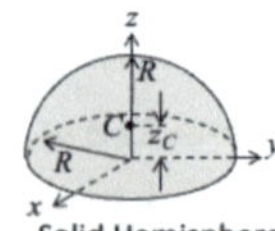
Solid Hemisphere

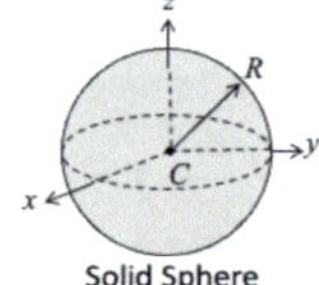
Solid Sphere

8/22/22 SC05B- 20

Mass Moments of Inertia of Solids (cont.)

Shape	CM location	I_x	I_y	I_z	Int. Volume
Rectangular prism	½(A, B, H)	$m(B^2 + H^2)/12$	$m(A^2 + H^2)/12$	$m(A^2 + B^2)/12$	ABH
Thin rod, length L (cylinder, $R << L$)	½ L	$mL^2/12$	~ 0	~ 0	~ 0

Rectangular prism

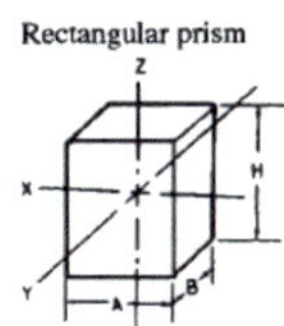

Thin rod

Mounted Solar Array

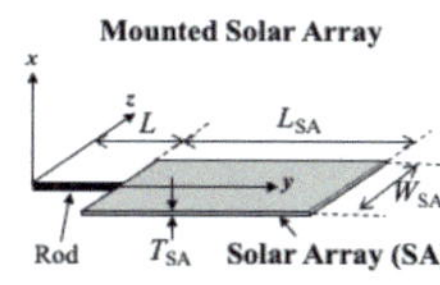

Calculate MoI of mounted solar array using Parallel Axis Theorem. The rod is offset $L/2$; the SA is offset $L + L_{SA}/2$. The SA's thickness is T_{SA}. Can neglect T_{SA}^2 terms compared to W_{SA}^2 & L_{SA}^2.

Mass Properties Topics

- Mass properties & coordinate systems
- Center of mass of assembly of items
- Moment of Inertia of individual items
- Principal axes & planes of symmetry
- Selected MoI values
- Examples of Moment of Inertia Calculation
  - Non-symmetric SC
  - Axisymmetric Bodies (2): rod without & with mass joining masses
  - Launch vehicle
- Propellants: dealing with liquids
- References

Example 1: Spacecraft MoI Calculation (1 of 8)

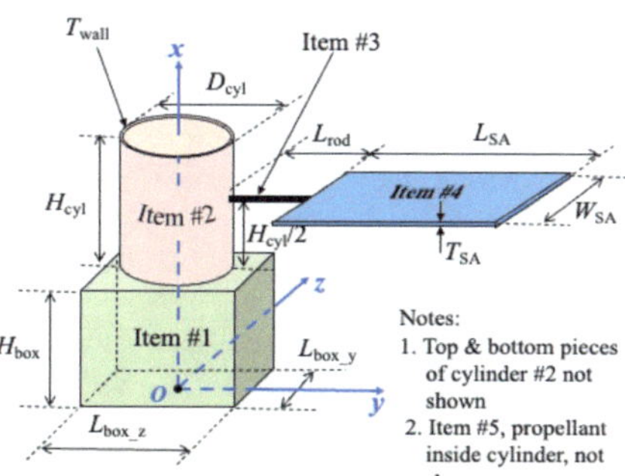

1. Calculate spacecraft total mass m_{tot} (kg)
2. Calculate $\boldsymbol{G} = (x_{cg}, y_{cg}, z_{cg})$, location of the CM relative to origin $\boldsymbol{O}$ at the bottom center of Item 1 (m)
3. Calculate the inertia tensor $[\mathbf{I}_{ij}]$ about G (kg-m²)

Spacecraft MoI Example 1 (2 of 8)

- Item 1: box contains electronics, $\rho_1 = 320$ kg/m³
- Item 2: a hollow graphite-epoxy cylinder/propellant tank (no top or bottom). $\rho_2 = 1{,}580$ kg/m³
- Item 2.1 is the cylinder's bottom cover, Item 2.2 is the top cover. $\rho_{2.1} = \rho_{2.2} = 1{,}580$ kg/m³
- Item 3 is the negligible diameter support rod for the solar array (SA)
- Item 4 is the SA, aligned the z-axis
- Item 5 (not shown) is solid propellant to fill #2, $\rho_5 = 1{,}000$ kg/m³
- Dimensions/masses are in the table. All items in non-symmetric spacecraft are of uniform density.

Example 1 Spacecraft Geometry & Properties (3 of 8)

Box (Item #1):		
Volume:	2.70	m³
height:	1.20	m
length:	1.50	m
depth:	1.50	m
Density ρ:	320.00	kg/m³
Cylinder (Item #2):		
ρ_GrE:	1,580.0	kg/m³
diameter:	1.00	m
height:	1.50	m
Top & bottom thick t_{wall}:	0.01	m
effective radius	0.495	m
Rod (item 3): mass	5.0	kg
length:	0.50	m
radius:	~0.0	m
Solar Panel (Item 4) mass	50.0	kg
volume (ea.)	0.18	m³
length:	3.50	m
width:	1.00	m
thickness:	0.05	m
Propellant #5,1=yes, 0=no		1
Volume	1.116	m³
Propellant density	1,000	kg/m³
diameter:	0.98	m
height:	1.48	m

Spacecraft Example 1 CM Calculations (4 of 8)

SC Item no	Item Name	Mass (kg)	x dist (m)	y dist (m)	z dist (m)
1	Avionics	864.00	0.600	0.000	0.000
2	Hollow Cylinder without ends	72.73	1.950	0.000	0.000
2.1	Cyl. bottom end	12.41	1.205	0.000	0.000
2.2	Cylinder top end	12.41	2.695	0.000	0.000
3	Rod	5.00	1.950	0.750	0.000
4	Solar Panel	50.00	1.950	2.750	0.000
5	Propellant	1116.36	1.950	0.000	0.000
totals	∑ mass:	2132.91	1.4031	0.0662	0.0000
			X_{cg}	Y_{cg}	Z_{cg}

Spacecraft Example 1
1st Moments & CM Location (5 of 8)

SC Item no	M_x (kg-m)	M_y (kg-m)	M_z (kg-m)	x_G (m)	y_G (m)	z_G (m)
1	518.40	0.00	0.00	-0.803	-0.066	0.000
2	141.82	0.00	0.00	0.547	-0.066	0.000
2.1	14.95	0.00	0.00	-0.198	-0.066	0.000
2.2	33.44	0.00	0.00	1.292	-0.066	0.000
3	9.75	3.75	0.00	0.547	0.684	0.000
4	97.50	137.50	0.00	0.547	2.684	0.000
5	2176.90	0.00	0.00	0.547	-0.066	0.000
totals	2992.77	141.25	0.00			

Spacecraft Example 1
MoI Comments (6 of 8)

On previous table, note:

- Must find CM location first
- Then use CM for new offset calculation
- Calculate each part's MoI about part's centroid (labeled as J_{0i})
- Calculate $m_i x_i^2$ for each part, sum $J_{0i} + m_i x_i^2$
- Note that usually $J_{0i} << m_i x_i^2$: almost don't need to calculate J_{0i}

Item Matrices about *O* (7 of 8) (kg-m²)

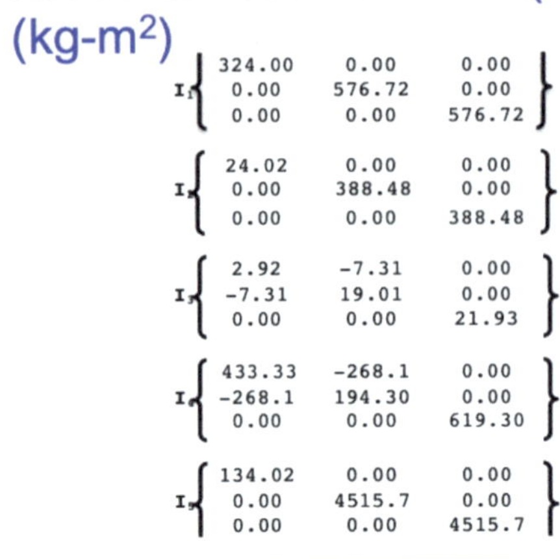

$$I_1 \begin{Bmatrix} 324.00 & 0.00 & 0.00 \\ 0.00 & 576.72 & 0.00 \\ 0.00 & 0.00 & 576.72 \end{Bmatrix}$$

$$I_2 \begin{Bmatrix} 24.02 & 0.00 & 0.00 \\ 0.00 & 388.48 & 0.00 \\ 0.00 & 0.00 & 388.48 \end{Bmatrix}$$

$$I_3 \begin{Bmatrix} 2.92 & -7.31 & 0.00 \\ -7.31 & 19.01 & 0.00 \\ 0.00 & 0.00 & 21.93 \end{Bmatrix}$$

$$I_4 \begin{Bmatrix} 433.33 & -268.1 & 0.00 \\ -268.1 & 194.30 & 0.00 \\ 0.00 & 0.00 & 619.30 \end{Bmatrix}$$

$$I_5 \begin{Bmatrix} 134.02 & 0.00 & 0.00 \\ 0.00 & 4515.7 & 0.00 \\ 0.00 & 0.00 & 4515.7 \end{Bmatrix}$$

∑ Inertia Matrices about *O* & *G* (8 of 8)

I_{total} about *O*

$$\begin{Bmatrix} 918.29 & -275.44 & 0.00 \\ -275.44 & 5694.25 & 0.00 \\ 0.00 & 0.00 & 6122.17 \end{Bmatrix}$$

I_{total} about *CG*

$$\begin{Bmatrix} 908.94 & -77.24 & 0.00 \\ -77.24 & 1495.0 & 0.00 \\ 0.00 & 0.00 & 1913.5 \end{Bmatrix}$$

Principal moments about *CG*

$$\begin{Bmatrix} 898.9 & 0 & 0 \\ 0 & 1,505 & 0 \\ 0 & 0 & 1,913.5 \end{Bmatrix}$$

Note: principal moments preferred because commanded rotations are *uncoupled*. They are obtained by eigenvalue analysis (not shown).

Axisymmetric Body Examples

- Assume long, skinny bodies such as LVs to be axisymmetric
- Only calculate *two* MoIs
  - About long axis
  - Perpendicular to long axis
- Three examples follow
  - Simple MoI: Massless rod + spheres
  - Simple MoI: Rod + spheres
  - Launch vehicle

MoI Calculation Examples

1. Pitch & roll MoI of 2 kg sphere centered on rod

2. Pitch & roll MoI for "dumbbell" with two 1 kg spheres mounted on ends of rod (same mass)

MoI Example 1: One 2 kg, 14.1 cm Sphere on 1 kg, 1 m Rod

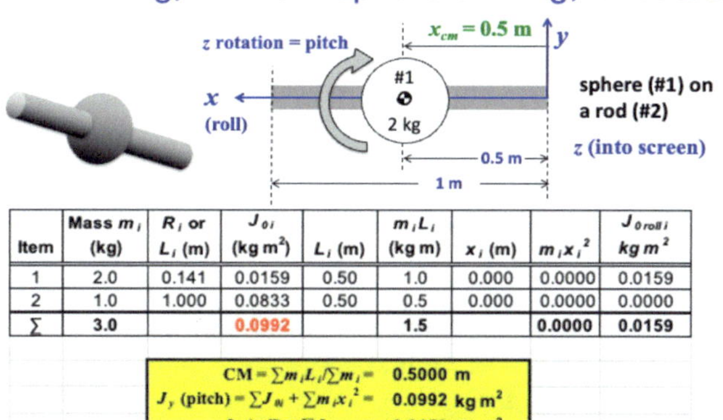

Item	Mass m_i (kg)	R_i or L_i (m)	J_{0i} (kg m²)	L_i (m)	m_iL_i (kg m)	x_i (m)	$m_ix_i^2$	$J_{0\,roll\,i}$ kg m²
1	2.0	0.141	0.0159	0.50	1.0	0.000	0.0000	0.0159
2	1.0	1.000	0.0833	0.50	0.5	0.000	0.0000	0.0000
∑	**3.0**		**0.0992**		**1.5**		**0.0000**	**0.0159**

$\text{CM} = \sum m_iL_i / \sum m_i =$ 0.5000 m

$J_y \text{ (pitch)} = \sum J_{0i} + \sum m_i x_i^2 =$ 0.0992 kg m²

$J_x \text{ (roll)} = \sum J_{0\,roll\,i} =$ 0.0159 kg m²

8/22/22

MoI Example 2: *Two* 1 kg, 10 cm Spheres on ends of 1 kg, 1 m Rod – same mass, **SIX TIMES** the pitch MoI!!

z rotation = pitch
x_{cm} = 0.5 m
y
x (roll)
#2 1 kg
#1 1 kg
sphere (#1 & 2) on a rod (#3)
0.5 m
1 m

Item	Mass m_i (kg)	R_i or L_i (m)	$J_{0\,pitch\,i}$ (kg m²)	L_i (m)	m_iL_i (kg m)	x_i (m)	$m_ix_i^2$	$J_{0\,roll\,i}$ kg m²
1	1.0	0.100	0.0040	0.00	0.0	-0.500	0.2500	0.0040
2	1.0	0.100	0.0040	1.00	1.0	0.500	0.2500	0.0040
3	1.0	1.000	0.0833	0.50	0.5	0.000	0.0000	0.0000
∑	**3.0**		**0.0913**		**1.5**		**0.5000**	**0.0080**

$\text{CM} = \sum m_iL_i / \sum m_i =$ 0.5000 m

$J_y \text{ (pitch)} = \sum J_{0i} + \sum m_i x_i^2 =$ 0.5913 kg m²

$J_x \text{ (roll)} = \sum J_{0\,roll\,i} =$ 0.0080 kg m²

8/22/22

Mass Properties Topics

- Mass properties & coordinate systems
- Center of mass of assembly of items
- Moment of Inertia of individual items
- Principal axes & planes of symmetry
- Selected MoI values
- Examples of Moment of Inertia Calculation
  - Non-symmetric SC
  - Axisymmetric Bodies (2)
  - Launch vehicle
- Propellants: dealing with liquids
- References

8/22/22

Example 4: Calculation of Mass Properties of Launch Vehicle

- Assume axisymmetric, so only *two* MoIs: about long axis, and perpendicular to long axis
- Use bottom of aft skirt as reference location
- Assemble table of masses and each mass's location

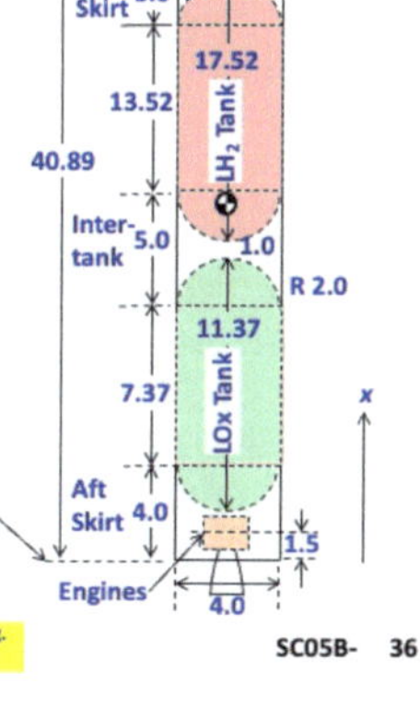

8/22/22

CM Calculation (Tanks Empty)

Item	Mass (kg)	Location (m)	Moment (kg-m)
Payload Fairing $h = 2D$	689	35.56	24507
Payload (5T w/PAF)	4250	34.40	146189
Payload Attach Fitting	750	33.15	24860
Forward Skirt $D = 2R$	501	31.40	15743
LH_2 Fwd Dome/Insul.	272	30.89	8388
LH_2 Cyl with Insulation	1836	23.14	42480
LH_2 Aft Dome/Insul.	272	15.37	4175
Intertank	836	13.87	11594
LOx Fwd Dome/Insul.	403	12.37	4993
LOx Cyl. with Insulation	1488	7.69	11436
LOx Aft Dome/Insul.	403	3.00	1210
Aft Skirt	669	2.00	1337
Thrust Structure	497	2.00	994
Engines	2236	1.00	2236
Gimbals	71	2.00	142
Avionics	744	31.40	23365
Wiring	1047	20.45	21400
LH_2 fuel	0	23.14	0
LOx oxidizer	0	7.69	0
Totals	16963	kg	345049
CM Location above ref.	20.34	m	

8/22/22

Launch Vehicle Shell Thicknesses

- For tanks, need the thickness t to calculate tank's mass, which is used to calculate MoI of tank
- For a shell of uniform thickness, $m_{shell} = \rho(Vol_{shell\ material}) = (\rho\ A_{shell}\ t)$
- Thickness $t = m_{shell}/(\rho A_{shell})$
- Use mass & area from CM calculations along with material density to find thickness

8/22/22

Launch Vehicle LOx Tank Thickness

- $D = 4$ m, $h = 11.37$ m, $m_{lox\_tank} = 2134$ kg,
  $A_{lox\_tank} = \pi D h_{LOx\_cyl} + \pi D^2$
  $= \pi D[(h_{LOx\_tank} - ½ D - ½ D) + D]$
  $= 143$ m^2
  $\rho_{lox\_tank} = 2{,}700$ kg/m^3
- Assume that the tank is of uniform thickness t
- $t = m_{lox\_tank}/(\rho A_{lox\_tank})$
  $= 2{,}134$ kg/[(2,700 kg/m^3)(143 m^2)] = 0.0055 m
- Use to determine masses of tank domes and cylinder

LV LOx Tank (cont.)

- Cylinder mass = shell mass + insulation mass
  $m_{LOx\_cyl} = m_{LOx\_shell} + m_{LOx\_shell\_ins}$
  $= A_{LOx\_cyl}[\rho\, t_{cyl}$ + insulation density]
  $= \pi$(4 m)(7.37 m)[2,700 kg/m^3(0.0055 m) + 1.123 kg/m^2]
  = 1,488 kg
- $m_{LOx\_dome} = m_{LOx\_domeshell} + m_{LOx\_dome\_ins}$
  $= A_{LOx\_dome}[\rho\, t_{dome}$ + insulation density]
  $= 2\pi$(2 m)^2[2,700 kg/m^3(0.0055 m) + 1.123 kg/m^2]
  = 403 kg

LV LOx Tank (cont.)

- $J_{x0\_LOx\_cyl} = m_{LOx\_shell}(R^2/2 + h_{LOx\_shell}{}^2/12)$
  = 1488 kg [(2 m)^2/2 + (17.52 m – 4 m)^2/12]
  = 9717 kg m^2
- $J_{x0\text{-}LOx\_dome} = 5\, m_{LOx\_dome} R^2/12$
  = 5 (403 kg) (2 m)^2/12
  = 672 kg m^2
- These values found in lines 9 – 11 in the MoI spreadsheet below

MoI of Assembled Structure

- Vehicle is assembly of cylinders, domes, boxes, engines, thrust structure, payload, PLF, PAF, etc.
- Vehicle MoI = ∑ component MoIs
- Effect of part's separation from CM is incorporated by knowing separation distance from the part's center to the CM
- We use "parallel axis" theorem: $J_i = J_{i0} + m_i L_i^2$, where J_{i0} = MoI of item i about its own centroid, m_i = item i mass, and L_i is separation of item i's centroid from CM of assembly
- Pitch inertia assembly $J_{pitch} = J_y = \sum J_y$ of all parts
- Roll inertia assembly $J_{roll} = \sum J_x$ of all parts

MoI of Launch Vehicle

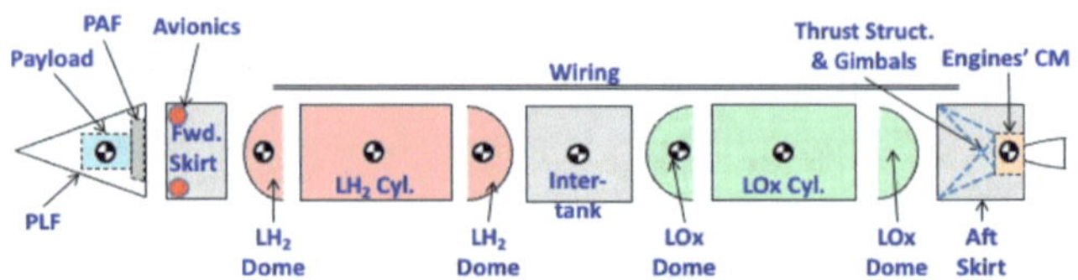

1. Use mass and dimensions to determine J_{0xi} & J_{0zi}, MoIs about own centroid, for each vehicle part (use CM list)
2. Determine L_i & R_i, part's centroid location relative to CM
3. Compute offset contribution to determine part's inertias: $J_{xi} = J_{0xi} + m_i R_i^2$, $J_{yi} = J_{0yi} + m_i L_i^2$
4. Vehicle inertias: for roll, $J_x = \sum(J_{0i} + m_i R_i^2)$: for pitch/yaw, $J_y = \sum(J_{0yi} + m_i L_i^2)$, for all parts

Notes On Calculation of LV MoI

- Insulated parts must include insulation mass; usually assume uniform areal distribution (const. mass/area)
- If offset L_i from vehicle CM > item's characteristic 'dimension' or length, may assume that inertia of items negligible ('point mass') compared to inertia component due to offset (i.e. $J_{0i} << m_i L_i^2$). Almost always true.
- Items such as wiring assumed uniform along vehicle length (modeled as thin solid rod)
- **MoI about long axis *several orders of magnitude less* than MoI perpendicular to long axis**

LV Wiring MoI Calculation

- Assume: wiring essentially a point mass in the z-direction (long dir.)
- Wiring consists of two bundles on outside of tank, each ½ of the mass, a radius R from center

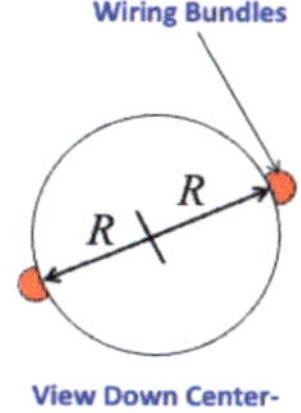

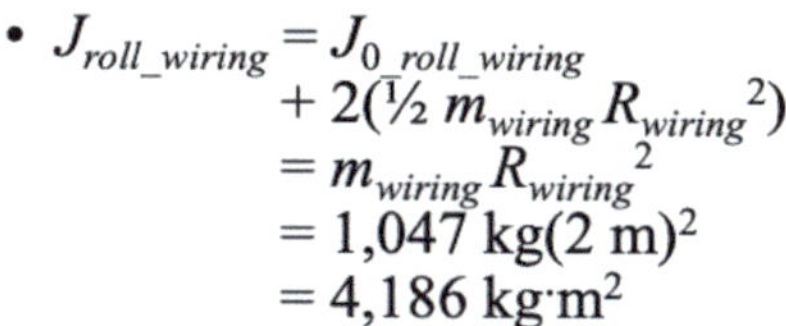

- $$J_{roll\_wiring} = J_{0\_roll\_wiring} + 2(\tfrac{1}{2}\, m_{wiring} R_{wiring}^2) = m_{wiring} R_{wiring}^2 = 1{,}047\ \text{kg}(2\ \text{m})^2 = 4{,}186\ \text{kg·m}^2$$
- In line 17 in the MoI spreadsheet

Mass Property Calculations: Launch Vehicle

Item	Mass (kg)	Location (m)	Moment (kg-m)	Thickness (m)*	Dist from CM (m)	J_o (kg m²)	$m\,x_{CM}^2$ (kg m²)	J_{pitch}, J_{yaw} (kg m²)	J_{roll} (kg m²)
Payload Fairing $h = 2D$	689	35.56	24507	0.0049	15.22	3139	159693	162832	2756
Payload (5T w/PAF)	4250	34.40	146189	-	14.06	0	839716	839716	0
Payload Attach Fitting	750	33.15	24860	-	12.81	0	123001	123001	0
Forward Skirt $D = 2R$	501	31.40	15743	0.0049	11.06	1379	61292	62671	2006
LH_2 Fwd Dome/Insul.	272	30.89	8388	0.0029	10.55	453	30219	30671	724
LH_2 Cyl with Insulation	1836	23.14	42480	0.0029	2.79	31654	14341	45995	7344
LH_2 Aft Dome/Insul.	272	15.37	4175	0.0029	-4.97	453	6699	7152	724
Intertank	836	13.87	11594	0.0049	-6.47	3412	34948	38360	3343
LOx Fwd Dome/Insul.	403	12.37	4993	0.0055	-7.97	672	25610	26282	1076
LOx Cyl. with Insulation	1488	7.69	11436	0.0055	-12.65	9717	238210	247927	5951
LOx Aft Dome/Insul.	403	3.00	1210	0.0055	-17.34	672	121332	122005	1076
Aft Skirt	669	2.00	1337	0.0049	-18.34	2228	224888	227117	2674
Thrust Structure	497	2.00	994	-	-18.34	0	167200	167200	0
Engines	2236	1.00	2236	-	-19.34	0	836462	836462	0
Gimbals	71	2.00	142	-	-18.34	0	23832	23832	0
Avionics	744	31.40	23365	-	11.06	0	90969	90969	0
Wiring	1047	20.45	21400	-	0.11	145867	12	145880	4186
LH_2 fuel	0	23.14	0	-	2.80	0	0	0	0
LOx oxidizer	0	7.69	0	-	-12.65	0	0	0	0
Totals	16963	kg	345049			199647	2.998E+6	3.198E+6	31860
CM Location above ref.	20.34	m				6%	of total	kg m²	kg m²
						uniform LV	2.364E+6	Pitch MoI	Roll MoI

* Thickness does not include insulation

Mass Properties Topics

- Mass properties & coordinate systems
- Center of mass of assembly of items
- Moment of Inertia of individual items
- Principal axes & planes of symmetry
- Selected MoI values
- Examples of Moment of Inertia Calculation
  - Non-symmetric SC
  - Axisymmetric Bodies (2)
  - Launch vehicle
- **Propellants: dealing with liquids**
- References

That's An Empty Launch Vehicle – What About Propellants?

- Solid propellants are … solid
  Add appropriate shapes to MoI calculations (cylinders, domes, etc.)
- Liquid propellants:
  - Often can neglect J_{0i} calculations for liquids, since most have low viscosity and do not act as rigid body (and the J_{0i}s usually have small effect on J_{tot})
  - Must include $m_i L_i^2$ effect of propellant mass whether solid or liquid

Mass Properties Topics

- Mass properties & coordinate systems
- Center of mass of assembly of items
- Moment of Inertia of individual items
- Principal axes & planes of symmetry
- Selected MoI values
- Examples of Moment of Inertia Calculation
  - Non-symmetric SC
  - Axisymmetric Bodies (2)
  - Launch vehicle
- Propellants: dealing with liquids
- **References**

References

1. **Orbital Mechanics for Engineering Students**, H. D. Curtis, 3rd ed., Elsevier, 2014 (Ch. 9)
2. **Handbook of Equations for Mass & Area Properties of Various Shapes**, NAVWEPS Report 7827, U.S. Naval Ordnance Test Station, April 1982
3. **How to Calculate Mass Properties – An Engineer's Practical Guide**, R. Boynton & K. Wiener, Space Electronics Inc., undated. Available: www.space-electronics.com/mass-properties-analysis

Additional References

- Heineman Jr., W., "*Fundamental Techniques of Weight Estimating and Forecasting for Advanced Manned Spacecraft and Space Stations,*" NASA TN-D-6349, 1971.
- Heineman Jr., W., "*Mass Estimation and Forecasting for Aerospace Vehicles Based on Historical Data*," NASA JSC-26098, 1994.
- MacConochie, I. O., and P. J. Klich, "*Techniques for the Determination of Mass Properties of Earth-to-Orbit Transportation Systems*," NASA TM-78661, 1978.
- Myers, J., "*Handbook of Equations for Mass and Area Properties of Various Geometrical Shapes*," NAVWEPS report 7827, April 1962, http://www.dtic.mil/dtic/tr/fulltext/u2/274936.pdf, accessed Feb. 2018. doi:10.21236/ad0274936
- NASA, "*WAATS - A Computer Program for Weights Analysis of Advanced Transportation Systems*," NASA CR-2420, 1974.

Spacecraft Types, Attitude Sensing, Navigation/Orbit Determination

22-Aug-22

(52)

SC Types & Sensing Topics

- Spacecraft types & descriptions
- Stabilization & control schemes
- Pointing requirements & missions
- Attitude determination
- Rotations in space
- Sensor types
- Sensor accuracy

22-Aug-22

Spacecraft Design Process

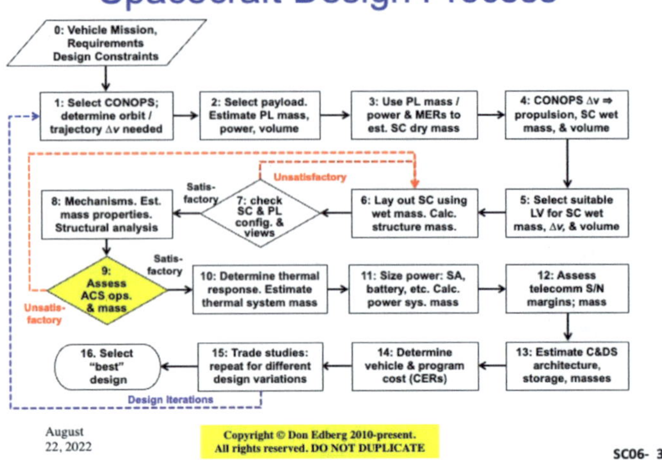
0: Vehicle Mission, Requirements Design Constraints
1: Select CONOPS; determine orbit / trajectory Δv needed
2: Select payload. Estimate PL mass, power, volume
3: Use PL mass / power & MERs to est. SC dry mass
4: CONOPS Δv ⇒ propulsion, SC wet mass, & volume
5: Select suitable LV for SC wet mass, Δv, & volume
6: Lay out SC using wet mass. Calc. structure mass.
7: check SC & PL config. & views
Unsatisfactory
Satis-factory
8: Mechanisms. Est. mass properties. Structural analysis
9: Assess ACS ops. & mass
Satis-factory
Unsatis-factory
10: Determine thermal response. Estimate thermal system mass
11: Size power: SA, battery, etc. Calc. power sys. mass
12: Assess telecomm S/N margins; mass
13: Estimate C&DS architecture, storage, masses
14: Determine vehicle & program cost (CERs)
15: Trade studies: repeat for different design variations
16. Select "best" design
Design Iterations

August 22, 2022

SC06- 3

SC Types & Sensing Topics

- Spacecraft types & descriptions
- Stabilization & control schemes
- Pointing requirements & missions
- Attitude determination
- Rotations in space
- Sensor types
- Sensor accuracy

22-Aug-22

SC06- 4

Types of Spacecraft Stabilization

(Green fill = *passive*, power not required)

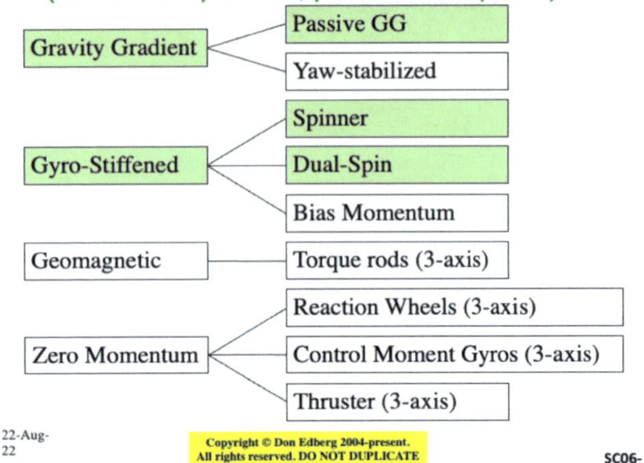
Gravity Gradient
Passive GG
Yaw-stabilized
Gyro-Stiffened
Spinner
Dual-Spin
Bias Momentum
Geomagnetic
Torque rods (3-axis)
Zero Momentum
Reaction Wheels (3-axis)
Control Moment Gyros (3-axis)
Thruster (3-axis)

22-Aug-22

SC06- 5

Passive *Gravity-Gradient* Stabilization

- Send out a long boom with a mass attached
- Tends to point towards (or away from!) Earth center PASSIVELY (i.e. same side of moon always faces Earth!)
- Passively controls two axes (pitch, roll), but **not yaw**
- Simple, passive, low cost
- Only good for LEO <1000 km
- Poor accuracy: Only good to at best ±5° pointing
- No yaw stability
- No maneuvering
- Examples: *Cerise* (left), *ORBCOM, GEOSAT*
- Recommended for: nadir-pointing

22-Aug-22

SC06- 6

Yaw-Stabilized Gravity-Gradient Spacecraft

- Add a momentum wheel ("gyro") with spin axis parallel to uncontrolled yaw
- Gyro provides yaw (& pitch) stiffnesses

End Mass
Body Rotates At 1 Rev / Orbit
V
Wheel Rate Ω
Orbit Normal
Down

Fig. 7.14 Gravity gradient stabilization with momentum wheel.
Ref: Griffin

22-Aug-22

SC06- 7

Spin-Stabilized Spacecraft ("Spinner")

- Gyroscopic stability provides "stiffness" for spin axis
- SC tends to keep pointing in same direction despite disturbances

Required hardware:

- Star, horizon, or sun sensor
- Thruster for changing spin axis
- TVC not needed for burns (spin averages out)

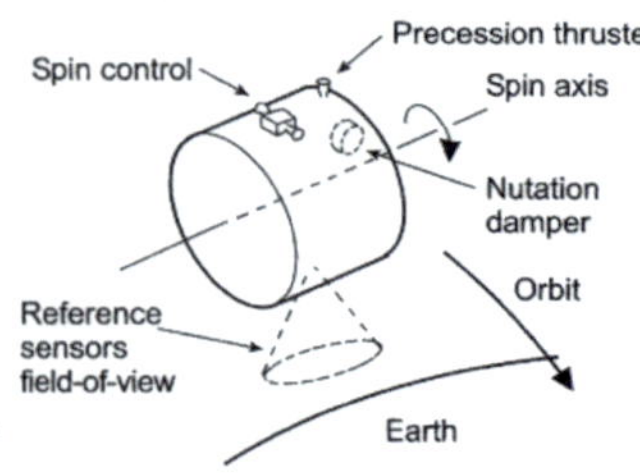
Precession thrusters
Spin control
Spin axis
Nutation damper
Orbit
Reference sensors field-of-view
Earth

Ref: Northrop-Grumman Hdbk

22-Aug-22

SC06- 8

Spin-Stabilized Spacecraft

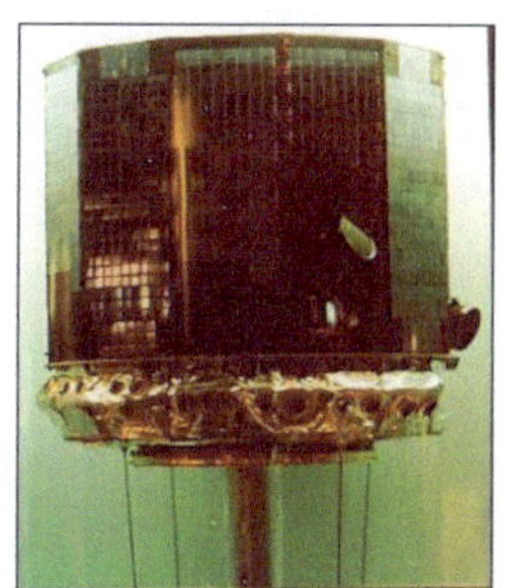

Pluses:
- Low cost
- Gives scan motion
- Averages out burn misalignments

Minuses:
- Low accuracy, poor maneuverability, lower power capability

Examples: *Explorer 1*, *Pioneer Venus*, *Intelsat I*, *II*, *III*

Typical spin: 30-80 RPM

22-Aug-22

SC06- 9

Momentum-Bias SC Schematic

- Spinning flywheel provides two-axis spin stabilization (like football)
- Third axis (roll) controlled by changes in spin speed
- Thrusters or other methods remove excess momentum

Arrays and appendages not shown

22-Aug-22

Ref: AIAA

SC06-10

Momentum-Bias Spacecraft

Courtesy: SS Loral

- Instead of spinning the entire SC, use an *internally-mounted* flywheel to provide gyroscopic stability
- Simple, cheaper than 3-axis
- 0.1° – 3° pointing accuracy
- Poor accuracy in wheel axis, limited maneuvering, TVC needed
- Examples: *Intelsat V* (shown), *Seasat*, *RCS Satcom*

22-Aug-22

SC06-11

Dual-Spin Spacecraft

Platform

Rotor

ANTENNA

ANTENNA REFLECTOR

ANTENNA FEEDS

THERMAL RADIATOR

CONTROL THRUSTER (4)

PROPELLANT TANK (4)

APOGEE KICK MOTOR

DESPUN REPEATER SHELF

TRAVELING WAVE TUBE AMPLIFIER

BATTERY PACK

SOLAR CELL ARRAYS

HS 376 SPACECRAFT CONFIGURATION

Ref: Wiesel, Boeing

22-Aug-22

SC06-12

Dual-Spin Spacecraft Schematic

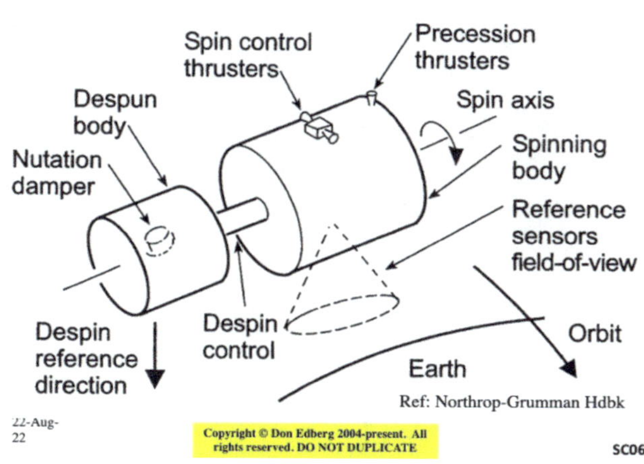

Ref: Northrop-Grumman Hdbk

22-Aug-22

SC06-13

Dual-Spin (cont.)

- Spin gives scan motion, pointing, gyro control for burns
- De-spun section gives inertial pointing (geo, planetary pointing)
- Expensive, complex mechanisms & moving parts, dynamics, lower power capability
- Moving parts (slip rings, bearing, drive motors) can fail
- Examples: *Galileo*, *Intelsat VI, Boeing 376*

22-Aug-22

SC06-14

Three-Axis Stabilized Spacecraft

- High pointing accuracy (*except for geomagnetic*), rapid maneuvering, large solar arrays
- Expensive, complex, zero-g feed for propellants needed, higher mass & power
- Examples: *Hubble*, *GPS TDRS* (left), *Magellan*, *Intelsat VIII*

22-Aug-22

Geomagnetic

- Electromagnets ("torque rods") in SC poiny "north" using external **B** field
- Low power
- Poor pointing performance
- Often used in cubesats

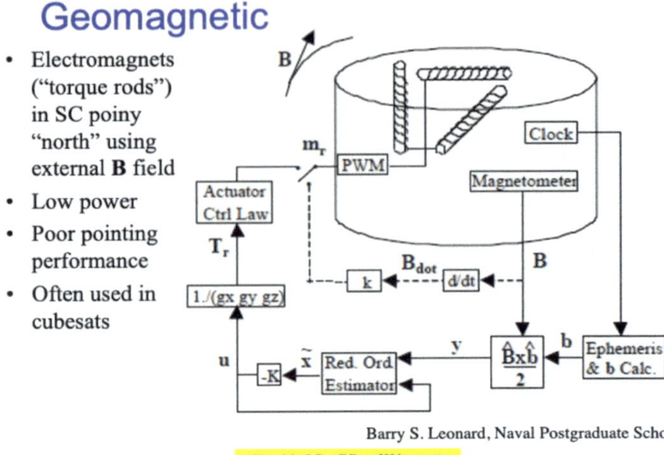

Barry S. Leonard, Naval Postgraduate School

22-Aug-22

Three-Axis *Reaction-Wheel* Spacecraft

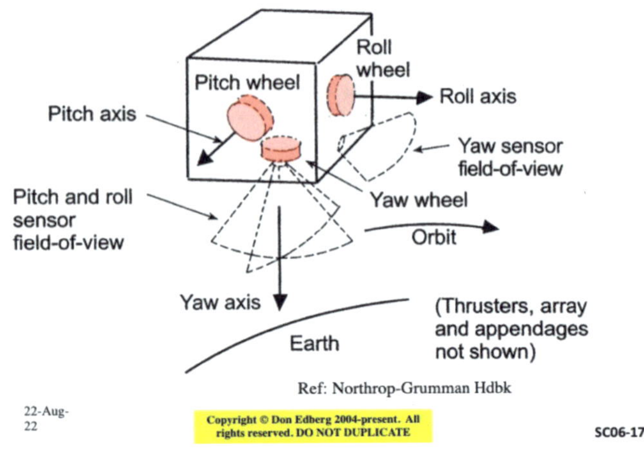

Ref: Northrop-Grumman Hdbk

22-Aug-22

Three-Axis *Control-Moment Gyro* SC

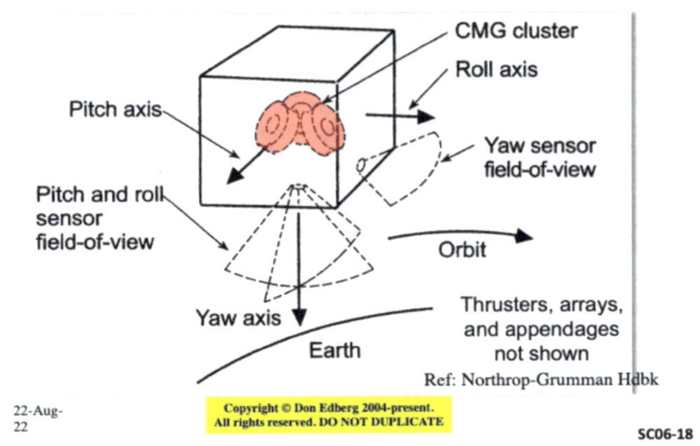

Ref: Northrop-Grumman Hdbk

22-Aug-22

Three-Axis *Thruster* Spacecraft

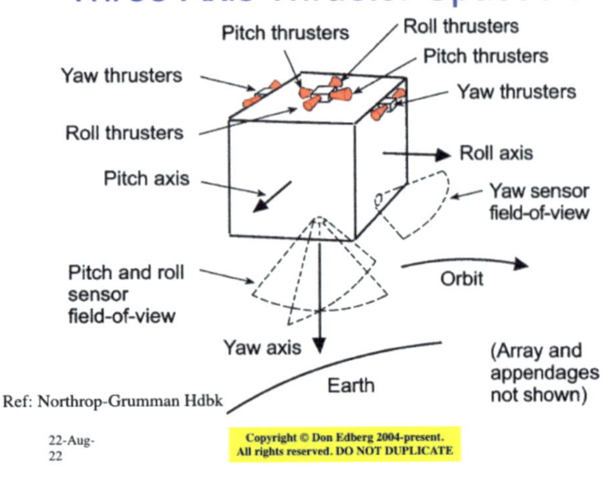

Ref: Northrop-Grumman Hdbk

22-Aug-22

SC Types & Sensing Topics

- Spacecraft types & descriptions
- Stabilization & control schemes
- Pointing requirements & missions
- Attitude determination
- Rotations in space
- Sensor types
- Sensor accuracy

22-Aug-22

SC Type vs. Mission Suitability

Spacecraft Type	Typical Missions	Pointing Accuracy	Maneuver Rates	Propellant Use	Space Examples
Gravity Gradient	Experimental only	5 – 10°	NA	NA	*Cerise*
Spin	Scanning sensors	0.1° – 1.0°	Low	High	*Pioneers, Explorer V*
Dual Spin	Comsats, Scientific	0.05° – 0.15°	Low	High	DSCS-II, HS376, *Intelsat, GLL*
Zero Momentum Spin	Scanning Sensors	0.05° – 0.15°	High	Low	*Vela, DSP*
Pitch Momentum Bias	Earth Probing sensors/ antennas	0.05° – 0.15°	Low	Low	*FLTSATCOM, TDRS, Satcom, Eurostar*
Thruster RCS	Interplanetary, earth-orbiting	0.1° – 0.5°	High Only	Medium	*Shuttle, Apollo/LM, Mariner, Voyager*
Reaction Wheel RCS	Scientific	0.001° – 0.50°	Low	Low	*OGO, HEAO, GRO, Nimbus*
CMG: Control-Moment Gyro	Orbiting Platforms	0.001° – 0.50°	High	Low	*Skylab, ISS, Mir*

Pointing Requirements

From mission, i.e.:

- Nadir-pointing or telescope pointing
- Scanning
- Etc.

Typical values:

- Optics, cameras, telescopes: 0.001° to 0.1°, or better
- High-gain antennae: 0.1° to 0.5°
- Solar arrays: 4° to 10°

Pointing Accuracy

Depends on three factors:

1. Sensor accuracy (how well you measure)
2. Attitude determination (how do you figure out what direction you're pointing)
3. Control accuracy (how well you can position your spacecraft)

Note that only TWO distinct objects are needed for three-axis orientation!

Maneuvering Also Important

- How often?
- How fast?
- Angular rates required?
- Control system response times

These also may define some of the propulsion system requirements

SC Types & Sensing Topics

- Spacecraft types & descriptions
- Stabilization & control schemes
- Pointing requirements & missions
- **Attitude determination**
- Rotations in space
- Sensor types
- Sensor accuracy

Attitude Determination vs. Guidance/Navigation (GN)

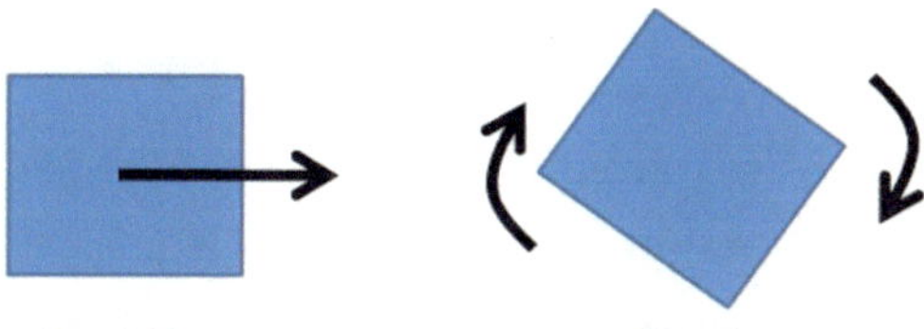

Translation of center of mass

Rotation about center of mass

GN

ADCS

GN: discussed in Telecommunication lecture SC10

ADCS: discussed in this lecture

Attitude Determination & Control Process

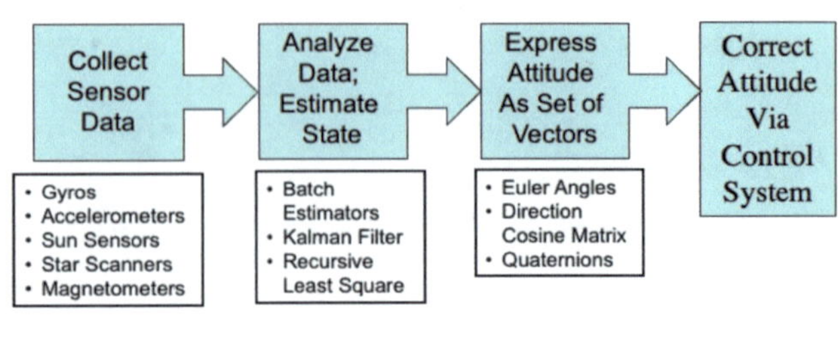

Common Coordinate System

Similar to airplane coordinates, facing forward:

- Forward = velocity direction = +x-axis
- "Up" (*zenith*) away from attracting body = +z-axis
  nadir = –z-axis
- Out left "wing" = +y-axis
- θ = pitch ("theta," about "wings")
- ψ = yaw ("psi" about vertical)
- ϕ = roll ("phi," about velocity dir.)

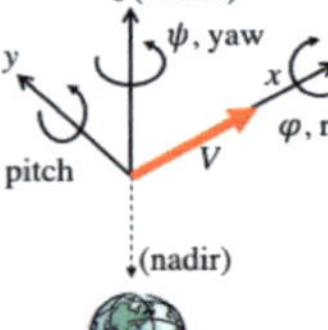

SC Types & Sensing Topics

- Spacecraft types & descriptions
- Stabilization & control schemes
- Pointing requirements & missions
- Attitude determination
- Rotations in space
- Sensor types
- Sensor accuracy

Rotations

- Relation between reference frame and spacecraft frame defined by three angles
- Euler angles determined by both magnitude and sequence of rotations
- A *direction cosine matrix* (DCM) specifies orientations after these rotations

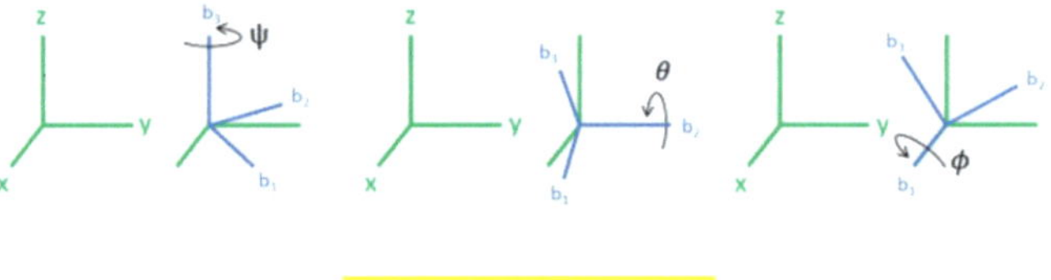

Rotation Computations

$$DCM = \begin{bmatrix} \cos\theta\cos\psi & \cos\theta\sin\psi & -\sin\theta \\ -\cos\phi\sin\psi + \sin\phi\sin\theta\cos\psi & \cos\theta\cos\psi + \sin\phi\sin\theta\sin\psi & \sin\phi\cos\theta \\ \sin\phi\sin\psi + \cos\phi\sin\theta\cos\psi & -\sin\phi\cos\psi + \cos\phi\sin\theta\sin\psi & \cos\phi\cos\theta \end{bmatrix}$$

- DCM requires 16 multiplies, 4 additions, 6 trig functions
- Reverse operation: multiply by $[DCM]^T$
- Large memory required
- Computationally intensive
- Alternatives?

Quaternions

- Contain same information as Euler angles
- Any series of rotations can be represented as a *single* rotation about *one fixed axis*
  - e_1 is the rotation amount
  - (e_2, e_3, e_4) describe the *direction* of the rotation vector
- Hence "quaternions" for 'four numbers'
- Non-singular
- Evaluation requires 16 multiplies, 12 adds, **0** trig functions (significant advantage over DCM)

Attitude Representations Compared

Method	Advantages	Disadvantages
Euler angles (yaw, pitch, roll)	• No redundant parameters • Clear physical interpretation • Convenient product rule using DCM (Direction Cosine Matrix)	• Singularities • 29 trig functions and 27 multiplies slow computation • Gimbal lock
Quaternions	• Rapid computation: 15 multiplies, 12 adds. No singularities or trig functions.	• Taught infrequently in schools

MATLAB codes & info on ACS operations:

- Convert between DCM, Euler angles, Quaternions, and Euler vectors: `www.mathworks.com/matlabcentral/fileexchange/20696-function-to-convert-between-dcm-euler-angles-quaternions-and-euler-vectors`
- Quaternion class: `www.mathworks.com/matlabcentral/fileexchange/33341-quaternion-m`

Courtesy: Brown T5.6

22-Aug-22 SC06-33

SC Types & Sensing Topics

- Spacecraft types & descriptions
- Stabilization & control schemes
- Pointing requirements & missions
- Attitude determination
- Rotations in space
- Sensor types
- Sensor accuracy

22-Aug-22 SC06-34

On-Board Sensors & Sensing

- Linear motion
  - Acceleration
- Angular
  - Attitude
  - Rate gyros
  - Sun sensor
  - Horizon sensor
  - Magnetometer
  - Star sensor/tracker
- Linear & angular combined in IMU
- Spatial position & attitude
  - GPS, GPS DCP

22-Aug-22 SC06-35

Linear Acceleration

$F = kx = ma$

m, k, accel, x, 0 (Rest)

- Measured by accelerometers
- Measure position of a "proof mass" on a spring
- Deflection proportional to acceleration being experienced

22-Aug-22 SC06-36

Stabilized Platform for Attitude Info

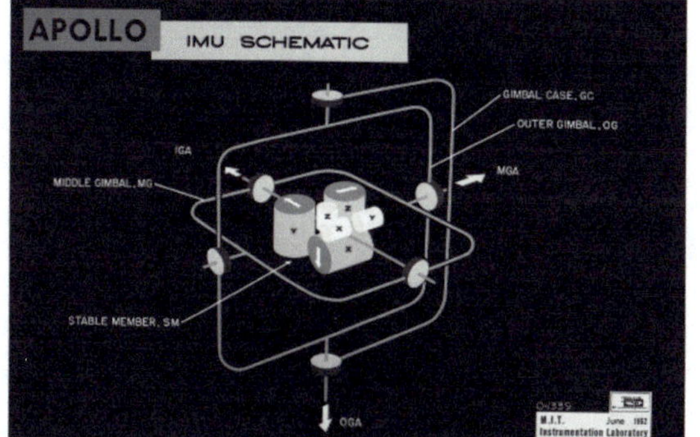

- Assembly uses torque motors to maintain orientation of platform regardless of motion
- Resolvers measure motion relative to reference position
- Expensive, complex, not commonly used

Courtesy MIT

8/22/22 SC06-37

Apollo Stabilized Platform Video

The IMU - The Heart of Apollo Guidance

Apollo IMU Draper Lab.mp4

8/22/22 SC06-38

Angular Position Sensor: Laser Gyroscope

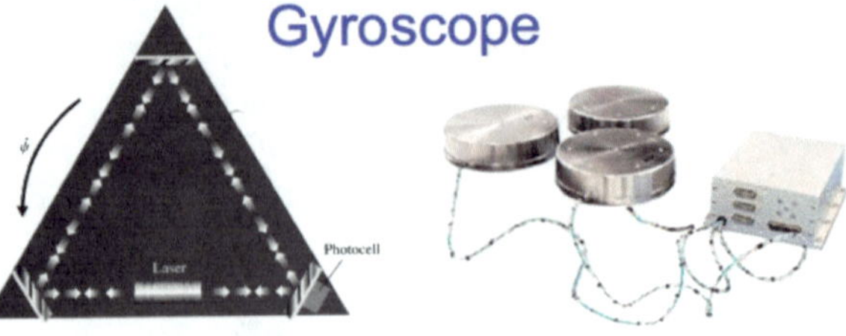

FIGURE 6.4
The laser gyroscope.

- Works as a gyro, but no moving parts
- More reliable
- Less power consumption
- Radiation tolerant

Left: Wiesel.
Right: Northrop-Grumman FOG 2500
High precision interferometric Fiber-Optic Gyros: 3.8 lbm, 5W @ 5V.

22-Aug-22

SC06-40

Hemispherical Resonator Gyro

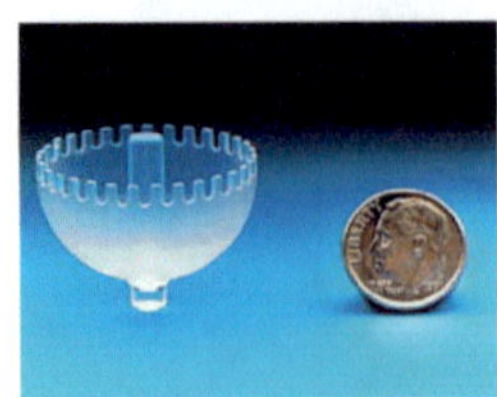

Courtesy: Northrop-Grumman

- Sense movement of a standing wave in a fused-silica shell
- Like a wineglass 'singing' as you slide finger around the rim
- Null points in the wave precess when the unit is rotated
- Vibrating sensor shell is only moving part

22-Aug-22

SC06-41

Inertial Measurement Unit (IMU)

Inertial Labs IMU-P
Accels: up to ±40 g_0
Gyros: up to ±2,000 deg/s
Mass: 0.070 kg
Size: 39 x 45 x 22 cm
Power: 0.8 W at 5 VDC

- Set of three (one per axis) silicon MEMS accelerometers sense accelerations experienced during orbit insertion & Δv adjust maneuvers
- Set of three (one per axis) fiber-optic gyros sense angular rates
- Integration of these values provides position and attitude information

8/22/22

SC06-42

2016 Mars Lander Crash Sparked By Bad Inertial Data

LYON, France: European Space Agency (ESA) engineers believe they've found the root cause of the Oct. 19 crash of the *Schiaparelli* Mars lander – erroneous information from the IMU resulting in an incorrect altitude reading.

Shortly after the lander's parachute deployed, the IMU "saturated" – the point at which it measured maximum rotation rates – and stayed that way for about 1 sec. When fed into the navigation system, the data generated an estimated altitude that was below ground level, according to ESA.

This triggered a premature release of the parachute and backshell. Braking thrusters then fired for the minimum preselected time of about 3 sec., instead of the nominal 60 sec. Finally, on-ground systems were activated as if *Schiaparelli* had already landed, whereas it was still 3.7 km (12,000 ft.) from the surface

The scenario has been "clearly reproduced in computer simulations of the control system's response to the erroneous information," ESA says.

Courtesy: Aviation Week

22-Aug-22

SC06-43

Angular Position Sensor: Sun Sensor

Sun's rays
Entry slit
Reticle slits
Photo - detectors
Binary outputs

Figure 10.7 Sun sensor with digital output

Left: Fortescue.
Right: Mariner 10, Cosmosphere

SC06-44

Multiple Sun Sensor Module

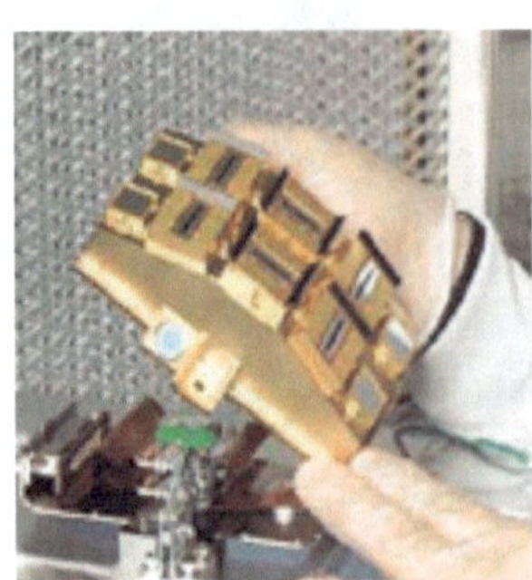

- Example: *Goodrich 13-517* coarse sun sensor assembly
- Mount several on periphery of SC to locate sun in any SC attitude
- *MER* had FIVE sun sensors around its cruise stage

22-Aug-22

SC06-45

Angular Position Sensor: Conical Scanning Horizon Sensor

Useful when near a planet or object large enough to be visible as a disc

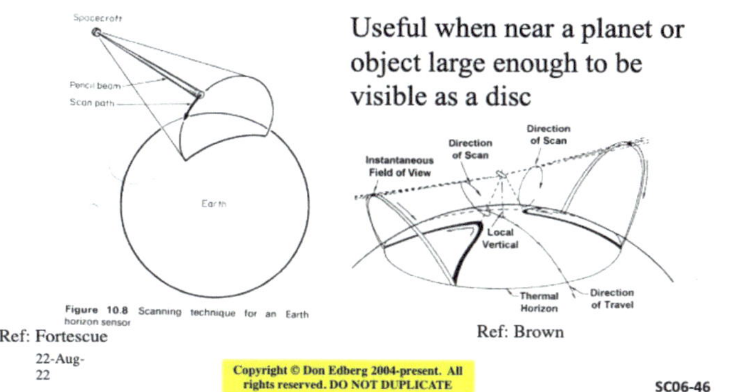

Ref: Fortescue

Ref: Brown

22-Aug-22

SC06-46

Horizon or Nadir Sensor

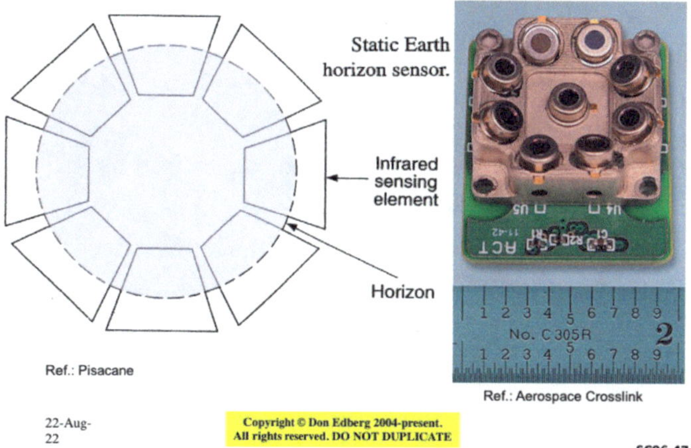

Ref.: Pisacane

Ref.: Aerospace Crosslink

22-Aug-22

SC06-47

Magnetometers

- Vector sensors: provide both direction & magnitude of magnetic field
- Reliable, lightweight, low power, no moving parts
- Limited to low altitudes above Earth due to field weakness (decreases as $1/r^3$)
- Not accurate inertial sensors
  - magnetic field not known accurately
  - influenced by field of SC

22-Aug-22

SC06-48

Star Sensor (courtesy: Jena Optronik)

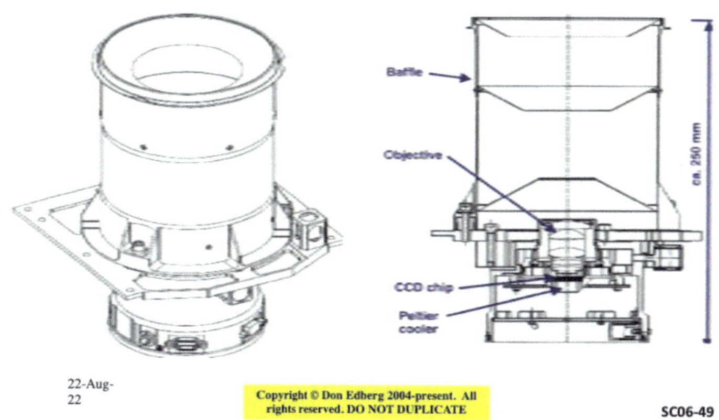

22-Aug-22

SC06-49

Medium Accuracy Star Sensor Typical Characteristics

- Accuracy 0.025° Line of sight
- Mass 3.7 kg (head, electronics, harness)
- Power 14 W/8 W (cooler on/off)
- FoV 18° × 13°
- Update rate 8 Hz
- Baffle 25-30° (sun exclusion angle)
- Temperature –25 to +35 C

22-Aug-22

SC06-50

Angular Position Sensor: Star Tracker

Ball Aerospace CT-631

- Like camera with CCD
- Electronically search and track over limited FoV
- Provides horiz. & vert. position, converts to position error
- Stray light is problem, keep >30-60° from sun
- Can be fixed or gimbaled

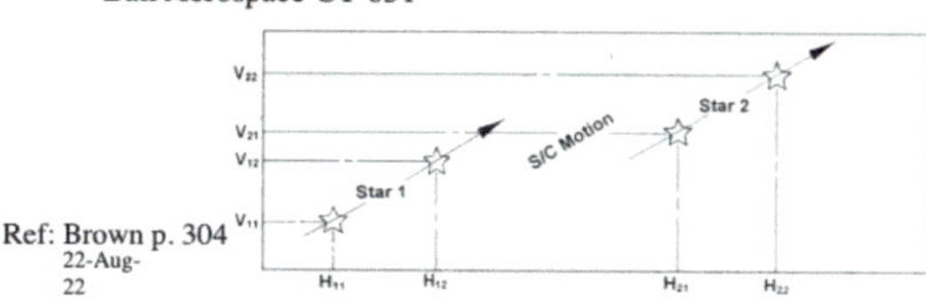

Ref: Brown p. 304

22-Aug-22

SC06-51

Angular Position Sensor: Star Cameras

- Inertial sensor
- Looks for patterns of bright stars
- On-board processor deduces position by pattern matching
- Requires no *a priori* pointing knowledge
- First used on *Clementine*: OCA camera FoV 28.9° × 43.4°, accuracy 96 arc-sec, 1.2 kg

22-Aug-22

SC06-52

GPS Attitude Measurement

- Multiple GPS antennae mounted at precise locations on spacecraft
- GPS Differential Carrier Phase (DCP) measurements fed back to attitude control system
- Only works near Earth (!)

22-Aug-22

SC06-53

SC Types & Sensing Topics

- Spacecraft types & descriptions
- Stabilization & control schemes
- Pointing requirements & missions
- Attitude determination
- Rotations in space
- Sensor types
- Sensor accuracy

22-Aug-22

SC06-54

Typical Accuracy of Sensors

Reference Object	Potential Accuracy
Star trackers	1 arc sec
Sun sensors	1 arc min (= 60 arc sec)
Earth Horizon sensor	6 arc min
Magnetometer	>30 arc min

22-Aug-22

SC06-55

Measures of Angular Resolution

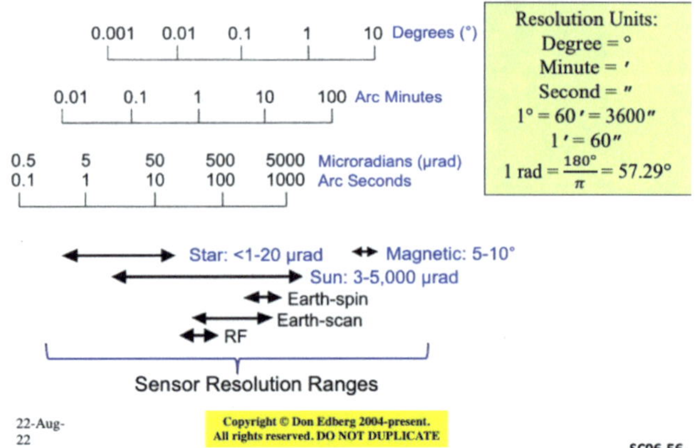

Resolution Units:
Degree = °
Minute = ′
Second = ″
1° = 60′ = 3600″
1′ = 60″
$1 \text{ rad} = \frac{180°}{\pi} = 57.29°$

22-Aug-22

SC06-56

Spacecraft Dynamics & Control Systems

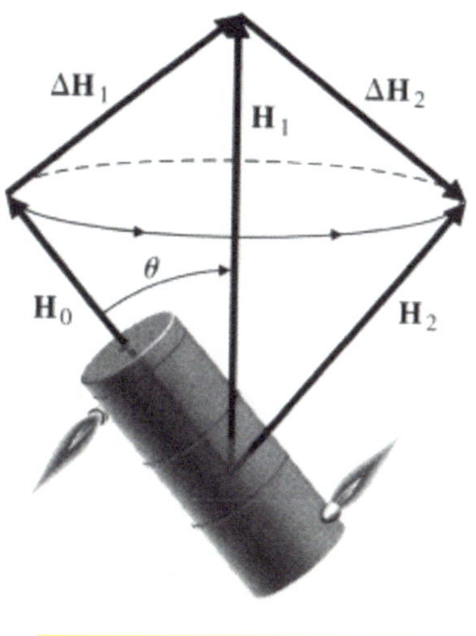

August 22, 2022

(62)

Dynamics & Control Topics

- Guidance/Navigation vs. Attitude Control
- Spacecraft dynamic characteristics
  - Non-Spinning
  - Spinning
- Moments of Inertia and spin
- Actuators
  - Thrusters
  - Reaction wheels
  - Control-moment gyros
  - Magnetic torquers
  - Mother Nature
- Attitude control systems: needs & operations
  - Disturbances
- Design considerations, modes of operation

August 22, 2022

Spacecraft Design Process

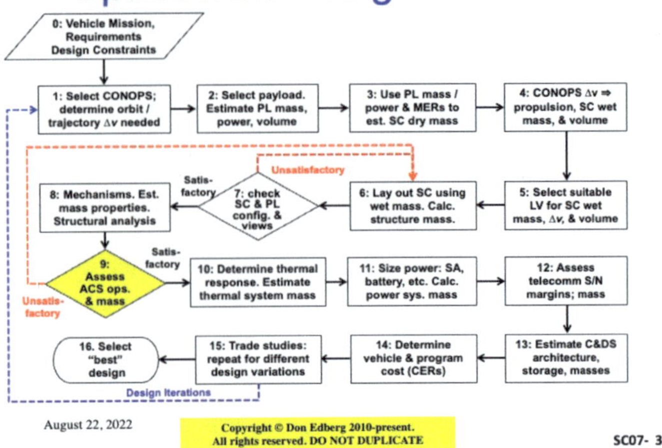

0: Vehicle Mission, Requirements Design Constraints
1: Select CONOPS; determine orbit / trajectory Δv needed
2: Select payload. Estimate PL mass, power, volume
3: Use PL mass / power & MERs to est. SC dry mass
4: CONOPS Δv ⇒ propulsion, SC wet mass, & volume
5: Select suitable LV for SC wet mass, Δv, & volume
6: Lay out SC using wet mass. Calc. structure mass.
7: check SC & PL config. & views
Unsatisfactory
Satis-factory
8: Mechanisms. Est. mass properties. Structural analysis
9: Assess ACS ops. & mass
Satis-factory
Unsatis-factory
10: Determine thermal response. Estimate thermal system mass
11: Size power: SA, battery, etc. Calc. power sys. mass
12: Assess telecomm S/N margins; mass
13: Estimate C&DS architecture, storage, masses
14: Determine vehicle & program cost (CERs)
15: Trade studies: repeat for different design variations
16. Select "best" design
Design Iterations

August 22, 2022

SC07- 3

The Key to Guidance

Click for recording of actual audio from an Air Force training tape on how guided missiles locate themselves and get to the target: `missile guidance.wav`

August 22, 2022

SC07- 4

Guidance/Navigation vs. **Attitude Control System (ACS)**

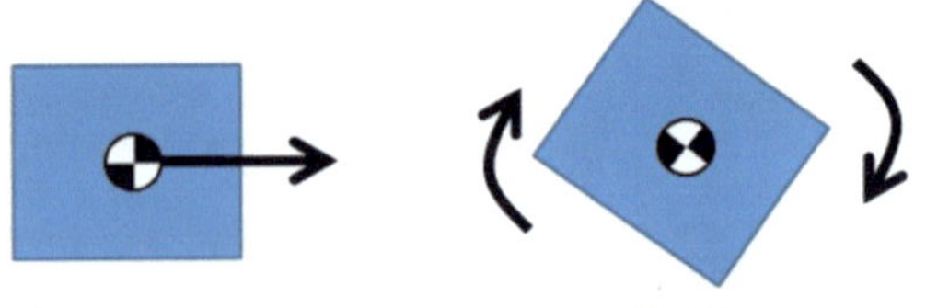

Translation of center of mass

Rotation about center of mass

Guidance /Navigation

- Ensures vehicle follows a certain trajectory
  - "Slow" control

Attitude Control System:

- **Ensures vehicle stability during flight ("fast control")**
- **Controls headings to maintain trajectory**

August 22, 2022

SC07- 5

Dynamics & Control Topics

- Guidance/Navigation vs. Attitude Control
- Spacecraft dynamic characteristics
  - Non-Spinning
  - Spinning
- Moments of Inertia and spin
- Actuators
  - Thrusters
  - Reaction wheels
  - Control-moment gyros
  - Magnetic torquers
  - Mother Nature
- Attitude control systems: needs & operations
  - Disturbances
- Design considerations, modes of operation

August 22, 2022

SC07- 6

Spacecraft Dynamics

- Non-spinning spacecraft
  - Maneuvering & propellant consumption
- Spinning spacecraft
  - Stability of spin
  - Changing spin axis orientation
- Vibrations & damping

August 22, 2022

SC07- 7

Dynamics & Control Topics

- Guidance/Navigation vs. Attitude Control
- Spacecraft dynamic characteristics
  - Non-Spinning
  - Spinning
- Moments of Inertia and spin
- Actuators
  - Thrusters
  - Reaction wheels
  - Control-moment gyros
  - Magnetic torquers
  - Mother Nature
- Attitude control systems: needs & operations
  - Disturbances
- Design considerations, modes of operation

August 22, 2022

SC07- 8

Non-Spinning Spacecraft

- Simple view: they obey $T = I\alpha$ (torque = moment of inertia × angular acceleration)
  MoI = moment of inertia, varies with different rotation axes
- Torque supplied by:
  - Thrusters
  - Reaction Wheels (RWAs)
  - Control moment gyros (CMGs). Russians call these "gyrodynes"

August 22, 2022

SC07- 9

Dynamics & Control Topics

- Guidance/Navigation vs. Attitude Control
- Spacecraft dynamic characteristics
  - Non-Spinning
  - Spinning
- Moments of Inertia and spin
- Actuators
  - Thrusters
  - Reaction wheels
  - Control-moment gyros
  - Magnetic torquers
  - Mother Nature
- Attitude control systems: needs & operations
  - Disturbances
- Design considerations, modes of operation

August 22, 2022

SC07-10

Spinning Spacecraft Dynamics

Spin dynamics "similar" to linear dynamics:

- Force **F** = rate of change of linear momentum **p** (vector quantities)

$$\mathbf{F} = \frac{d\mathbf{P}}{dt}$$

Analogy to rotational systems:

- Moment or torque **M** = rate of change of angular momentum **H**

$$\mathbf{M} = \frac{d\mathbf{H}}{dt}$$

- Spinning top example

August 22, 2022

SC07-11

Repointing A Spinning Spacecraft

ΔH_1 H_1 ΔH_2 θ H_0 H_2

Ref: Wiesel

- SC has initial angular momentum $\mathbf{H}_0$
- Apply a torque $\Delta\mathbf{H}_1$ to start precession, SC "wobbles"

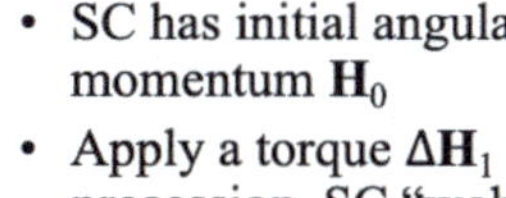

- Wait for one-half of a precession cycle, *body* axis rotates 2θ
- Apply second torque $\Delta\mathbf{H}_2$ to stop precession
- Spacecraft now spins with new spin axis direction

August 22, 2022

SC07-12

Spinning Spacecraft Thruster Locations

Spin Axis

Spin-Despin Thrusters

Pulsing Thrusters Change Angular Momentum for **Precession Control**

Ref: Brown

August 22, 2022

SC07-13

Precession Movie

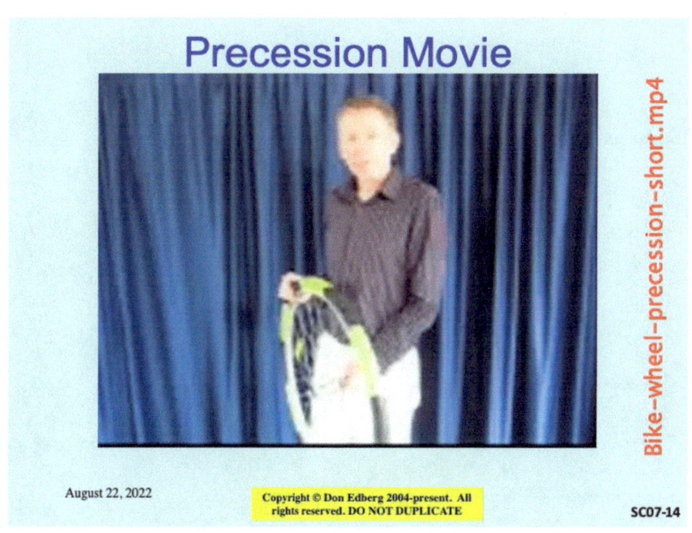

Bike–wheel–precession–short.mp4

August 22, 2022

SC07-14

Time Interval for Precession

- Assume axisymmetric spacecraft (moments of inertia A, A, and C, $C > A$) spinning at a rate of ω rad/s
- The time required to rotate an angle θ is

$$t = \pi\left(\frac{A - C}{C}\right)\frac{\cos\theta}{\omega}$$

- The total "cost" is $\Delta H_{\text{tot}} = 2|\mathbf{H}_0|\tan\theta$
- So 180° rotation ($\theta = 90°$) is *impossible* using a single pair of burns!

Large-Angle Rotations

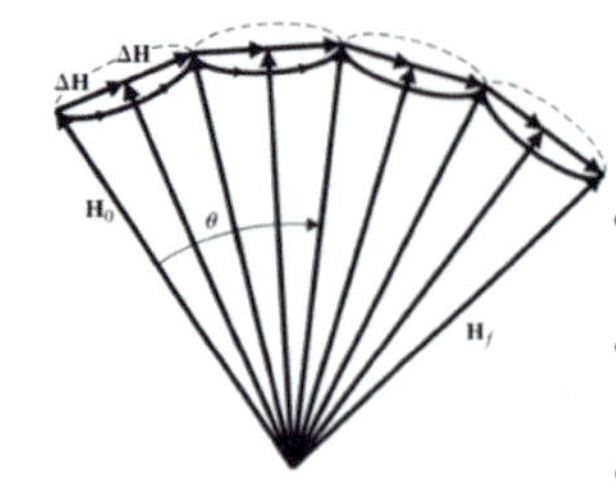

FIGURE 5.11
Large-angle reorientation by repeated impulses.

- Large angle rotations take large Δvs
- Solution is to break into smaller rotations
- Timing of each successive maneuver is critical

Ref: Wiesel

Dual-Spin Control Method Similar

Ref: AIAA

DUAL SPIN CONTROL

Dynamics & Control Topics

- Guidance/Navigation vs. Attitude Control
- Spacecraft dynamic characteristics
  - Non-Spinning
  - Spinning
- Moments of Inertia and spin
- Actuators
  - Thrusters
  - Reaction wheels
  - Control-moment gyros
  - Magnetic torquers
  - Mother Nature
- Attitude control systems: needs & operations
  - Disturbances
- Design considerations, modes of operation

Moments of Inertia (*I* matrix)

$$I = \begin{bmatrix} I_{xx} & I_{xy} & I_{xz} \\ I_{yx} & I_{yy} & I_{yz} \\ I_{zx} & I_{zy} & I_{zz} \end{bmatrix}$$

Note: the matrix is symmetric

$I_{ij} = I_{ji}$

Moments of Inertia

$$\begin{cases} I_{xx} = \sum_i m_i\left(y_i^2 + z_i^2\right) \\ I_{yy} = \sum_i m_i\left(x_i^2 + z_i^2\right) \\ I_{zz} = \sum_i m_i\left(x_i^2 + y_i^2\right) \end{cases}$$

"Products of Inertia" $$\begin{cases} I_{xy} = -\sum_i m_i x_i y_i \\ I_{xz} = -\sum_i m_i x_i z_i \\ I_{yz} = -\sum_i m_i y_i z_i \end{cases}$$

The *mass properties* group keeps track of all component masses and their locations

When Masses Change Location:

- If things move, the *parallel axis theorem* is invoked:

$$I_P = I_G + \begin{bmatrix} m\left(y_{G/P}^2 + z_{G/P}^2\right) & -mx_{G/P}y_{G/P} & -mx_{G/P}z_{G/P} \\ -mx_{G/P}y_{G/P} & m\left(x_{G/P}^2 + z_{G/P}^2\right) & -my_{G/P}z_{G/P} \\ -mx_{G/P}z_{G/P} & -my_{G/P}z_{G/P} & m\left(x_{G/P}^2 + y_{G/P}^2\right) \end{bmatrix}$$

Here I_P is the (unknown) moment of inertia about P, I_G is the moment of inertia about the CM, and $x_{G/P}$, $y_{G/P}$, and $z_{G/P}$ are the distances from the mass center to P.

Inertia Matrix Properties

Can find a set of directions so inertia matrix is symmetric. These are called the "Principal Moments of Inertia" and the "Principal Directions."

$$I = \begin{bmatrix} A & 0 & 0 \\ 0 & B & 0 \\ 0 & 0 & C \end{bmatrix}$$

If there is a plane of symmetry, the inertia becomes very simple. The spacecraft is said to be *axisymmetric*.

$$I = \begin{bmatrix} A & 0 & 0 \\ 0 & A & 0 \\ 0 & 0 & C \end{bmatrix}$$

August 22, 2022

SC07-21

Spin & Spacecraft Geometry

Low Inertia direction (minimum Moment of Inertia, or Minor Principal MoI)

High Inertia direction (maximum MoI or Major Principal MoI)

Prolate ("long & skinny"): $C < A$

Oblate ("short & fat") $C > A$

August 22, 2022

SC07-22

Preferred Spin Axis

- Spacecraft can be set to spin about principal axis
- They "prefer" to spin around maximum MoI axis
- Prolate spacecraft spinning about min MoI tend to eventually "tumble" (*Explorer 1*)
- Oblate spacecraft are preferred — they spin stably around max MoI axis

August 22, 2022

SC07-23

Explorer I & Whip Antennas

Photos: D. Edberg

August 22, 2022

SC07-24

Explorer 1 Movie

Movie: Explorer 1-1958.mp4

August 22, 2022

SC07-25

Inertia Axes

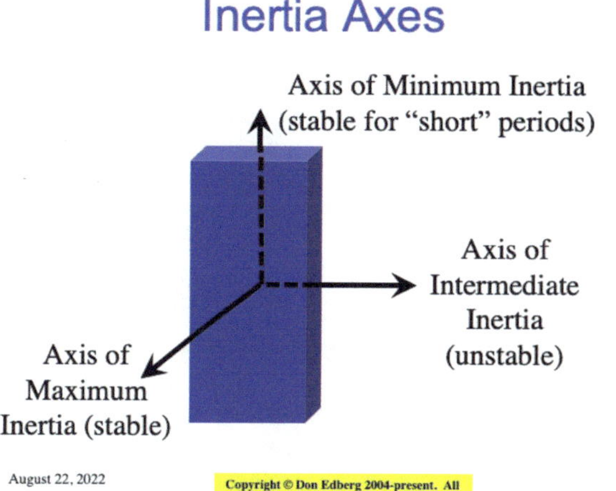

August 22, 2022

SC07-26

Major, Minor, Intermediate Spin

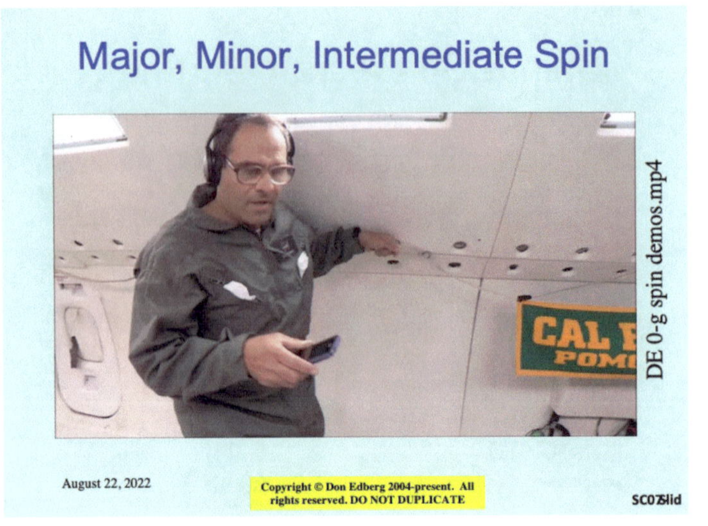

DE 0-g spin demos.mp4

August 22, 2022

Spin Movie — Intermediate Axis

RichardGarriottSpaceVideoBlogRotational Inertia.mp4

August 22, 2022

Another Spin Movie

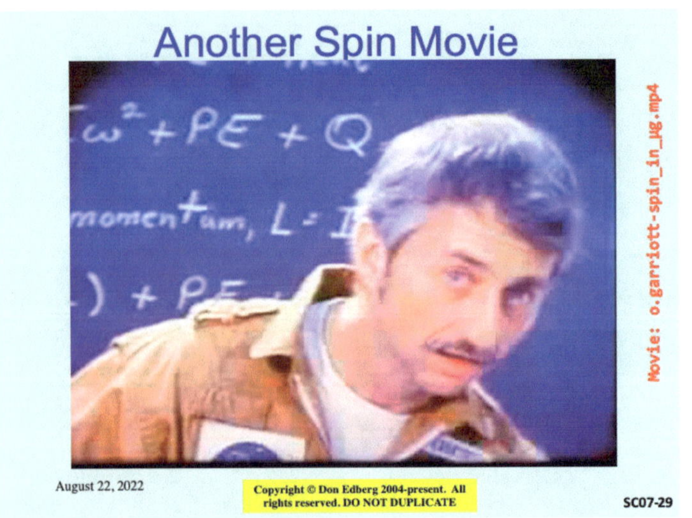

Movie: o.garriott-spin_in_μg.mp4

August 22, 2022

Stopping a Spinning Spacecraft

- During solid rocket motor firings SC spun for
  - Stability
  - To "average out" thrust imbalances
- After firing, need to stop spin
- "Yo-Yo" de-spin mechanism used (angular momentum conserved)

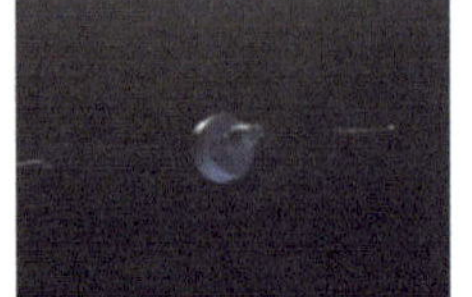

Courtesy: NASA/MER

August 22, 2022

"Yo-Yo" De-spinner

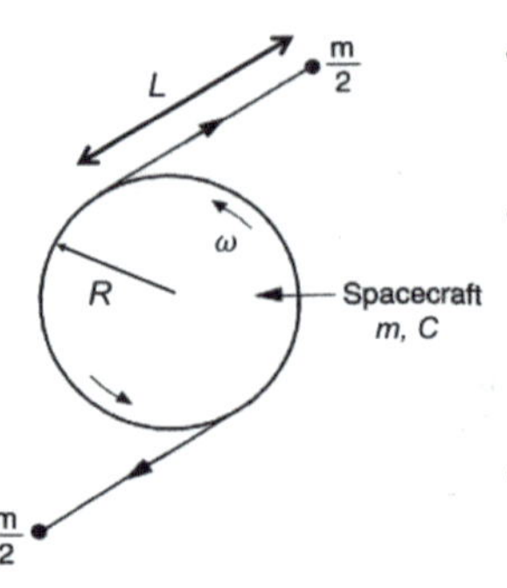

- Angular momentum transferred to weights that are jettisoned
- To completely de-spin: $L = [R^2 + (C/m)]^{1/2}$ where R = radius, C = spin axis inertia, and m = mass of balls
- Length independent of spin rate ω!

Ref: Pisacane (fig.), Wiesel (analysis)

August 22, 2022

Yo-Yo Despinner Operation

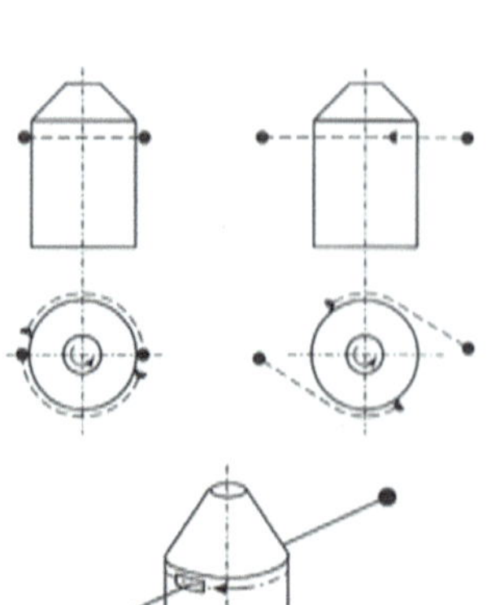

Courtesy: Hdbk. Space Technology

August 22, 2022

Conservation of Angular Momentum

August 22, 2022

SC07-33

Tuned-Mass Damper

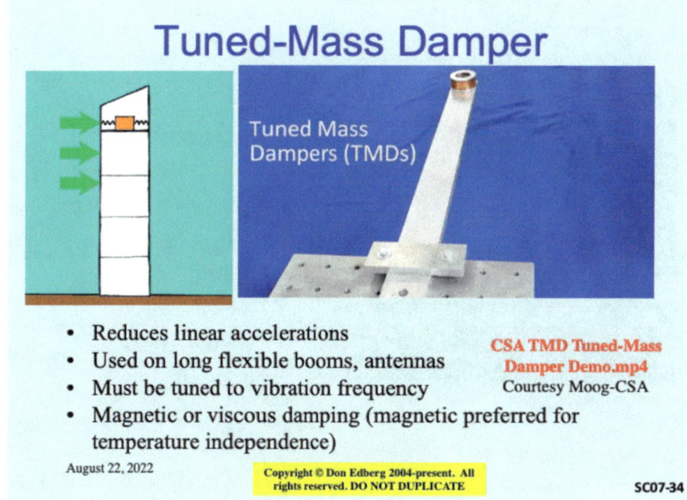

- Reduces linear accelerations
- Used on long flexible booms, antennas
- Must be tuned to vibration frequency
- Magnetic or viscous damping (magnetic preferred for temperature independence)

CSA TMD Tuned-Mass Damper Demo.mp4
Courtesy Moog-CSA

August 22, 2022

SC07-34

Vibration Damping

- Tuned-mass damping used on HST
- Reduced SA vibrations for better astronomy

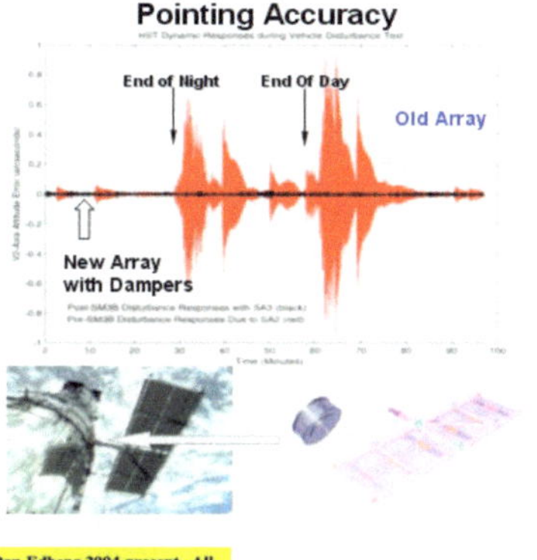

Courtesy CSA engineering

August 22, 2022

SC07-35

Dynamics & Control Topics

- Guidance/Navigation vs. Attitude Control
- Spacecraft dynamic characteristics
  - Non-Spinning
  - Spinning
- Moments of Inertia and spin
- Actuators
  - Thrusters
  - Reaction wheels
  - Control-moment gyros
  - Magnetic torquers
  - Mother Nature
- Attitude control systems: needs & operations
  - Disturbances
- Design considerations, modes of operation

August 22, 2022

SC07-36

Spacecraft Control Actuators

Actuators needed to *change rotational attitude* of spacecraft

Types of Actuators

- Thrusters
- Reaction Wheels
- Control-Moment Gyros
- Magnetic Torquers
- (Gravity Gradient)
- (Solar torque)

August 22, 2022

SC07-37

Dynamics & Control Topics

- Guidance/Navigation vs. Attitude Control
- Spacecraft dynamic characteristics
  - Non-Spinning
  - Spinning
- Moments of Inertia and spin
- Actuators
  - Thrusters
  - Reaction wheels
  - Control-moment gyros
  - Magnetic torquers
  - Mother Nature
- Attitude control systems: needs & operations
  - Disturbances
- Design considerations, modes of operation

August 22, 2022

SC07-38

Thruster Control

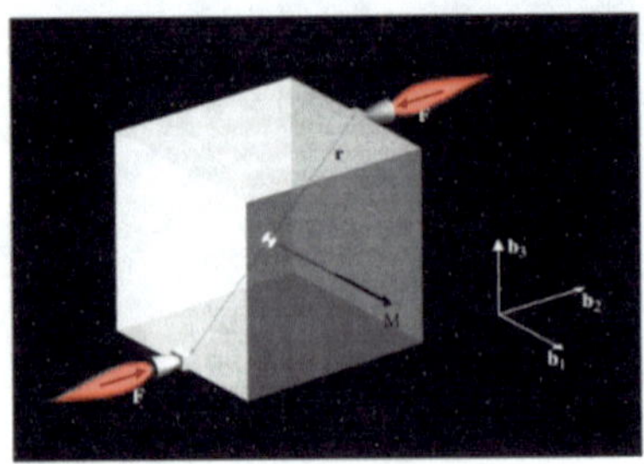

FIGURE 5.12
Thruster-controlled spacecraft.

- Direct control of spacecraft
- Requires propellant
- Discrete amount of control (Minimum Impulse Bit, MIB)
- Can be cross-coupling

Ref: Wiesel

August 22, 2022 SC07-39

Thrusters (cont.)

- Moment generated per thruster: $\mathbf{M} = \mathbf{r} \times \mathbf{F}$
- Change in momentum $\Delta\mathbf{H} = \mathbf{M}\Delta t = 2\mathbf{r} \times \mathbf{F}\, \Delta t$
- Stop rotation by firing in opposite direction
- Want to align thrusters with principal axes
- If steady disturbance, get "limit cycle" motion: spacecraft rotates to end of tolerance, thrusters fire, rotates to opposite end, thrusters fire, etc.
- Thrusters may be monopropellant or cold gas

August 22, 2022 SC07-40

Three-Axis Control Spacecraft Thruster Arrangement (12 Needed)

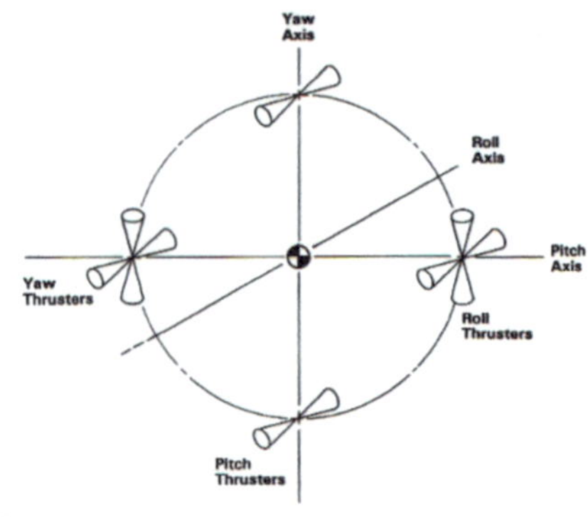

Fig. 4.11 Typical thruster installation for three-axis spacecraft.

Ref: Brown

August 22, 2022 SC07-41

Multiple Thrusters May Control Different Motions

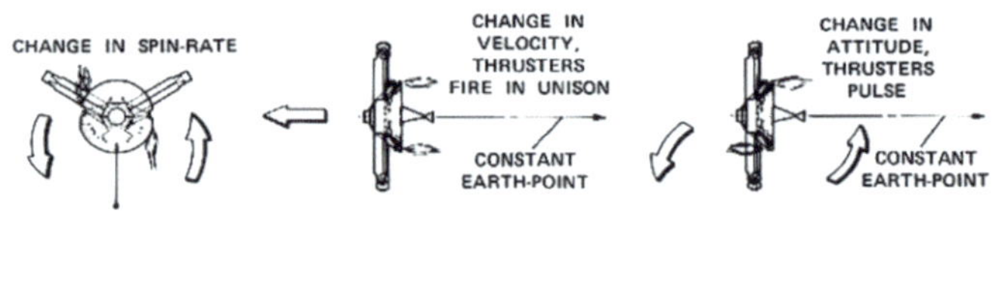

Courtesy JPL

August 22, 2022 SC07-42

One-Axis Rotational Maneuver

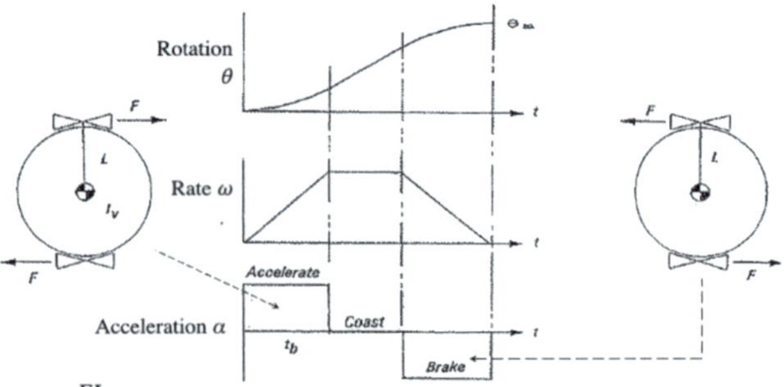

$\theta_{rotation} = \frac{nFL}{I_{rot}} t_b (t_b + t_c)$ where n = #thrusters, F = thrust, L = moment arm, t_b = burn time, t_c = coast time, I_{rot} = rotation inertia. Propellant consumed: $m_p = \frac{nFt_b}{I_{sp}}$

August 22, 2022 SC07-43

"Bang-Bang" Control

- Thrusters either on or off (no partial settings)
- Direction from sign of error
- Add "deadband" to cycle thrusters, save propellant
- Thrusters pulse only when *outside* of deadband angle
- Reduced deadband increases propellant use

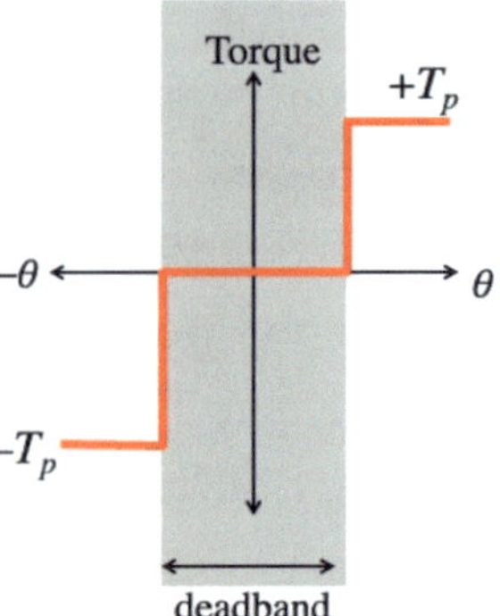

August 22, 2022 SC07-44

Minimum Impulse Bit *p*

- Smallest increment of impulse ("fine-ness") a thruster can deliver called p (N·s or lb_f·s)
- Depends on how fast valves can open and shut, and size of thruster components (plumbing, nozzles, etc.)
- Usually p must be measured with hardware
- Typical value: $p = 10^{-3}$ N·s

August 22, 2022

SC07-45

Propellant Consumption vs. MIB

- For a bang-bang spacecraft:
  Propellant consumption $\sim \frac{p^2 R}{I\theta}$, where

p = thruster minimum impulse bit (MIB)

R = thruster distance from mass center

I = spacecraft moment of inertia

θ = angular window width

We want to use small MIB thrusters (small p) to minimize propellant consumption

Table of thrusters in Propulsion section.

August 22, 2022

SC07-46

20 ms (0.020 s) Thrust Pulse

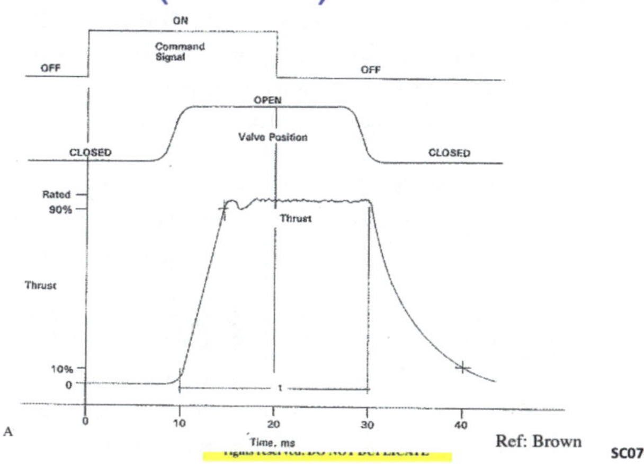

Ref: Brown

SC07-47

Soyuz ACS Thruster Video

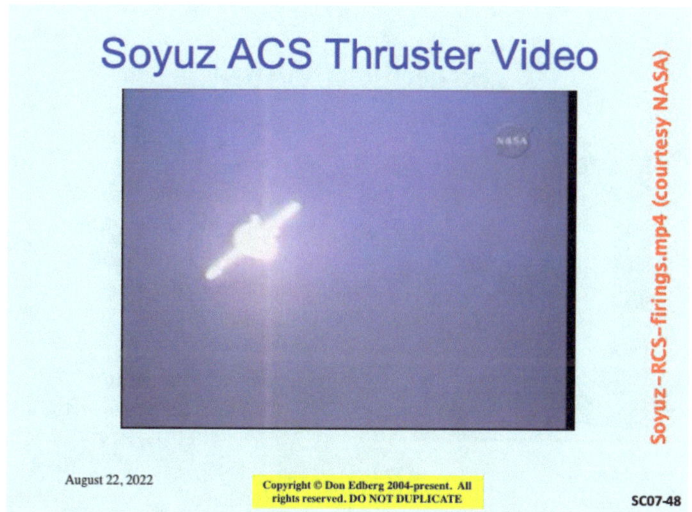

August 22, 2022

SC07-48

Starliner ACS Thruster Video

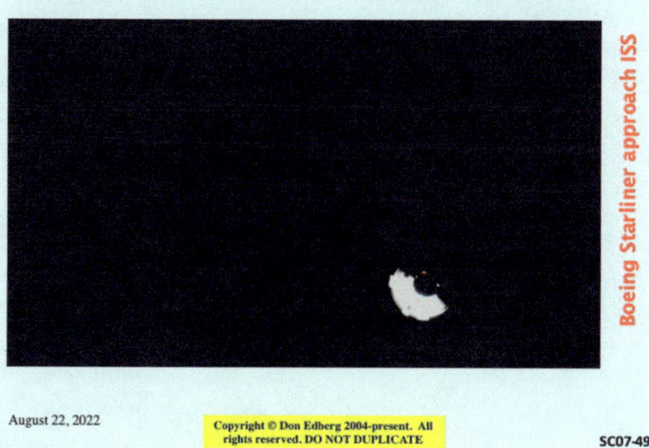

August 22, 2022

SC07-49

"Cross-Coupled" Effect of Thrusters

- Thrusters usually not perfectly symmetric
- Mass center not centered between thrusters
- Firing of thrusters produces unwanted:
  - Rotations in 2nd and/or 3rd axes
  - Change in translational velocity
- Both create unwanted change in attitude or rates (affects rendezvous, etc.)

August 22, 2022

SC07-50

X-37: Cross-coupling

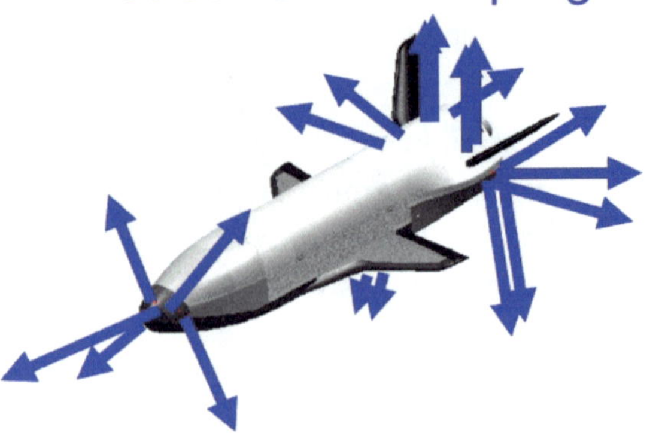

August 22, 2022 SC07-51

Dynamics & Control Topics

- Guidance/Navigation vs. Attitude Control
- Spacecraft dynamic characteristics
  - Non-Spinning
  - Spinning
- Moments of Inertia and spin
- Actuators
  - Thrusters
  - Reaction wheels
  - Control-moment gyros
  - Magnetic torquers
  - Mother Nature
- Attitude control systems: needs & operations
  - Disturbances
- Design considerations, modes of operation

August 22, 2022 SC07-52

Reaction Wheel

(RWA = Reaction Wheel Assembly)

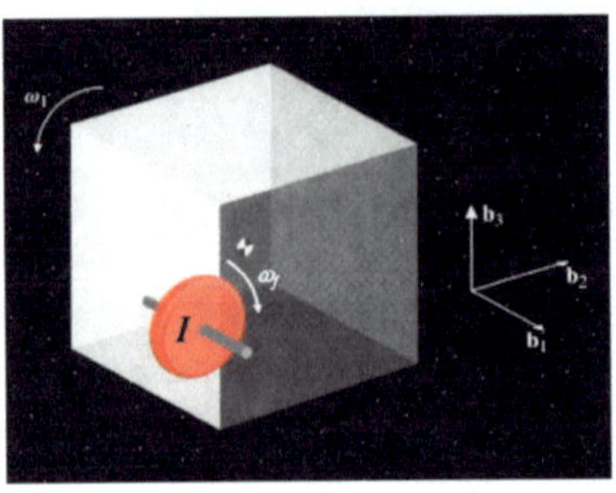

- Mount a flywheel with inertia I on an electric drive motor
- If spacecraft and wheel are non-rotating, total momentum $\mathbf{H} = 0$.
- Now, spin motor to create speed ω_f with respect to spacecraft

Ref: Wiesel

August 22, 2022 SC07-53

Reaction Wheel (cont.)

- Torque created by speeding up flywheel to ω_f causes spacecraft (with inertia = A) to rotate in opposite direction with rotation rate ω_1
- Speed of flywheel in *inertial* space: $(\omega_f - \omega_1)$
- No external torque:
  $H_{tot} = 0 = I(\omega_f - \omega_1) - A\omega_1$, so $\omega_1 = \frac{I\omega_f}{A+I}$
- $\omega_1 << \omega_f$ since $A >> I$: fine control of rotation rates (and pointing angles) possible. Accelerating/decelerating RWA provides angular rate in *same* axis as wheel spin axis

August 22, 2022 SC07-54

RWA (cont.)

- Can get << 0.001° spacecraft pointing accuracy
- Reaction wheels can accelerate either direction
- Applied torque determined by acceleration α
- No steady rotation speed, varies as needed (typically kept above some minimum to avoid "stiction")
- Example: Honeywell HR12 has momentum capacity of 12 N·m·s, mass 7 kg, 6,000 RPM max wheel speed, 22 W continuous power

August 22, 2022 SC07-55

Example: 180° Rotation Maneuver with Honeywell HR12

- HR12 has momentum capacity H = 12 N·m·s, produces torque T = 0.20 N·m
- Time for wheel to accelerate to maximum ω_f is
  $$t = \frac{H}{T} = \frac{12\ \text{N}\cdot\text{m}\cdot\text{s}}{0.20\ \text{N}\cdot\text{m}} = 60\ \text{s}$$
- If SC has inertia A = 2000 kg·m², applied torque produces SC rotational acceleration $\alpha_{SC} = \frac{T}{A} = \frac{0.20\ \text{N·m}}{2{,}000\ \text{kg·m}^2} = 1\times10^{-5}\,\frac{\text{rad}}{\text{s}^2}$; rotation rate $\omega = \alpha_{SC}\, t = \left(1\times10^{-5}\,\frac{\text{rad}}{\text{s}^2}\right)60s = 0.006\,\frac{\text{rad}}{\text{s}} = \frac{0.344°}{\text{s}}$
- In spin up, average rotation rate ω_{ave} = ½ ω = 0.172°/s
- In 60 s spin-up, SC turns 10.32°, same during spin-down, leaving 156.36°, time needed = 463 s.
- Total = spin up 60 + coast 463 + spin down 60 s = 583 s ≈ 9 m 53 s

August 22, 2022 SC07-56

Reaction Wheel Data (Courtesy Honeywell)

Characteristic	Mini-wheel	HR 0610	HR 12	HR 14	HR 16	HR 4820	HR 2010	HR 2020	HR 2030	HR 4520
Angular momentum, N-m-s	0.2 to 1.0	4 to 12	12 to 50	20 to 75	75 to 150	65	33.2 to 68.4	27	19.5 to 45.6	60.75
Output torque, N-m	>0.028	0.07 to 5	0.1 to 0.2	0.1 to 0.2	0.1 to 0.2	0.14	0.1	0.13	0.21	0.135
Wheel rpm, ±	9000	6000	6000	6000	6000	6000	6000	6500	6000	5400
Power, W[b]	>6	<15	22	22	22	20	17	35	20	35
Bus voltage, dc	12 to 34	14 to 35	23 to 57	23 to 57	23 to 57	22 to 36	27 to 44	70	27.7 to 31.3	51
Mass, kg	1.3	3.6 to 5.0	7.0	8.5	12	10.2	9.2 to 10.9	7.9	8.9 to 11.2	11.1
Integral electronics	Y	Y	Y	Y	Y	Y	N	Y	Y	Y
Diameter, mm[c]	108	267	316	368	418	405	406	300	305	406
Height, mm	54	12.0	159	159	178	214	235	172	191	215
Op temperature,[d]										
Low	−25	−15	−30	−30	−30	−15	−15	−13	−15	−24
High	+60	+60	70	+70	+70	+71	+70	+75	+80	+61

[a]Reproduced with permission of Honeywell International, Inc.
[b]Power values are steady-state power at maximum wheel speed, W.
[c]Dimensions are overall envelope, mm.
[d]Temperature ranges are qualification limits, operating, °C.

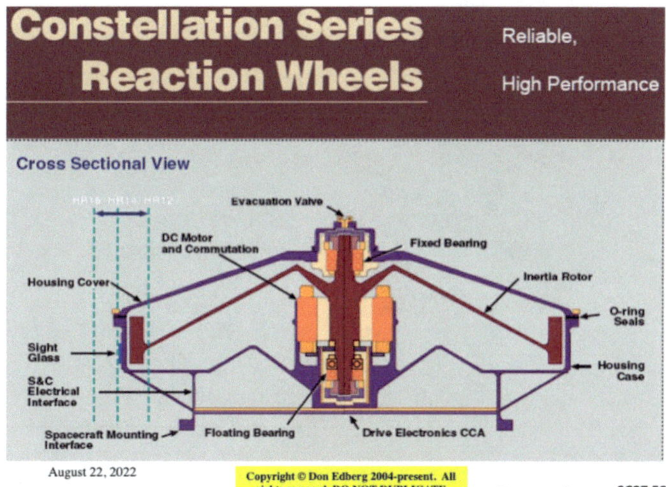

Courtesy Honeywell

Reaction Wheel Assembly

Honeywell HR-061 Reaction Wheel

Other RWA Considerations

- Steady disturbance torque causes steady increase in RWA spin, to cancel
- Eventually RWA reaches critical RPM and must be de-spun ("desaturated")
- Desaturation means thrusters fired simultaneously when RWA wheel braked
- The higher the disturbance torque, the more often the RWA requires desaturation
- Desaturation can cause unwanted velocity changes due to cross-coupling

Time Required to Saturate Example: *Magellan*

- Assume steady solar torque applied: $T_{solar} = 2.7 \times 10^{-6}$ N·m
- RWA can store maximum angular momentum H_w= 27 N·m·s at 4000 RPM maximum wheel speed
- Time to saturate (reach max wheel speed):

$$t_{sat} = \frac{H_w}{T} = \frac{27 \text{ N}\cdot\text{m}\cdot\text{s}}{27{\times}10^{-7} \text{ N}\cdot\text{m}} = 10^{+7} \text{ s} = 115 \text{ days}$$

- Must then desaturate (next)

Time Required for Desaturation Example: *Magellan*

- Saturated RWA: max angular momentum storage H_w= 27 N·m·s
- Thrusters: F = 0.889 N; arm L = 2.134 m
- Wheel ang. mom. = thruster ang. mom.: $H_w = 2(FL)t$ (Note: number of thrusters = 2)
- Solve: $t = \frac{H_w}{2FL} = \frac{27 \text{ N·m·s}}{2(0.889 \text{ N})2.134 \text{ m}} = 7.116 \text{ s}$
- 8.6 g propellant consumed, if I_{sp} = 150 s

RWA Arrangement

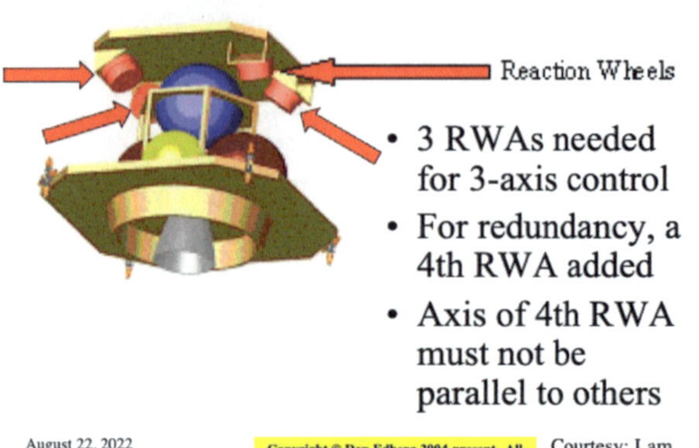

- 3 RWAs needed for 3-axis control
- For redundancy, a 4th RWA added
- Axis of 4th RWA must not be parallel to others

August 22, 2022 Courtesy: Lam SC07-63

RWA Demonstration: *Cubli*

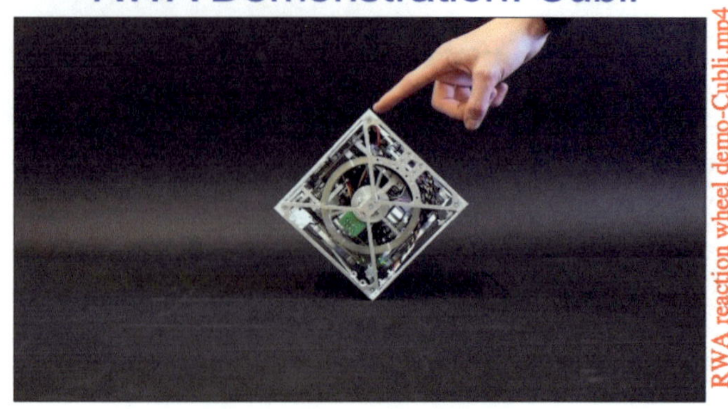

RWA reaction wheel demo-Cubli.mp4

August 22, 2022 SC07-64

Dynamics & Control Topics

- Guidance/Navigation vs. Attitude Control
- Spacecraft dynamic characteristics
  - Non-Spinning
  - Spinning
- Moments of Inertia and spin
- Actuators
  - Thrusters
  - Reaction wheels
  - Control-moment gyros
  - Magnetic torquers
  - Mother Nature
- Attitude control systems: needs & operations
  - Disturbances
- Design considerations, modes of operation

August 22, 2022 SC07-65

Control-Moment Gyro (CMG)

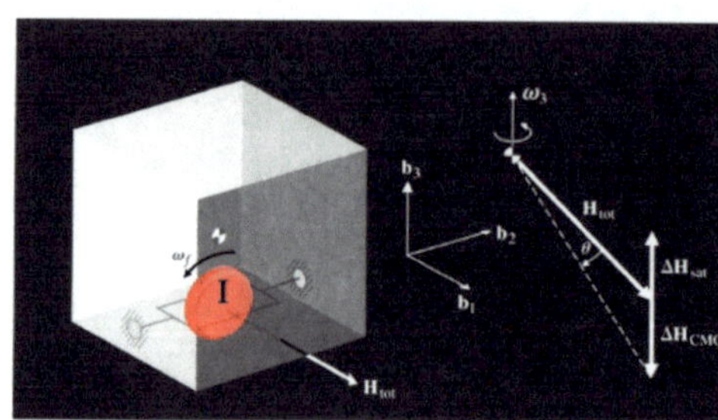

FIGURE 5.14
The control-moment gyroscope

- A gimbaled flywheel spins continuously
- Pivot up/down
- Provides torque in *third* axis (vertical)
- Right-hand rule to calculate

Ref: Wiesel

August 22, 2022 SC07-66

Control-Moment Gyro (CMG)

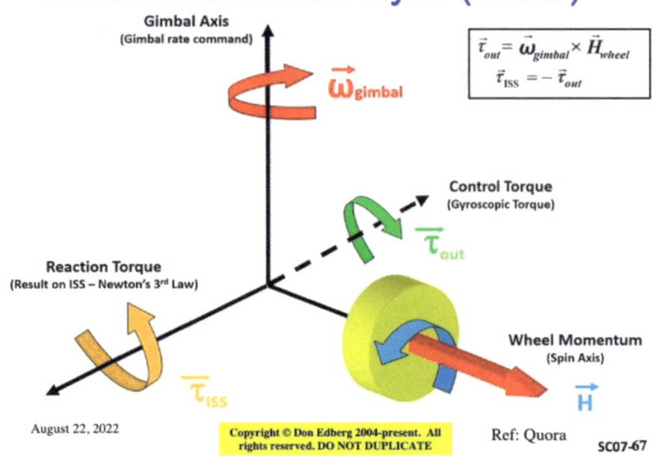

August 22, 2022 Ref: Quora SC07-67

CMG (cont.)

- Flywheel (inertia I) on electric motor
- Gimbal allows the spinning flywheel to be rotated around another axis (see previous)
- If SC is non-rotating, total momentum $\mathbf{H}_{total} = I\omega_f$ ($\mathbf{b}_1$ direction)
- If rotate gimbal an angle θ with respect to the spacecraft, $\Delta\mathbf{H}_{CMG} = -I\omega_f \sin\theta$ ($\mathbf{b}_3$ dir)
- Total $\mathbf{H}$ cannot change, so $\Delta\mathbf{H}_{SC} = \Delta\mathbf{H}_{CMG}$, so $\omega_3 = \frac{I\omega_f \sin\theta}{C}$ where C is SC moment of inertia in $\mathbf{b}_3$ dir.

August 22, 2022 SC07-69

CMG Summary

- CMG *does not* rotate spacecraft around spin axis *or* gimbal axis
- CMG *does* cause rotation about third axis
- CMG operates at constant rotational speed
- Acts in reverse of a rate gyro
- Typically more torque than RWA
- Example: *EADS CMG 15-45S*, 15 kg, 45 N·m of torque

ISS CMG courtesy NASA; below Honeywell CMG

Control-Moment Gyro Movie

Movie: cmg-man-wheel0.mpg

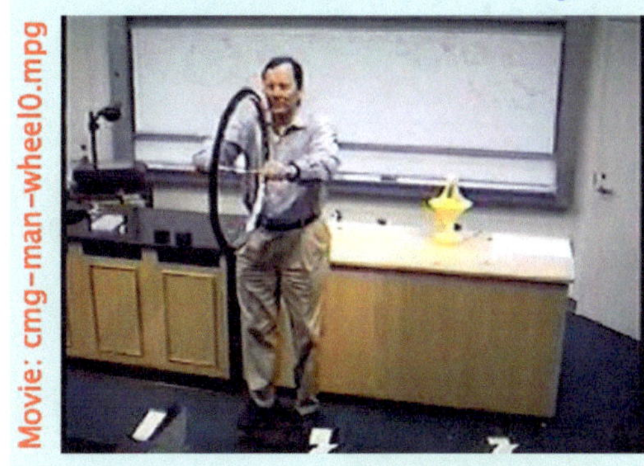

- Flywheel angular momentum in *y*-axis ("left wing")
- Apply torque to axle of spinning flywheel in roll axis ("forward")
- Rotate in yaw axis ("zenith")

Control-Moment Gyro Demo

Movie: cmg-maneuver.mp4

Dynamics & Control Topics

- Guidance/Navigation vs. Attitude Control
- Spacecraft dynamic characteristics
  - Non-Spinning
  - Spinning
- Moments of Inertia and spin
- Actuators
  - Thrusters
  - Reaction wheels
  - Control-moment gyros
  - Magnetic torquers
  - Mother Nature
- Attitude control systems: needs & operations
  - Disturbances
- Design considerations, modes of operation

Magnetic Torquers ("Torque Rods" or "Magnetorquers")

- Electromagnet tries to align with local magnetic lines, just like a compass
- Electric – no propellant
- Best in LEO where fields strongest
- Poor accuracy, only good to $\pm 30°$ rotation
- In GEO, fields can reverse
- Many issues with sign errors! (TOMS-EP, TIMED, TERRIERS, etc.)

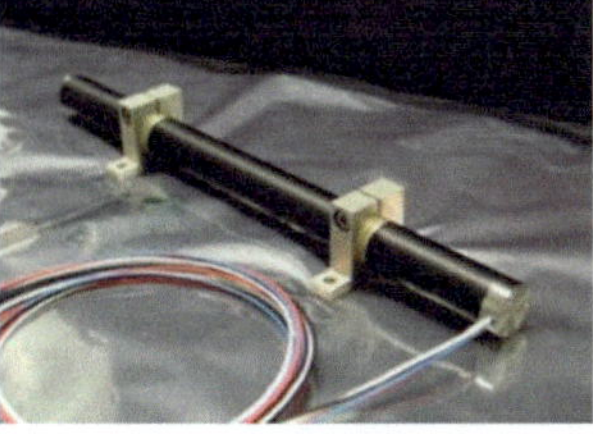

Source: Microcosm

How To Desaturate With *No* Thrusters!

- Use torque rods to apply torque via external magnetic fields
- HST uses torque rods for RWA momentum desaturation (no on-board thrusters due to risk of contamination)

Stabilite Magnetic Control System

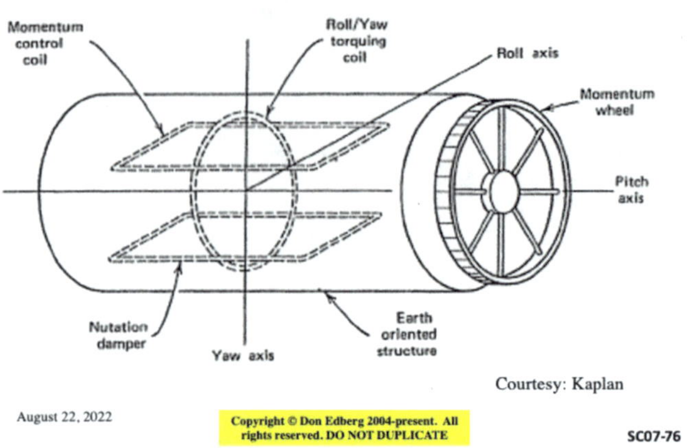

Courtesy: Kaplan

August 22, 2022

SC07-76

Dynamics & Control Topics

- Guidance/Navigation vs. Attitude Control
- Spacecraft dynamic characteristics
  - Non-Spinning
  - Spinning
- Moments of Inertia and spin
- Actuators
  - Thrusters
  - Reaction wheels
  - Control-moment gyros
  - Magnetic torquers
  - **Mother Nature**
- Attitude control systems: needs & operations
  - Disturbances
- Design considerations, modes of operation

August 22, 2022

SC07-77

Solar Pressure Torque

- Differential pitch commanded on SAs.
- Produced roll torque
- Used for roll control on *Mariner 10*, Mercury

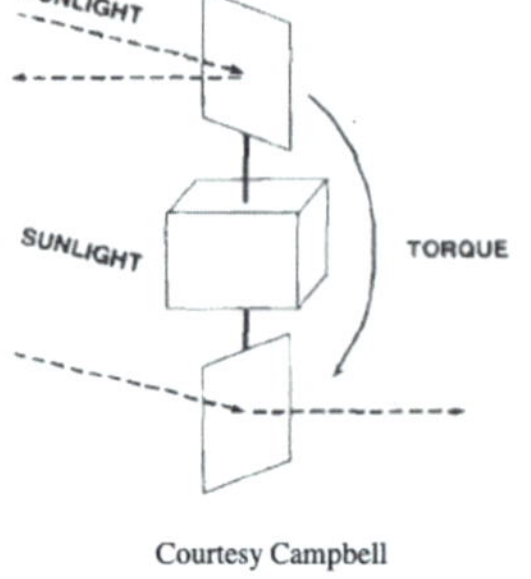

Courtesy Campbell

August 22, 2022

SC07-78

Dynamics & Control Topics

- Guidance/Navigation vs. Attitude Control
- Spacecraft dynamic characteristics
  - Non-Spinning
  - Spinning
- Moments of Inertia and spin
- Actuators
  - Thrusters
  - Reaction wheels
  - Control-moment gyros
  - Magnetic torquers
  - Mother Nature
- **Attitude control systems: needs & operations**
  - Disturbances
- Design considerations, modes of operation

August 22, 2022

SC07-79

Why An Attitude Control System? 1: *Operations*

LEASAT Launch Sequence

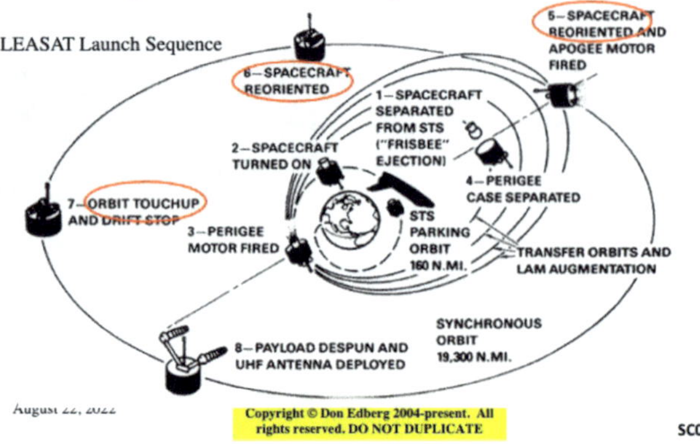

August 22, 2022

SC07-80

Dynamics & Control Topics

- Guidance/Navigation vs. Attitude Control
- Spacecraft dynamic characteristics
  - Non-Spinning
  - Spinning
- Moments of Inertia and spin
- Actuators
  - Thrusters
  - Reaction wheels
  - Control-moment gyros
  - Magnetic torquers
  - Mother Nature
- Attitude control systems: needs & operations
  - **Disturbances**
- Design considerations, modes of operation

August 22, 2022

SC07-81

Why An Attitude Control System? 2: *Disturbances*

- Aerodynamic torques
- Magnetic fields
- Gravity gradient torques
- Solar torques
- Venting, etc.

Refer to "Orbital Environment" charts to calculate magnitudes of torques

August 22, 2022

SC07-82

Dynamics & Control Topics

- Guidance/Navigation vs. Attitude Control
- Spacecraft dynamic characteristics
  - Non-Spinning
  - Spinning
- Moments of Inertia and spin
- Actuators
  - Thrusters
  - Reaction wheels
  - Control-moment gyros
  - Magnetic torquers
  - Mother Nature
- Attitude control systems: needs & operations
  - Disturbances
- Design considerations, modes of operation

August 22, 2022

SC07-83

Attitude Control System Block Diagram

Figure 10.2 Block diagram for an attitude control system

ACS = Attitude Control System
RCS = Reaction Control System

Functions of ACS:

- Attitude sensing
- Attitude control
- Control software
- Interface with ground control

August 22, 2022

SC07-84

ACS Selection Criteria

Depends on:

- Pointing requirements, including:
  - Initial orientation
  - Normal operations
  - Backup operations
  - Instrument pointing
- Orientation and TVC for any burns
- Slew rates
- Solar array size and pointing

August 22, 2022

SC07-85

Possible ACS "Modes"

Mode	Description	Sensors/Actuators
Orbit Insertion Mode	None (provided by launch vehicle)	n/a
Acquisition Mode	Determines initial attitude & aligns spacecraft for mission	Star tracker & sun sensor, RWAs
Normal, On-Station Mode	Maintain solar panels within 1° of sun vector	Sun sensor, star tracker, gyros, RWAs, thrusters
Slew Mode	Slew maneuver for attitude recovery	Thrusters, RWAs, CMGs
Contingency/ Safe Mode	Ensures power and thermal constraints are met	Sun sensor, star tracker, RWAs
SC Bus Undervoltage	Disconnect Non-Essential Loads (DNEL)	Sun sensor, star tracker, RWAs

August 22, 2022

SC07-86

Spacecraft Thermal Control

Courtesy regardnews.com

8/22/22

Topics

- Need for thermal control & requirements
- Energy balance
- Calculating temperatures
- Thermal control
  - Passive heating/cooling
  - Active heating/cooling
- Analysis examples
- Radiation with view factors
- Simplified whole-spacecraft thermal analysis
- Transient thermal analysis
- Thermal system design methodology

8/22/22

Spacecraft Design Process

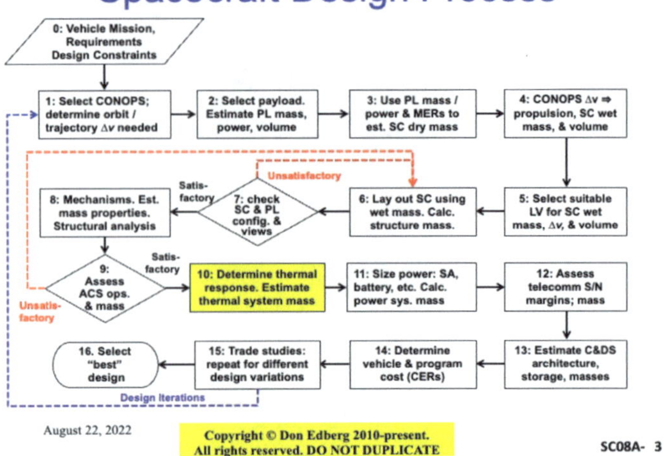

0: Vehicle Mission, Requirements Design Constraints
1: Select CONOPS; determine orbit / trajectory Δv needed
2: Select payload. Estimate PL mass, power, volume
3: Use PL mass / power & MERs to est. SC dry mass
4: CONOPS Δv ⇒ propulsion, SC wet mass, & volume
Unsatisfactory
8: Mechanisms. Est. mass properties. Structural analysis
Satis-factory
7: check SC & PL config. & views
6: Lay out SC using wet mass. Calc. structure mass.
5: Select suitable LV for SC wet mass, Δv, & volume
Satis-factory
9: Assess ACS ops. & mass
Unsatis-factory
10: Determine thermal response. Estimate thermal system mass
11: Size power: SA, battery, etc. Calc. power sys. mass
12: Assess telecomm S/N margins; mass
16. Select "best" design
15: Trade studies: repeat for different design variations
14: Determine vehicle & program cost (CERs)
13: Estimate C&DS architecture, storage, masses
Design Iterations

August 22, 2022

Topics

- **Need for thermal control & requirements**
- Energy balance
- Calculating temperatures
- Thermal control
  - Passive heating/cooling
  - Active heating/cooling
- Analysis examples
- Radiation with view factors
- Simplified whole-spacecraft thermal analysis
- Transient thermal analysis
- Thermal system design methodology

8/22/22

Need For Thermal Control

- Vehicle & payload must meet thermal requirements for survival & proper operation

Thermal issues:

- Spacecraft's attitude & orbit (sun vs. shade) can greatly affect temperature
- *Batteries & liquid propellants* typically have narrow temperature ranges, as do optically-stable equipment such as mirrors, support structure, etc.
- Solar arrays *lose efficiency* as temperature rises
- Sensors may require *low* temperatures
- High power dissipating components may create local "*hot spots*," and sun-free faces may be required

8/22/22

Temperature Requirements

Temperatures must typically be maintained within the following ranges:

Item	Low Temp. (°C)	High Temp (°C)
Batteries	+5	+20
Hydrazine propellant	+7	+35
Electronics	0	+40
Solar Arrays	–150	+100
Structures	–45	+65
IR sensors	–200	–80

AFT = "Allowable Flight Temperature"

8/22/22

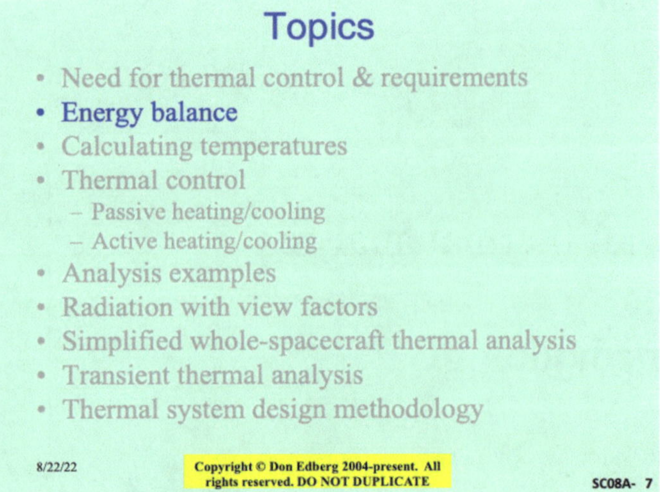

Topics
Need for thermal control & requirements
Energy balance
Calculating temperatures
Thermal control
Passive heating/cooling
Active heating/cooling
Analysis examples
Radiation with view factors
Simplified whole-spacecraft thermal analysis
Transient thermal analysis
Thermal system design methodology
8/22/22
Copyright © Don Edberg 2004-present. All rights reserved. DO NOT DUPLICATE
SC08A- 7

Spacecraft Heat Balance

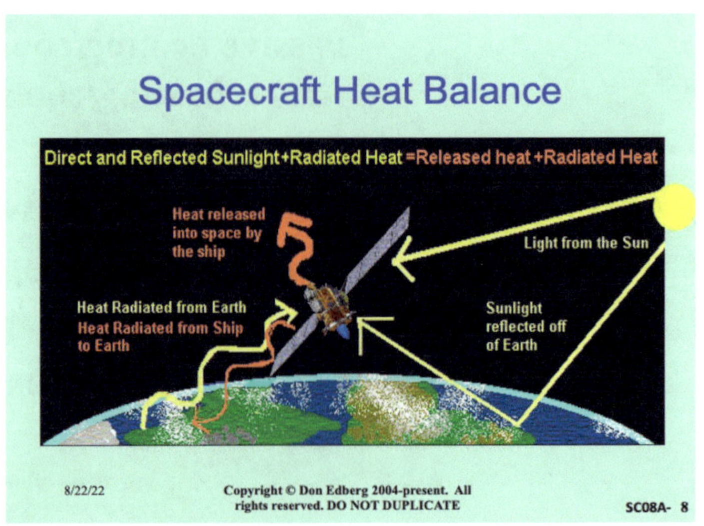

Direct and Reflected Sunlight+Radiated Heat =Released heat+Radiated Heat
Heat released into space by the ship
Light from the Sun
Heat Radiated from Earth
Heat Radiated from Ship to Earth
Sunlight reflected off of Earth

8/22/22

Definition of Heat Flux $\dot{q}$

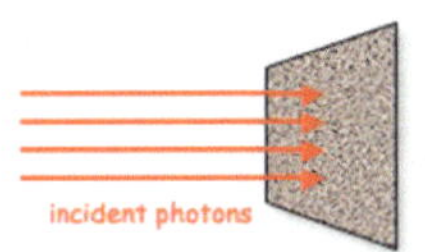

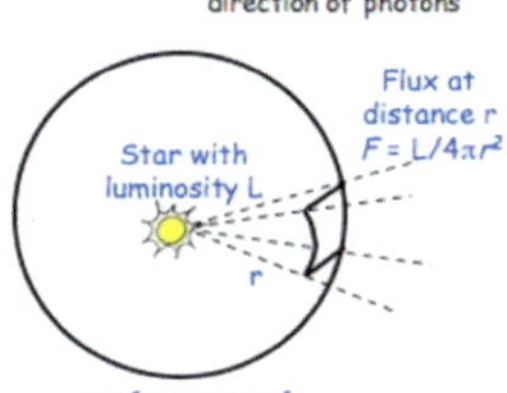

- Flux is the amount of energy crossing a unit area in unit time
- Heat flux (and all electromagnetic) goes as $\frac{1}{r^2}$, where r is the radial distance

Courtesy Swinburne University

Energy From Sun & Earth

- G_{sun} = incident radiation from sun, ON EVERY EXPOSED SURFACE!

  Mean value $G_{sun} = \left[1{,}371\ \frac{\mathrm{W}}{\mathrm{m}^2}\right] \cdot \left(\frac{r}{1\ \mathrm{AU}}\right)^2$,

  where r = distance from sun;

  1 AU = mean Earth-sun distance (elliptical orbit)
  - Max G_{sun} =1,387 W/m^2 at 03 January (perihelion)
  - Min G_{sun} = 1,320 W/m^2 at 04 July (aphelion)
- G_{sun} varies as $\frac{1}{r^2}$

  the further you are from sun, the less received
- G_{earth} = Earth infrared radiation = 237 W/m^2 in LEO

Energy Received by Spacecraft

- Amount of energy falling on spacecraft from sun is

$$\dot{Q}_{sun} = G_{sun} A_{exposed\ to\ sun}$$

- Amount of energy reaching spacecraft from Earth includes two parts:

  Earth's own IR radiation

$$\dot{Q}_{Earth} = G_{Earth} A_{exposed\ to\ Earth}$$

  and 2) earth *reflection* of sun energy

$$\dot{Q}_{from\ sun\ reflected\ by\ Earth} = G_{sun} \rho_{Earth} A_{exposed\ to\ Earth}$$

- Amount actually received depends on *absorptivity* of SC, and *view factors*

Three Energy Transfer Terms

- **Absorptivity** α = fraction of incident radiation that is absorbed
  - For a "black body," α = 1: all energy is absorbed
- **Emissivity** ε = fraction of energy that is *emitted* by radiation
  - For a black body, ε = 1: 100% of energy is emitted
- **Reflectivity** ρ = **Albedo** a = fraction of incident radiation that is *reflected*
  - Earth albedo a = 0.306 (from average earth temp.)

Energy Transfer Out of Spacecraft

- No atmosphere, no convection.
- No mass in contact, no conduction.
- Only energy out is through *radiation*
  - Radiate to empty space
  - Radiate to neighboring objects

Heat Lost to Radiation

Heat radiated $\dot{q} = \varepsilon\sigma\left(T^4 - T_{target}^4\right)$

ε = emissivity

$\sigma = 5.67 \times 10^{-8}$ W/(m^{2}K^4)
known as Stefan-Boltzmann constant

T & T_{target} are the ABSOLUTE temps of the radiating surface and target surface (in K).

- $T_{target} \approx 0$ for space.
- T_{target} NOT negligible for Earth: 200–230 K.

Emissivity ε

- **Emissivity** or **emittance** ε describes *how well* a surface radiates energy
- $\varepsilon = 1$ is a perfect emitter (a blackbody)
  $\varepsilon \approx 0.95$ for some black paints (shuttle's bottom)
- $\varepsilon = 0$ is a surface that doesn't emit
  Polished metal (gold, silver) has $\varepsilon \approx 0.05$
  Polished playground slide in the sun = ***HOT***!

D. Edberg photo

8/22/22 SC08A-15

Equilibrium SC Thermal Analysis

- Energy in = energy out (in equilibrium)

So $\dot{Q}_{\text{absorbed from sun}} + \dot{Q}_{\text{absorbed from nearby objects}}$

$+$ internal heat generation $= \dot{Q}_{\text{radiated}}$

Surface being irradiated by sun at equilibrium:

$$G_{sun}\alpha_s = \varepsilon_{IR}\sigma T^4$$

Assumes: radiation to $T_{\text{target}} = 0$;
No radiation from other objects;
Ignores view factors (perpendicular for now);
No internal heat generation

8/22/22 SC08A-16

Topics

- Need for thermal control & requirements
- Energy balance
- **Calculating temperatures**
- Thermal control
  - Passive heating/cooling
  - Active heating/cooling
- Analysis examples
- Radiation with view factors
- Simplified whole-spacecraft thermal analysis
- Transient thermal analysis
- Thermal system design methodology

8/22/22 SC08A-17

Equilibrium Temp of Irradiated Surface

Solve to get

$$T = \sqrt[4]{\frac{G_{sun}\alpha_s}{\varepsilon_{IR}\sigma}} = \underbrace{\sqrt[4]{\frac{G_{sun}}{\sigma}}}_{\text{Constants}} \cdot \underbrace{\sqrt[4]{\frac{\alpha_s}{\varepsilon_{IR}}}}_{\textit{Designer can change}}$$

Use $\frac{\alpha_s}{\varepsilon_{IR}}$ ratio for temperature predictions

Note: lower $\left(\frac{\alpha_s}{\varepsilon_{IR}}\right) \Longrightarrow$ lower temperatures

8/22/22 SC08A-18

Equilibrium Temperature in Sun

(1 AU, insulated flat plate in vacuum)

Material	Absorptivity	Emissivity	α/ε Ratio	Eq. Temp, K
Gold	0.299	0.023	13.0	749
6061-T6 Aluminum as received	0.379	0.0383	9.9	699
6061-T6 Aluminum polished	0.20	0.034	5.88	614
Steel (polished)	0.357	0.135	2.64	503
Black paint	0.975	0.874	1.12	406
White enamel	0.252	0.853	0.30	291
Aluminized Teflon®	0.163	0.80	0.20	264
Aluminum anodized	0.14†	0.84†	0.17	252
OSR (optical solar reflector, quartz over silver)	0.077	0.79	0.10	220

†http://www.solarmirror.com/fom/fom-serve/cache/43.html Courtesy Brown

SC08A-19

Thermal Control

Temperatures may be controlled by :

1. Passive surface coatings
   - Adjust $\frac{\alpha_s}{\varepsilon_{IR}}$ ratio
2. Passive control devices: no energy required
   - Modulate $\frac{\alpha_s}{\varepsilon_{IR}}$ ratio depending on current temperature
3. Active control devices
   - Electrical power required

8/22/22 SC08A-20

Topics

- Need for thermal control & requirements
- Energy balance
- Calculating temperatures
- **Thermal control**
  - **Passive heating/cooling**
  - Active heating/cooling
- Analysis examples
- Radiation with view factors
- Simplified whole-spacecraft thermal analysis
- Transient thermal analysis
- Thermal system design methodology

8/22/22

SC08A-21

Passive Thermal Control Elements

"Passive" = no energy required

- Radiation Control
  - Absorb/Emit α/ε Control
    - Coating
    - Painting
    - Plating
    - Optical Solar Reflector
    - Al/Ag-coated Teflon®
    - Al-coated Kapton®
  - Cut Off
    - Insulation
      - MLI Blanket
    - Shut Off
      - Hood
      - Shutter / Pinwheel
- Conductivity Control (only used *within* a spacecraft)
  - Increase
    - Heat Pipe
  - Decrease
    - Spacer
  - Increase heat capacity
    - Heat sink doubler

8/22/22

SC08A-22

Passive Thermal Control by $\frac{\alpha_S}{\varepsilon_{IR}}$ Ratio

Use paints and coatings to provide desired energy balance – options:

- Make "too hot" items black to radiate better (i.e. black tiles on bottom of Space Shuttle)
- Use white or reflective coatings to reduce absorption
- Consult chart (next)
- "Rotisserie mode" to equalize temperatures

8/22/22

SC08A-23

Thermal Properties of Materials

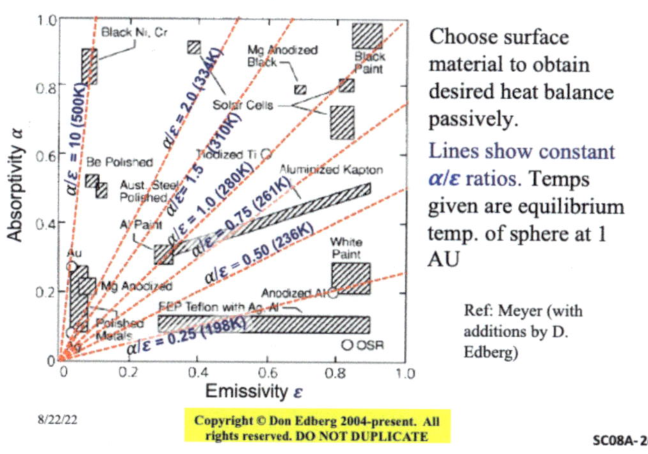

Choose surface material to obtain desired heat balance passively.

Lines show constant α/ε ratios. Temps given are equilibrium temp. of sphere at 1 AU

Ref: Meyer (with additions by D. Edberg)

8/22/22

SC08A-24

Passive Coatings: Easiest & Simplest

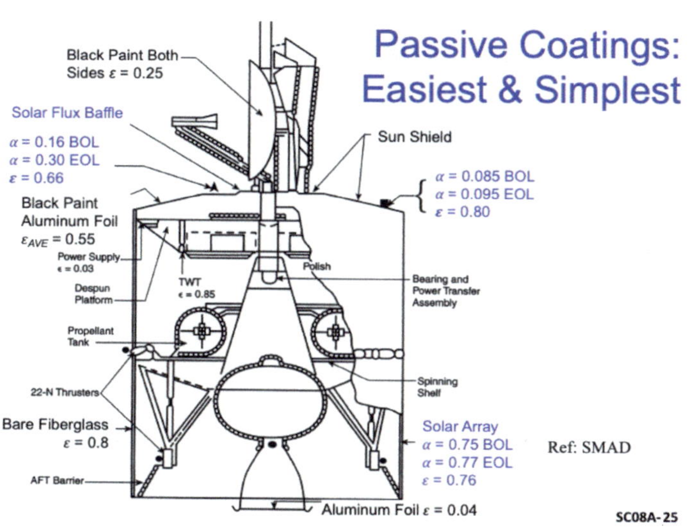

Ref: SMAD

SC08A-25

Thermal Paint Absorptivity (α) Changes With Thruster Deposits

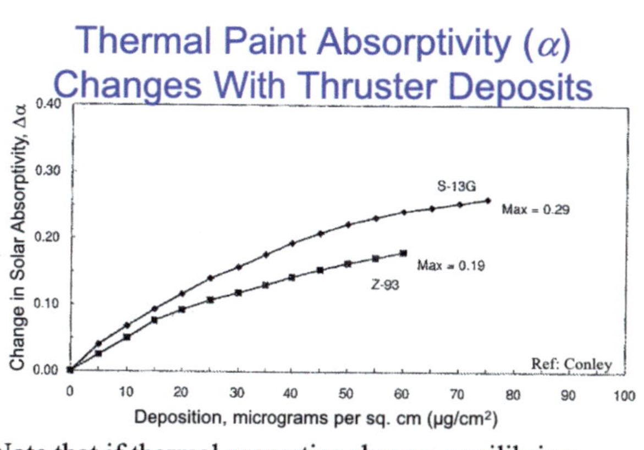

Note that if thermal properties change, equilibrium temperature changes: can show $\Delta T = \frac{T}{4} \cdot \frac{\Delta(\alpha/\varepsilon)}{\alpha/\varepsilon}$

8/22/22

SC08A-26

Passive Coatings

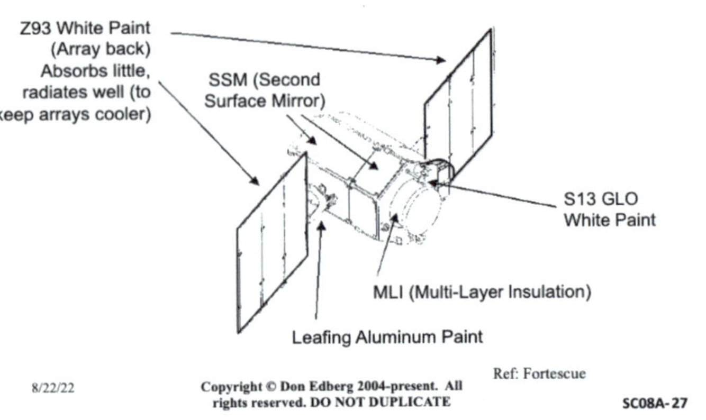

Ref: Fortescue

8/22/22 SC08A-27

Multi-Layer Insulation (MLI)

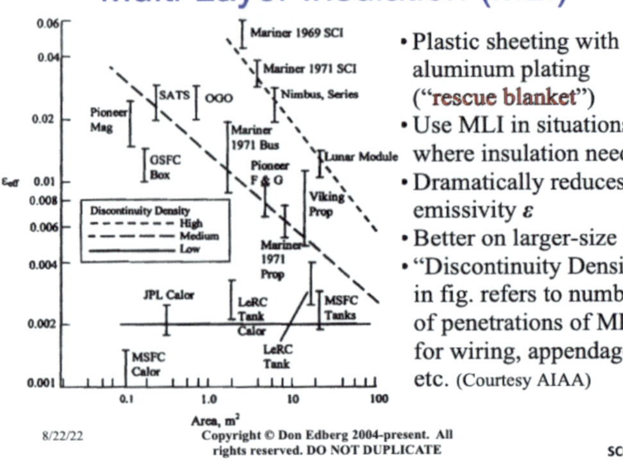

- Plastic sheeting with aluminum plating ("rescue blanket")
- Use MLI in situations where insulation needed
- Dramatically reduces emissivity ε
- Better on larger-size SC
- "Discontinuity Density" in fig. refers to number of penetrations of MLI for wiring, appendages, etc. (Courtesy AIAA)

8/22/22 SC08A-28

Multi-Layer Insulation (MLI)

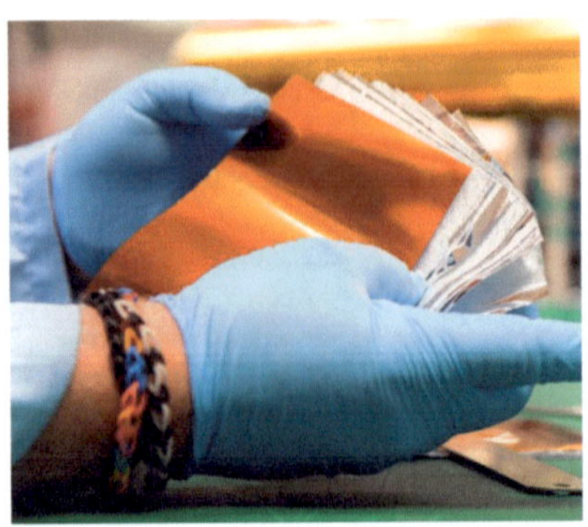

Photo: ESA

Photo: Fortescue

Multilayer Insulation

- Typical: 30 alternating layers of aluminized Mylar® (orange) and separator cloth (white)
- About 5 mm (0.2 in) thickness
- Holes punched to allow air to vent during ascent

Added 210604

8/22/22 SC08A-29

How Does MLI Work?

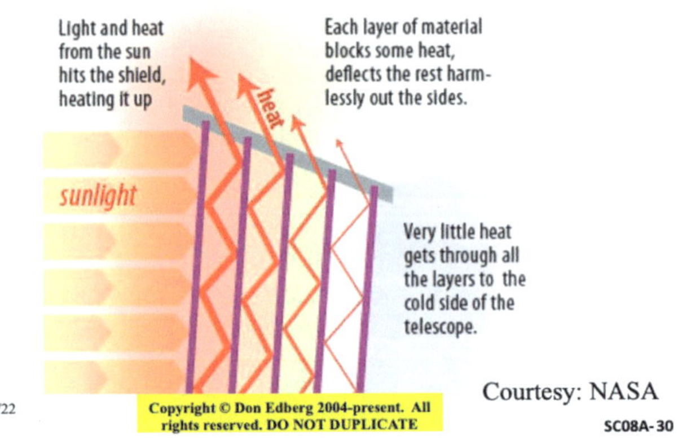

Courtesy: NASA

8/22/22 SC08A-30

Cassini MLI Blankets

Courtesy: NASA

8/22/22 SC08A-31

Variable-Property Devices

These devices change emissive and/or absorptive properties in response to temperature:

- Fins
- Louvers
- Shutters
- Pinwheels

8/22/22 SC08A-32

Bimetallic Fins

- Fins automatically open at high temps to radiate heat
- Fins close when temperature drops sufficiently

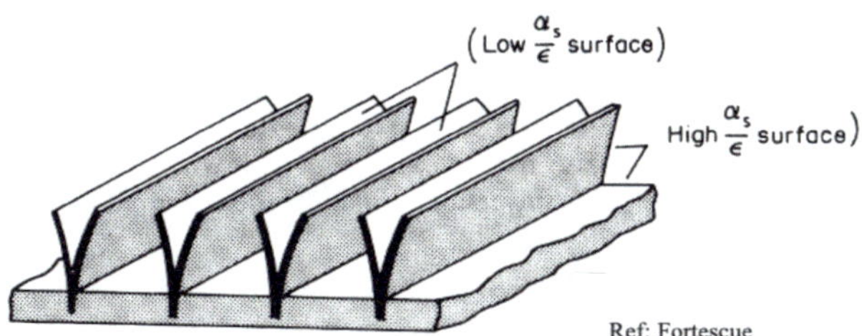

Ref: Fortescue

Figure 12.8 Active thermal control system with bimetallic fins. (*After* J. A. Wiebelt)

Operation of Thermal Louvers

Low ε_{IR} surface

radiator

vanes

High ε_{IR} surface

bimetallic spring

Louver Closed

Louver Open

Mariner 10, Cosmosphere (Edberg)

Ref: Pisacane

Louver Operation & Performance

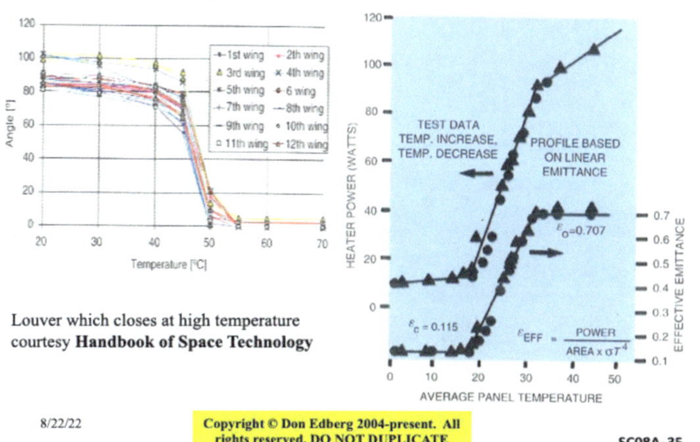

Louver which closes at high temperature courtesy **Handbook of Space Technology**

Movable Shutters

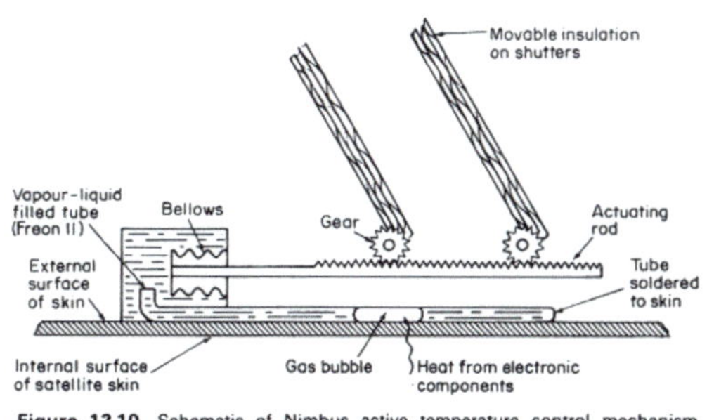

Figure 12.10 Schematic of Nimbus active temperature control mechanism (General Electric)

Shutters are another scheme to modulate radiation to control temperature

Ref: Fortescue

Shutters

- Open and close to control radiation when needed

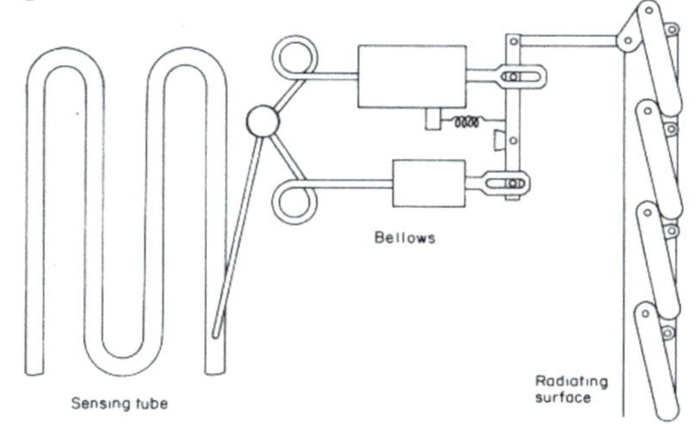

Figure 12.9 Schematic of gas-bellows-actuated dynamic temperature controller

Ref: Fortescue

Pinwheels: "Rotary Shutters"

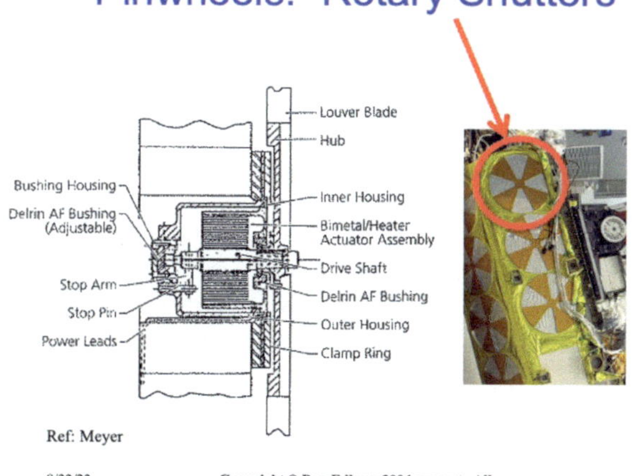

Ref: Meyer

If Coatings, MLI, or Passive Shutters Are Not Enough…

Go to other schemes:

- Heat Pipes
- Radioisotope thermal decay
- *Cold Plates/Radiators*
- *Heating films or heaters*
- *Thermoelectrics*
- Combinations

Active thermal control elements: require *external power*

8/22/22 SC08A-39

Heat Pipe

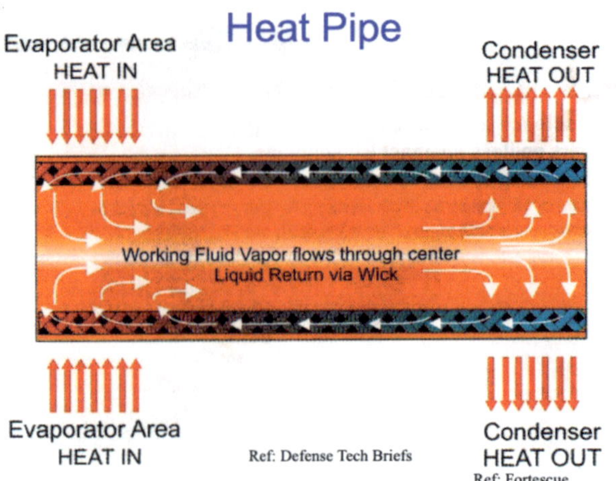

Ref: Defense Tech Briefs

Ref: Fortescue

8/22/22 SC08A-40

Radioactive Heating Units (RHUs)

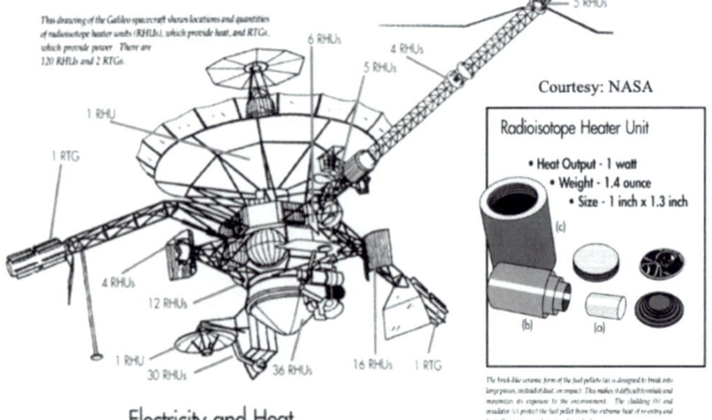

Courtesy: NASA

8/22/22 SC08A-41

Topics

- Need for thermal control & requirements
- Energy balance
- Calculating temperatures
- **Thermal control**
  - Passive heating/cooling
  - **Active heating/cooling**
- Analysis examples
- Radiation with view factors
- Simplified whole-spacecraft thermal analysis
- Transient thermal analysis
- Thermal system design methodology

8/22/22 SC08A-42

Cold Plates

- Heat conducted from heat load (hot equipment) into cold plate
- Liquid coolant (water) pumped through cold plate to extract heat via convection
- Heat dissipated in a radiator in a separate location
- Active: power required for pump

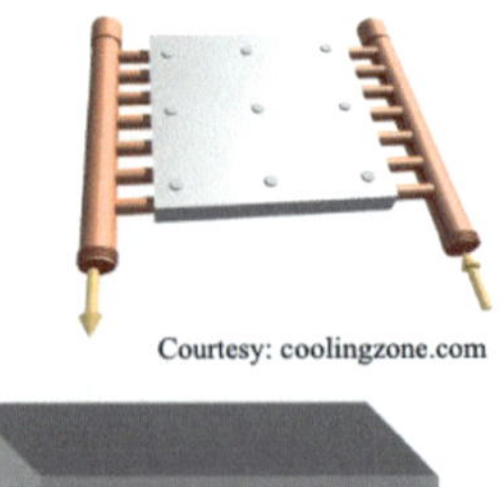

Courtesy: coolingzone.com

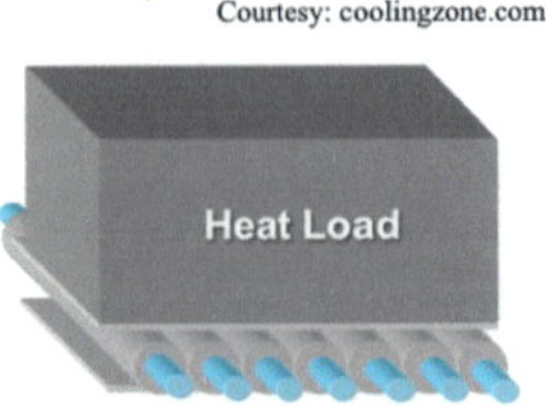

Courtesy: Quora

8/22/22 SC08A-43

Getting Rid of the Heat on ISS

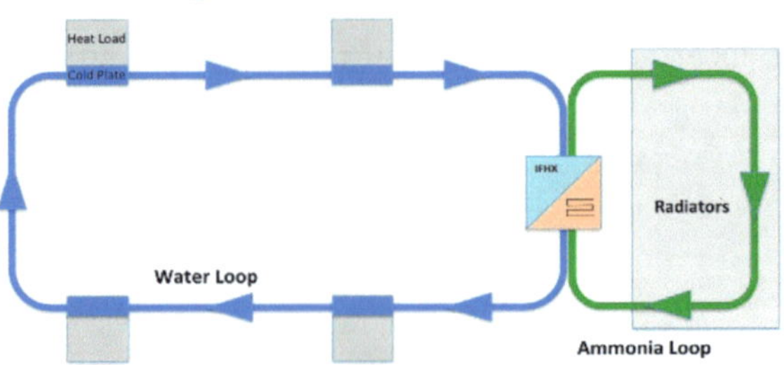

- Water flows through vehicle picking up heat from cold plates; then enters Interface Heat Exchanger (IFHX) at surface of vehicle
- Ammonia loop outside vehicle also enters the IFHX. Inside IFHX, heat from water loop conducts into ammonia loop
- Warmed ammonia flows into radiators; heat is radiated to space
- Cold ammonia flows back to IFHX to pick up more heat

Courtesy: Quora

8/22/22 SC08A-44

Radiators on ISS

Photo: Quora

8/22/22

SC08A-45

Radiators Dissipate Excess Heat

Courtesy: NASA

Add a waste heat term, q_w, to previous eq. to represent heat dissipated by radiator:

$$G_{sun}\alpha_s \cos\theta + \frac{q_w}{A_R} = \varepsilon_{IR}\sigma T^4$$

- A_R = radiator area (m^2)
- θ = angle of the solar vector

Assume value of T, solve to get q_w, or assume q_w and solve for T

8/22/22

SC08A-46

Spacelab Equipment Cooling

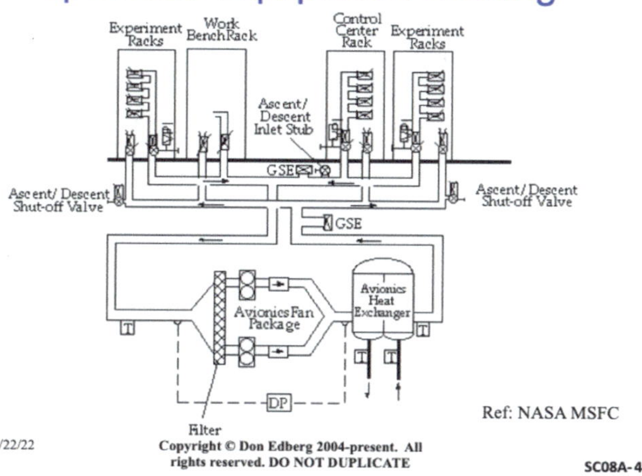

Ref: NASA MSFC

8/22/22

SC08A-47

Electrical Heaters

- Use electrical heaters to heat local areas using i^2R (resistance) heating
- Like a hot pad on an injured limb

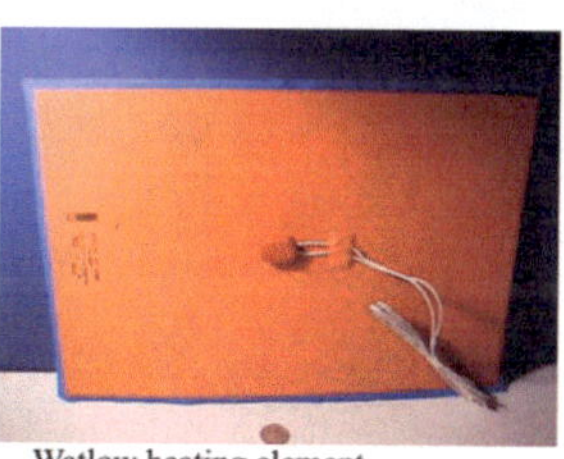

Watlow heating element

8/22/22

SC08A-48

Thermoelectric Cooling

- Use thermoelectric device(s) as a "heat pump"
- Get VERY LOW temperatures
- More stages = lower temps
- Very localized
- Great for IR sensors, etc.

HEAT ABSORBED FROM DEVICE BEING COOLED
CERAMIC
COLD EXCHANGER CONNECTED TO THIS SURFACE
CONDUCTOR
P
N
CONDUCTOR
HEAT DISSIPATED INTO HOT EXCHANGER
HEAT SINK CONNECTED TO THIS SURFACE
V
CERAMIC ELECTRICAL INSULATOR
DC POWER SOURCE

Multi Stage TECs

Courtesy: Marlow

8/22/22

SC08A-49

"Cold-Biasing"

- A spacecraft component is intentionally made to have a lower-than-desired temperature, then heat is added using thermostatically-controlled heaters
- Often used for hydrazine propellant lines to ensure freezing does not occur

8/22/22

SC08A-50

Topics

- Need for thermal control & requirements
- Energy balance
- Calculating temperatures
- Thermal control
  - Passive heating/cooling
  - Active heating/cooling
- **Analysis examples**
- Radiation with view factors
- Simplified whole-spacecraft thermal analysis
- Transient thermal analysis
- Thermal system design methodology

8/22/22 SC08A-51

Equilibrium Thermal Calculation: Example 1

Cube-shaped, nadir-facing polar-orbit earth SC:

$$Q_{solar} + Q_{albedo} + Q_{EarthIR} + P_{internal} = \varepsilon_{radiator} A_{radiator} \sigma \left(T^4_{radiator} - T^4_{space}\right)$$

- Note: $Q_{solar} = \alpha\, q_{solar}\, A$; $Q_{albedo} = \alpha\, q_{albedo}\, A$; $Q_{EarthIR} = \varepsilon\, q_{EarthIR}\, A$
- Assume 1 m cube; $T_{space} = 0$; internal power $P_{internal} = 150$ W; all faces insulated ($Q_{solar} = 0$) except nadir-pointing bottom face

8/22/22 SC08A-52

Thermal Example 1 (cont.)

- Averaged nadir heat fluxes:
  - Solar = 0 W/m^2 (all faces insulated)
  - Albedo (reflected) = 41.3 W/m^2
  - Planet IR = 60.6 W/m^2
- All other faces: negligible fluxes (insulated)

Then the equation becomes

$$[\alpha(0 + 41.3\ \text{W/m}^2) + \varepsilon(60.6\ \text{W/m}^2)]A + 150\ \text{W} = \varepsilon\sigma A T^4$$

If $\alpha/\varepsilon = 0.15/0.85$ & desired $T = 300$ K, result is $\boldsymbol{A = 0.38}$ **m^2**; insulate 0.62 m^2 on bottom

8/22/22 SC08A-53

Topics

- Need for thermal control & requirements
- Energy balance
- Calculating temperatures
- Thermal control
  - Passive heating/cooling
  - Active heating/cooling
- Analysis examples
- **Radiation with view factors**
- Simplified whole-spacecraft thermal analysis
- Transient thermal analysis
- Thermal system design methodology

8/22/22 SC08A-54

Radiation Calcs More Complicated…

In real systems:

- Must worry about "view factors"
  - They describe how well one surface "talks to" or transmits energy to another surface
  - Solve simple examples by hand
  - More complex problems solved using tables or specialized software
- Internal electronics, solar arrays also generate heat energy
- Also must account for *transient* conditions...

8/22/22 SC08A-55

Radiation Heat Transfer via *View Factors*

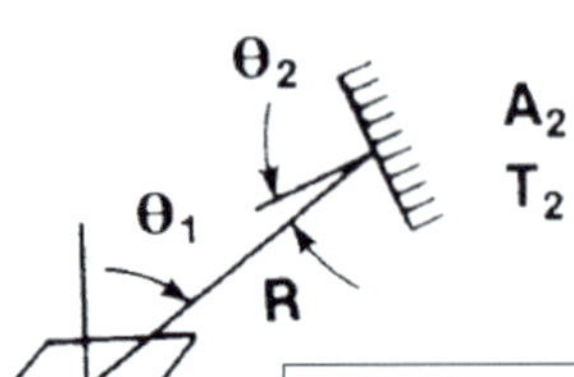

Q_{12} is the radiative exchange from A_1 to A_2:

$$Q_{12} = \sigma F_{1\text{-}2} A_1 (T_1^4 - T_2^4) = \sigma F_{2\text{-}1} A_2 (T_1^4 - T_2^4),$$

where $F_{1\text{-}2}$ is the view factor of surface 1 by surface 2, (the fraction of photons leaving 1 that intercept 2).

Note that $\Sigma F_{i\text{-}2} = 1$

Courtesy G. Nacouzi

8/22/22 SC08A-56

View Factor Calculation

- View Factors F_{ij} defined by

$$F_{12}A_1 = \iint \frac{\cos\theta_1 \cos\theta_2}{\pi R^2} dA_1 dA_2$$

View factor between diffuse surfaces A_1 and A_2 is $F_{A1\text{-}A2}$ = F_{12} = [Radiative energy leaving A_1 & intercepted by A_2] ÷ [Radiative energy leaving A_1 in all directions]

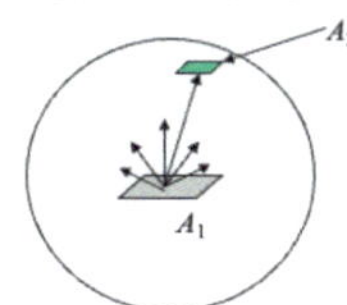

A_2 surface element is on hemisphere of radius R

Courtesy G. Nacouzi

8/22/22 SC08A-57

View Factor Calculation Estimation

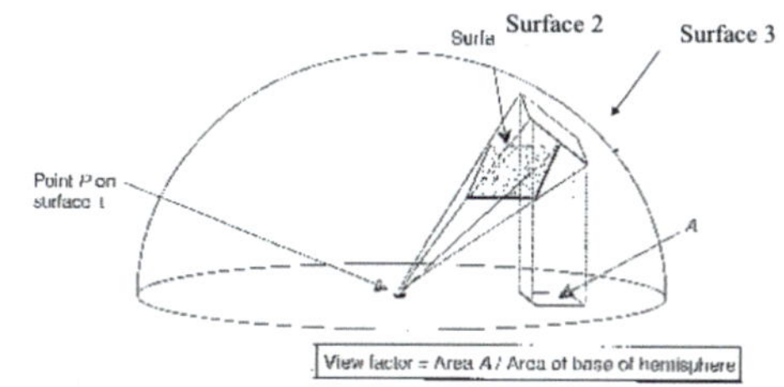

The view factor for flat surfaces can be approximated by:
1- Calculate the projected area of surface 2 on the hemisphere, and
2- Obtain normal component of that surface with respect to source
These steps are accomplished using "simple" trigonometry

Courtesy G. Nacouzi

8/22/22 SC08A-58

Energy Balance in Earth Orbit with View Factors

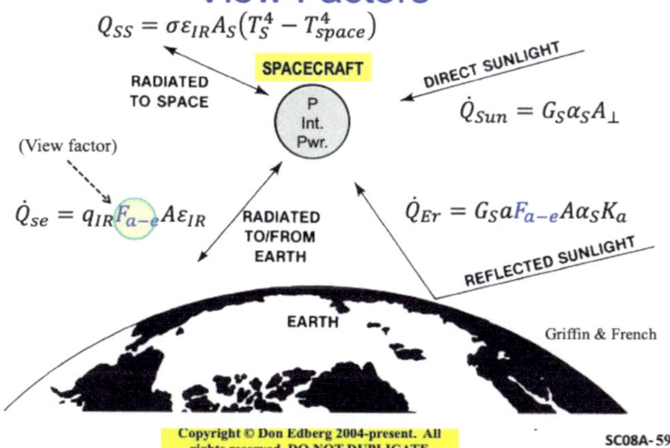

Griffin & French

 SC08A-59

Worst-Case Temperatures

- Worst-case HIGH temperature: in direct sun, sun reflection & IR from planet underneath
- Worst-case LOW steady temperature: in *shadow* (no sun), minimal IR from cold asteroid underneath

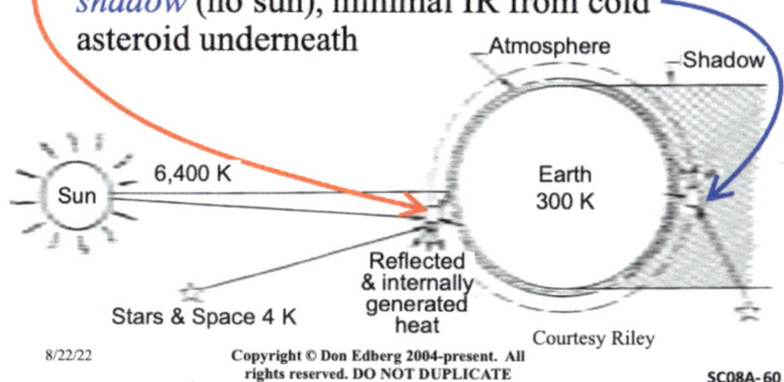

Courtesy Riley

8/22/22 SC08A-60

Solar Array Temperature Equilibrium With View Factors & Internal Power

- Energy absorbed + dissipated* = emitted

$$\underbrace{G_s A\alpha_{s-t}}_{\text{Energy from Sun}} + \underbrace{q_{IR}F_{a-e}A\varepsilon_{IR-b}}_{\text{IR energy emitted by Earth}} + \underbrace{G_s a F_{a-e} A\alpha_{s-b}K_a}_{\text{Sun energy reflected off Earth}}$$

$$\underbrace{+\eta G_s A}_{\text{SA power dissipation}} = \underbrace{\sigma\varepsilon_{IR-b}AT^4}_{\text{Heat radiated by SA bottom}} + \underbrace{\sigma\varepsilon_{IR-t}AT^4}_{\text{Heat radiated by SA top}}$$

*Assumes: 1. no heat conduction through SA stem;
2. plane of SA perpendicular to sun.

Variables defined next slide.

8/22/22 SC08A-61

Definitions of Variables

T = temperature
G_s = energy from sun (W/m^2)
q_{IR} = infrared radiation from Earth (W/m^2)
A = solar array area (m^2)
α_{s-t}, α_{s-b} = solar array absorptivity, t = top, b = bottom
ε_{IR-t}, ε_{IR-b} = solar array emissivity, top/bottom
a = Earth albedo, or reflectivity ρ (both ≈ 0.30)
F_{a-e} = view factor, array to Earth
K_a = light reflection collimation term
σ = Stefan-Boltzmann constant, 5.67×10^{-8} W/(m^{2}K^4)
η = solar array efficiency

8/22/22 SC08A-62

Calculation of Terms

- The plate (SA) ↔ *large* sphere (Earth) view factor $F_{a\text{-}e}$ may be calculated from

$$F_{a-e} = \left(\frac{R_e}{R_e + H}\right)^2$$

where H = altitude, R_e = earth radius.

- The collimation term K_a is calculated from

$$K_a = 0.657 + 0.54\frac{R_e}{R_e + H} - 0.196\left(\frac{R_e}{R_e + H}\right)^2 = 0.657 + \frac{0.54}{1 + H/R_e} - \frac{0.196}{(1 + H/R_e)^2}$$

Ref.: Brown

8/22/22 SC08A-63

Example 2: Solar Array (SA) Temperature

Solve the SA temp eqn. for T_{MAX} and T_{MIN}:

$$T_{max} = \sqrt[4]{\frac{G_s(\alpha_{s-t} + a\, F_{a-e}\, \alpha_{s-b}\, K_a - \eta) + q_{IR} F_{a-e}\, \varepsilon_{IR-b}}{\sigma(\varepsilon_{IR-b} + \varepsilon_{IR-t})}}$$

$$T_{min} = \sqrt[4]{\frac{q_{IR}\, F_{a-e}\, \varepsilon_{IR-b}}{\sigma(\varepsilon_{IR-b} + \varepsilon_{IR-t})}}$$ (no direct or reflected sun)

Given parameters for 28% efficiency SA, 500 km orbit:

$a = 0.3$, $q_{IR} = 237$ W/m^2, $G_s = 1371$ W/m^2,

$\eta = 0.28$, $\alpha_{s\text{-}t} = 0.805$, $\alpha_{s\text{-}b} = 0.3$, $\varepsilon_{IR\text{-}b} = \varepsilon_{IR\text{-}t} = 0.8$

Result: T_{max} **= +49.9 C** and T_{min} **= –67.3 C**

8/22/22 SC08A-64

Topics

- Need for thermal control & requirements
- Energy balance
- Calculating temperatures
- Thermal control
  - Passive heating/cooling
  - Active heating/cooling
- Analysis examples
- Radiation with view factors
- Simplified whole-spacecraft thermal analysis
- Transient thermal analysis
- Thermal system design methodology

8/22/22 SC08A-65

"Spherical Spacecraft" Thermal Analysis (Ref: Brown)

- Simple but useful for initial design process
- Assumes complex spacecraft is a lumped-mass with radiation properties of a sphere
- Assumes equilibrium
- View factors can be calculated easily or looked up on graphs
- Must be refined later to actual geometry, transient conditions

8/22/22 SC08A-66

Preliminary Thermal System Design Process Using Sphere Geometry

1. Temperature limits from on-board equipment
2. Establish internal electric power dissipation (power levels at worst-case hot & cold situations)
3. Determine equivalent sphere radius matching spacecraft surface area. Assume SC is isothermal at uniform temperature (approximation!)
4. Select surface properties. Calculate *view factors.*
5. Compute worst-case *hot* temp using direct solar, earth IR, reflection, etc. If too hot, calculate radiator area needed
6. Compute worst-case *cold*, compare to spec. If radiator temp too low, get electric heat power needed.

Courtesy Brown

8/22/22 SC08A-67

Equivalent Sphere Radius: Need SC Average Density

- Example: SC has dry mass m_{dry} = 3,200 kg & dry density ρ_{dry} = 160 kg/m^3; propellant mass m_{prop} = 12,800 kg & prop bulk density ρ_{bulk} = 1,280 kg/m^3

$$\rho_{ave} = \frac{m_{dry} + m_{prop}}{m_{dry}/\rho_{dry} + m_{prop}/\rho_{bulk}} = \frac{3{,}200 + 12{,}800}{3{,}200/160 + 12{,}800/1{,}280} = 533.3 \text{ kg/m}^3$$

$$Vol_{sphere} = \frac{m_{total}}{\rho_{ave}} = \frac{16{,}000 \text{ kg}}{533.3 \text{ kg/m}^3} = 30 \text{ m}^3$$

$$Vol_{sphere} = \frac{4}{3}\pi r^3, \text{ so } r_{sphere} = \sqrt[3]{\frac{3Vol_{sphere}}{4\pi}} = 1.928 \text{ m}$$

8/22/22 SC08A-68

Sphere SC Temperature Equilibrium With View Factors & Internal Power

- Energy absorbed – dissipated = radiated

$$\underbrace{G_s \pi R^2 \alpha_s}_{\text{Energy from Sun}} + \underbrace{q_{IR}\varepsilon_{IR}F_{s\text{-}e}A_s}_{\text{IR energy emitted by Earth}} + \underbrace{G_s a F_{s\text{-}e}A_s\alpha_s K_a}_{\text{Sun energy reflected off Earth}} + \underbrace{P}_{\text{SC power dissipation}}$$

$$= \underbrace{\sigma\varepsilon_{IR}A_s T^4}_{\text{Heat radiated by SC}}$$

$F_{s\text{-}e}$ is view factor, sphere-to-earth

A_s= spacecraft *surface* area (= $4\pi R^2 = \pi D^2$, sphere)

πR^2 = spacecraft *cross-sectional* area (sphere)

P = power dissipated inside spacecraft

8/22/22 SC08A-69

Sphere SC Temperature Calculation

- Solve the SC temp eqn. for T_{max} and T_{min}:

$$T_{max} = \sqrt[4]{\frac{G_s\alpha_s\left(\frac{1}{4} + aK_aF_{s-e}\right) + q_{IR}\varepsilon_{IR}F_{s-e} + \frac{P}{4\pi r^2}}{\sigma\varepsilon_{IR}}}$$

$$T_{min} = \sqrt[4]{\frac{q_{IR}\varepsilon_{IR}F_{s-e} + \frac{P}{4\pi r^2}}{\sigma\varepsilon_{IR}}} \text{ (no direct or reflected sun)}$$

- G_s, a, q_{IR}, σ, orbital constants are known
- SC design gives r, α_s, ε_{IR}, P, etc.
- Calculate view factor $F_{s\text{-}e}$ & collimation term K_a from previous formulae

8/22/22 SC08A-70

Example 3: Spacecraft Temp in LEO

Spacecraft at 700 km altitude LEO

White paint: $\alpha_s = 0.316$, $\varepsilon_{IR} = 0.8$,

On-board power dissipation $P = 170$ W,

equivalent sphere diameter $R = 1.035$ m,

G_s, a, q_{IR} for Earth:

- Resulting worst-case temperatures:
  $T_{max} = 264$ K = –8° C
  $T_{min} = 186$ K = –87° C (**COLD**!!)

8/22/22 SC08A-71

Topics

- Need for thermal control & requirements
- Energy balance
- Calculating temperatures
- Thermal control
  - Passive heating/cooling
  - Active heating/cooling
- Analysis examples
- Radiation with view factors
- Simplified whole-spacecraft thermal analysis
- **Transient thermal analysis**
- Thermal system design methodology

8/22/22 SC08A-72

Simple Thermal Calculations *Do Not Consider TRANSIENT Situations*

- Into/out of eclipse conditions, etc.
- Input actual geometry, masses, & thermal properties to thermal modeler
- Determine radiation view factors & thermal balances with thermal software
- Use temperature distributions to calculate thermal stresses / loads with thermal/structural analysis software (i.e. NASTRAN®)

8/22/22 SC08A-73

Calculation of Eclipse Time Fraction

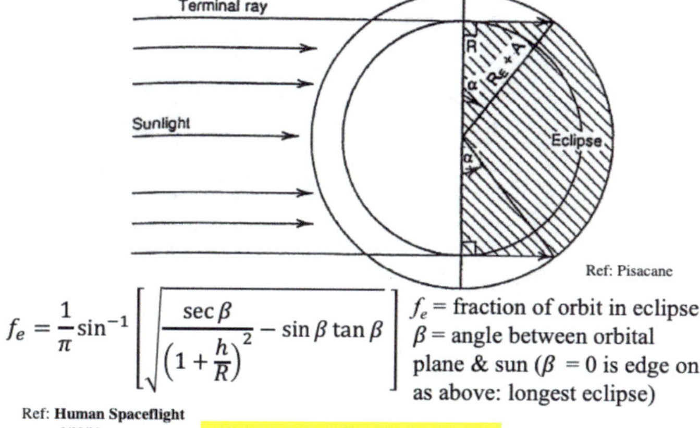

Ref: Pisacane

$$f_e = \frac{1}{\pi}\sin^{-1}\left[\sqrt{\frac{\sec\beta}{\left(1+\frac{h}{R}\right)^2} - \sin\beta\tan\beta}\right]$$

f_e = fraction of orbit in eclipse
β = angle between orbital plane & sun (β = 0 is edge on as above: longest eclipse)

Ref: **Human Spaceflight**

8/22/22 SC08A-74

Eclipse Time Graph (Earth)

Maximum eclipse time all angles of inclination

Minimum eclipse time for inclination = 0

Inclination = 10°

Inclination = 20°

Inclination = 30°

Inclination = 40°

Orbital period

Eclipse time (min): 0, 5, 10, 15, 20, 25, 30, 35, 40

Orbital period (min): 90, 100, 110, 120, 130, 140, 150, 160, 170

Altitude (km): 0, 400, 800, 1200, 1600, 2000, 2400, 2800, 3200, 3600

Ref: Pisacane

8/22/22

SC08A-75

Transient (Time Varying) Conditions

- Actual situation: rate of change of stored thermal energy = input energy – output

$$MC_p \frac{dT}{dt} = P + G_{sun}\alpha_s A_i + IR_e + G_{sun-e} - \varepsilon_{IR}A_R\sigma T^4$$

| rate of change of heat energy in (lumped mass) SC | heat energy direct from sun | Earth (e) IR radiation & sun reflection | heat energy radiated by spacecraft |

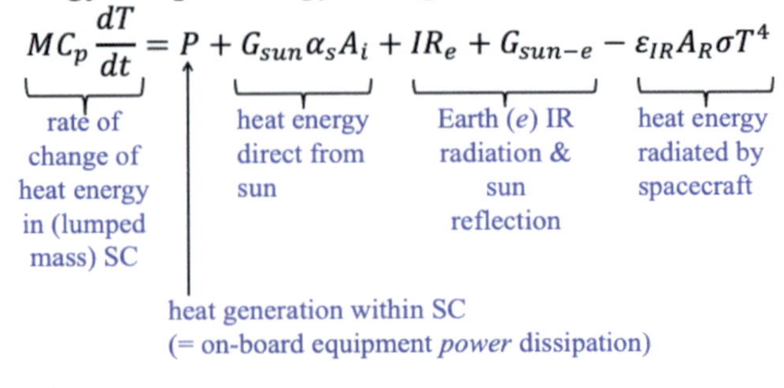

heat generation within SC
(= on-board equipment *power* dissipation)

8/22/22

SC08A-76

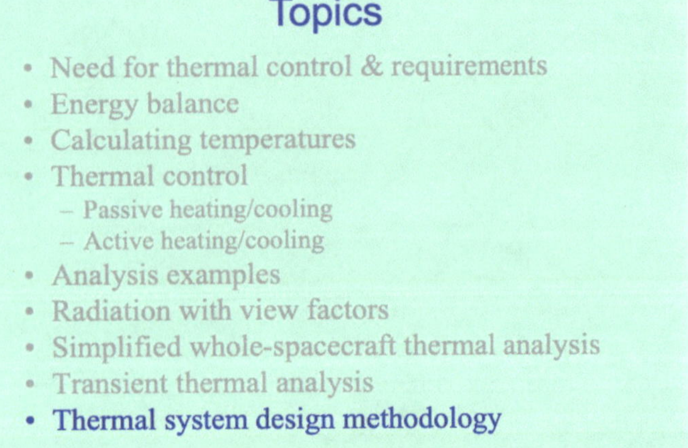

Thermal Control System Design Methodology

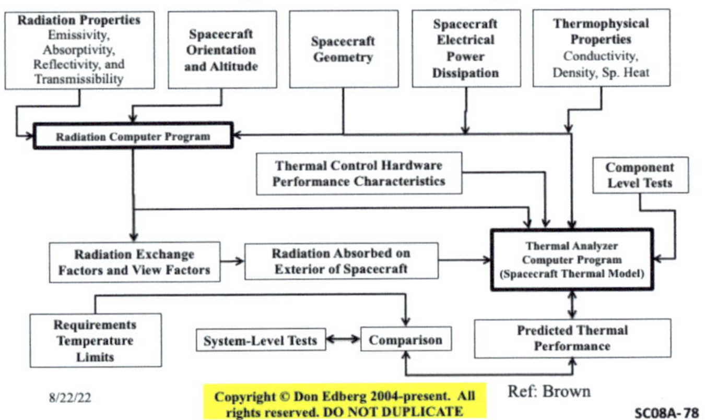

8/22/22

Ref: Brown

SC08A-78

Pioneer Finite-Element Modeling

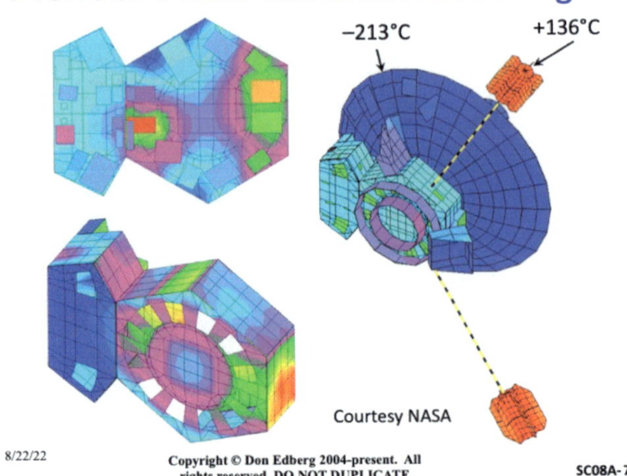

8/22/22

SC08A-79

Thermal Margins

- Specified for each component
- Need to keep propellants, batteries, electronics, instruments within a given temperature range
- Margin about max and min temperatures added for safety.

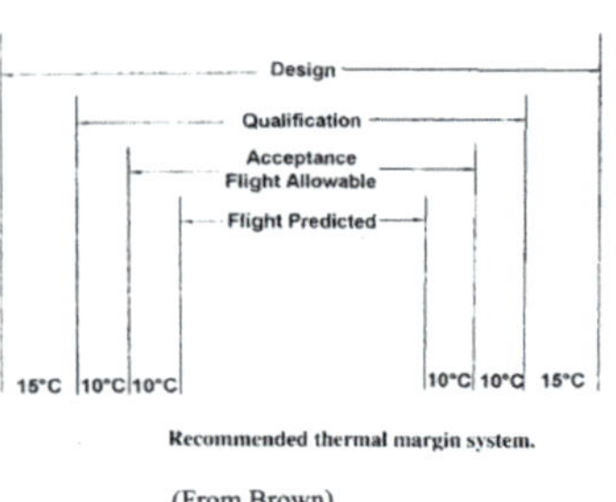

Recommended thermal margin system.

(From Brown)

8/22/22

SC08A-80

Predicted Temps, *TerraSAR-X*

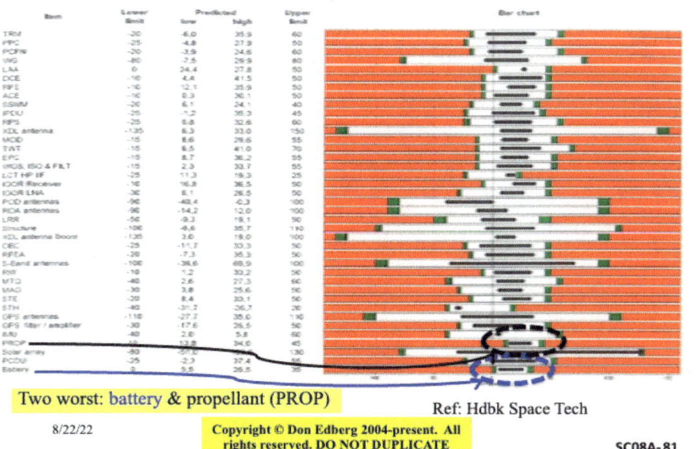

Ref: Hdbk Space Tech

Sample Thermal Transient Case

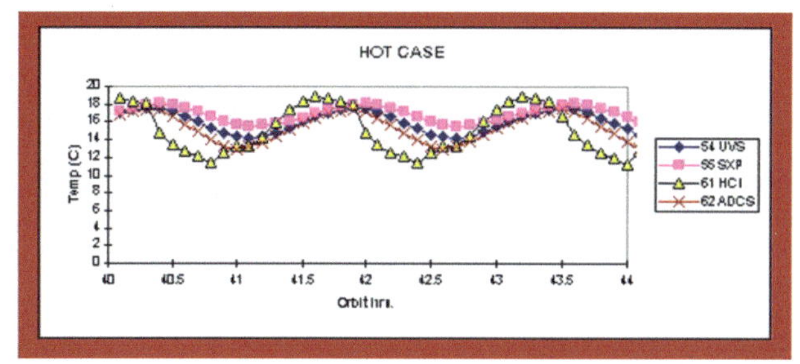

Ref: U. Co. SNOE

Thermal Control Hardware Masses
(Brown T7.6)

MLI (12 layers, with separators)	0.03 kg/m^2
Heat pipes (Aluminum, 12.7 mm dia.)	0.33 kg/m
Louvers	7.3 kg/m^2
Thermostat (250 W rating)	0.03 kg
Foam insulation	64 kg/m^3
Heaters (installed weight)	2.0 kg/m^2
OSR (optical solar reflector, installed)	1.0 kg/m^2
Radiators (Aluminum, installed)	2.71 kg/m^2
Paint	0.24 kg/m^2
Phase-change devices (Q in Wh)	$0.076 + 0.141Q$ kg

Thermal subsystem cost = 0.04 × total SC cost

Venus, Earth, Mars Thermal Data

Parameter	Venus	Earth	Mars
Solar constant G_s (W/m^2):	2,611.0	1,367.6	589.2
Albedo a (no units):	0.750	0.306	0.250
Infrared flux G_{planet} (W/m^2):	163	237	110
Equilibrium temp. T_e (K):	232	254.3	210.1
Ave. surface temp. T_s (K):	735	288	~210
Surface pressure P (Pa):	9,210,000	101,325	636
Atmosphere Scale Height (km)	15.9	8.5	11.1

Courtesy NASA

Inhabited Spacecraft: Environmental Control & Life Support Systems (ECLSS)

August 22, 2022

Courtesy: NASA

ECLSS Topics

- Human requirements
- Atmosphere functions:
  - Oxygen mixture control
  - Contaminant removal & atm. revitalization
  - Temperature & humidity control
  - Pressure integrity
- Water recovery & management
- Waste management
- Fire detection & suppression
- Radiation protection
- Food storage, preparation (growth?)
- Other functions
  - Crew exercise
  - Extra-vehicular activity (EVA) suits & tools
  - Food storage & prep
  - Personal hygiene, sleep, waste facilities
  - Clothing washing
  - Long-term missions?

August 22, 2022

Spacecraft Design Process

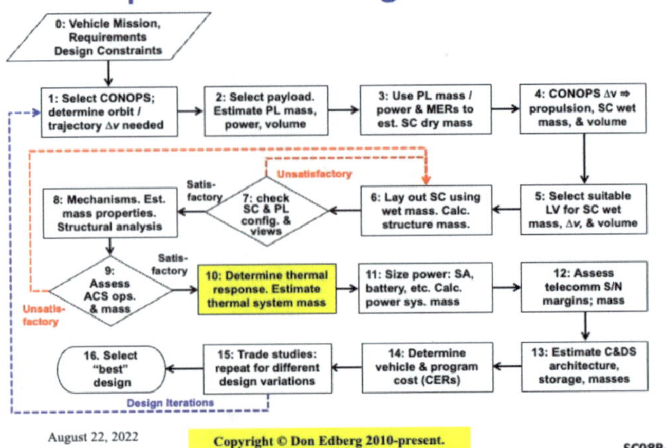
0: Vehicle Mission, Requirements Design Constraints
1: Select CONOPS; determine orbit / trajectory Δv needed
2: Select payload. Estimate PL mass, power, volume
3: Use PL mass / power & MERs to est. SC dry mass
4: CONOPS Δv ⇒ propulsion, SC wet mass, & volume
5: Select suitable LV for SC wet mass, Δv, & volume
6: Lay out SC using wet mass. Calc. structure mass.
Unsatisfactory
7: check SC & PL config. & views
Satis-factory
8: Mechanisms. Est. mass properties. Structural analysis
9: Assess ACS ops. & mass
Satis-factory
Unsatis-factory
10: Determine thermal response. Estimate thermal system mass
11: Size power: SA, battery, etc. Calc. power sys. mass
12: Assess telecomm S/N margins; mass
13: Estimate C&DS architecture, storage, masses
14: Determine vehicle & program cost (CERs)
15: Trade studies: repeat for different design variations
16. Select "best" design
Design Iterations

Environmental Control and Life Support Systems (ECLSS)

Why ECLSS? **Human habitation imposes *many* additional requirements on spacecraft** compared to robotic payload. ECLSS functions:

- Atmosphere oxygen, humidity, temperature
- Water recovery & management
- Waste management
- Food storage, preparation (growth?)
- Fire detection & suppression
- Human "facilities"
- Other functions

ECLSS Topics

- Human requirements
- Atmosphere functions:
  - Oxygen mixture control
  - Contaminant removal & atm. revitalization
  - Temperature & humidity control
  - Pressure integrity
- Water recovery & management
- Waste management
- Fire detection & suppression
- Radiation protection
- Other functions
  - Crew exercise
  - Extra-vehicular activity (EVA) suits & tools
  - Food storage & prep
  - Personal hygiene, sleep, waste facilities
  - Clothing washing
  - Long-term missions?

Human Requirements

NEEDS
Oxygen = 0.84 kg
Food Solids = 0.62 kg
Water in Food = 1.15 kg
Food Prep Water = 0.76 kg
Drink = 1.62 kg
Metabolized Water = 0.35 kg
Hand/Face Wash Water = 4.09 kg
Shower Water = 2.73 kg
Urinal Flush = 0.49 kg
Clothes Wash Water = 12.50 kg
Dish Wash Water = 5.45 kg
Total = 30.6 kg
EFFLUENTS
Carbon Dioxide = 1.0 kg
Respiration/Perspiration Water = 2.28 kg
Food Preparation, latent water = 0.036 kg
Urine = 1.50 kg
Urine Flush Water = 0.5 kg
Feces Water = 0.091 kg
Sweat Solids = 0.018 kg
Urine Solids = 0.059 kg
Feces Solids = 0.032 kg
Hygiene Water = 12.58 kg
Clothes Wash Water:
Liquid = 11.9 kg
Latent = 0.6 kg
Total = 30.6 kg

Courtesy NASA

Note: values are per person per day based on an average metabolic rate of 136.7 W/person (11,200 BTU/person/day) and 0.87 respiration quotient (the molar ratio of CO_2 generated to O_2 consumed). Values are higher for greater activity levels and for larger-than-average people.

ISS Consumable Budget

Consumable	Design Load (kg/person-day)
Oxygen	0.85
Water (drinking)	1.6
Water (in food)	1.15
Water (clothes/dishes)	17.9
Water (sanitary)	7.3
Water (food prep)	0.75
Food solids	0.62

ISS ECLSS Functions

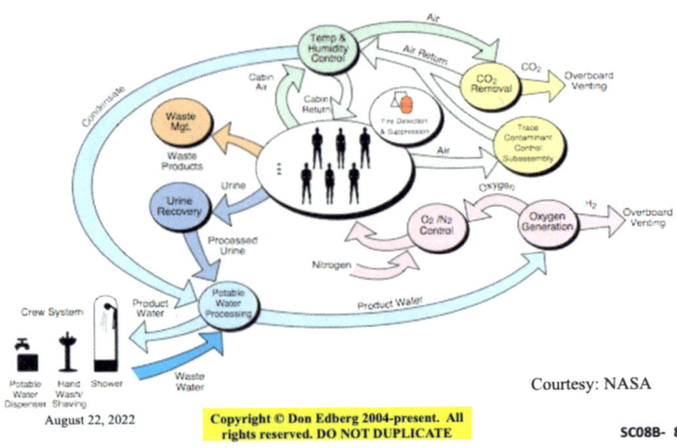
Temp & Humidity Control
Air
Air Return
CO₂ Removal
CO₂
Overboard Venting
Cabin Air
Cabin Return
Condensate
Waste Mgt.
Waste Products
Fire Detection & Suppression
Trace Contaminant Control Subassembly
Air
Oxygen
Urine
Urine Recovery
O₂/N₂ Control
Oxygen Generation
H₂
Overboard Venting
Processed Urine
Nitrogen
Potable Water Processing
Product Water
Crew System
Potable Water Dispenser
Hand Wash/ Shaving
Shower
Waste Water

Courtesy: NASA

ISS ECLS System

- Five main components:
  - Atmospheric control
  - Water supply
  - Food supply
  - Sanitation equipment
  - Fire detection & suppression
- Primary life support systems located in the *Zvezda* service module
  - Monitored & controlled by laptops

How Things Work Together

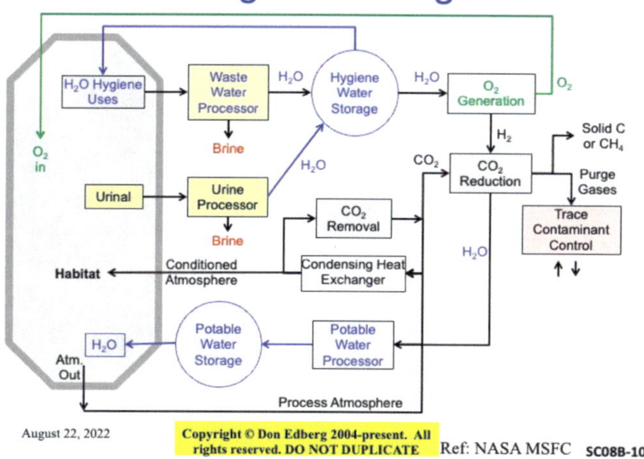

Ref: NASA MSFC

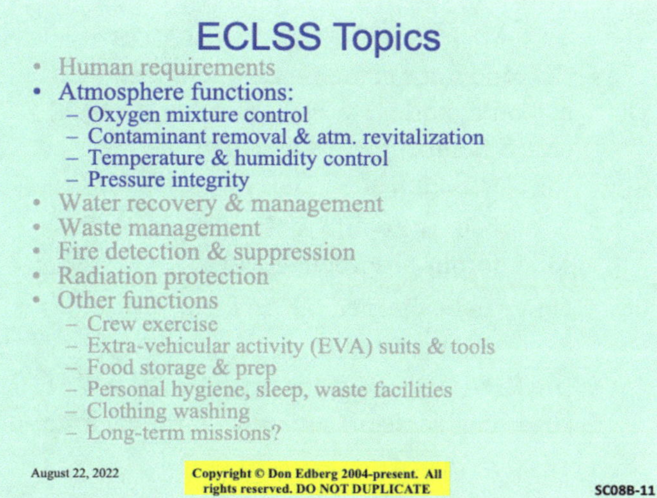

Physiological Effects of Oxygen Concentration

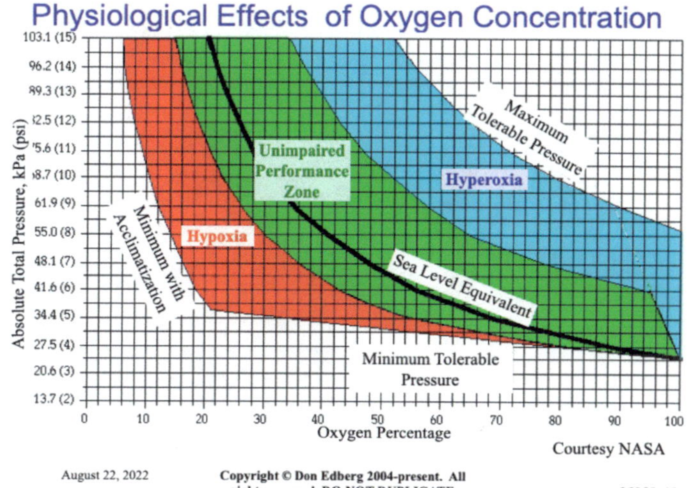

Courtesy NASA

Atmospheric Mixture Control

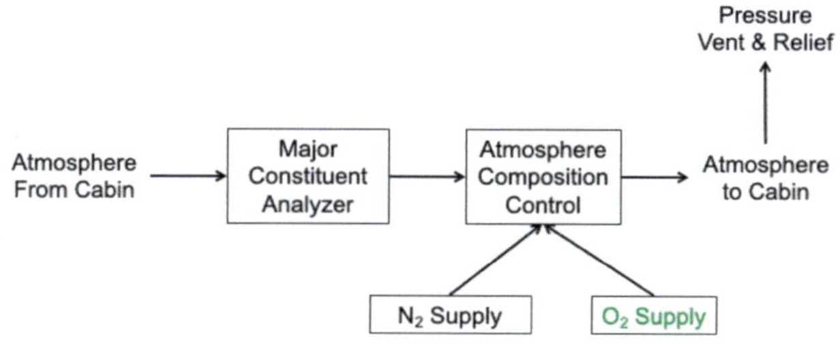

Atmosphere Revitalization

- Monitors/controls major constituents (proper composition)
- Generates, stores, & supplies O_2
- Controls pressure
- Removes CO_2
- Monitor & control trace contaminants

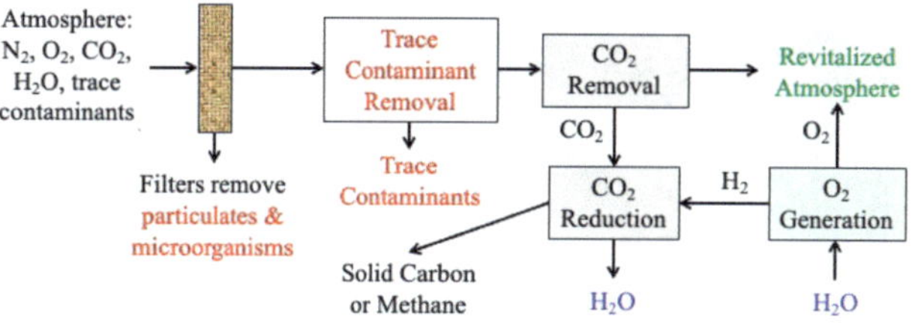

ISS *Elektron* Oxygen Supply System

- Decomposes H_2O electrochemically. Mfr.: NIIKHIMMASH
- Uses a liquid loop with 30% potassium hydroxide (KOH) in solution, gas lines with a pressure regulator
- The decomposition of 4 kg of H_2O/h yields 100 L of O_2/h at a pressure of 160 mm Hg, can support four crew members/day

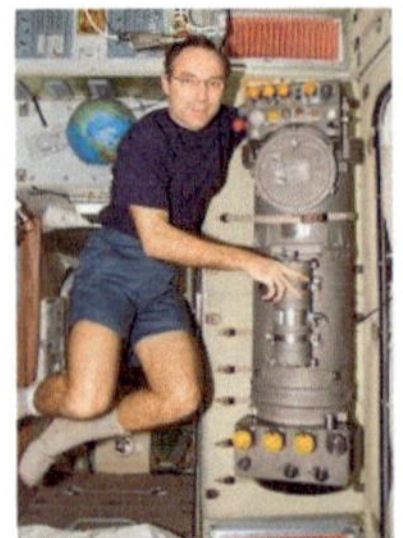

August 22, 2022 SC08B-15

Elektron O_2 Generator Operation Diagram

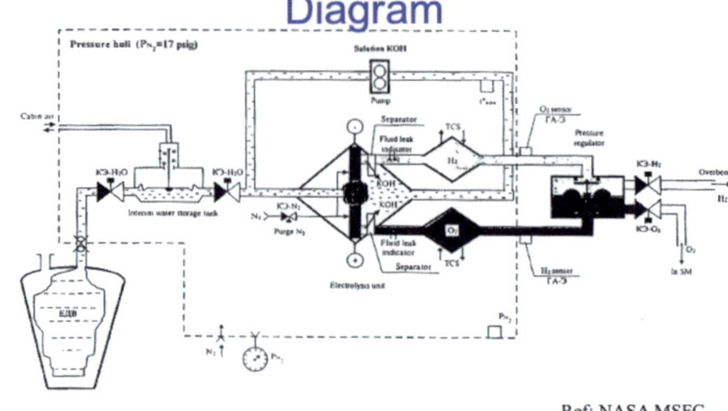

Fig. A-3. Elektron unit flow diagram

Ref: NASA MSFC

August 22, 2022 SC08B-16

ISS *TGK* Solid-Fuel Oxygen Generators

- Backup to Elektron, similar to airliner oxygen masks
- Solid potassium perchlorate ($KClO_4$) thermally decomposed into potassium chloride (KCl) and O_2 @ 400 C.
- Consists of a replaceable cartridge with igniter, striker mechanism, filter, dust-collection filter, & fan
- 1 cartridge yields 600 L of oxygen
- Outer surface temperature may reach 50 C (122 °F)
- One of these almost started a fire on *Mir*

Photo: Quora

August 22, 2022 SC08B-17

Air Purification Systems

- CO_2 and trace contaminants scrubbed by carbon dioxide removal assembly (CDRA) on US Segment and Vozdukh ('air') in Russian Segment
- Absorbed by Zeolite molecular sieves, a solid porous absorbent material
- Unit can be cleaned out and reset by exposure to vacuum.
- If CDRA or *Vozdukh* is off-line, CO_2-absorbing canisters are utilized for back-up

August 22, 2022 SC08B-18

Backup: CO_2-Absorbing Canisters on ISS

- Lithium hydroxide canisters remove CO_2 from the atmosphere: $2\ LiOH \cdot H_2O + CO_2 \rightarrow Li_2CO_3 + 3\ H_2O$
- One canister removes 1600 liters of CO_2
- On average, one crew member produces 480 L/day of CO_2
- Not reusable; usage must be budgeted.
- Backup for CDRA & Vozdukh
- May be used for short-duration missions
- Apollo 13 issue!

August 22, 2022 SC08B-19

Gas Analysis System

- Continuously monitors partial pressures of oxygen, carbon dioxide, water vapor and hydrogen content
- One analyzer each in *Zvezda* and *Zarya*

Climate Control System

- Controls temperature
- Controls humidity
- Provides ventilation
- Cools equipment in inhabited areas

August 22, 2022 SC08B-21

FVP Trace Contaminants Control Unit

- Removes contaminants from the atmosphere
- Consists of two activated-charcoal cartridges, a catalytic oxidizer cartridge, filter, fan, and valves
- Charcoal beds absorb high molecular-weight trace contaminants
- Catalyst oxidizes CO to CO_2, and H_2 to H_2O
- Charcoal cartridges regenerated every 20 days by heating to 200 °C and exposing to vacuum for 12 h

August 22, 2022 SC08B-22

Human Comfort "Box"

Courtesy NASA

August 22, 2022 SC08B-23

ISS Temperature & Humidity Control

- Atmosphere composition N_2 & O_2
- Temperature ~ 23 C, humidity around 40-60%
- Liquid-air heat exchangers provide temperature & humidity control

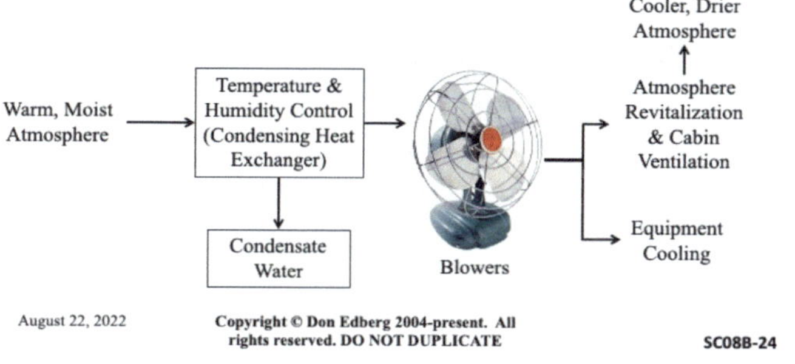

August 22, 2022 SC08B-24

Spacelab Cabin Ventilation

Ref: NASA MSFC

August 22, 2022 SC08B-25

Pressure Integrity Monitoring System

- Monitors the total atmospheric pressure
- Transmits a warning signal to the *Caution and Warning panel* if pressure drops below acceptable levels (i.e. if there is a hull breach)
- Used to pressurize and equalize atmosphere in docking chambers, such as during a *Soyuz* or *Progress* docking

August 22, 2022 SC08B-26

ECLSS Topics

- Human requirements
- Atmosphere functions:
  - Oxygen mixture control
  - Contaminant removal & atm. revitalization
  - Temperature & humidity control
  - Pressure integrity
- **Water recovery & management**
- Waste management
- Fire detection & suppression
- Radiation protection
- Other functions
  - Crew exercise
  - Extra-vehicular activity (EVA) suits & tools
  - Food storage & prep
  - Personal hygiene, sleep, waste facilities
  - Clothing washing
  - Long-term missions?

August 22, 2022 SC08B-27

- Stores
- Distributes
- Recovers
- Monitors quality

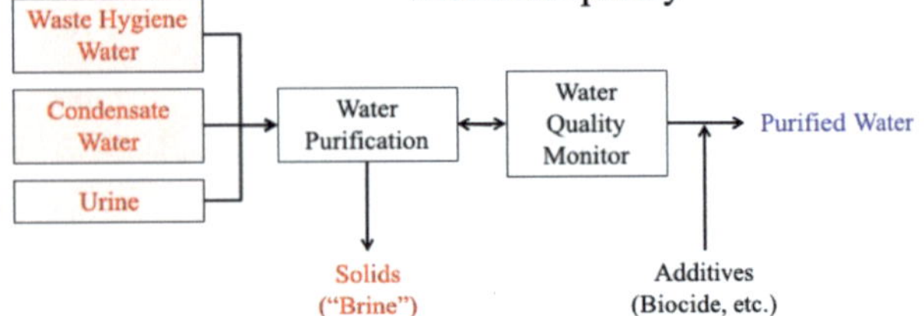

ECLSS Topics

- Human requirements
- Atmosphere functions:
  - Oxygen mixture control
  - Contaminant removal & atm. revitalization
  - Temperature & humidity control
  - Pressure integrity
- Water recovery & management
- **Waste management**
- Fire detection & suppression
- Radiation protection
- Other functions
  - Crew exercise
  - Extra-vehicular activity (EVA) suits & tools
  - Food storage & prep
  - Personal hygiene, sleep, waste facilities
  - Clothing washing
  - Long-term missions?

Waste System Deals With:

- Solid
- Liquid
- Gaseous
- Waste elimination (0-g toilet?)

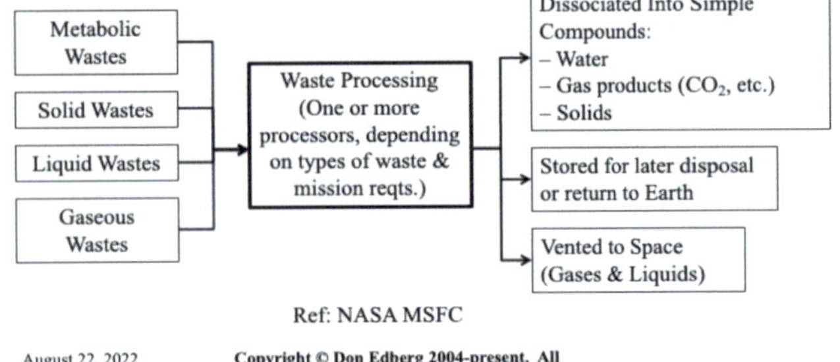

Ref: NASA MSFC

Waste & Hygiene Compartment

- A.K.A. toilet
- Pre-treated waste (liquid/solid) stored in disposable containers

Courtesy: ESA

ECLSS Topics

- Human requirements
- Atmosphere functions:
  - Oxygen mixture control
  - Contaminant removal & atm. revitalization
  - Temperature & humidity control
  - Pressure integrity
- Water recovery & management
- Waste management
- **Fire detection & suppression**
- Radiation protection
- Other functions
  - Crew exercise
  - Extra-vehicular activity (EVA) suits & tools
  - Food storage & prep
  - Personal hygiene, sleep, waste facilities
  - Clothing washing
  - Long-term missions?

ISS Fire System

- Detects fires with one or more sensors
- Suppresses with foam
- Gas masks supplied to crew
- Post-fire cleanup

ISS Fire Detection & Suppression

- Modules have IR or induction smoke detectors
- Portable fire extinguishers produce non-toxic foam
  - Cannot be used in the U.S. segments
  - Foam conducts electricity, poses an electric shock hazard from US 120 VAC equipment, while Russian segment is 28 VDC)
- Rebreather-type gas masks are provided for protection against smoke inhalation; produce O_2 via a chemical reaction for up to 140 minutes

August 22, 2022 SC08B-34

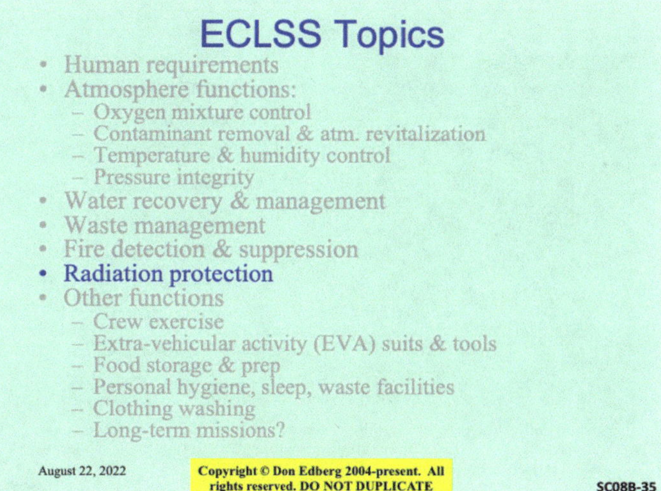

ECLSS Topics

- Human requirements
- Atmosphere functions:
  - Oxygen mixture control
  - Contaminant removal & atm. revitalization
  - Temperature & humidity control
  - Pressure integrity
- Water recovery & management
- Waste management
- Fire detection & suppression
- **Radiation protection**
- Other functions
  - Crew exercise
  - Extra-vehicular activity (EVA) suits & tools
  - Food storage & prep
  - Personal hygiene, sleep, waste facilities
  - Clothing washing
  - Long-term missions?

August 22, 2022 SC08B-35

Radiation Exposure

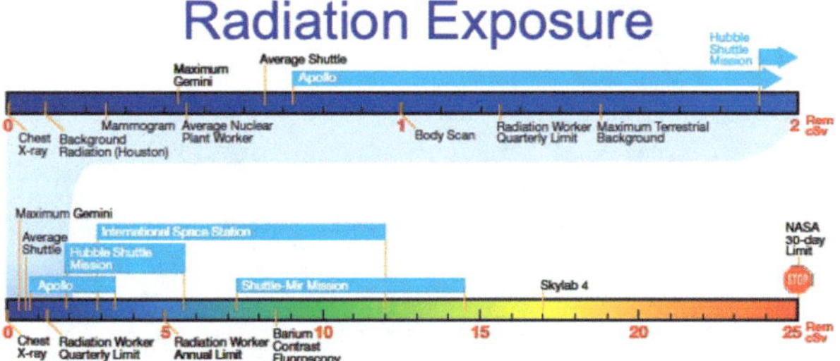

- Radiation exposure depends on location & altitude
- No EVAs during high radiation periods (i.e. SAA)
- Must provide protection for long missions (water works fairly well)
- *Sievert* unit (Sv): a dose of ionizing radiation of 1 J/kg of recipient biological mass

August 22, 2022 SC08B-36

Best Way to Protect from Radiation

- Minimize dosage by passing quickly through and weakest zones (i.e. Apollos 8, 10-17)
- Apollo lunar mission astronauts got a radiation dose of about 0.5 rad, slightly less than a chest CT scan.
- Mission total of 4.5 rad per person
- CT scan ~ 0.7 rad

Courtesy: Quora

August 22, 2022 SC08B-37

ECLSS Topics

- Human requirements
- Atmosphere functions:
  - Oxygen mixture control
  - Contaminant removal & atm. revitalization
  - Temperature & humidity control
  - Pressure integrity
- Water recovery & management
- Waste management
- Fire detection & suppression
- Radiation protection
- **Other functions**
  - Crew exercise
  - Extra-vehicular activity (EVA) suits & tools
  - Food storage & prep
  - Personal hygiene, sleep, waste facilities
  - Clothing washing
  - Long-term missions?

August 22, 2022 SC08B-38

Other Environmental Functions

- Crew exercise
- Extra-Vehicular Activity (EVA)
  - Suits with oxygen, cooling/heating
  - Specialized tools
- LONG-TERM missions
  - Food storage & preparation
  - Clothing washing etc.
  - Plant growth facilities
  - Hyperbaric chamber with airlock

August 22, 2022 SC08B-39

Crew Exercise

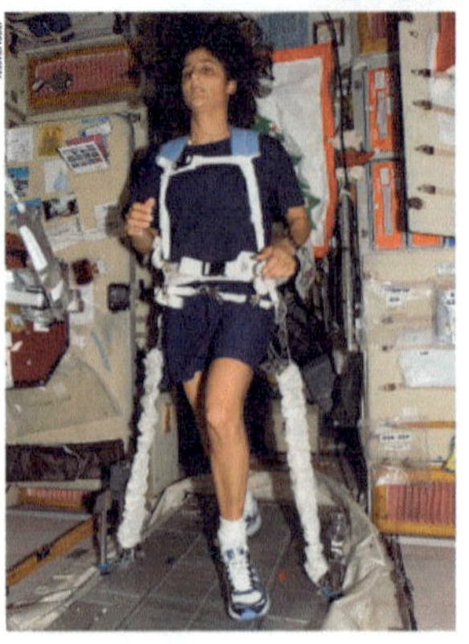

- Bone mass decreases in microgravity environment, perhaps do to reduced need
- Regular exercise that puts mechanical loads on the skeleton helps to alleviate loss
- Artificial gravity is long-term solution

EVA (Extra-Vehicular Activity)

- Pressure suit must provide pressure, cooling, etc.
  - Orlan (Russian), EMU (USA)
  - Cooling undergarment pumps cooling water
  - Waist vs. rear-entry systems

Apollo EVA Suit

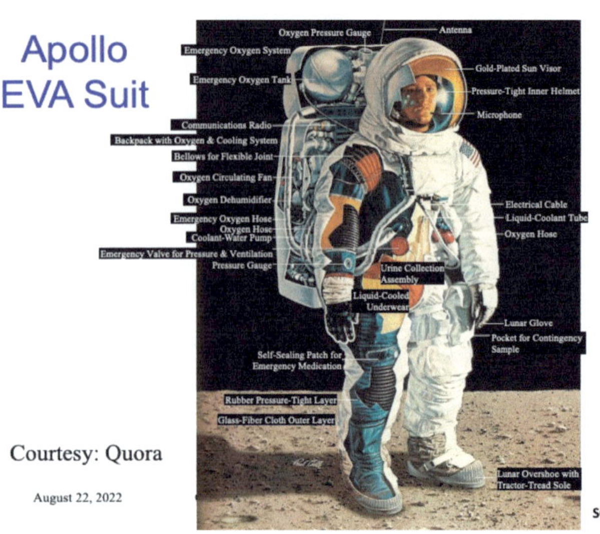

Courtesy: Quora

EVA Tools (courtesy NASA)

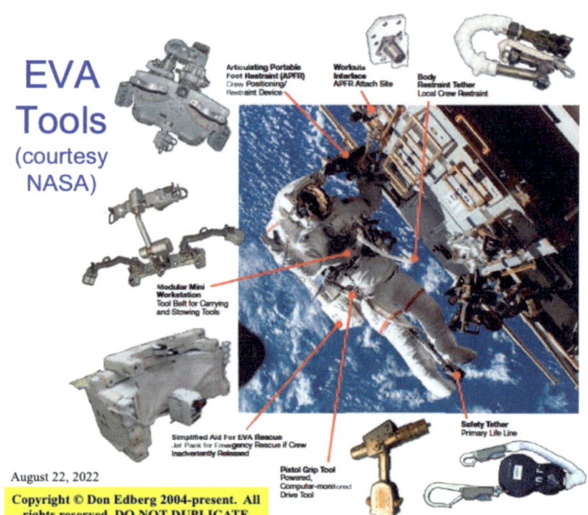

Long-Term Missions

- So-called "open-loop" missions require continuous resupply
  - Lots of up-mass needed, costly
  - Cannot do long missions (such as Mars)
- Situation improves with partial "recycling" (partially closed-loop)
- Best, but hardest, is fully closed-loop (= self sufficient, minimum mass)

Other functions:

- Exercise?
- Artificial gravity?
- Acoustics?
- Privacy? Isolation from loved ones?

ISS ECLSS Hardware Location (Feb. 2010)

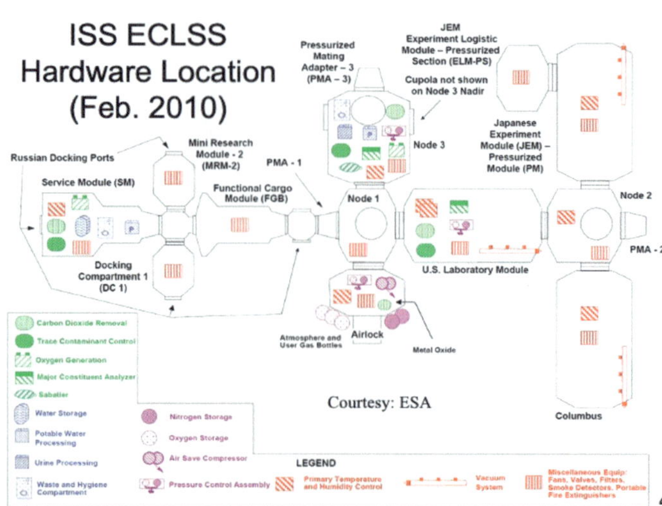

Courtesy: ESA

Spacecraft Power: Power Sources & Storage, Delivery Systems

Courtesy Boeing, Wikipedia

August 22, 2022

Spacecraft Power Topics

- Power system elements, definitions, & power required
- Sources of power
  - Chemical cells / batteries
    - Definitions, chemistries
    - Series vs. parallel, strings
    - Battery design & life estimation, sizing example
  - Fuel cells
  - Photovoltaic / solar cells
    - Considerations: power & mechanical
    - Deployment
    - Losses & efficiency, sizing
  - Solar Dynamic
  - Nuclear thermal: RTG
  - Nuclear reactors
- Power control systems
- Power allocation to subsystems, operational modes
- Two spacecraft riddles
- Mass estimation

August 22, 2022

Spacecraft Design Process

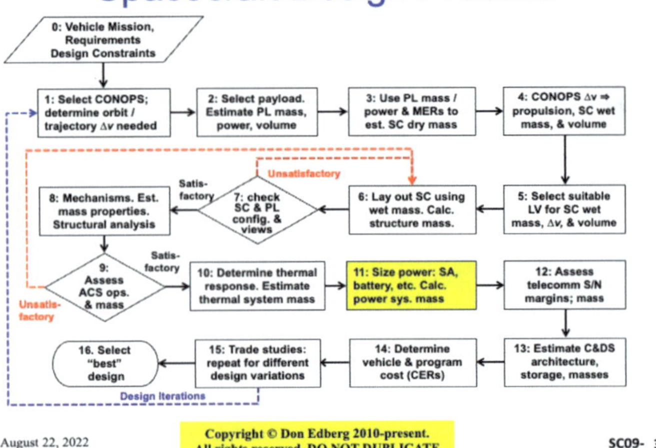
0: Vehicle Mission, Requirements Design Constraints
1: Select CONOPS; determine orbit / trajectory Δv needed
2: Select payload. Estimate PL mass, power, volume
3: Use PL mass / power & MERs to est. SC dry mass
4: CONOPS Δv ⇒ propulsion, SC wet mass, & volume
5: Select suitable LV for SC wet mass, Δv, & volume
6: Lay out SC using wet mass. Calc. structure mass.
7: check SC & PL config. & views
Unsatisfactory
Satis-factory
8: Mechanisms. Est. mass properties. Structural analysis
9: Assess ACS ops. & mass
Satis-factory
Unsatis-factory
10: Determine thermal response. Estimate thermal system mass
11: Size power: SA, battery, etc. Calc. power sys. mass
12: Assess telecomm S/N margins; mass
13: Estimate C&DS architecture, storage, masses
14: Determine vehicle & program cost (CERs)
15: Trade studies: repeat for different design variations
16. Select "best" design
Design Iterations

Spacecraft Power Topics

- Power system elements, definitions, & power required
- Sources of power
  - Chemical cells / batteries
    - Definitions, chemistries
    - Series vs. parallel, strings
    - Battery design & life estimation, sizing example
  - Fuel cells
  - Photovoltaic / solar cells
    - Considerations: power & mechanical
    - Deployment
    - Losses & efficiency, sizing
  - Solar Dynamic
  - Nuclear thermal: RTG
  - Nuclear reactors
- Power control systems
- Power allocation to subsystems, operational modes
- Two spacecraft riddles
- Mass estimation

Power System Elements

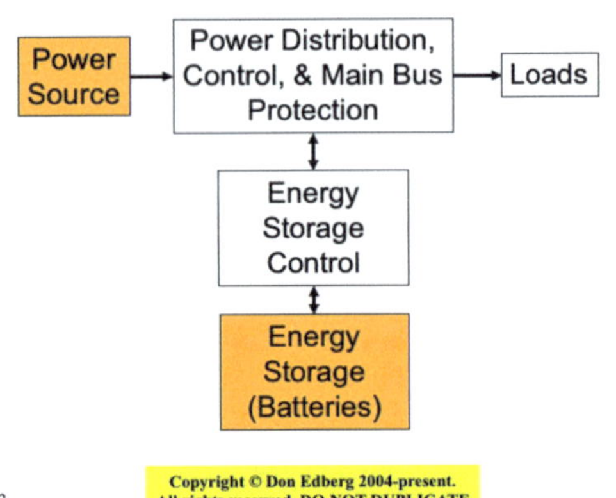
Power Source
Power Distribution, Control, & Main Bus Protection
Loads
Energy Storage Control
Energy Storage (Batteries)

Power System Definitions

- EPS = Electrical Power System
- PCS = Power Control System
- SA = Solar Array
- RTG = Radioisotope Thermal Generator
- ASRG = Advanced Stirling Radioisotope Generator
- DoD = Depth of Discharge
- DET = Direct Energy Transfer
- PPT = Peak Power Tracking

Estimating Power Required

Preliminary value of total power needed (P_{total}) depends on type of spacecraft (per Brown):

- General: $P_{total} = 1.13\ P_{payload} + 122$ W
- Comsats: $P_{total} = 1.13\ P_{payload} + 56$ W
- Metsats: $P_{total} = 1.96\ P_{payload}$
- Planetary: $P_{total} = 332.93 \ln(P_{payload}) - 1047$ W

Add 90% margin for new designs!

Subsystem power $P_{subsystem} = P_{total} - P_{payload}$

Note: 40% battery capacity (energy storage) margin recommended at Phase C/D start

Brown

Power Sources

Power source selection

- Chemical sources:
  - Batteries
  - Fuel Cells
- Solar Energy
  - Photovoltaic
  - Solar Dynamic
- Radioisotope Thermal Generators
- Nuclear Reactors

Select Power Source Based on Mission & Spacecraft Type

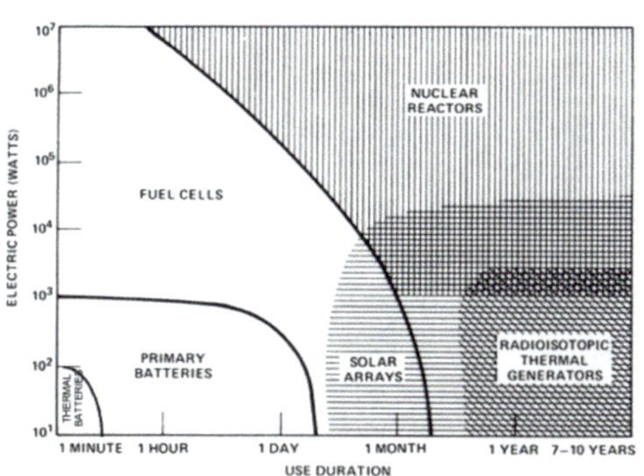

Fig. 10.2 Operating regimes of spacecraft power sources.

Electrochemical Power

- Batteries: Primary vs. Secondary
- Battery types (Ni-Cd, Ni-MH, Ag-Zn, Li)
- Fuel Cells

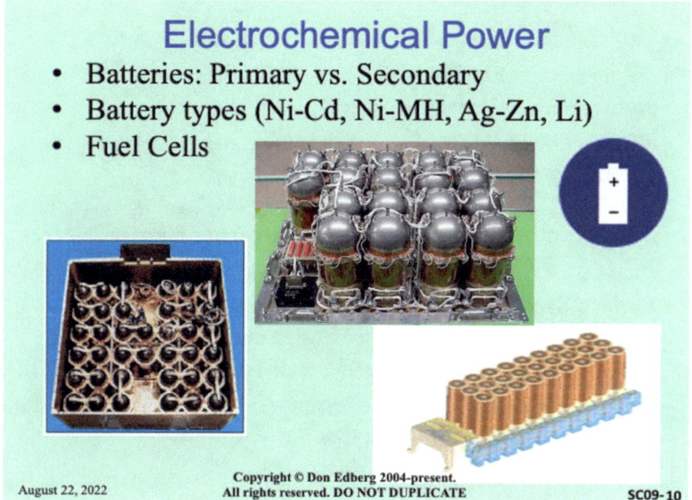

Battery Requirements

- Storage for many spacecraft (primary power source for LVs)
- Battery must supply all power needs and not experience complete discharge
- Must be charged during non-eclipse periods
- (Thermal constraints)

Spacecraft Power Topics

- Power system elements, definitions, & power required
- Sources of power
  - Chemical cells / batteries
    - Definitions, chemistries
    - Series vs. parallel, strings
    - Battery design & life estimation, sizing example
  - Fuel cells
  - Photovoltaic / solar cells
    - Considerations: power & mechanical
    - Deployment
    - Losses & efficiency, sizing
  - Solar Dynamic
  - Nuclear thermal: RTG
  - Nuclear reactors
- Power control systems
- Power allocation to subsystems, operational modes
- Two spacecraft riddles
- Mass estimation

Cell & Battery Definitions

Definition	Description	Units
Average discharge voltage, V_{avg}	V_{avg} = # series cells × $V_{avg\text{-}cell}$	V
Charge capacity, C_{chg}	C_{chg}: electric charge stored in battery	Ah
Energy capacity, E_{bat}	$E_{bat} = C_{chg} \times V_{avg}$; energy stored in battery	J or Wh
Specific energy, e_{bat}	e_{bat} = energy stored in battery *per unit mass*	J/kg or Wh/kg (higher better)
Cell Energy Density	Energy stored in battery per unit volume	Wh/L or Wh/m³
Depth of Discharge, DoD	DoD = percent of battery capacity used during discharge cycle	Percent
Charge rate R_{chg}	R_{chg} = rate battery can accept charge*	Amps

* Too high causes rapid temp increase & battery failure; charge voltage must be greater than battery's open-circuit voltage in order to charge (rule of thumb ~ 25% higher)
Units: V = volts, A = amperes (amps), J = joule, V·A = W = watt = J/s, h = hour

Ideal Cell Properties

- Cell mistakenly called a "battery"
- Example cell has a capacity of 1 Ah = 1,000 mAh @ 3.6 V
- Ideal current supplied at a voltage of 3.6 V:
  - 1 A = 1,000 mA (= 3.6 W) for 1 hour or
  - 0.5 A = 500 mA (= 1.8 W) for 2 hours or
  - 0.25 A = 250 mA (= 0.9 W) for 4 hours.
- Energy (not power) rating of (1 Ah) × 3.6 V = 3.6 watt-hours (Wh)
- In reality, these values reduced by internal impedance at high current levels

"Real-Life" Cell Description

- All lls have internal resistance R_{int} that reduces delivered voltage V_{out} at higher current levels: $V_{out} = V_{OC} - iR_{int}$
  - A car's headlights dim when the starter cranks the engine
- Low R_{int} minimizes internal resistive power loss $P_{resistive} = i^2 R_{int}$ and delivers more power
- Large currents induce significant voltage loss at battery terminals, reducing effective energy capacity and favoring higher-voltage systems (which reduce i)
- Higher voltages also reduce wiring mass: ISS has 100 V system (most vehicles use 28 V)

Courtesy splung.com

Cell Chemistry

Primary (non-rechargeable)

- Thermal batteries
- Alkaline, etc.
- Silver-Zinc
- Some Lithiums

Secondary (rechargeable)

- Lithium-ion, Lithium-metal, Lithium-polymer
- Nickel-Cadmium or NiCd (oldest)
- Nickel-Metal Hydride (Nickel Hydrogen, NiH_2)

Below for older Nickel chemistry only:

- "Memory": NiCd & NiH batteries suffer from life limitations driven by repeated charging cycles & DoD
- "Reconditioning" (deep discharge & charge) can extend NiCd & NiH battery life… NOT LITHIUMS!!!

Cell Characteristics

Courtesy **Human Spaceflight Mission Analysis & Design**

Characteristic	NiCd	NiH_2	NiMH	Li-Ion
Specific Energy (Wh/kg)	35	49 IPV 60 CPV	60	85-175
Energy Density (Wh/L)	45	25 IPV 40 CPV	86	160
Cell Specific Energy (Wh/kg)	39	60 – 80	–	100 – 200
Cell Energy Density (Wh/L)	137	64	–	260
Operating Temp. (°C)	–40 to +70	–5 to +20	–5 to +20	–40 to +70
Self-Discharge, 20° C, per month	10 – 25%	6%* *Ref. says 60%. Typo?	6%*	0.167%
Cell's Discharge Voltage $V_{discharge}$	1.2	1.2 – 1.3	1.2 – 1.3	3.7

Lithium-Chemistry Cells

- Higher operating voltage: 3.6 V reduces cell count
- High energy density: $125 - 200\ \frac{\text{Wh}}{\text{kg}}$ reduces battery mass
- High volume efficiency: $300\ \frac{\text{Wh}}{\text{L}}$ reduces battery volume required
- High-cycle usage: cell life is more than 10,000 charge/discharge cycles
- Can be shaped for convenience

Quallion QL075KA Lithium-Ion Cell

Battery design for long cycle life (>100,000 cycles)

Electrical Characteristics	
Nominal Capacity	72 Ah
Nominal Voltage	3.6 V
Maximum Recommended Continuous Discharge Current	72 A
Operating Discharge Temperature	0°C to 42°C

Physical Characteristics	
Height	173.7 mm
Width	80.9 mm
Thickness	56.2 mm
Weight	1820 g
Volume	790 cc
Gravimetric Energy Density	142 Wh/kg
Volumetric Energy Density	330 Wh/L

Spacecraft Power Topics

- Power system elements, definitions, & power required
- Sources of power
  - Chemical cells / batteries
    - Definitions, chemistries
    - **Series vs. parallel, strings**
    - Battery design & life estimation, sizing example
  - Fuel cells
  - Photovoltaic / solar cells
    - Considerations: power & mechanical
    - Deployment
    - Losses & efficiency, sizing
  - Solar Dynamic
  - Nuclear thermal: RTG
  - Nuclear reactors
- Power control systems
- Power allocation to subsystems, operational modes
- Two spacecraft riddles
- Mass estimation

Ideal Cells in *Series*

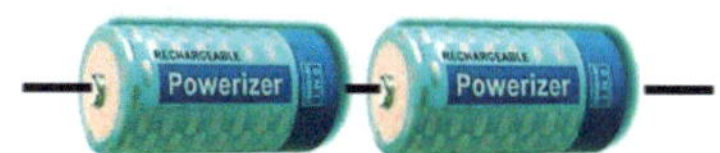

- Two cells in series produces a battery with an energy capacity of 1 Ah = 1,000 mAh at 2 × 3.6 V = 7.2 V
- It can supply 1 A for 1 hour, or 0.5 A for 2 hours, or 0.25 A for 4 hours, at voltage 7.2 V
- Energy rating of 1 Ah × 7.2 V = 7.2 Wh

 # SERIES cells = required voltage ÷ cell voltage

August 22, 2022 SC09-21

Ideal Cells in *Parallel*

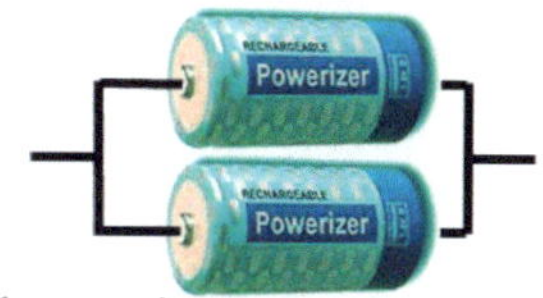

- Two cells in parallel produce a battery with an energy capacity of 2 × 1 Ah = 2 Ah = 2,000 mAh at 3.6 V
- It can supply 2 A for 1 hour, or 1 A for 2 hours, or 0.5 A for 4 hours, at voltage 3.6 V
- Energy rating of 2 Ah × 3.6 V = 7.2 Wh

 # PARALLEL cells = req'd current ÷ cell current

August 22, 2022 SC09-22

Batteries vs. Cells vs. Strings

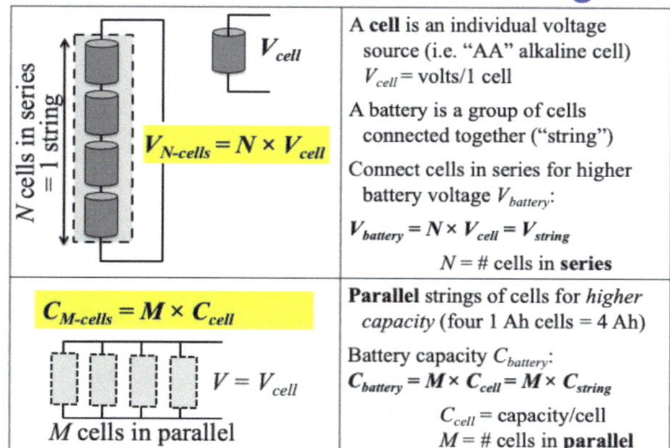

A **cell** is an individual voltage source (i.e. "AA" alkaline cell) V_{cell} = volts/1 cell

A battery is a group of cells connected together ("string")

Connect cells in series for higher battery voltage $V_{battery}$:

$V_{battery} = N \times V_{cell} = V_{string}$

N = # cells in **series**

Parallel strings of cells for *higher capacity* (four 1 Ah cells = 4 Ah)

Battery capacity $C_{battery}$:

$C_{battery} = M \times C_{cell} = M \times C_{string}$

C_{cell} = capacity/cell

M = # cells in **parallel**

August 22, 2022 SC09-23

Strings of Cells in Parallel

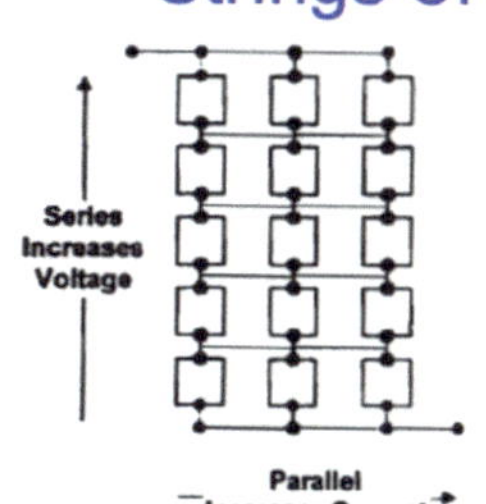

Remember: connect cells in **series** for more **voltage**, **parallel** for more **capacity**.

Courtesy: Brown (L); Hdbk of Space Flight (2xR)

August 22, 2022 SC09-24

Spacecraft Power Topics

- Power system elements, definitions, & power required
- Sources of power
  - Chemical cells / batteries
    - Definitions, chemistries
    - Series vs. parallel, strings
    - **Battery design & life estimation, sizing example**
  - Fuel cells
  - Photovoltaic / solar cells
    - Considerations: power & mechanical
    - Deployment
    - Losses & efficiency, sizing
  - Solar Dynamic
  - Nuclear thermal: RTG
  - Nuclear reactors
- Power control systems
- Power allocation to subsystems, operational modes
- Two spacecraft riddles
- Mass estimation

August 22, 2022 SC09-25

Battery Design Analysis

The battery energy capacity required, E_b, is determined by the average eclipse time energy (average power × time, or $P_e t_e$) divided by the total system efficiency η and depth of discharge (DoD):

$$E_b = \frac{P_e t_e}{\eta \cdot DoD}$$

- η = total system efficiency

August 22, 2022 SC09-26

Battery Depth of Discharge (*DoD*)

- *DoD* is fraction or percent of battery energy storage used — how much you take out divided by how much is there:

  $DoD = \frac{P_L t_d}{E_{batt}} = \frac{P_L t_d}{C_{chg} V_{avg}}$, where

  P_L = Load power (W)

  t_d = Discharge time (h)

  C_{chg} = Charge capacity (Ah = amp-hours)

  V_{avg} = average discharge voltage (usually 28V)
- Charging is specified in terms of charge *rate* C_R:

  $C_R = \frac{C_{chg}}{t_c}$, where t_c is charging time in hours (h)

Battery Life

Depends on:

- Depth of Discharge (DoD).
  Smaller DoD preferred for longer battery life
- DoD measured in percent (80% discharged = 0.8)
- Number of discharge cycles
- Temperature

Must factor Depth of Discharge (DoD) into design of power system for life of SC

Smaller DoD means *larger* battery capacity needed

Lithium & Nickel Chemistry Battery Life vs. DoD

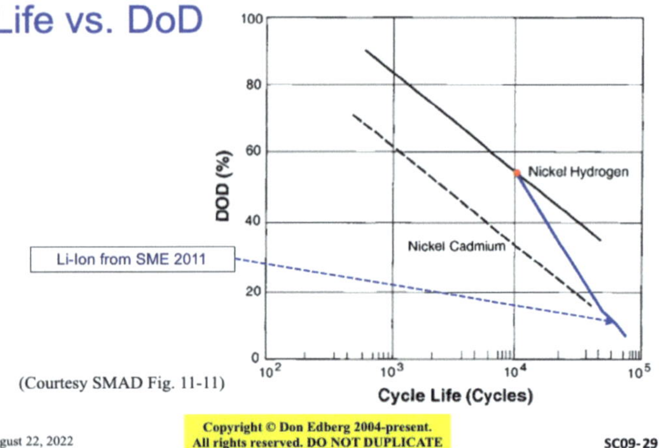

(Courtesy SMAD Fig. 11-11)

Li-Ion Battery Capacity Fading

According to the **Handbook of Space Technology**, the following mathematical relationship is not applicable to all Li-Ion cell types, but is appropriate to estimate the capacity fading of 20 Ah cells and larger:

Capacity fading CF (in %) = calendar loss + cycling loss

$CF = k_a\sqrt{\text{mission duration (years)}} + k_c\sqrt{\text{\# discharge cycles}}$

k_a depends on temperature θ in °C:

$$k_a = 0.0009\theta^2 - 0.0129\theta + 0.1533$$

k_c depends on percent Depth of Discharge (DoD, 0-100%):

$$k_c = 0.005\ \text{DoD}(\%) + 0.021$$

Lithium Battery Sizing Example 1 of 2

- Power system to provide 567 W eclipse power (28V for 38 min), power loss 3%. 10,000 cycles required at 25° C. Cells: 28 Ah capacity.
- Li: chart @ 10K cycles: DoD = 55%
- Capacity $C = \frac{\text{power}\times\text{time}}{\text{DoD(voltage)}(1-\text{loss})} = \frac{567\ \text{W}\times\frac{38}{60}\text{h}}{0.55(28\ \text{V})(0.97)}$

  C = 24.04 Ah; one cell (= 28 Ah) sufficient
- Energy capacity E_b = 28 Ah × 28 V = 784 Wh
- Mass = 784 Wh/(130 Wh/kg) = **6.03 kg**

Lithium Battery Sizing Example 2 of 2

- 28 volts required, Li cells produce ~3.6 volts: 28 V/(3.6 V/cell) = 7.78 cells ≈ 8 cells in series (need integer number of cells)
- 8-cell string produces 28.8 V
- One cell provides proper capacity, therefore battery consists of 1 string of 8 cells in series (=8s1p configuration)
- For one battery-out capability: add second string: 2 parallel strings, 8 cells in series = 8s2p. Mass = **12.1 kg**

Spacecraft Power Topics

- Power system elements, definitions, & power required
- Sources of power
  - Chemical cells / batteries
    - Definitions, chemistries
    - Series vs. parallel, strings
    - Battery design & life estimation, sizing example
  - **Fuel cells**
  - Photovoltaic / solar cells
    - Considerations: power & mechanical
    - Deployment
    - Losses & efficiency, sizing
  - Solar Dynamic
  - Nuclear thermal: RTG
  - Nuclear reactors
- Power control systems
- Power allocation to subsystems, operational modes
- Two spacecraft riddles
- Mass estimation

August 22, 2022 SC09-33

Fuel Cells

- Cell converts chemicals (usually H_2 and O_2) directly into electricity
- 500 Wh/kg at 2600 W power
- ~$3000/kW for commercial, space higher

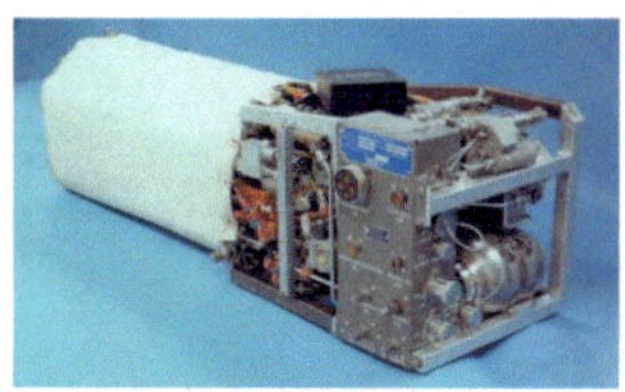

August 22, 2022 SC09-34

Fuel Cell Fundamentals

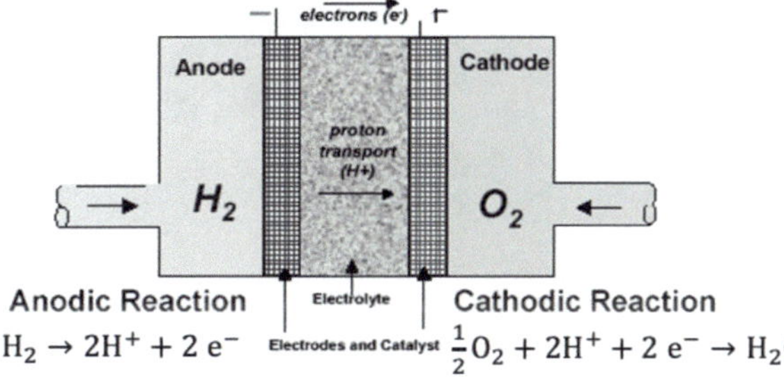

Anodic Reaction

$$H_2 \rightarrow 2H^+ + 2\,e^-$$

Cathodic Reaction

$$\frac{1}{2}O_2 + 2H^+ + 2\,e^- \rightarrow H_2O$$

August 22, 2022 SC09-35

Fuel Cell Facts

- High power levels available: shuttle developments have produced specific power of 275 W/kg, ~12 kW with start up time of ~ 15 minutes and instantaneous shutdown
- Up to 35% electrical efficiency
- Must store fuel cell chemicals as cryogenics for volume efficiency
- Thermal insulation possibilities limit mission duration
- Used on manned flights, water is a byproduct

Apollo CSM fuel cell photo: J Humphries

August 22, 2022 SC09-36

Apollo Fuel Cell System

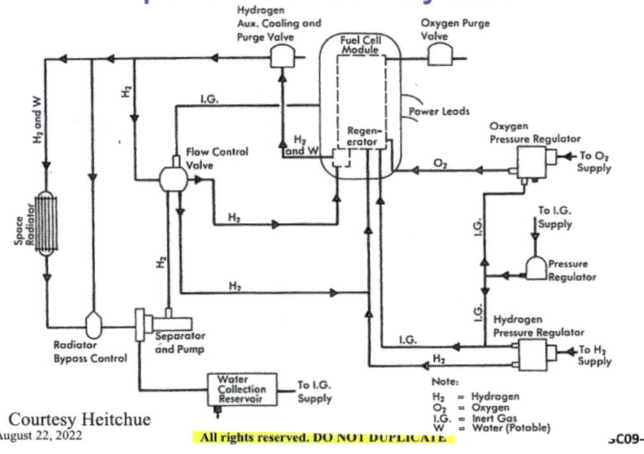

Courtesy Heitchue

August 22, 2022 SC09-37

Fuel Cell Info: Space Shuttle

- Reactor: 14" × 15" × 40", 255 lb_m
  28 VDC output; continuous power 7 kW, peak power 12 kW
- LOx tank is 36.8 in dia.
  empty weight 201 lb_m, capacity 781 lb_m
- LH_2 tank is 45.5 in dia.
  empty weight 216 lb_m, capacity 92 lb_m
- Nominal consumables usage:
  4 lb_m/h LOx, 0.6 lb_m/h LH_2 at 220 A
- 0.339 kg/kWh or 2950 Wh/kg reactants

August 22, 2022 SC09-38

Spacecraft Power Topics

- Power system elements, definitions, & power required
- Sources of power
  - Chemical cells / batteries
    - Definitions, chemistries
    - Series vs. parallel, strings
    - Battery design & life estimation, sizing example
  - Fuel cells
  - **Photovoltaic / solar cells**
    - Considerations: power & mechanical
    - Deployment
    - Losses & efficiency, sizing
  - Solar Dynamic
  - Nuclear thermal: RTG
  - Nuclear reactors
- Power control systems
- Power allocation to subsystems, operational modes
- Two spacecraft riddles
- Mass estimation

August 22, 2022 SC09-39

Photovoltaic Power Generation: Convert Light to Electricity! (sample)

August 22, 2022 SC09-40

Triple-Junction Compound Solar Cell Composition (Photovoltaic = *PV*)

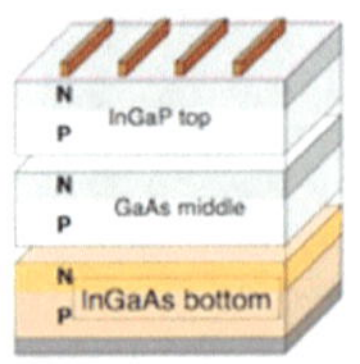

InGaP: Indium Gallium Phosphide
GaAs: Gallium Arsenide
InGaAs: Indium Gallium Arsenide

Courtesy Sharp

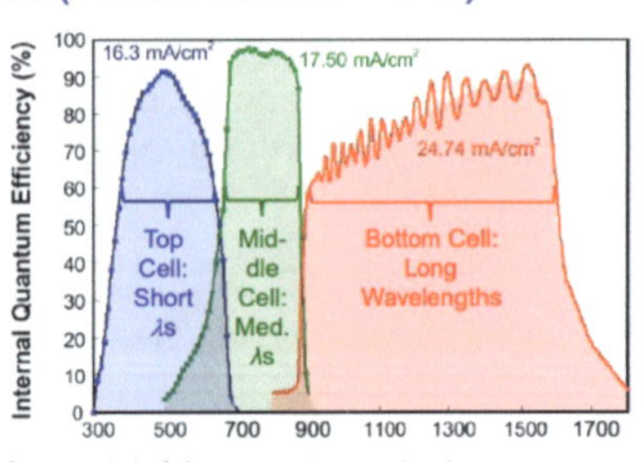

Spectrolab 3J **Wavelength (nm)**

August 22, 2022 SC09-41

Typical Solar Array (SA) Construction

Ref: Aerospace *Crosslink*

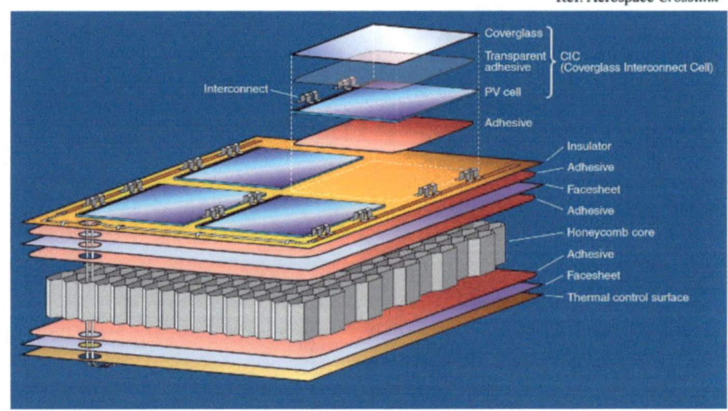

August 22, 2022 SC09-42

Spacecraft Power Topics

- Power system elements, definitions, & power required
- Sources of power
  - Chemical cells / batteries
    - Definitions, chemistries
    - Series vs. parallel, strings
    - Battery design & life estimation, sizing example
  - Fuel cells
  - Photovoltaic / solar cells
    - **Considerations: power & mechanical**
    - Deployment
    - Losses & efficiency, sizing
  - Solar Dynamic
  - Nuclear thermal: RTG
  - Nuclear reactors
- Power control systems
- Power allocation to subsystems, operational modes
- Two spacecraft riddles
- Mass estimation

August 22, 2022 SC09-43

Solar Power Considerations

- Solar intensity AKA "insolation" = incident solar radiation varies like $1/r^2$: less with more distance
- Solar cell power now effective to ~ Jupiter's distance from sun (2011 Juno mission, courtesy NASA)
- Sun powers photovoltaic, solar dynamic

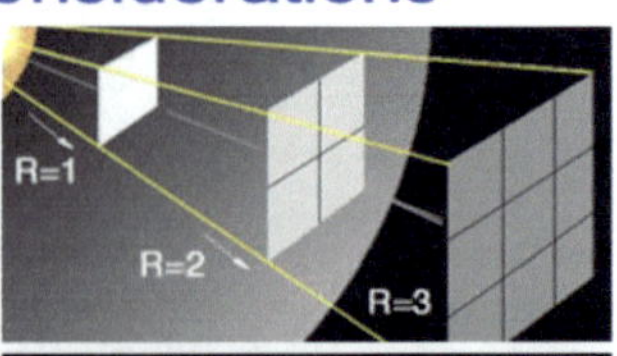

August 22, 2022 SC09-44

Solar Array Mechanical Issues

- Deployment: body-mount vs flexible vs. fold-out vs. roll-out
- Launch locks (for deployables)
- Sun tracking (one or two degrees of freedom, rotary joint or joints needed)
- Power transfer over moving joint
- Mass & stiffness

Body Mounted Arrays

- Used with spinners
- Array mounted on outside surface
- *Advantages*:
  1. Lighter
  2. Back side of cells shielded
  3. No deployment mechanism
  4. Smaller thermal cycle magnitude
- *Disadvantages*
  1. Cells' illumination angles rotate reducing output to $\sim 1/\pi = 32\%$ loss of total area
  2. Area limited to body surface

Spacecraft Power Topics

- Power system elements, definitions, & power required
- Sources of power
  - Chemical cells / batteries
    - Definitions, chemistries
    - Series vs. parallel, strings
    - Battery design & life estimation, sizing example
  - Fuel cells
  - Photovoltaic / solar cells
    - Considerations: power & mechanical
    - **Deployment**
    - Losses & efficiency, sizing
  - Solar Dynamic
  - Nuclear thermal: RTG
  - Nuclear reactors
- Power control systems
- Power allocation to subsystems, operational modes
- Two spacecraft riddles
- Mass estimation

Flexible SA Concepts 1, 2&3, 4, 5

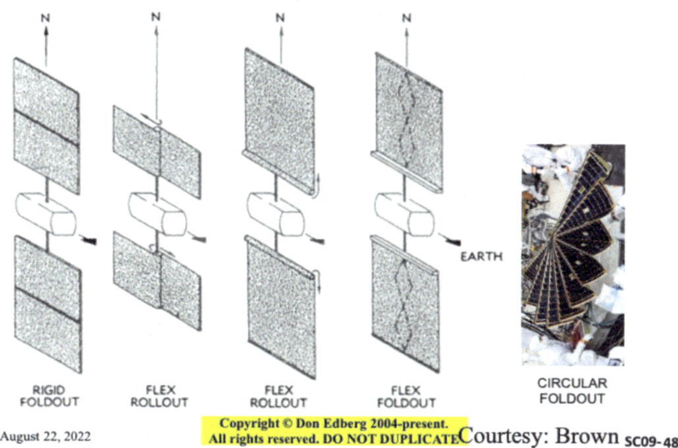

Courtesy: Brown

Solar Array Hinge & Flex Cable

Courtesy: Conley

Rigid Planar Array – Fold-Up Concept 1

- *Advantages*: active pointing toward sun, no area constraints
- *Disadvantages*: deployment mechanism, articulation, power transfer through articulation mechanism

Movie: `SMOS_SA_Deployment.mpg`

Flexible Rolled Solar Array (FRSA) Example: Hubble Space Telescope, Concept 2

- *Advantages*: minimal substrate weight, flexible stowage
- *Disadvantages*: reliability, original HST arrays needed replacement

Hubble FRSA

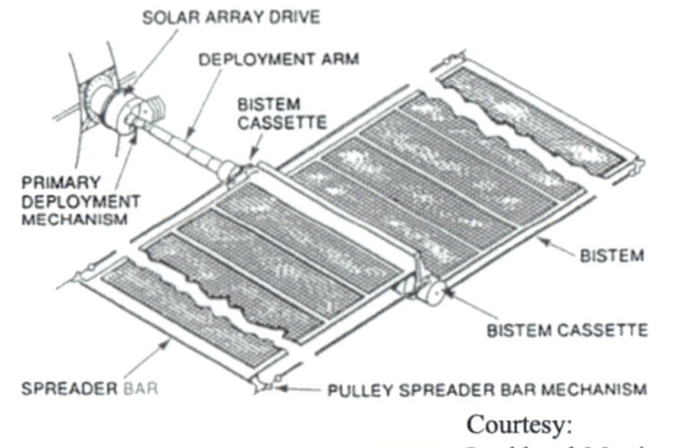

Courtesy: Lockheed-Martin, via Brown

ISS Folding Solar Array, Concept 4: (note extendable boom in center)

Courtesy: NASA

Rotational Folding Solar Array, Concept 5

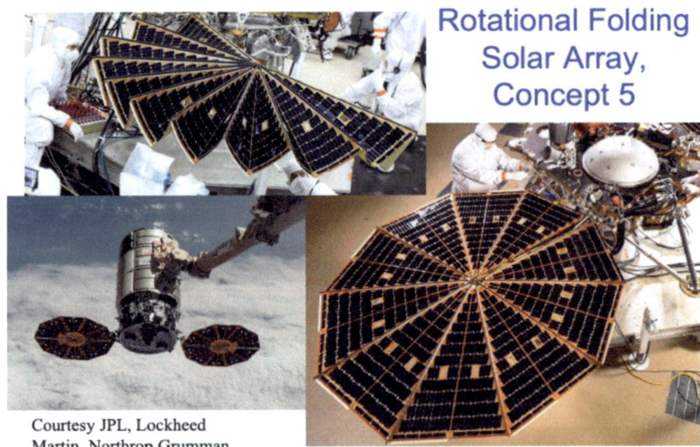

Courtesy JPL, Lockheed Martin, Northrop Grumman

Spacecraft Power Topics

- Power system elements, definitions, & power required
- Sources of power
  - Chemical cells / batteries
    - Definitions, chemistries
    - Series vs. parallel, strings
    - Battery design & life estimation, sizing example
  - Fuel cells
  - Photovoltaic / solar cells
    - Considerations: power & mechanical
    - Deployment
    - Losses & efficiency, sizing
  - Solar Dynamic
  - Nuclear thermal: RTG
  - Nuclear reactors
- Power control systems
- Power allocation to subsystems, operational modes
- Two spacecraft riddles
- Mass estimation

Solar/PV Cells, Modules, & Arrays: Design & Sizing

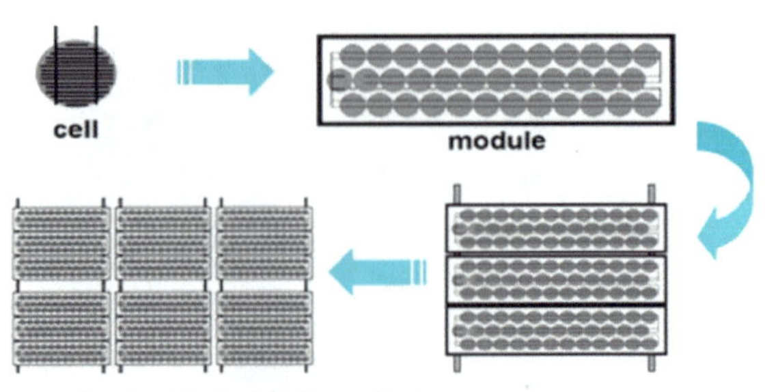

Courtesy: Florida Solar Energy Center

Solar Array Power/Sizing Requirements

Solar arrays must provide:

- Adequate power to charge batteries enough to run SC in eclipse
- Sufficient power to cover all losses:
  - Transfer efficiencies
  - Charging efficiency
  - Battery transfer efficiency
  - Operating efficiency (multiple factors)

Ideal Energy Balance (no losses)

$$P_{solar\ array}\, T_{day} = P_{SC\text{-}day}\, T_{day} + P_{SC\text{-}eclipse}\, T_{eclipse}$$

Energy coming in from the solar array during the day … … must cover both the daytime energy used AND the *eclipse* / night-time energy used.

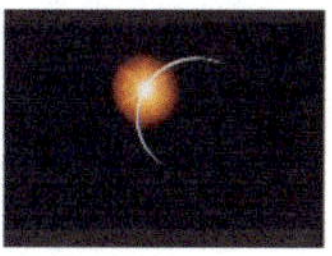

Real Energy Balance: Losses!

With **losses**, it's more complex: use *this* relation:

$$P_{solar\ array} = \frac{P_{SC-eclipse}T_{eclipse}}{X_{A-B}X_{B-L}T_{day}} + \frac{P_{SC-day}}{X_{A-L}}$$

where

$P_{solar\ array} = P_{SA}$ = solar array power output
$P_{SC\text{-}day}$ = spacecraft power, day
$P_{SC\text{-}eclipse}$ = spacecraft power, eclipse
T_{day}, $T_{eclipse}$ = Day & night durations of orbit
$X_{A\text{-}B}$ = Solar **A**rray to **B**attery transfer efficiency
$X_{B\text{-}L}$ = **B**attery to night **L**oads transfer efficiency
$X_{A\text{-}L}$ = Solar **A**rray to day **L**oads transfer efficiency

Power Transmission Efficiencies

System Type	$X_{A\text{-}B}X_{B\text{-}L}$: Solar array through batteries to night/eclipse loads transfer efficiency	$X_{A\text{-}L}$: Solar array to day loads transfer efficiency
Direct Energy Transfer (DET)	0.65	0.85
Peak-Power Tracking (PPT)	0.60	0.80

Differences between DET and PPT are due to the fact that PPT requires a power converter between the SAs and the loads (courtesy SMAD)

Efficiency Drops With *Higher* Temp.

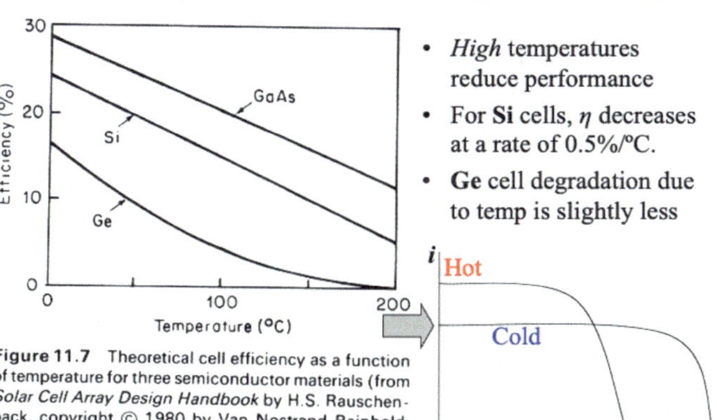

- *High* temperatures reduce performance
- For **Si** cells, η decreases at a rate of 0.5%/°C.
- **Ge** cell degradation due to temp is slightly less

Figure 11.7 Theoretical cell efficiency as a function of temperature for three semiconductor materials (from *Solar Cell Array Design Handbook* by H.S. Rauschenback, copyright © 1980 by Van Nostrand Reinhold. All rights reserved)

Efficiency *Climbs* With *Lower* Temp.

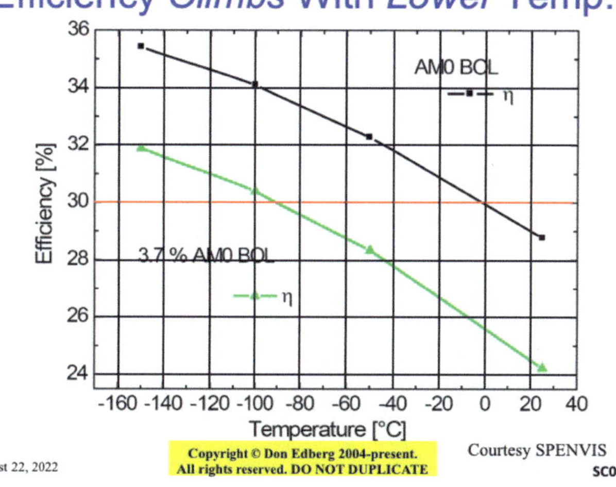

Courtesy SPENVIS

Efficiency *Climbs* With *Lower* Temp.

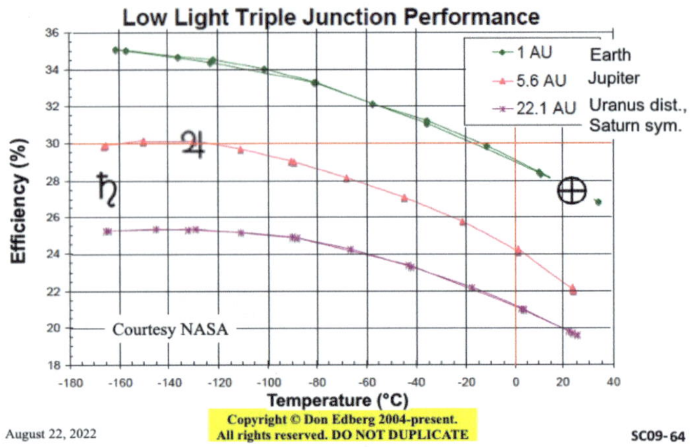

Solar Cell Power

i
Short Circuit Current, 0 volts
Max Power Point (MPP)
$P = i \times V$
V
Open Circuit Voltage, 0 amps

- ***i-V*** curve illustrates current-voltage behavior
- Cell efficiency $\eta = \frac{P_{out}}{P_{in}} = \frac{\text{output power}}{\text{incident power}}$
- Peak power occurs at max product of iV [current × voltage]

Power from Sun

- Solar power depends on inverse square of distance from sun
- Common measurement is the Astronomical Unit A.U., the mean Earth-Sun distance
- Use solar distance rating factor η_l:

$$\eta_l = \left(\frac{\text{dist. from sun in A.U.}}{1\ \text{A.U.}}\right)^2$$

- Longer distances = less power from sun and vice versa; Mars was considered limit for solar power *until Juno probe to Jupiter*

PV Power vs. Sun Angle

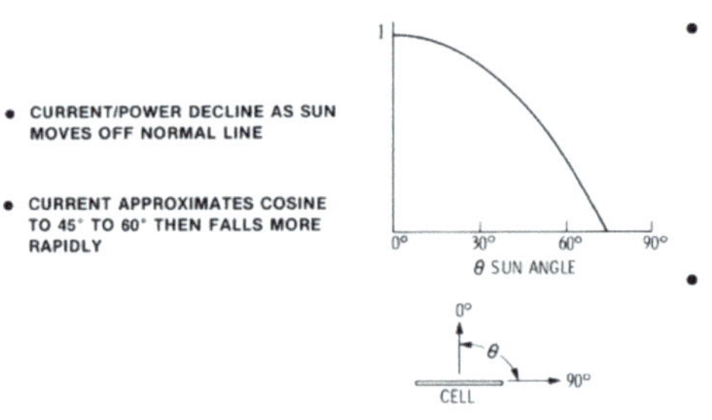

Fig. 10.10 Effect of sun angle on solar array power.

- Power ~ cos θ (θ = angle with sun, drops faster when $\theta > 45°$)
- 5° error still provides 99.6% of max

Solar Cell Degradation From Radiation

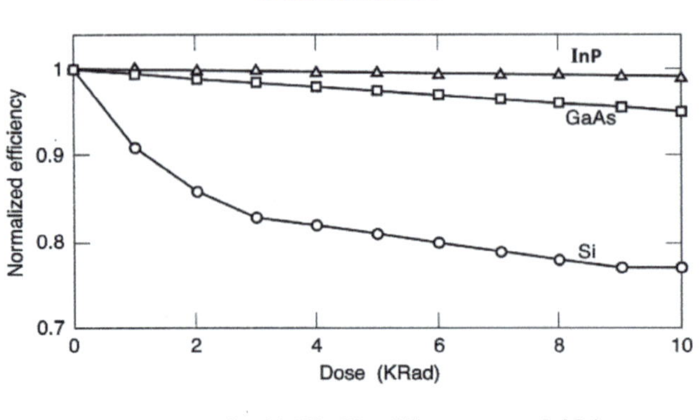

Ref: Rainey

Other PV Loss Sources

Output also affected by

- UV discoloration
- Dust, outgassing, contamination, etc.
- Duration of use (thermal cycling?), loss of
  - 3.75%/year, Si cells
  - 2.75%/year, Ga-As cells
- Micrometeoroid hits
- Packing Factor

Solar Array Packing Factor

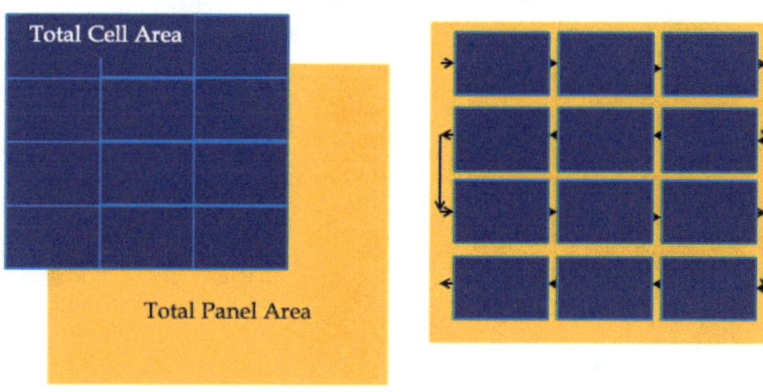

Packing Factor = Total Cell Area ÷ Total Panel Area, usually ~80 – 90%

Solar Cell Concentrators

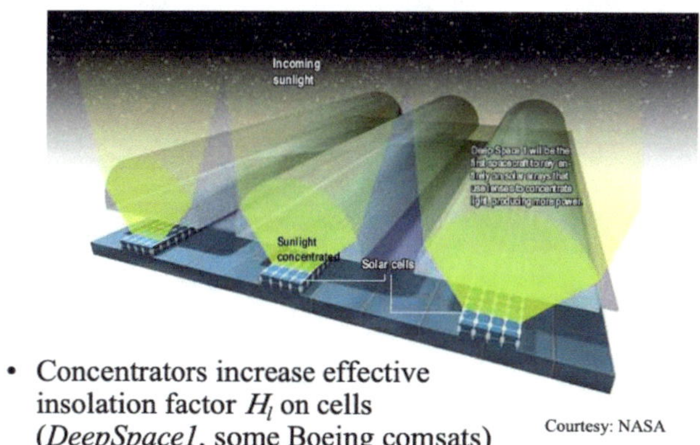

- Concentrators increase effective insolation factor H_l on cells (*DeepSpace1*, some Boeing comsats)

Courtesy: NASA

Solar Array Sizing Example (1 of 5)

- Solar array required power P_{SA} = 1,200 W EOL (end-of-life) power at 1 AU, Peak temp 115 °C
- 76 mm × 37 mm *Spectrolab* $GaInP_2$/GaAs/Ge Dual Junction Solar Cells
- $\eta = 0.215$, or 21.5% efficiency, cell temp = 28 °C

$$P_{ideal\ cell} = \begin{pmatrix}\text{solar}\\ \text{flux}\end{pmatrix} \times \begin{pmatrix}\text{dist.}\\ \text{correct.}\end{pmatrix} \times \begin{pmatrix}\text{cell}\\ \text{area}\end{pmatrix} \times \begin{pmatrix}\text{cell}\\ \text{efficiency}\end{pmatrix}$$

$$= G_s H_l A_{cell} \eta$$

$$= 1370\ \frac{\text{W}}{\text{m}^2} \times 1 \times 0.002812\ \frac{\text{m}^2}{\text{cell}} \times 0.215$$

$$= 0.83\ \frac{\text{W}}{\text{cell}}\ (\text{@ 1 AU from sun, 28 °C})$$

SA Sizing Example (2 of 5)

At Earth = 1 AU, so $\eta_l = 1.0$

Other assumed losses:

- 92% packing factor $\eta_p = 0.92$
- 30% loss due to radiation effects ($\eta_{rad} = 0.7$)
- Temperature loss = $\frac{0.5\%}{°\text{C}}(115 - 28)$°C = 43.5%. Remaining = 100–43.5=56.5%; efficiency $\eta_t = 0.565$
- UV loss creates efficiency $\eta_{UV} = 0.98$
- Thermal cycling loss creates $\eta_{cy} = 0.99$
- Losses due to cell mismatch $\eta_m = 0.975$
- Losses due to resistance in interconnects $\eta_i = 0.98$
- Losses due to contamination $\eta_{con} = 0.99$

SA Sizing Example (3 of 5)

- $P_{cell} = P_{ideal}\ (\eta_l\ \eta_{rad}\ \eta_{UV}\ \eta_{cy}\ \eta_m\ \eta_i\ \eta_{con}\ \eta_t)$
  $= 0.83$ W/cell × (1.0 × 0.70 × 0.98 × 0.99 × 0.975 × 0.98 × 0.99 × 0.565)

$$P_{cell} = 0.3006\ \frac{\text{W}}{\text{cell}}$$

- Total # cells = $\frac{1{,}200\ \text{W}}{0.3006\ \text{W/cell}} = 3{,}992$ cells
- Cells/m² = $\frac{(1\ \text{m}^2)\eta_p}{A_{cell}} = \frac{(1\ \text{m}^2)0.92}{0.002812\ \text{m}^2} = 327$
- Total array area $A_{SA} = \frac{3{,}992\ \text{cells}}{327\ \text{cells/m}^2} = \mathbf{12.2\ m^2}$

Solar Array Electrical (4 of 5)

- Cell open circuit voltage ≈ 2.2 V
- Cell peak-power voltage ≈ 2.05 V
- Need voltage 1.1-1.15 × Li battery voltage to charge, so array has to deliver 1.1 to 1.15 × 28.8 V = 31.68 to 33.12 V
- # series cells = $\frac{33.12\ \text{V}}{2.05\ \text{V/cell}} = 16.15$ cells
- Approx. 16 series solar cells produce required charge voltage: # strings= $\frac{\#\text{cells}}{\#\text{cells/string}} = \frac{3{,}992}{16} =$ 249.5, **250 strings of 16 cells (16s250p)**

SA Sizing Example (5 of 5)

- Select geometry/deployment, calc. mass (estimation values given at end of section):
- Cannot use body mount, power too limited
- Deployable: use $\frac{4\ \text{kg}}{\text{m}^2}$ for rigid fold-out:
  - Panel mass = $12.2\ \text{m}^2 \times \frac{4\ \text{kg}}{\text{m}^2} = 48.8\ \text{kg}$
  - Attachment mass = 15% of 48.8 kg = 7.32 kg
  - Drive mechanism mass (Larson & Pranke, in back): $m_{drive} = (-0.014\ m_{array} + 20.6)m_{array}/100 = 9.72\ \text{kg}$
- Total mass = 65.84 kg

Solar Dynamic

- Sun provides thermal energy
- Drives Brayton, Rankine, or Stirling engines which in turn drive electrical generator
- Attractive at higher power (>100 kW)

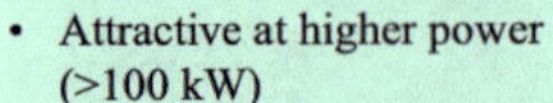

- Reduced reliability (moving parts)
- Vibration?
- Waste heat dissipation
- Possible ACS issues

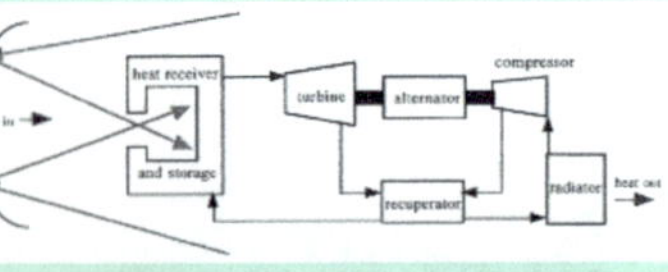

Nuclear Thermal Power: Radioisotope Thermal Generator

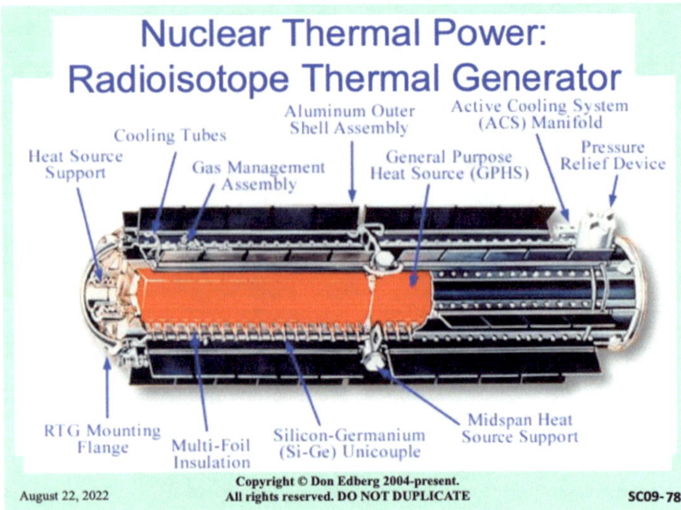

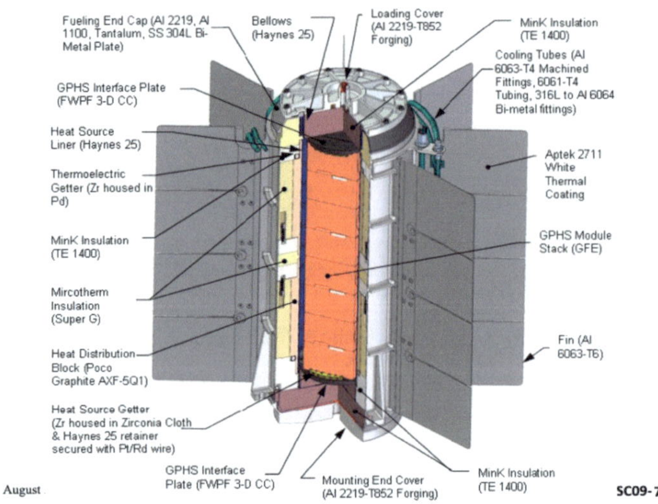

Radioisotope Thermal Generator

- Radioactive decay produces heat; thermoelectric effect converts directly into electricity (~7% efficient)

Pu^{238} glows red hot: 65 W

- Provides continuous long-term power
- Power level drops due to half-life of material used (e.g. Sr^{90} half-life ~ 28 years)
- Thermoelectric elements degrade with age

RTG Comments

- RTGs eliminate need for Sun on distant missions
- RTGs are environmentally "handicapped"
- RTGs are costly & difficult to handle
- RTG radiation affects instruments and electronics, keep as far away as possible
- Plutonium shortage
- *Only use if solar will not suffice*

Courtesy space.com

Heat Source Pellet Details

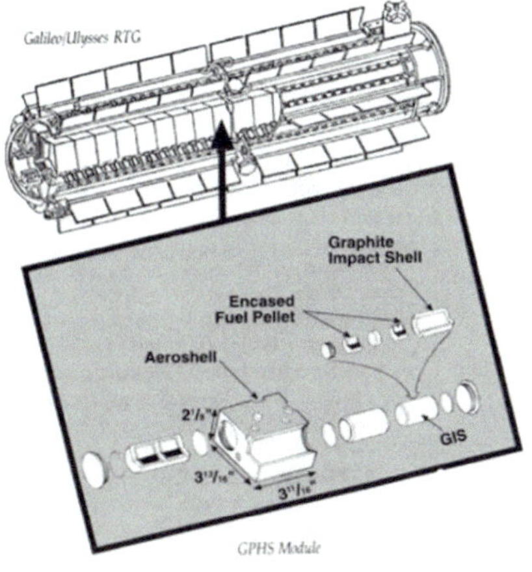

Each of the RTG's 18 modules contains four Pu-238 fuel pellets enclosed in three layers of protection - - the metal encasing the pellets, the graphite shell, and the aeroshell.

How A Thermoelectric Device Produces Electricity

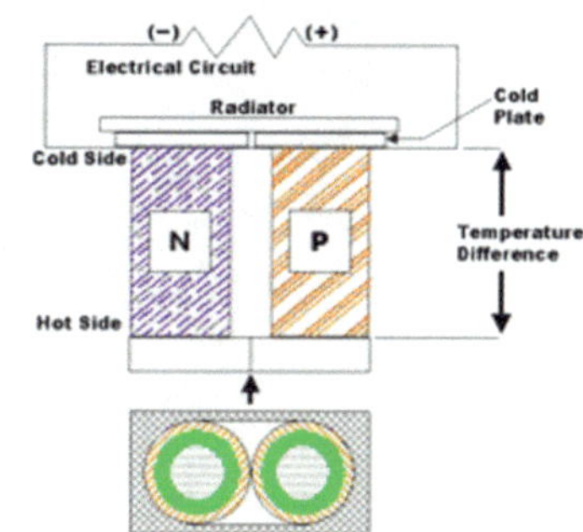

- Thermoelectric unicouple is a semi-conductor device with "N" and "P" type material in legs
- Heat applied at hot junction and cooling side produces electrical potential difference between materials ("Seebeck Effect")
- Connecting cold side terminals through a resistive load causes current to flow in electrical circuit

RTG Cross-Section

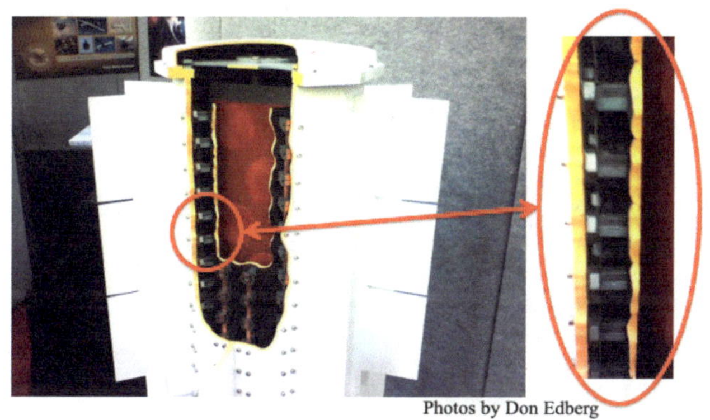

Photos by Don Edberg

SNAP 19 RTG

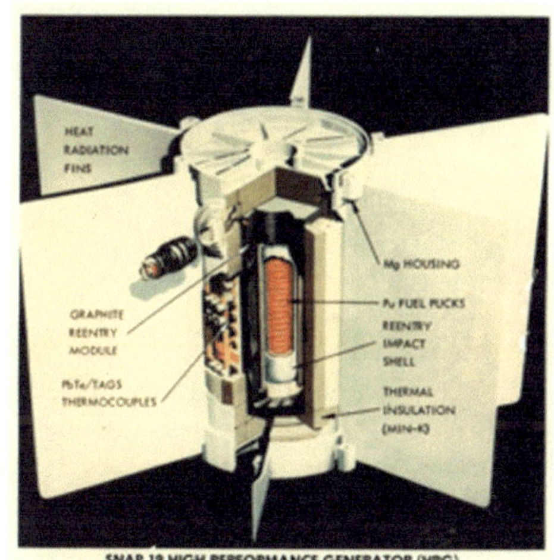

SNAP 19 HIGH PERFORMANCE GENERATOR (HPG)

Advanced Stirling Radioisotope Generator: ASRG (NASA program cancelled Dec. 2013)

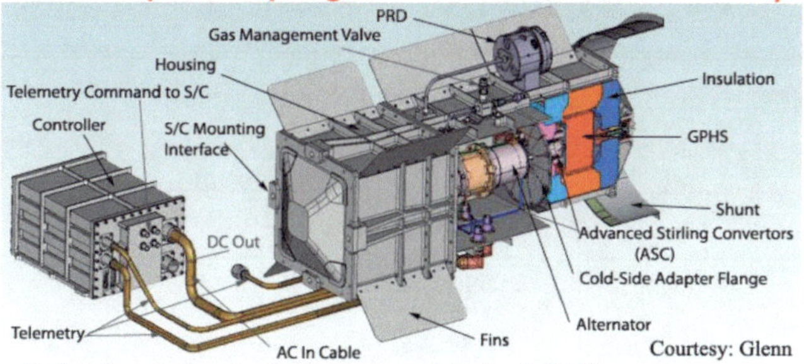

Courtesy: Glenn

- Reduced mass of nuclear heat source, combined with Stirling engine
- Higher conversion efficiency compared to RTG: 26% vs. ~6%
- Power output 130 W BOL, Mass 32 kg
- Still in development. NASA cancellation: //futureplanets.blogspot.com/2013/12/the-asrg-cancellation-in-context.html

Future RTG Info (email 09/2018)

Name	Power/Unit @ Launch (W)	Mass (kg)	Cost (2018 USD)	Expected Service	Used On
MMRTG	110	45	$77M/1 unit; $94M /2 units; $117M/3 units	Current	Curiosity, M2020
eMMRTG	150	44	$77M/1 unit; $94M /2 units; $117M/3 units	2024	
Next-Gen RTG	400-500	62	TBD	2028	

Courtesy A. Johnson, JPL

MMRTG Info

Multi-Mission Radioisotope Thermoelectric Generator
(expanded view)
GPHS (General Purpose Heat Source) module
Stack of eight GPHS modules
Thermoelectric module (thermocouples)
Radiator fins

Courtesy Quora

NASA, DOE May Face Plutonium Supply Shortage (SPACE 10/13/17)

- Oct. 4 2017 Government Accountability Office (GAO) report found that the long-term supply for an isotope used to power space missions may be in jeopardy: "… while there is sufficient plutonium-238 in stockpiles now for missions planned through the mid-2020s, scaling up production of the isotope faces a number of technical issues."
- In Congressional testimony, GAO Director of Acquisition & Sourcing Management Shelby Oakley said that the DOE "faces challenges in hiring and training the necessary workforce, perfecting and scaling up chemical processing, and ensuring the availability of reactors that must be addressed or its ability to meet NASA's needs could be jeopardized."
- Oakley said that the DOE plan lacked concrete steps toward the goal of producing 1.5 kilograms of plutonium-238 a year by 2025
- Mars 2020 rover and the New Frontiers Mission anticipated to launch in 2025 require the isotope.
- The administration's recently announced focus on lunar exploration may create additional demand and outstrip current stockpiles.

Nuclear Reactors: The "Other" Nuclear Power Source

- Nuclear reactors provide very large power levels, up to megawatts
- Reaction generates heat which is converted to electricity
  - Thermoelectric
  - Rankine
  - Thermionics: heat converted to electricity by boiling e^- (electrons) from cathode and collecting at cooler anode.
    Cathode ~ 2,000 K, Anode ~ 1,000 K, η ~ 15%

Nuclear Reactors

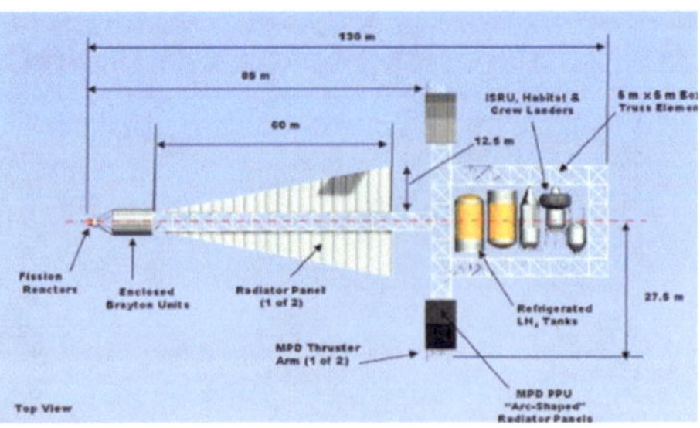

- Large radiators to reject heat required
- Heavy shielding required to protect spacecraft
  - Additional shielding for manned systems
  - Separation can reduce shielding (mass) requirement
- Usually considered for interplanetary missions where large power is required
- Although being considered for some current programs, political constraints limit its use

Courtesy nextbigfuture.com

Spacecraft Power Topics

- Power system elements, definitions, & power required
- Sources of power
  - Chemical cells / batteries
    - Definitions, chemistries
    - Series vs. parallel, strings
    - Battery design & life estimation, sizing example
  - Fuel cells
  - Photovoltaic / solar cells
    - Considerations: power & mechanical
    - Deployment
    - Losses & efficiency, sizing
  - Solar Dynamic
  - Nuclear thermal: RTG
  - Nuclear reactors
- Power control systems
- Power allocation to subsystems, operational modes
- Two spacecraft riddles
- Mass estimation

SC Power Control System (PCS) Principal Functions

Power Source
Power Distribution, Control, & Main Bus Protection
Energy Storage Control
Energy Storage (Batteries)
Loads

- Regulate voltage, isolate noise
- AC and/or DC conversion
- Control battery charge-discharge cycle
- Control solar array output in response to load
- Switch equipment on and off
- Keep memory alive
- Protect subsystems from power source faults, i.e. overcurrent protection
- Fire pyros

Power System Design Procedures

1. Establish power requirements from payload, SC config., & mission orbit/requirements
2. Select power source (solar, RTG, etc.) based on mission & SC type
3. Size power source (solar array, RTG, etc.)
4. Size battery system for eclipse
5. Determine power distribution system
6. Prepare mass & power estimates
7. Conduct trades; refine if needed

August 22, 2022

SC09-94

Dissipative Power Control Systems for Solar Arrays

- *Dissipative systems*, also called direct energy transfer systems (DET), shed power resistively
- Shunt regulator in parallel with SA, simple design

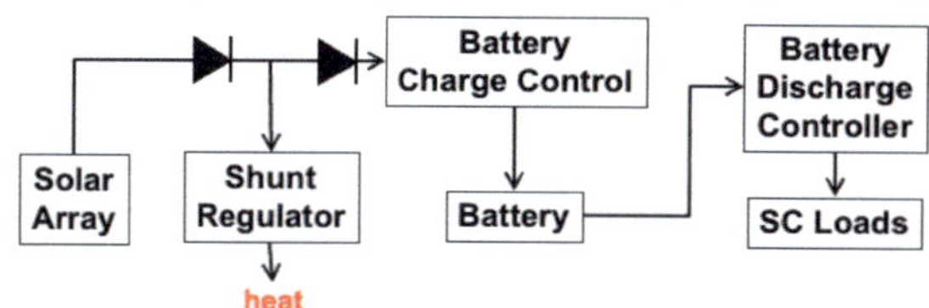

Courtesy: Griffin

August 22, 2022

SC09-95

Non-Dissipative Power Control Systems for Solar Arrays

- Also called peak power tracking systems (PPT)
- Array regulated through DC-DC converter
- Operates at its peak point based on required load
- As loads decrease, output shifts toward open circuit to yield lower current (and conversely)

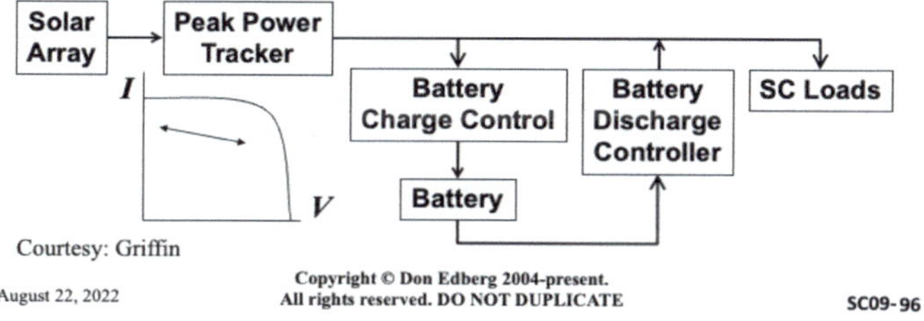

Courtesy: Griffin

August 22, 2022

SC09-96

Block Diagrams of Other Power Systems

DET
Solar Array
Shunt Reg.
Battery
Unregulated Bus with Direct Energy Transfer

DET
Solar Array
Shunt Reg.
Solar Array
Battery
Semi-Regulated Bus with Direct Energy Transfer

Solar Array
Peak Power Tracking
Battery
Unregulated Bus with Peak Power Tracking

DET
Solar Array
Shunt Reg.
Solar Array
Battery Charge Regulator
Battery Discharge Regulator
Hybrid
Unregulated
Battery
Regulated Bus with Direct Energy Transfer

August 22, 2022

Courtesy of Ley, et al.

SC09-97

What Are The ▶| Symbols?

- This symbol represents a *diode*
- Diodes allow flow of electrical current only one direction
  - Current can flow in the arrow direction, not against arrow
- Similar to a one-way or "flapper" valve in plumbing
- Diodes prevent battery discharging through SAs when not producing power (called "back bias"), or other circuitry

August 22, 2022

SC09-98

Typical CubeSat Power System

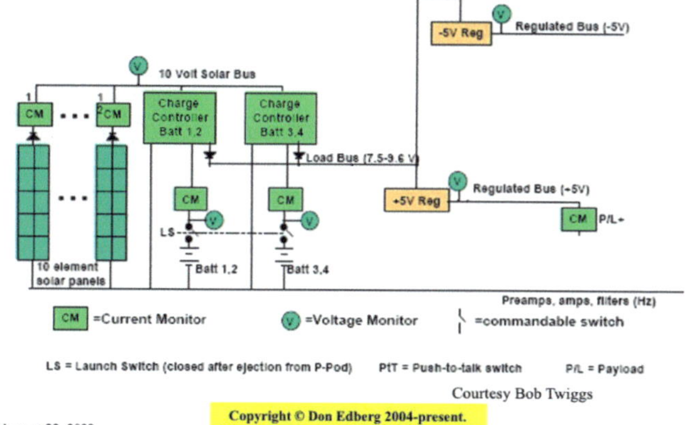

Courtesy Bob Twiggs

August 22, 2022

SC09-99

Spacecraft Power Topics

- Power system elements, definitions, & power required
- Sources of power
  - Chemical cells / batteries
    - Definitions, chemistries
    - Series vs. parallel, strings
    - Battery design & life estimation, sizing example
  - Fuel cells
  - Photovoltaic / solar cells
    - Considerations: power & mechanical
    - Deployment
    - Losses & efficiency, sizing
  - Solar Dynamic
  - Nuclear thermal: RTG
  - Nuclear reactors
- Power control systems
- **Power allocation to subsystems, operational modes**
- Two spacecraft riddles
- Mass estimation

Power Allocation

Initial power allocations from statistical studies of many conventional spacecraft

Subsystem	Comsats	Metsats	Planetary	Other
Thermal Control	30%	48%	28%	33%
Attitude Control	28	19	20	11
Power	16	5	10	2
Command & Data	19	13	17	15
Communication	0 (incl. in C&DS)	15	23	30
Propulsion	7	0	1	4
Mechanisms	0	0	1	5

(From Brown)

Establish Power Requirements For Each Operational Mode

- Launch
- Post separation/cruise
- In-sun or daytime
- Eclipse or nighttime
- Safe mode
- Etc.

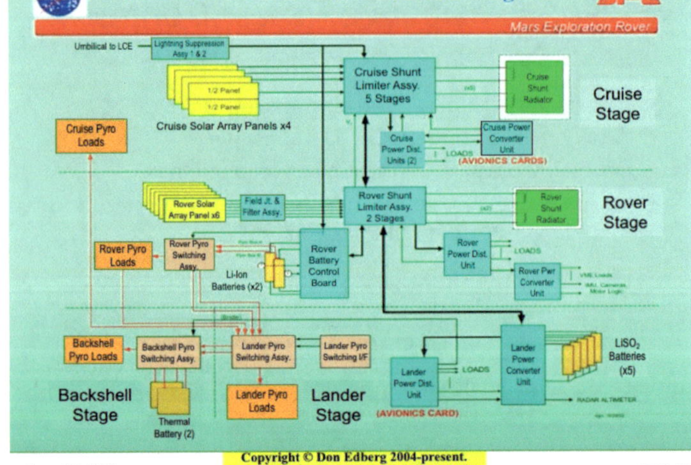

Magellan Power: Day vs. Night Ops

Characteristic	Cruise (near Earth)	Venus Mapping, Full Sun (daytime)*	Venus Playback, Eclipse (nighttime†)
Time in Mode (min)	—	37.2	19.5
Attitude Control (W)	138.8	159.8	142.8
Command & Data (W)	47.96	63.41	65.23
Power System (W)	42.7	50.18	53.68
Propulsion (W)	0.76	7.96	5.74
Thermal Control (W)	313.0	**126.58**	**239.75**
Communication (W)	43.8	61.80	110.90
Payload (W)	0.0	**248.78**	**18.54**
Total Power to Loads (W)	587.02	718.51	636.64
Total Wire Loss (W)	71.25	40.99	34.23
Total Power Required	**658.27**	**759.50**	**670.87**

†During nighttime, thermal power must be increased for additional heater power; *a factor of 2 is a good initial estimate.* Other systems remain near daytime levels

*Daytime allocations *must include battery charging power*

Courtesy: Brown

Spacecraft Power Topics

- Power system elements, definitions, & power required
- Sources of power
  - Chemical cells / batteries
    - Definitions, chemistries
    - Series vs. parallel, strings
    - Battery design & life estimation, sizing example
  - Fuel cells
  - Photovoltaic / solar cells
    - Considerations: power & mechanical
    - Deployment
    - Losses & efficiency, sizing
  - Solar Dynamic
  - Nuclear thermal: RTG
  - Nuclear reactors
- Power control systems
- Power allocation to subsystems, operational modes
- **Two spacecraft riddles**
- Mass estimation

Before We Finish, Two Spacecraft Riddles:

1. "What's critical to a high-technology space mission and yet is low technology?"
2. "What task is one of the last to be defined but must be one of the first delivered?"

"If you answered 'the cabling,' you're clearly an experienced spacecraft builder."
High Velocity Leadership: The *Mars Pathfinder* Approach to Faster, Better, Cheaper; Brian Muirhead & William Simon, HarperBusiness, 1999, ISBN 0887309747, p. 118

August 22, 2022 SC09-106

Don't Forget Cabling! Cabling mass is usually carried as subsystem level entry and includes Data and Power cabling.

Courtesy JPL

August 22, 2022 SC09-107

Spacecraft Power Topics

- Power system elements, definitions, & power required
- Sources of power
  - Chemical cells / batteries
    - Definitions, chemistries
    - Series vs. parallel, strings
    - Battery design & life estimation, sizing example
  - Fuel cells
  - Photovoltaic / solar cells
    - Considerations: power & mechanical
    - Deployment
    - Losses & efficiency, sizing
  - Solar Dynamic
  - Nuclear thermal: RTG
  - Nuclear reactors
- Power control systems
- Power allocation to subsystems, operational modes
- Two spacecraft riddles
- **Mass estimation**

August 22, 2022 SC09-108

Power System/Solar Array Mass Estimation

Type	Body-fixed / Spinner	Rigid fold-out	Flexible fold-out	Flex Rolled Up (FRUSA)
SA mass	3.4 kg/m^2	4.0 kg/m^2	3.6 kg/m^2	1.9 kg/m^2
SA Deploy & drive mechanism mass	—	$m_{drive} = (-0.014\, m_{array} + 20.6) m_{array}/100$ Courtesy Larson & Pranke $m_{deploy+1DoF\ mech.} = 15$ kg/panel (Brown Table 6.21) *Magellan*: deployment hinge + 1-axis rotation mech. = 13.8 kg/panel (Brown p. 346)		
Bus attach	—	15% of array		
Control Electronics	0.071 kg × (end of life power in watts)			
Cabling (includes data & power)	k × On-Orbit Dry Mass (k = 0.045 for Comsats; k = 0.068 for all others)			

Table based on Brown

August 22, 2022 SC09-109

Spacecraft Telecommunications & GN&C

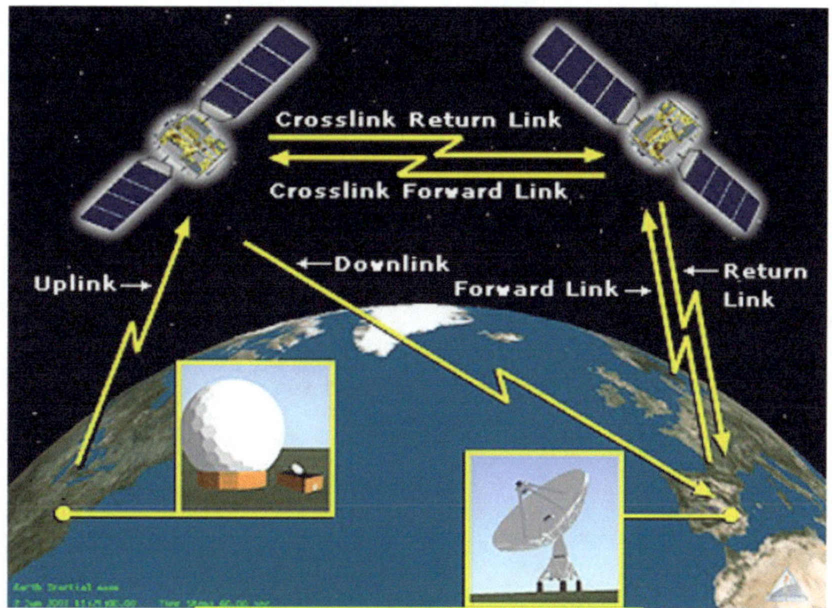

Courtesy AGI

August 23, 2022

SC10- 1

Telecommunications Topics

- Telecommunications Basics: Objectives and Definitions
- Long-range communication & the Link equation
- Antennas & performance
- Ground stations & weak signals
- Deep Space Considerations
- Ranging & Navigation

August 23, 2022

SC10- 2

Spacecraft Design Process

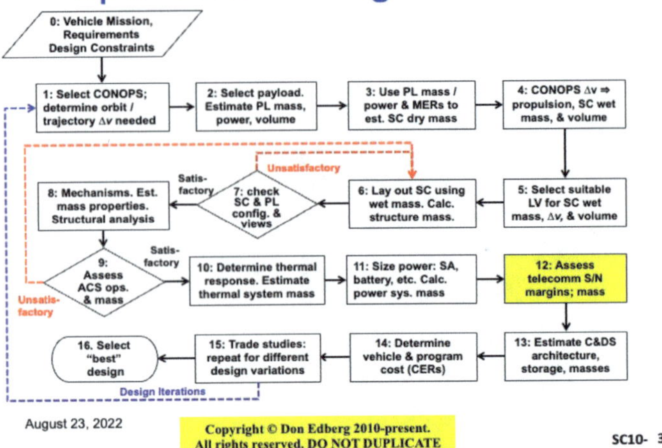

0: Vehicle Mission, Requirements Design Constraints
1: Select CONOPS; determine orbit / trajectory Δv needed
2: Select payload. Estimate PL mass, power, volume
3: Use PL mass / power & MERs to est. SC dry mass
4: CONOPS Δv ⇒ propulsion, SC wet mass, & volume
5: Select suitable LV for SC wet mass, Δv, & volume
6: Lay out SC using wet mass. Calc. structure mass.
7: check SC & PL config. & views
Unsatisfactory
Satis-factory
8: Mechanisms. Est. mass properties. Structural analysis
9: Assess ACS ops. & mass
Satis-factory
Unsatis-factory
10: Determine thermal response. Estimate thermal system mass
11: Size power: SA, battery, etc. Calc. power sys. mass
12: Assess telecomm S/N margins; mass
13: Estimate C&DS architecture, storage, masses
14: Determine vehicle & program cost (CERs)
15: Trade studies: repeat for different design variations
16. Select "best" design
Design Iterations

August 23, 2022 SC10- 3

Telecommunications Topics

- Telecommunications Basics: Objectives and Definitions
- Long-range communication & the Link equation
- Antennas & performance
- Ground stations & weak signals
- Deep Space Considerations
- Navigation & Ranging

August 23, 2022 SC10- 4

Objectives of Telecommunications

- Return data collected by spacecraft
- Provide SC system status & health
- Determine SC position & velocity
- Send SC commands from ground; control the various SC systems

August 23, 2022 SC10- 5

Telecomm Definitions

- *TM* = Telemetry = remote reception of spacecraft information
- *TT&C* = Telemetry, Tracking & Control
- *TC&R* = Telemetry, Command & Ranging
- *Ranging/Tracking* = Measure distance to spacecraft & spacecraft velocity
- *Downlink* = data coming DOWN from SC
- *Uplink* = data sent UP to SC
- *Crosslink* = data sent from one SC to another SC
- *Xmit* = transmit, *Xmitter* = transmitter, etc.

August 23, 2022 SC10- 6

Still More Definitions

- *bER* = bit Error Rate = E_b
- *bPS* = bits Per Second (data rate)
- *EDAC* = Error Detection and Correction
- *MbPS* = Million bits Per Sec (data rate)
- *SNR* = Signal-to-Noise Ratio
- *Hz* = hertz = 1 cycle or operation/second
- *MHz* = megahertz = 1 million cycles or operations/second (M = mega = 10^6)
- *GHz* = gigahertz = 1 billion cycles or ops per second (G = giga = 10^9)

August 23, 2022 SC10- 7

Antenna Definitions

- λ = wavelength, m or cm or mm
- *Gain* = amplification of system = output/input. Tends to increase with diameter or length.
- *HGA* = High-gain antenna, usually very directional
- *LGA* = Low gain antenna, usually not very directional
- *Isotropic* = equal radiation, all directions
- *EIRP* = *Effective Isotropic Radiated Power* is a measure of power density receiver gets EIRP = transmit power × antenna gain

August 23, 2022 SC10- 8

Decibels

- Used to describe *ratios*
- Based on "powers of ten" of a ratio $\frac{Q_2}{Q_1}$

  Decibels: $\text{dB} = 10\log_{10}\left(\frac{Q_2}{Q_1}\right)$

Example:

- $Q_2 = 100$ and $Q_1 = 10$,
  $\text{dB} = 10\log_{10}\left(\frac{100}{10}\right) = 10\log_{10}(10) = 10$
- Since $\log_{10}(2) \approx 0.3$,
  doubling $\approx +3$ dB
  (and halving ≈ -3 dB)

dBm, dBW Values Based on Reference Level

dBm = Power referenced to 1 mW

dBm Level	Power
0 dBm	1 mW
–3 dBm	0.5 mW
–10 dBm	0.1 mW
–20 dBm	0.01 mW or 10 μW

Received power levels are often in the picowatt = 10^{-12} watt range! *Galileo* transmitted 15-20 W; received at Earth, power = –170 dBm = 10 zeptowatts (10×10^{-21} watts!)

dBW = Power referenced to 1 watt

dBW level	Power
0 dBW	1 W
3 dBW	2 W
10 dBW	10 W
20 dBW	100 W
30 dBW	1 kW
50 dBW	100 kW

Transmit power levels are often in the 100 kW range or more.

Telecommunications Topics

- Telecommunications Basics: Objectives and Definitions
- Long-range communication & the Link equation
- Antennas & performance
- Ground stations & weak signals
- Deep Space Considerations
- Navigation & Ranging

Tele-command

Ground control commands spacecraft to:

- Initiate a process
- Adjust the SC altitude or attitude
- Switch transmitters on/off
- Upload new software
- *Whatever is needed to fix issues before they become disasters*

1981 *Oscar 9* AMSAT Incident

- *Oscar 9* AMSAT accidentally commanded to transmit on *both of* its frequencies
- Result: could not hear ground signals due to its on-board "shouting" transmitters
  - A power level of +64 dBW was needed to blast commands to the desensitized receiver on the SC
- Stanford's dish sent 71 dBW = 12.6 MW(!)
  - Turned off downlink
  - Communications restored

Long-Range Communication Considerations

Data/Message → Modulator → Transmitter → Receiver → Demodulator → Data/Message

Noise → Receiver

Transmission Time Issues

- Propagation speed of radio waves & light $\approx$ 300,000 km/s $\approx$ 186,000 mi/s = c
- For spacecraft at long range, transmission delays can be minutes or hours
  - Example: Mars one-way = 20 – 40 min
  - More for outer planets
- Autonomous operation and "safe modes" required

August 23, 2022

The Link Equation & Signal to Noise Ratio

- Error-free telecommunication requires a certain energy per bit (E_b)
- Signal-to-noise ratio: $SNR = \frac{S}{N} = \frac{E_b}{N_0}$
  (N_o is noise spectral density)
- We start by considering the *power flux density* F: the power P distributed over a sphere a distance R from the source: $F = \frac{P}{4\pi r^2}$

August 23, 2022

Signal-to-Noise Calculation (1 of 4)

- The antenna *gain* $G = 4\pi \frac{A\eta}{\lambda^2}$, where:
  A = physical area of antenna (m²)
  η = efficiency of antenna (about 0.55 for parabolic)
  λ = wavelength of signal (m)
- The product of the transmit power and the antenna gain is known as *effective isotropic radiated power*, EIRP
  EIRP = $P_t G_t$, where:
  P_t = Power output of transmitter (W)
  G_t = Gain of transmit antenna

August 23, 2022

S/N Ratio Calc. (2 of 4)

- The signal strength S received is the product of the power flux and the receiving area

$$S = \left(\frac{P_t G_t}{4\pi R^2}\right) A\eta$$

where R = distance between stations (m)

- Solving for A, substituting into the other eq.:

$$S = P_t G_t \left(\frac{\lambda}{4\pi R}\right)^2 G_r$$

Here G_r is the gain of the *receiving* antenna.

August 23, 2022

S/N Ratio Calc. (3 of 4)

- The term in parenthesis, $\frac{\lambda}{4\pi r^2}$, is known as the *path loss* L_p, and describes how the power drops as the signal spreads out with distance.
- The definition of *noise power*: $N = kT_s B$
  - k = Boltzmann constant, 1.38×10^{-23} J/K
  - T_s = Effective system temperature (lower better*)
  - B = receiving system bandwidth (Hz)
- Note: *lower system temperature means less noise.* **Why GPS & XM receivers take time to 'warm up': they're actually* ***cooling off!***

August 23, 2022

S/N Ratio Calc. (4 of 4)

- Bandwidth is the range of frequencies a receiver is designed to receive.
  Less bandwidth = less noise.
- Combining the noise & signal received eqs., the *signal-to-noise ratio S/N* is defined as:

$$\frac{S}{N} = \left(\frac{P_t G_t}{kB}\right)\left(\frac{\lambda}{4\pi R}\right)^2\left(\frac{G_r}{T_s}\right)$$

- Since the gain of the antennas $\sim \frac{1}{\lambda^2}$, increasing frequency = decreasing wavelength = increased *S/N* ratio: higher freqs preferred

August 23, 2022

Frequency Selection

- Higher frequency also means *more information transmission per unit time capability*
- Wavelength $\lambda = \frac{\text{speed of light}}{\text{frequency}} = \frac{c}{f}$
- Attenuation from traffic, weather/atmospherics
- Must be supported by ground station network

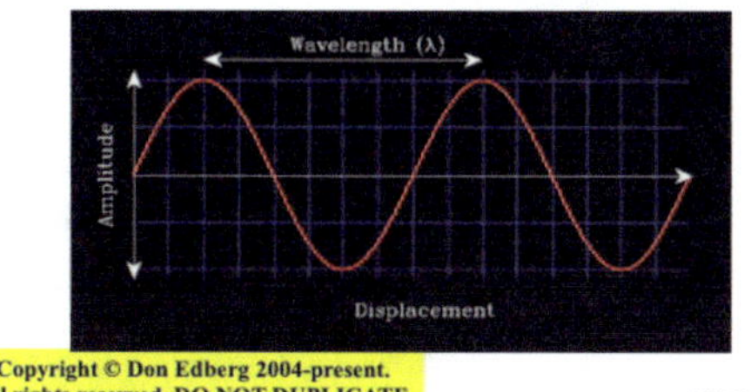

August 23, 2022

SC10-21

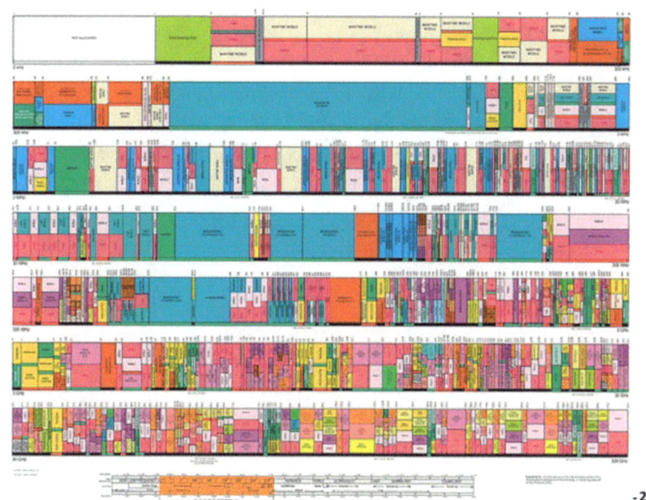

-22

Frequencies Used

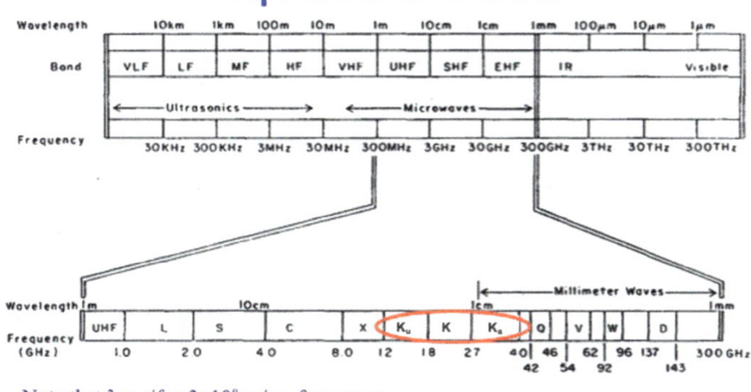

Note that $\lambda = c/f = 3\times10^8$ m/s ÷ frequency.
Example: K_u band, f = 12 GHz. $\lambda = c/f = 3\times10^8$ m/s ÷ 12E9 Hz = **0.025 m = 25 mm**

August 23, 2022

SC10-23

Telecommunications Topics

- Telecommunications Basics: Objectives and Definitions
- Long-range communication & **the Link equation**
- Antennas & performance
- Ground stations & weak signals
- Deep Space Considerations
- Navigation & Ranging

August 23, 2022

SC10-24

SNR, A.K.A. The Link Equation

The Link equation relates the *S/N* ratio to the different links of the chain:

$$SNR = \frac{E_b}{N_0} = \frac{P_t G_t G_r}{L_p k T_s L_c R_s}$$

- P_t = Power input to transmitting antenna (W)
- G_t = Gain of transmitting antenna
- G_r = Gain of receiving antenna
- L_p = Path loss (freq., distance, atmosphere dependent)
- k = Boltzmann constant, 1.38×10^{-23} Ws/K
- T_s = Effective system temperature (lower better*)
- L_c = Cable loss
- R_s = Symbol rate

Courtesy: Brown

August 23, 2022

SC10-25

Range & Path Loss Calculation

The slant range d for elevation angle ε is

$$d = \sqrt{(H + R_E)^2 - (R_E \cos \varepsilon)^2} - R_E \sin \varepsilon$$

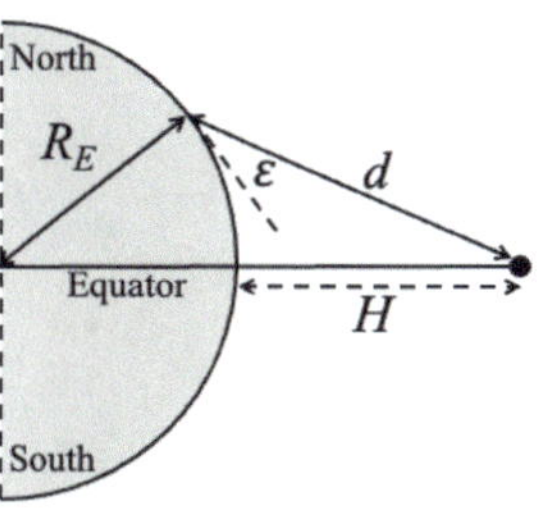

August 23, 2022

SC10-26

Example: Path Loss Calculation

For a 12 GHz GEO SC with 44° elevation:

$H + R_E = 42{,}164$ km, $R_E = 6{,}378$ km

$$d = \sqrt{(42{,}164)^2 - (6{,}378 \cos 44°)^2} - 6{,}378 \cos 44° = 37{,}483 \text{ km}$$

The path loss L_p at 12 GHz:

$$\lambda = \frac{c}{\nu} = \frac{3\times10^8 \text{m/s}}{12\times10^9/\text{s}} = 0.025 \text{ m}$$

$$L_p = 20 \log\left(\frac{4\pi d}{\lambda}\right)$$

$$= 20 \log\left[\frac{4\pi(37{,}483 \text{ km})}{0.025 \text{ m}}\right] = 205.5 \text{ dB}$$

August 23, 2022

SC10-27

Geostationary Path Loss

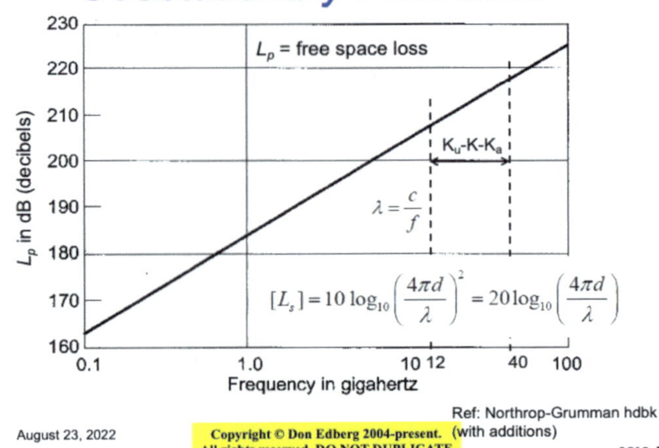

Ref: Northrop-Grumman hdbk (with additions)

August 23, 2022

SC10-28

Weather Attenuation Loss

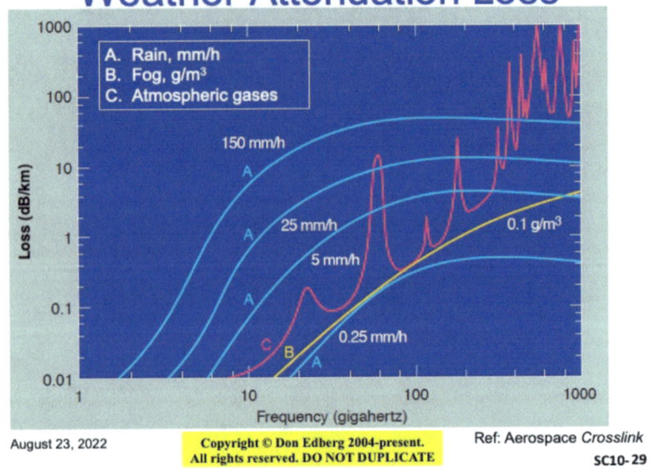

August 23, 2022

Ref: Aerospace *Crosslink*

SC10-29

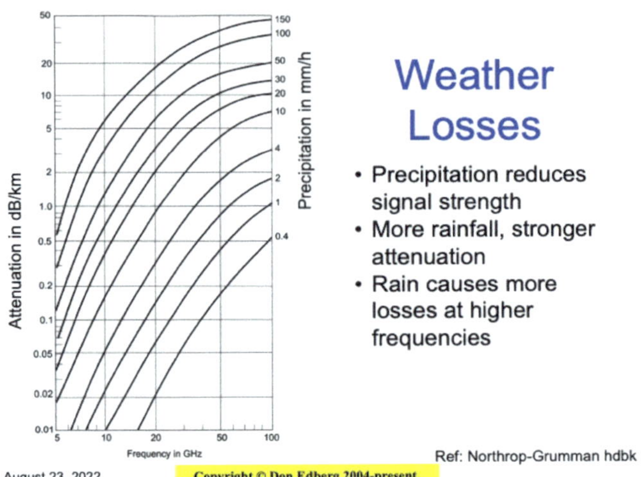

Weather Losses

- Precipitation reduces signal strength
- More rainfall, stronger attenuation
- Rain causes more losses at higher frequencies

Ref: Northrop-Grumman hdbk

August 23, 2022

SC10-30

Communications Blackout

- Solar Conjunctions
  - Superior conjunction: sun *between* SC & Earth
  - Inferior conjunction: sun *behind* SC
- No communication ±3° of conjunctions
- SC must be self-sufficient
- Play back data afterwards

Earth
Sun
Inferior Planet
Superior Planet

Courtesy: Swinburne University

August 23, 2022

SC10-31

The Link Budget, Pictorially

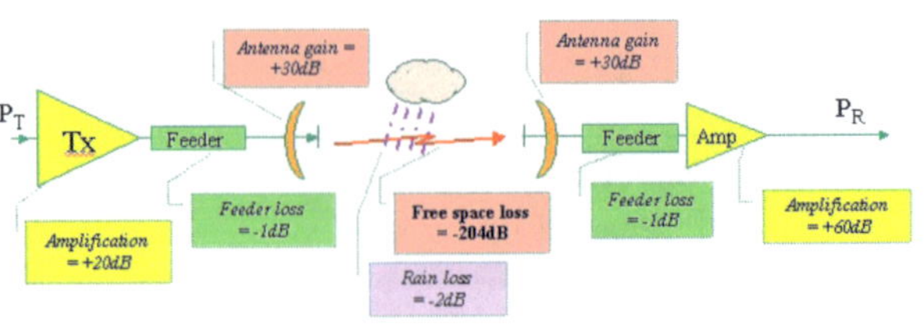

Courtesy: Satcom Online

August 23, 2022

SC10-32

The UPLINK Budget, Pictorially

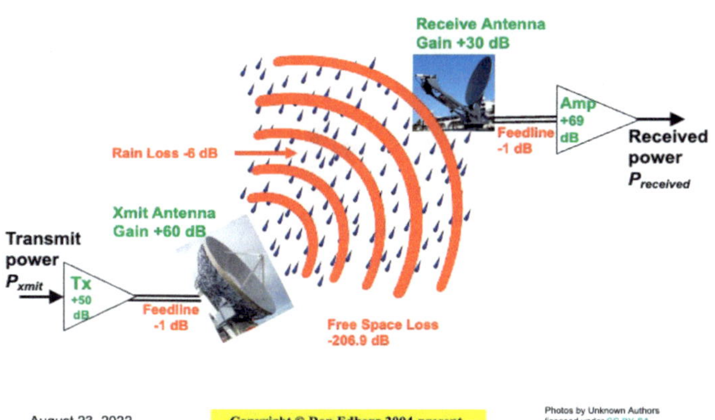

August 23, 2022

Photos by Unknown Authors licensed under CC BY-SA

Link Budget, Graphically

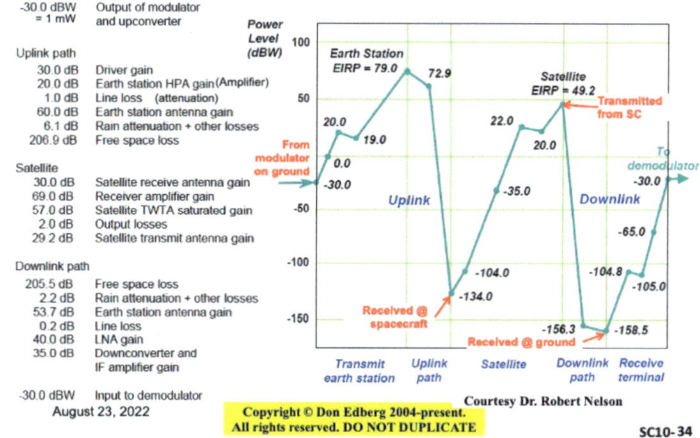

Courtesy Dr. Robert Nelson

August 23, 2022

Simple Link Budget Calculator

https://www.pasternack.com/t-calculator-link-budget.aspx

- Does not include weather losses such as rain attenuation

August 23, 2022

Telecommunications Topics

- Telecommunications Basics: Objectives and Definitions
- Long-range communication & the Link equation
- **Antennas & performance**
- Ground stations & weak signals
- Deep Space Considerations
- Navigation & Ranging

August 23, 2022

Parabolic Reflecting Antenna

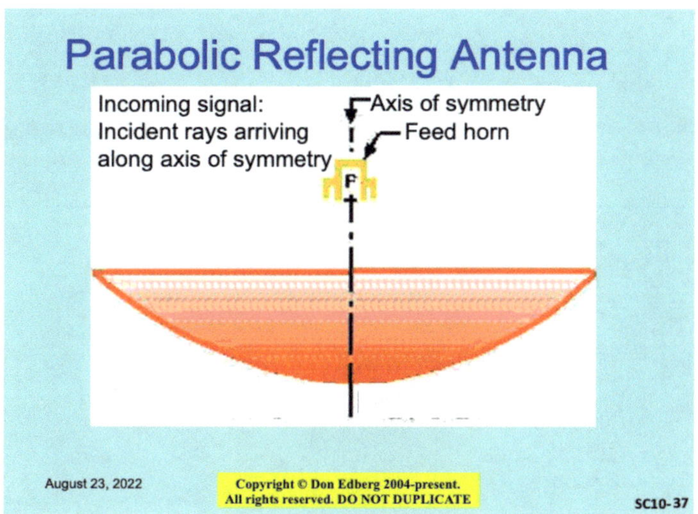

August 23, 2022

Feed and Focus

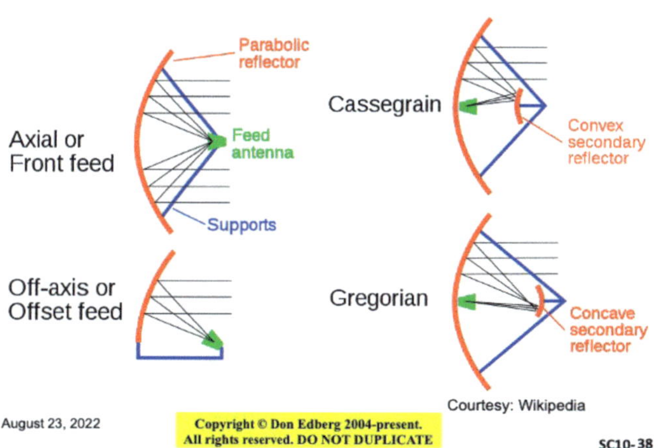

Courtesy: Wikipedia

August 23, 2022

Directional Antenna Radiation

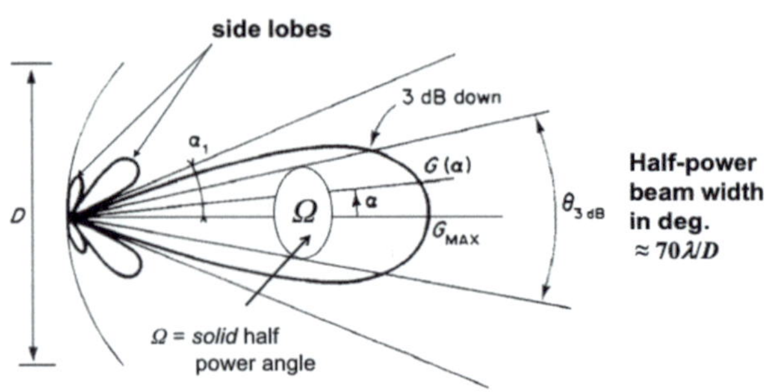

Half-power beam width in deg. $\approx 70\lambda/D$

Antenna calculator at http://www.q-par.com/antenna-calculator

Antenna Types

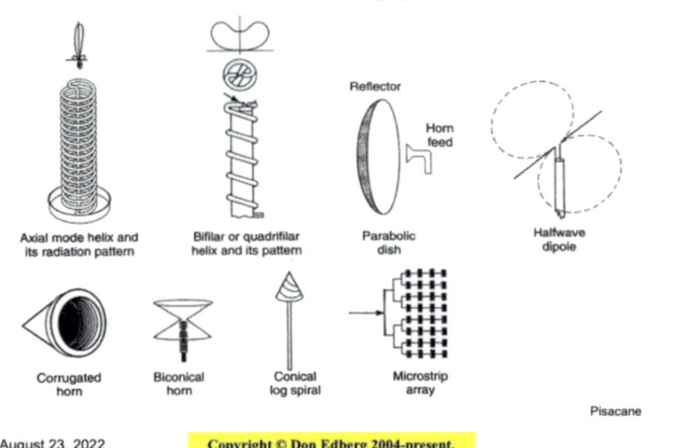

Pisacane

GPS-IIR-M

(courtesy Lockheed-Martin)

Global Positioning System (GPS) IIR-M satellite

Phased-Array or Microstrip Array Antenna

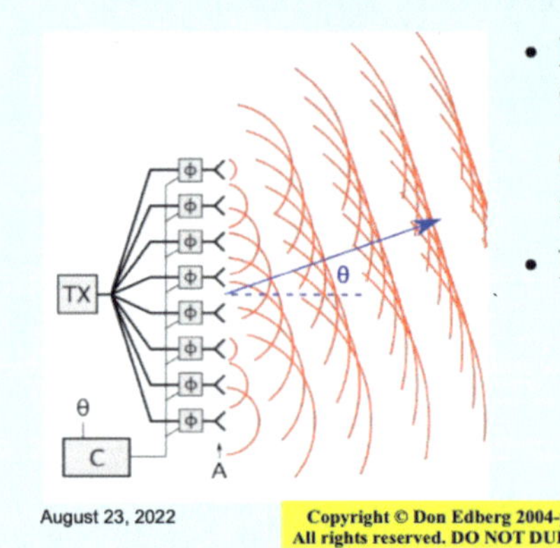

- Beam can be "steered" by changing timing of individual xmitters
- No moving parts

Courtesy Quora

Antenna Types, Gains, & Radiation Patterns

(λ = wavelength; dBi = dB relative to isotropic)

Courtesy: Brown

Configuration	Peak gain, dBi	Beam width, deg	Pattern
Half-wave dipole	1.64	—	
Planar array	$10\log\left(\frac{A}{\lambda^2}\right)+8$	—	
Turnstile	0.6	—	—
Horn	$20\log\left(\frac{D}{\lambda}\right)+7$ (Typically 5 to 20 dBi)	$\frac{72\lambda}{D}$	—
Bi-cone	$5\log\left(\frac{D}{\lambda}\right)+3.5$ (Typically 5 dBi)	Typically 45 × 360	
Helix	$10\log\left(\frac{D^2L}{\lambda^3}\right)+20.2$ (Typically 5 to 20 dBi)	$\frac{16.6}{\sqrt{D^2L/\lambda^3}}$	
Parabola	$20\log(f)+20\log(D)+17.8$ (Typically 10 to 65 dBi)	$\frac{65.3\lambda}{D}$	
Yagi	$\approx 12dBi$	—	

Parabolic Antenna Gain vs. Diameter

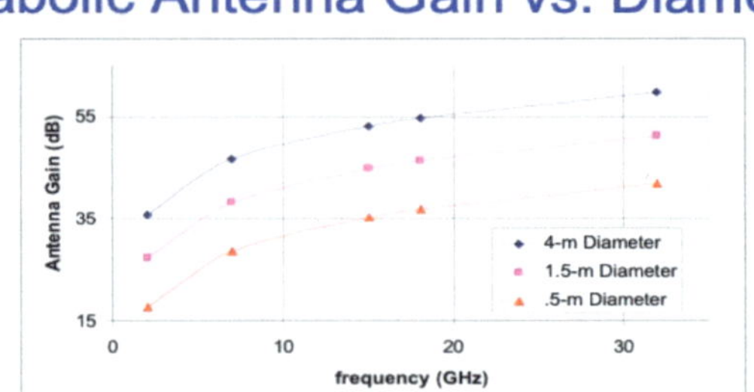

$$\text{Gain} = \frac{\text{Energy on target via directed antenna}}{\text{Energy on target via omni-directional antenna}}$$

Need Gain? How 'Bout 70 m Diameter?

August 23, 2022

SC10-45

Typical Spacecraft Telecomm Components

- High-gain and/or low gain antennas
- Solid-state recorder
- Superheterodyne receiver
- Wide band transmitter
- Encoder/decoder
- Power amplifier
  - Traveling-wave tube amplifier (TWTA)
  - Solid-state power amplifier
- Transponder (returns signal related to input)
- Cabling!

Heated Cathode (-)
Weak RF in
Magnetic Field
Amplified RF out
Electron Gun Anode (+)
Waveguide
Helix
Electron Beam (-)
Collector (+)

TWTA figure courtesy: ESA

August 23, 2022

SC10-46

Telecommunications Topics

- Telecommunications Basics: Objectives and Definitions
- Long-range communication & the Link equation
- Antennas & performance
- Ground stations & weak signals
- Deep Space Considerations
- Navigation & Ranging

August 23, 2022

SC10-47

Ground Stations

August 23, 2022

Courtesy ESA

SC10-48

Ground Station Functions

- Ranging, distance measuring of satellite
- Satellite command transmission
- Telemetry reception
- Antenna tracking
- Support launch & early orbit-phase test & housekeeping

August 23, 2022

SC10-49

Ground Station Block Diagram (Receiving)

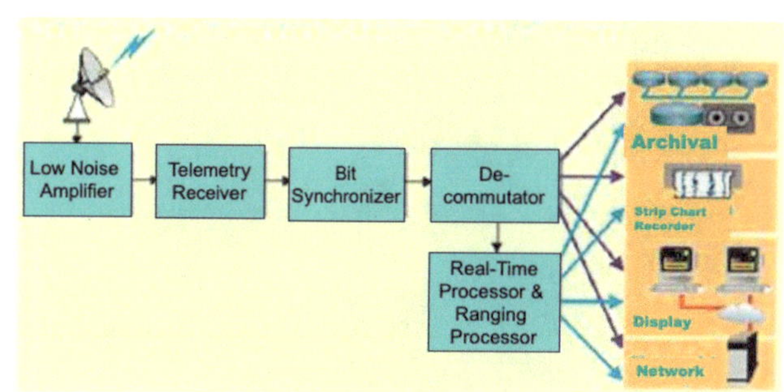

August 23, 2022

SC10-50

Telecommunications Topics

- Telecommunications Basics: Objectives and Definitions
- Long-range communication & the Link equation
- Antennas & performance
- Ground stations & weak signals
- **Deep Space Considerations**
- Navigation & Ranging

August 23, 2022

SC10-51

Receiving Low Signal Levels From Space (Power ~ $1/r^2$)

- Use LNAs (= Low Noise Amplifiers). Liquid helium cooling slows down electron motion in amplifier components: high SNR
- Bigger antennas
- Antenna arraying: use multiple ground station antennas together
  - Intercontinental arraying was used for the Galileo mission after difficulties with main antenna

Credit: ATNF/CSIRO

August 23, 2022

SC10-52

Telecommunication Challenges

Filename: mars telecomm.mp4 Courtesy: JPL

August 23, 2022

SC10-53

NASA's Deep Space Network (DSN)

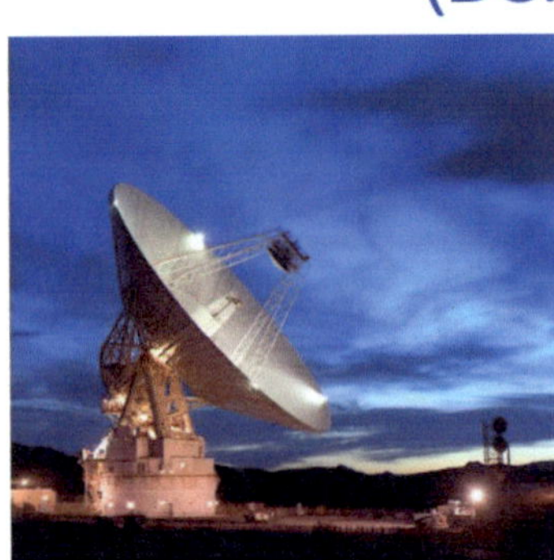

- Use for $r \geq 2 \times 10^6$ km (~1.2 million miles)
- Supported frequencies:
  - 2 GHz (S-band)
  - 13-15 GHz (K_U-band)
- Locations and sizes
  - 34 m & 70 m dishes at Goldstone (CA), Canberra (Aus.), Madrid (Spain)

August 23, 2022

SC10-54

Telecommunications & the DSN

Filename: Communicating_Through_Space.mpg Courtesy: JPL

August 23, 2022

SC10-55

Optical Communications

- Radio Frequency (RF) reaching limits for data capacity
- NASA press release 13-309, 22 Oct 2013: Lunar Laser Communication Demo on *LADEE* achieved 622 Mbps download using pulsed laser beam from Moon to Earth (equal to 30 HD video channels!)
- Also achieved 20 Mbps error-free upload from New Mexico to Moon
- Laser communications may enable increased image resolution and 3D video transmission from deep space

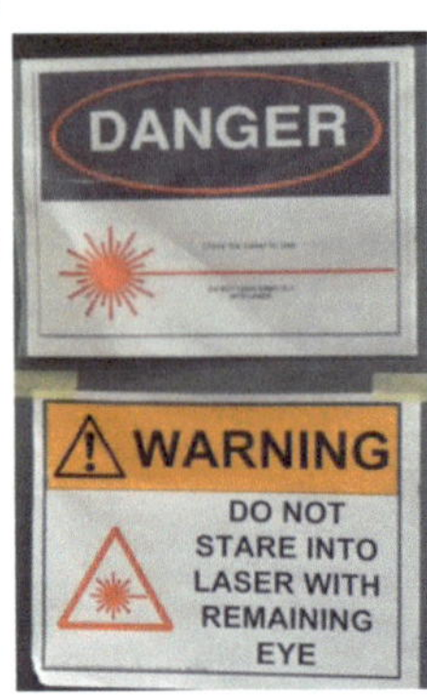

DANGER
WARNING
DO NOT STARE INTO LASER WITH REMAINING EYE

D Edberg photo

August 23, 2022

SC10-56

Laser Communications Relay Demonstration (LCRD, 11/2021)

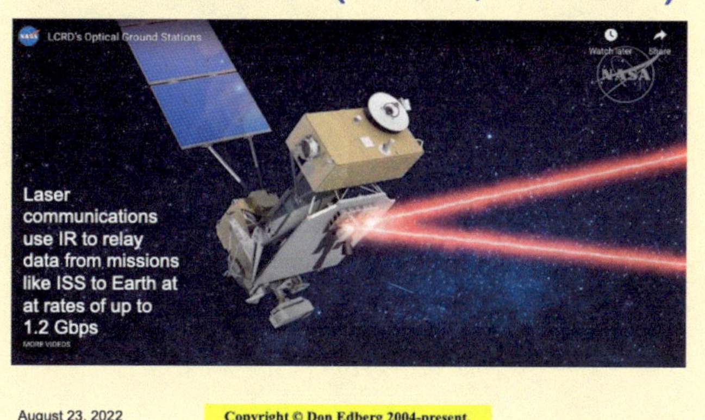

LCRD's Optical Ground Stations
Laser communications use IR to relay data from missions like ISS to Earth at at rates of up to 1.2 Gbps

August 23, 2022

SC10-57

Estimating System Performance: Jupiter Spacecraft Example

Given data:

- Parabolic antenna
- Diameter of 2.5 m
- Gain of 44 dB
- Transmitter power of 25 W
- Use data to create "Uplink Table"
  (Examples may be found in Brown, Tables 9.7, 9.8)

August 23, 2022

SC10-58

Downlink at Maximum Range

- Maximum range is the Earth and Jupiter at opposite sides of the sun

Carrier Link Performance	Value	Units
Modulation Index	68	degrees
Carrier Power/Total Power (P_c/P_t)	-8.53	dB
Received Carrier Power	-173.85	dB
Carrier Noise Bandwidth	20	Hz
	13.01	dB-Hz
Carrier Signal to Noise	29.19	dB
Carrier signal to noise required by DSN	10	dB
Carrier link margin	19.19	dB

August 23, 2022

SC10-59

Downlink at Maximum Range (cont.)

Data Link Performance	Value	Units
Data Power/Total Power(P_d/P_t)	-0.66	dB
Data Power Received	-165.98	dB
Data Symbol Rate	-43.01	dB-Hz
E_b/N_o achieved	7.06	dB
E_b/N_o required	4.1	dB
Data Link Margin	2.96	dB

~3 dB margin (factor of 2) considered acceptable.

August 23, 2022

SC10-60

Fixing the Problem if the Rates Cannot be Accommodated

- Increase the transmitting antenna area
- Improve the gain of transmitting antenna
- Increase the transmitter power
- Increase receiving antenna area, decrease receiver bandwidth or temperature (unlikely)
- Simplest: reduce data rate! Cutting data rate in half provides +3 dB
  - Example: Galileo spacecraft suffering antenna jam sent data at 160-1,000 bps instead of 134,000 bps to increase SNR, and accomplished 70% of mission

August 23, 2022

SC10-61

Telecommunications Topics

- Telecommunications Basics: Objectives and Definitions
- Long-range communication & the Link equation
- Antennas & performance
- Ground stations & weak signals
- Deep Space Considerations
- **Navigation & Ranging**

August 23, 2022

SC10-62

Guidance/Navigation vs. Attitude Control System* (ACS)

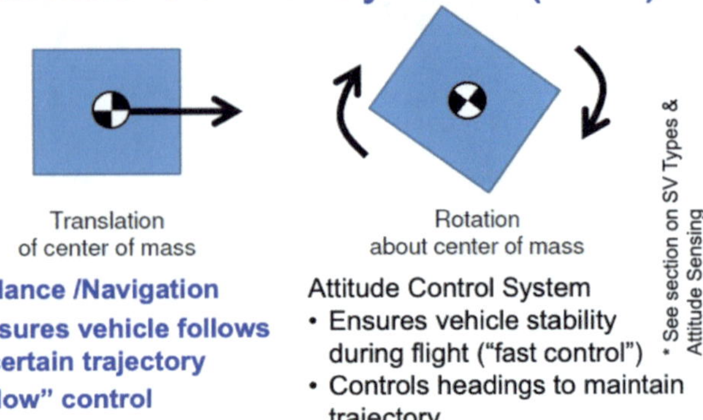

Translation of center of mass

Rotation about center of mass

Guidance /Navigation

- **Ensures vehicle follows a certain trajectory**
- **"Slow" control**

Attitude Control System

- Ensures vehicle stability during flight ("fast control")
- Controls headings to maintain trajectory

* See section on SV Types & Attitude Sensing

Guidance/Navigation: A.K.A. "Orbit Determination"

- Use Telecomm system to determine position in space and thus orbit/trajectory
- Begins with 'last known position'
- Incorporate "observables" to update the motion since last update
- Include large & small disturbances such as thruster firings, outgassing, solar radiation pressure, even SC-emitted radiation

Types of Navigation Info

- *Velocity measurement*: use Doppler effect to measure *radial* velocity component v_R with respect to ground station
  - No direct information on *tangential* velocity, can estimate from "cross-motion" of antenna dish

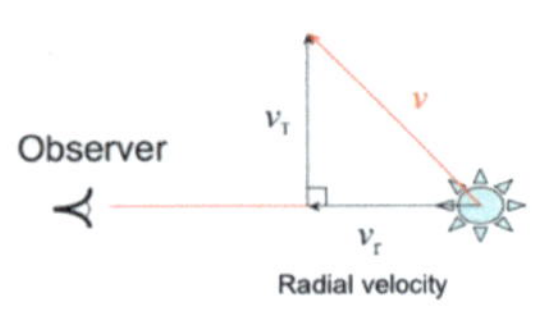

- *Angular measurement*: dish *direction* rate gives high-precision angular rate of SC with respect to ground, can translate into cross velocity v_T

Doppler During Flyover

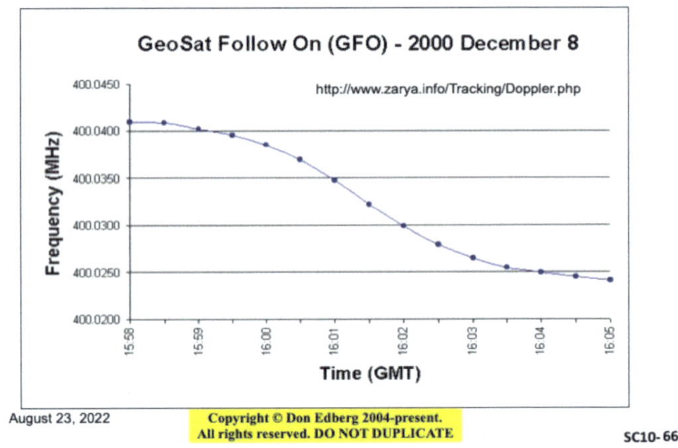

Range Measurement (Ranging)

Techniques use RF propagation delay to measure distance

- *Tone Ranging*: tones sent to SC & returned by SC's transponder
  - phase difference between sent & received tones measured; different tones resolve ambiguity.
- *PN Ranging*: tone modulated by known "pseudo-random" bit pattern
- *Regenerative Ranging*: removes uplink noise retransmitted in the turnaround process, improves SNR

Radial & Angular Position Accuracy

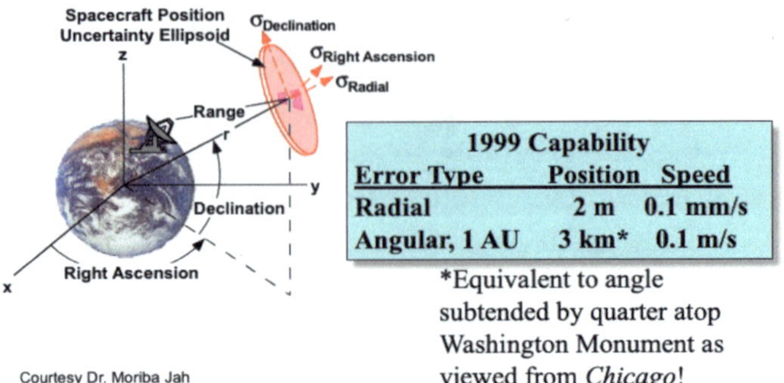

1999 Capability		
Error Type	Position	Speed
Radial	2 m	0.1 mm/s
Angular, 1 AU	3 km*	0.1 m/s

*Equivalent to angle subtended by quarter atop Washington Monument as viewed from *Chicago*!

Courtesy Dr. Moriba Jah

Position Quality Assessment

- Errors between predicted & observed quantities are known as *residuals*: statistics of residuals used to assess quality of OD process.
- Consistent residuals = course correction needed
- Random variation in residuals = statistical variation, vehicle on course as planned

August 23, 2022

SC10-69

Orbit Determination Example

- SC flying between Earth E & Mars M, currently at position A
- E's position is known exactly

"Observables" determined:

- EA, distance to SC, using radiometric techniques (ranging)
- Radial velocity v_{EA} along EA by Doppler
- Angle SEA by dish pointing

August 23, 2022

SC10-70

Observables Used to Predict Position

- Similar techniques used to get out-of-plane positions & velocities
- Rates, linear & angular positions used to extrapolate motion of SC from A to B
- In case its course needs correction, thrusters fired
- … and trajectory eventually leads to M = Mars!

August 23, 2022

SC10-71

The Center of the Universe?

At JPL Space Flight Operations Facility!

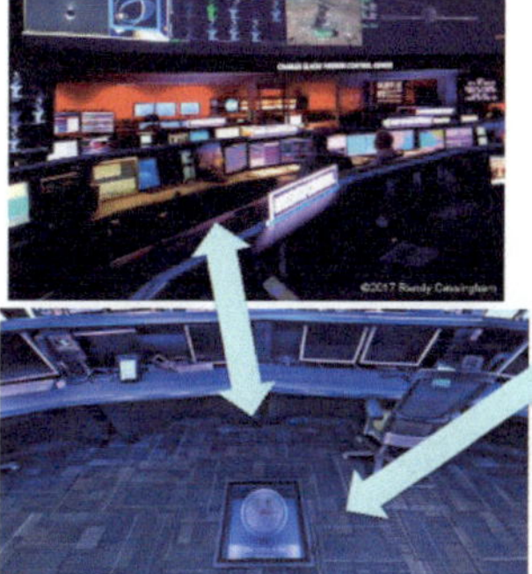

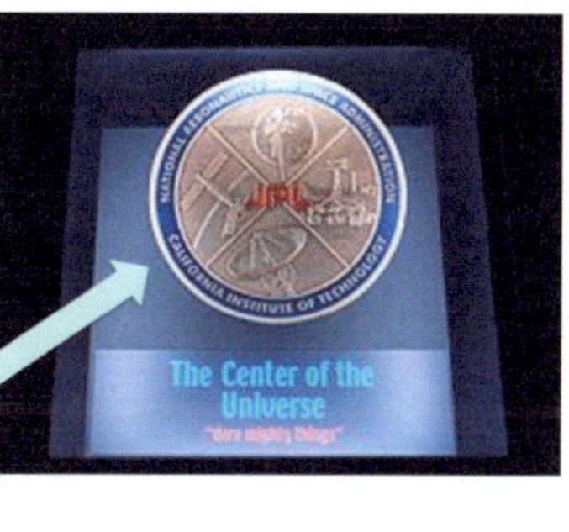

Courtesy: thisistrue, wikipedia

August 23, 2022

SC10-72

Command & Data System

August 23, 2022

(50)

Command & Data System (CDS) Topics

- Command & Data System definitions & functions
- Data downlinking & telemetry
- Sampling, and data conversion errors
- Data Encoding
- Fault protection & safing
- Hardware
- Software
- Mass & power estimation

August 23, 2022

Spacecraft Design Process

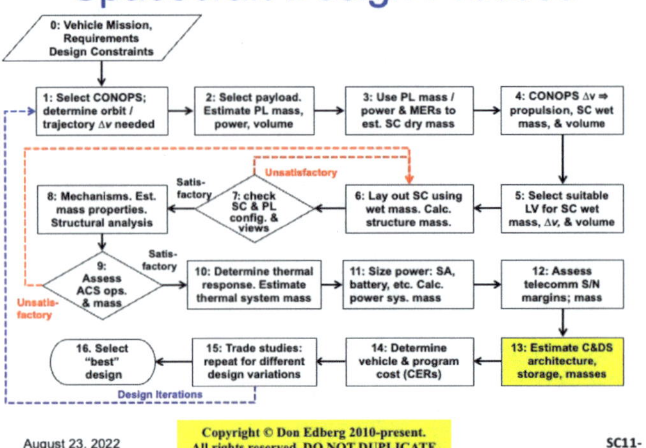

0: Vehicle Mission, Requirements Design Constraints
1: Select CONOPS; determine orbit / trajectory Δv needed
2: Select payload. Estimate PL mass, power, volume
3: Use PL mass / power & MERs to est. SC dry mass
4: CONOPS Δv ⇒ propulsion, SC wet mass, & volume
5: Select suitable LV for SC wet mass, Δv, & volume
6: Lay out SC using wet mass. Calc. structure mass.
7: check SC & PL config. & views
Unsatisfactory
Satis-factory
8: Mechanisms. Est. mass properties. Structural analysis
9: Assess ACS ops. & mass
Satis-factory
Unsatis-factory
10: Determine thermal response. Estimate thermal system mass
11: Size power: SA, battery, etc. Calc. power sys. mass
12: Assess telecomm S/N margins; mass
13: Estimate C&DS architecture, storage, masses
14: Determine vehicle & program cost (CERs)
15: Trade studies: repeat for different design variations
16. Select "best" design
Design Iterations

C&DS Topics

- Command & Data System definitions & functions
- Data downlinking & telemetry
- Sampling, and data conversion errors
- Data Encoding
- Fault protection & safing
- Hardware
- Software
- Mass & power estimation

Command & Data System (CDS)

- Sometimes called "Command & Data Handling" (C&DH) system or "On-Board Data Handling" (OBDH) system
- Serves as spacecraft's central nervous system
- Manages three data streams:
  - Payload/science data (downlink)
  - Engineering data (downlink)
  - Commands (uplink)

C&DS Primary Functions

- Process uploaded commands
  - Validate, decode, & distribute commands
- Schedule onboard activity
  - Resource management & coordination
- Data functions (payload & telemetry)
  - Collect from subsystems
  - Provide durable storage
  - Format & analyze
- Communication and messaging
  - Ground and subsystems

C&DS Functional Requirements

- Time-critical commands should *not* be needed for normal operation
- Critical events require *two* commands each
- Fault protection system to allow ground commanding of enable/disable states
- NEVER turn off all receivers at once
- No "toggle" commands: be able to command to known states *without knowing current state*
- Be able to telemeter contents of any memory

CDS Functional Reqts. (cont.)

- On-board memory reprogrammable
- All enable/disable states must be telemetered
- Commandable variation in engineering data rates desirable for certain measurements during critical events
- Engineering data stream must provide rapid, unambiguous determination of state at start of downlink & must show fault protection response

Typical CDS System

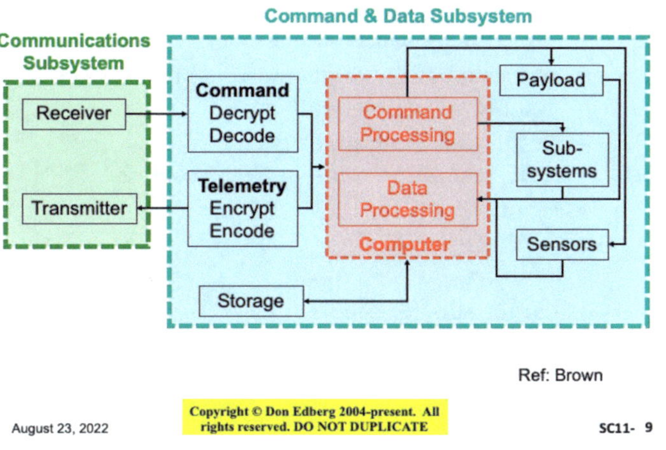

Ref: Brown

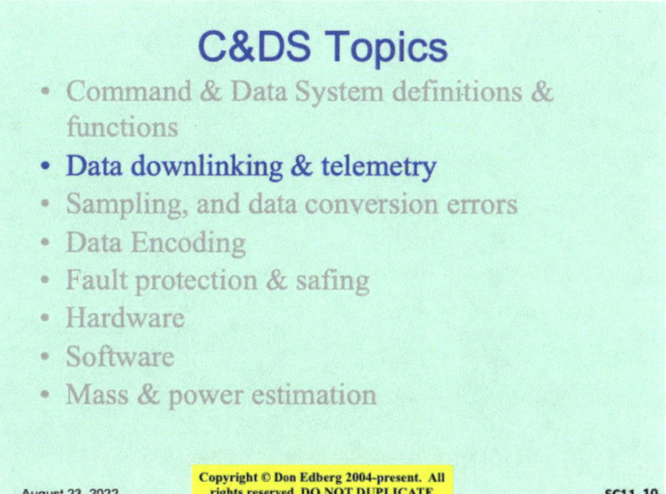

Ground Control Needs Downlinked Data

- Telemetry = automatically measure & transmit data by wire or radio or other means from remote sources to receiving stations for recording & analysis
- Science / payload data
  - What the SC was sent to do
  - Downlinked at highest possible rate
- Engineering data
  - Includes accelerations, component temperatures, pressures, voltages, currents, wheel speeds, rotational positions, ACS sensor outputs, fuel levels
  - Status measurements such as valve, switch, relay, and safe/arm positions
  - Information is used by ground ops to provide spacecraft subsystem status and health
  - Usually downlinked at medium rate, less than Science or Payload rate

Downlink Process for Science SC

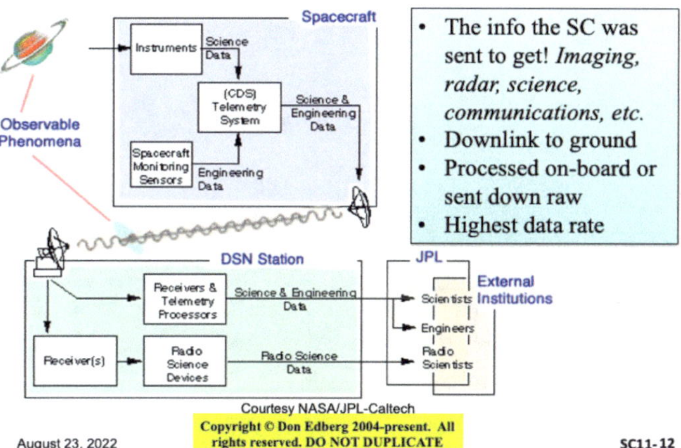

- The info the SC was sent to get! *Imaging, radar, science, communications, etc.*
- Downlink to ground
- Processed on-board or sent down raw
- Highest data rate

Courtesy NASA/JPL-Caltech

Mars Exploration Rover: Power System (Engineering) Telemetry

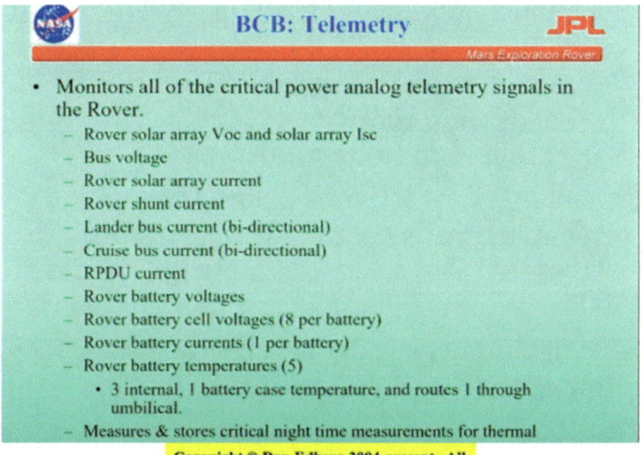

BCB: Telemetry

JPL — Mars Exploration Rover

- Monitors all of the critical power analog telemetry signals in the Rover.
  - Rover solar array Voc and solar array Isc
  - Bus voltage
  - Rover solar array current
  - Rover shunt current
  - Lander bus current (bi-directional)
  - Cruise bus current (bi-directional)
  - RPDU current
  - Rover battery voltages
  - Rover battery cell voltages (8 per battery)
  - Rover battery currents (1 per battery)
  - Rover battery temperatures (5)
    - 3 internal, 1 battery case temperature, and routes 1 through umbilical.
  - Measures & stores critical night time measurements for thermal

Telemetry Process

- Transducers convert physical quantities (e.g. vibration, temperature) into electrical signals
- Signals are digitized
- Extra bits added – "overhead" – for information such as instrument time, date, etc.
- Data transmitted by radio (or laser)
- Ground stations receive the transmissions
- Computers extract, display, process, store the data

C&DS Topics

- Command & Data System definitions & functions
- Data downlinking & telemetry
- **Sampling, and data conversion errors**
- Data Encoding
- Fault protection & safing
- Hardware
- Software
- Mass & power estimation

August 23, 2022 SC11-15

System Resolution: Number of Bits

- Analog-to-Digital converters (ADC) sample data and convert to digital numbers
- The value of the digitized signal is proportional to input level ÷ full-scale level
- Resolution of digitized data:
  ADC resolution = full-scale level ÷ $2^{[ADC\ \#\ bits]}$
- Example: 5 V full scale, 10-bit system resolution: $5\ V/2^{10} = 5\ V/1024 = 0.00488\ V = 4.88\ mV \approx 5\ mV$
- Larger number of bits:
  - Higher resolution 🙂
  - More bits that need to be transmitted 🙁

8/23/22 SC11-16

TM System Operation

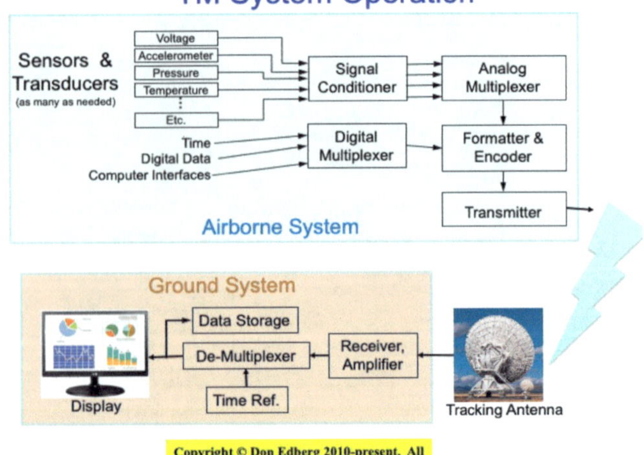

8/23/22 SC11-17

Sampling Data & Multiplexing – Commutation

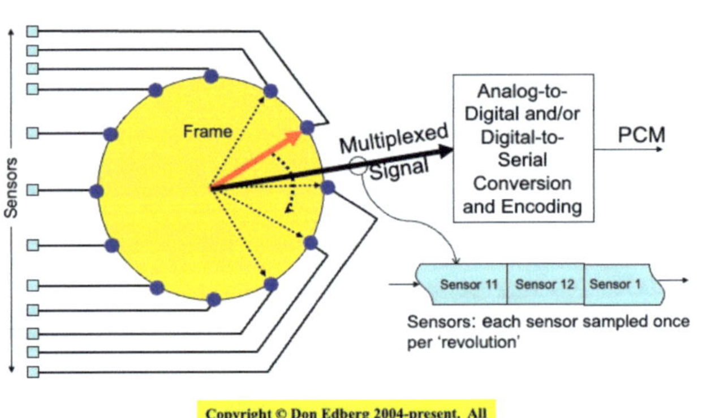

August 23, 2022 SC11-18

Framing of Data

- Data are split into "frames" that are sent in "blocks"

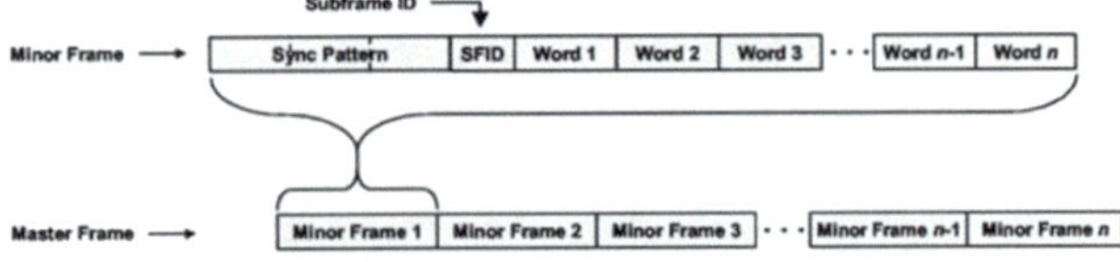

- Sync patterns separate strings of data
- Computers de-commutate to keep track of what's what

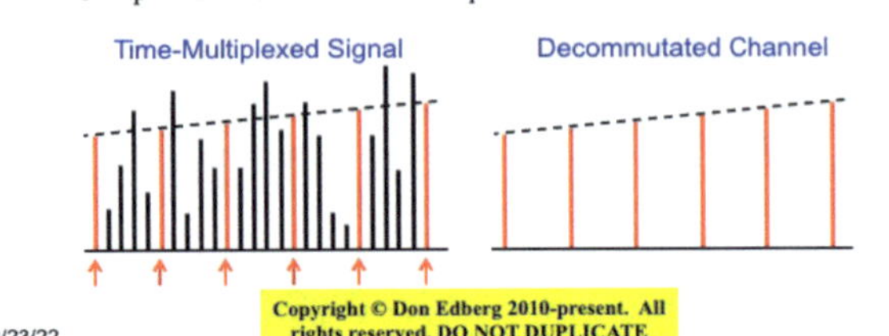

8/23/22 SC11-19

Data Encoding

- Shannon's Law: $R = B\log_2(1 + SNR)$
  where:
  R = rate of symbols transmitted, bit/s
  B = bandwidth, Hz
  SNR = Signal-to-noise ratio
- Shannon's Law imposes a hard limit on the performance of a communication channel: to increase symbol rate, need more bandwidth & SNR
- Practical limits usually less due to redundancy, error correction, etc. (overhead)

8/23/22 SC11-20

Coding = More Reliability

- Coding adds redundancy bits to the data for bit error detection & correction
- Improves signal-to-noise ratio

Types of coding:

- Reed-Solomon
- Viterbi
- Forward-Error Correction (FEC)
- Turbo

8/23/22 SC11-21

Coding Systems Comparison

Ref: SMAD fig. 13-9

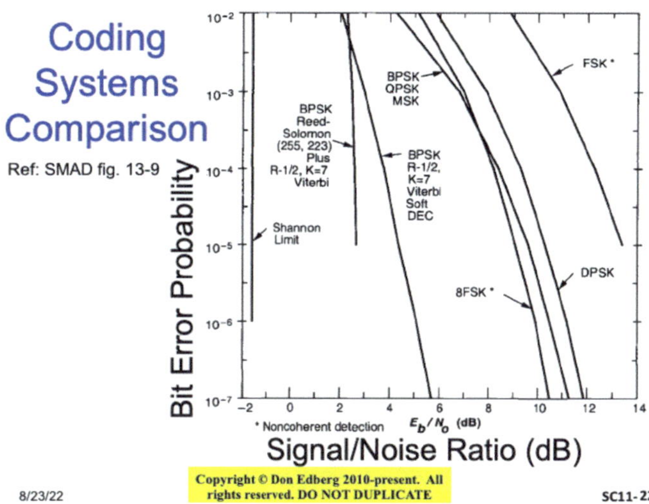

8/23/22 SC11-22

Minimum Sampling Rate & Conversion Errors

- Data sampling frequency f_s:
  $f_s \geq 2.2 \times$ max frequency component, or *aliasing* occurs
- Conversion induces quantization or rounding error
  - Use more bits to minimize
- Some missions may require encryption
  - Tends to slow system down

August 23, 2022 SC11-23

Aliasing Example

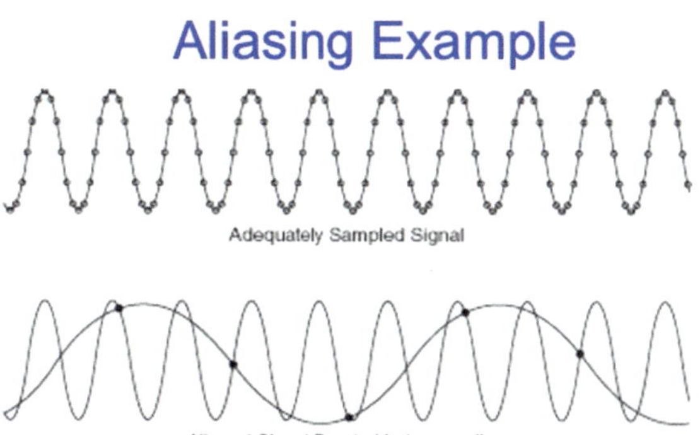

Actual higher-frequency signal (above) appears as low-frequency signal (below)

August 23, 2022 SC11-24

Aliasing Movie

RMAX aliasing P1030406.MOV

8/23/22 SC11-25

Aliasing Time Records (cont.)

- Say we have a signal between 5 & 10 Hz, or $\frac{1}{2}f_s$ & f_s, where $f_s = 10\text{ Hz} = \frac{1}{T} = \frac{1}{\text{sampling interval}} = \frac{1}{0.1\text{ s}}$
  7 Hz appears as (10 – 7) Hz = 3 Hz
  8 Hz appears as (10 – 8) Hz = 2 Hz
  Any signal (f') appears as ($f_s - f'$)
- The frequency is "folded" at $\frac{1}{2}f_s$
- $\frac{1}{2}f_s$ is known as the *Nyquist* frequency, f_N
- How to eliminate aliasing? Place low-pass filter ("anti-aliasing filter") in front of ADC that filters out frequencies above f_N

8/23/22 SC11-26

Aliasing Time Records

True spectrum (top plot, solid vertical lines)
Plot 1

×

Anti-aliasing filter (middle plot)
Plot 2

=

Filtered spectrum (Plot 3)

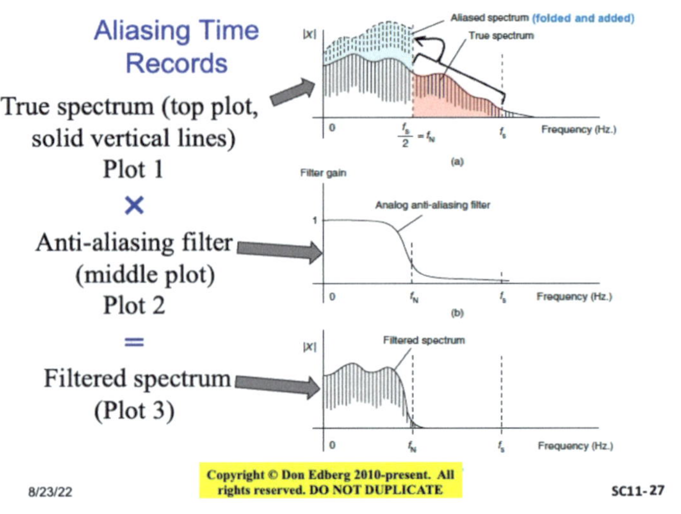

8/23/22

Quantization Error

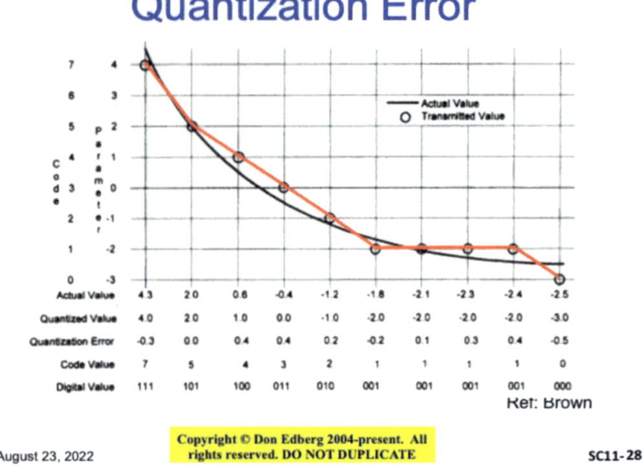

Actual Value	4.3	2.0	0.8	-0.4	-1.2	-1.8	-2.1	-2.3	-2.4	-2.5
Quantized Value	4.0	2.0	1.0	0.0	-1.0	-2.0	-2.0	-2.0	-2.0	-3.0
Quantization Error	-0.3	0.0	0.4	0.4	0.2	-0.2	0.1	0.3	0.4	-0.5
Code Value	7	5	4	3	2	1	1	1	1	0
Digital Value	111	101	100	011	010	001	001	001	001	000

Ref: Brown

August 23, 2022

Error Due To Sampling Rate

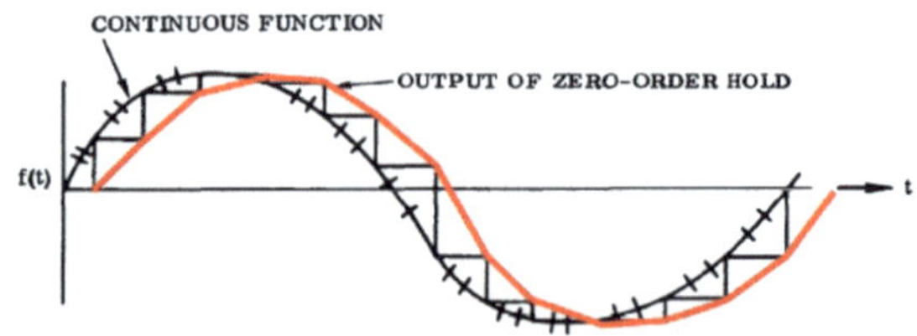

Samples Per Cycle	Maximum Amplitude Error
5	19.1%
10	4.9%
30	0.6%
50	0.2%

August 23, 2022

C&DS Topics

- Command & Data System definitions & functions
- Data downlinking & telemetry
- Sampling, and data conversion errors
- **Data Encoding**
- Fault protection & safing
- Hardware
- Software
- Mass & power estimation

August 23, 2022

Data Encoding

- Shannon's Law: $R_s = B \log_2(1 + SNR)$
  where:
  R_s = symbol rate or capacity, bit/s
  B = bandwidth, Hz
  SNR = Signal-to-noise ratio
- Shannon's Law imposes a hard limit on the performance of a communication channel
- Practical limits usually less due to redundancy, error correction, etc. (overhead)

August 23, 2022

Coding = More Reliability

- Coding adds redundancy bits to the data for bit error detection & correction
- Improves signal-to-noise ratio

Types of coding:

- Reed-Solomon
- Viterbi
- Forward-Error Correction (FEC)
- Turbo

August 23, 2022

Coding Systems Comparison

Ref: SMAD fig. 13-9

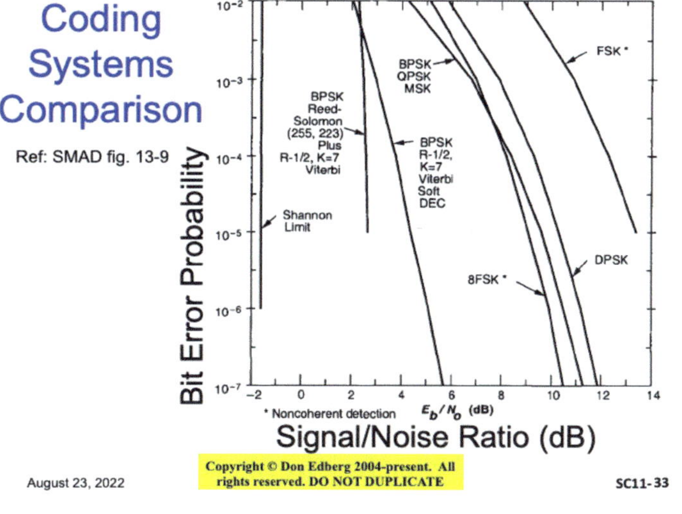

August 23, 2022

SC11-33

Data Rates

- Mission-critical communications use low bit rates for high reliability (BPS – KBPS range)
- Some earth observers use bit rates in the MBPS range (~400 MBPS for image transmission)
- Bit Error Rate (BER) limited by SNR

August 23, 2022

SC11-34

Commands: Uplink From Ground

- Low data-rate uplink (from ground)
- Commands dictate spacecraft configuration, attitude, and actions
- Accuracy is critical; errors can be dangerous or life-threatening

August 23, 2022

SC11-35

C&DS Topics

- Command & Data System definitions & functions
- Data downlinking & telemetry
- Sampling, and data conversion errors
- Data Encoding
- Fault protection & safing
- Hardware
- Software
- Mass & power estimation

August 23, 2022

SC11-36

Fault protection (FP) algorithms

- Normally run in one or more of the spacecraft's subsystems
- Ensure the ability to mitigate the impact of a mishap, and re-establish ability to contact Earth if an anomaly has caused a communications interruption
- A spacecraft may have many different FP algorithms running simultaneously with the ability to request CDS to take action

August 23, 2022

SC11-37

Example Fault Protection: *Command-Loss Timer*, CLT

- A software timer that's reset to a given value (i.e. 1 week) every time the SC receives a command
- If the timer gets all the way to zero, SC assumes it has experienced a failure in its command string
- The CLT FP then issues commands for actions, i.e. switching to redundant hardware to attempt to re-establish ability to receive commands

August 23, 2022

SC11-38

"Watchdog" Timer

- Independent circuit monitors 'heartbeat' of CPU
- Resets CPU if 'heartbeat' lost

Courtesy: James Cutler, Stanford

August 23, 2022 SC11-39

Example: "Watchdog" Timer: *Priority Inversion* Resetting Mars Pathfinder's Computer Several Times

Three Mars Pathfinder types of tasks executed as threads with priorities reflecting task urgency:

- High-priority threads, i.e. bus management task
- Medium priority threads, i.e. communications task
- Low priority threads, i.e. meteorological data gathering task

Bus connects all SC's components; management task moves data in / out of the information bus Access to bus is synchronized with 'mutexes' (short for "mutual exclusion", mechanisms used to prevent unauthorized access to resources currently in use). Using the mutexes guaranties that each bus thread waits until the previous one is finished.

All tasks use bus to publish data: when publishing, it acquires a mutex, writes to the bus, releases mutex. Normally, the threads system with priorities & bus management with mutexes works fine.

A problem known as 'priority inversion' can occur: assume a medium-priority communications task is scheduled while high-priority bus management task is blocked waiting for the low-priority meteorological data thread. The communications task, having higher priority than the meteorological task, would prevent it from running, consequently preventing the blocked bus management task from running. Since the communications task is a long-running task, a watchdog timer will eventually go off, noticing that the data bus task had not been executed for some time. The watchdog task will conclude that something had gone wrong, and initiate a total system reset.

Reset comparable to *home computer reboot*; clears registers so system can restart without errors

August 23, 2022 SC11-40

Opportunity Faults

- Low-power fault causes the rover to hibernate, assuming that it will wake up when there's more sunlight to let it recharge.
- Clock fault: the onboard clock is critical to operating while in hibernation. If the rover doesn't know the time, it doesn't know when to attempt to communicate. It can use clues like an increase in sunlight to make assumptions about the time.
- Uploss fault: if the rover hasn't heard from Earth in a long time, it goes into "uploss" fault – a warning that comm equipment may not be functioning. It begins to check the equipment and tries different ways to communicate with Earth.

Courtesy NASA

August 23, 2022 SC11-41

Safing

- Shutting down or reconfiguring components
  - Prevents damage either from within or from the external environment
- May include an automated search to re-establish Earth-pointing & regain communications
- May include Emergency Sun Reacquisition (ESR) to get solar arrays facing the sun
- May temporarily disrupt science & require flight team to perform additional work
- Provides strong & reliable protection for the spacecraft & its mission

August 23, 2022 SC11-42

Kepler & Safe Mode

- 3/22/2011: "There was likely a pretty big sigh of relief at NASA's Ames Research Center this week as satellite *Kepler* recovered from a glitch that took it offline for 144 hours." The event started on March 14 "after the spacecraft issued a network interface card (NIC) reset command to implement a computer program update." *Kepler* entered "safe mode" but resumed operations after it "passed a health and safety check."

August 23, 2022 SC11-43

MAVEN & 3-Month Safe Mode

- Mars Atmosphere & Volatile Evolution (MAVEN) SC carries two IMUs: primary & backup, responsible for keeping orientation
- IMU-1 issues in 2017 switched to backup unit
- Late 2021 IMU-2 starting to wear out; team returned to original IMU-1 Feb. '22
- Two weeks later SC couldn't seem to use either IMU to properly position itself, action needed
- 1st: stabilize the spacecraft via "heartbeat termination:" reboot main computer. When that didn't work, it went to backup computer (had never used the backup computer before)
- After >1 hour trying to revive IMU-1, computer swap which also put MAVEN on IMU-2, held.
- Spacecraft's sun-pointing was beginning to stray; team rushed to implement "all-stellar" mode using star trackers to replace IMUs
- Large loss of data on Mars UV, solar flares, & relay capability to/from Mars surface resulted from extended safe mode

August 23, 2022 SC11-44

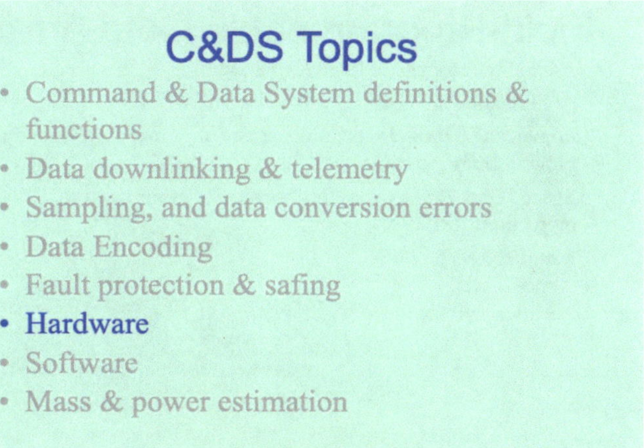

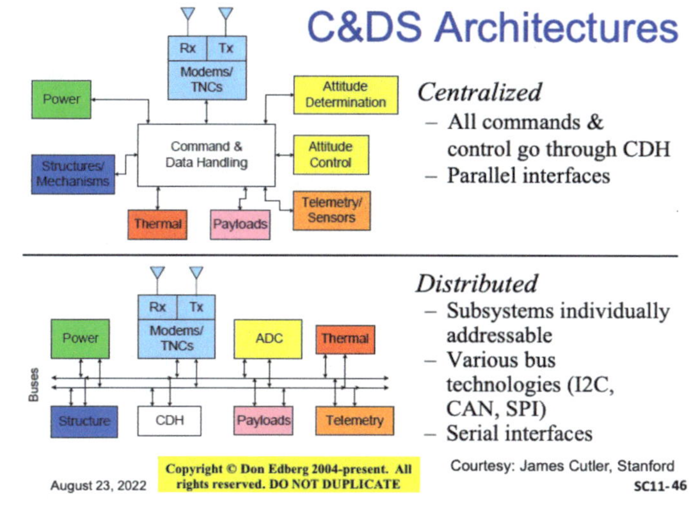

Rad-Hard Processor for Space Use

- BAE Aerospace RAD750 CPU
- "the rad-hard equivalent of the most advanced commercial microprocessor"
- 1 Mrad radiation-hardened
- Capable of running at 200+ MHz

New Rad-Hard Processor: BAE Aerospace RAD5545™

- SpaceVPX single-board computer
- Throughput 5.6 G-ops/s, 3.7 GFLOPS (>10 × RAD750®)
- 4 GB DDR3 SDRAM with error correction; 1 GB triple modular redundant non-volatile flash memory
- 35 W
- Total ionizing dose: 100 krad
- Single event upset: 0.001 SEUs/card-day, latchup immune
- Data from 2017 flyer

Ref: RAD5545 SpaceVPX_SBC.pdf

QinetiQ Space Computer in Thermal-Vac Test (May 2022)

- Onboard computer (OBC), power control & distribution unit (PCDU), remote terminal unit (RTU), & mass memory unit (MMU)
- Two weeks in simulated space environment

Source: ESA

Data Storage

- Solid-state recorders now universal
  - Faster
  - More reliable (no moving parts)
  - Lower power consumption

Historical note:

- Tape recorders no longer used, eliminating:
  - Motors, bearings, tape readers, & moving parts
  - Tape wear issues (GLL)

Solid State Mass Memory (SSMM)

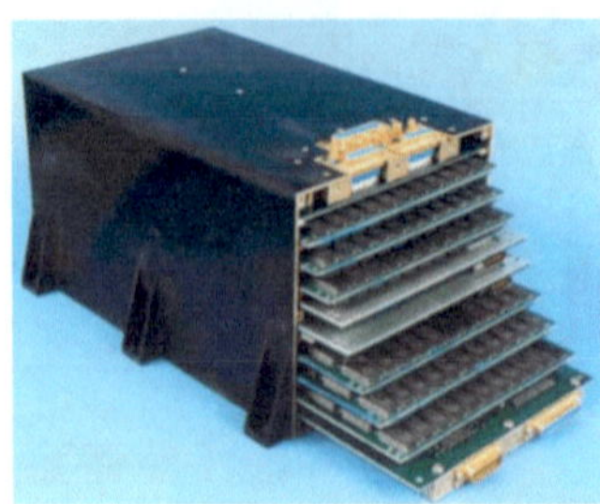

- Example: 3.1 Tbit SSMM device (Tbit = Terabit = 10^{12} bit)
- Mass 25 kg (est.)
- Power consumption: 20-140 W
- Radiation-hardened (= "**rad hard**")
- No moving parts
- Mfr.: Seakr Engineering

August 23, 2022 SC11-51

Field-Programmable Gate Array

- An integrated array of logic elements
- Abbreviation: FPGA
- Logic can be programmed into device after manufacture; retain programming permanently
- Space flight FPGAs have built-in single-event upset (SEU) protection
- Manufacturers: Atmel, Actel, Aeroflex, Xilinx

August 23, 2022 SC11-52

C&DS Topics

- Command & Data System definitions & functions
- Data downlinking & telemetry
- Sampling, and data conversion errors
- Data Encoding
- Fault protection & safing
- Hardware
- **Software**
- Mass & power estimation

August 23, 2022 SC11-53

Spacecraft Software

Building software is like building a cathedral:

1. It takes longer than originally expected;
2. It costs far more than originally budgeted;
3. When it is finally built, YOU PRAY!

August 23, 2022 SC11-54

Standard Software Tasks

1. Initialization after LV separation
2. Housekeeping: collect & store status information
3. Telemetry: send housekeeping to ground
4. Command handling
5. Maintain time clock
6. Supervision/safety
7. Autonomy

Subsystem control:

8. Attitude control
9. Navigation
10. Power management
11. Thermal control
12. Guidance control
13. Software uploads

August 23, 2022 SC11-55

Software Difficult To Create

- Often one-of-a-kind
- Usually multi-tasked, real-time, interrupt-driven
- Requires extreme reliability
- Must be remotely reconfigurable and maintainable
- Often designed while flight HW& mission operations still in flux
  - Interface definitions may occur late
  - ConOps may arrive late
  - Schedules are tightly coupled
- Flight HW & development tools greatly lag ground-based

August 23, 2022 SC11-56

Flight SW Development Issues

- Follows process similar to hardware development, beginning with requirements
- Timing of execution: each operation takes time. Allow for processing, non-synchronization, etc.
  - 1st space shuttle launch scrub
- Number of software lines of code (SLOC or LOC) used to estimate schedule, cost
  - Average cost/LOC **~$10-$25** for fully-tested line of code in 2008, GNC section of JPL*
  - Cost estimation discussed in later section
- "A professional programmer can produce ***ten lines*** of fully-tested, bug-free code per day." †

*Amy Attiyah, Supervisor, Navigation & Mission Design Sys. Engr., JPL
†Brian Muirhead, **High Velocity Leadership: The Mars Pathfinder Approach**, p. 115

August 23, 2022 SC11-57

Software (& Hardware) Design Flow

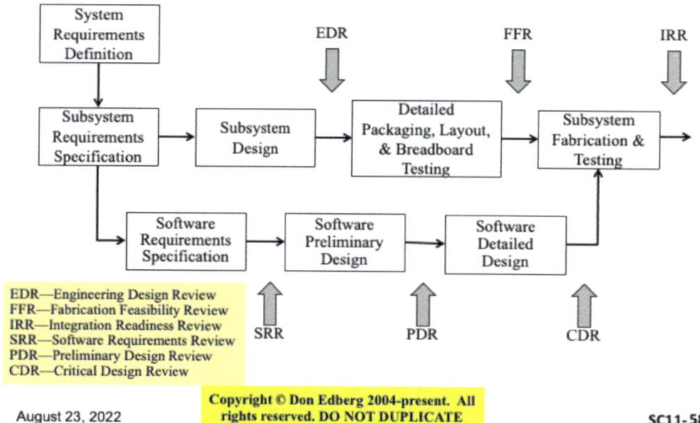

August 23, 2022 SC11-58

Software Complexity of Systems

System	Lines of Code (LOC)
Human comprehension	~5K
BIRD & TerraSAR satellites	90K
A310 control system	400K
Cellular phone	1 M
Automobile	2 M
Space Shuttle	3 M
X-47B UCAV (NGC)*	3.4 M
B-2 Stealth Bomber	4 M
A340 control system	20 M
Windows XP	45 M
Nuclear Missile Defense & Strategic Defense Initiative (est.)	25-100 M

Values courtesy Ley et al, except *Aviation Week

August 23, 2022 SC11-59

C&DS Topics

- Command & Data System definitions & functions
- Data downlinking & telemetry
- Sampling, and data conversion errors
- Data Encoding
- Fault protection & safing
- Hardware
- Software
- Mass & power estimation

August 23, 2022 SC11-60

CDS Mass & Power Estimation

Equipment	Mass (kg)	Power (W)
Telecomm, clock, signal cond., command proc.	30	20
Computer (only)	2	10
Science data processor	15	2 + 1/instrument
Engr. data processor	10	5
Data Storage (SSR)	0.25/Gbit	1/Gbit
Software	massless	negligible

Ref: Brown

August 23, 2022 SC11-61

Mass Properties Query (Dated humor)

NASA needed to calculate the spacecraft's mass; a very important factor. The spacecraft would have a computer. Since computers consist of hardware and software, they obtained the mass of the hardware and then went to the programmer responsible for loading the software.

- *Mass Properties*: "I need to know how much the software for the onboard computer weighs."
- *Programmer*: points at a large stack of punched cards.
- *Mass Properties*: "Wow! That looks heavy!"
- *Programmer*: "Don't worry. We only use the holes."

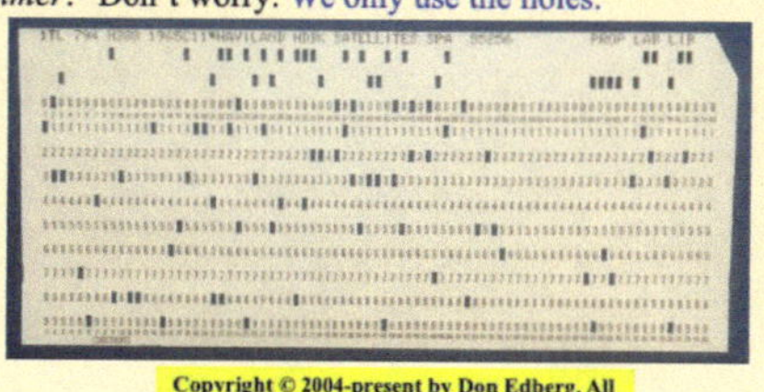

8/23/22 SC11-62

Blank slide (almost)

Spacecraft Testing

Source: Boeing

August 23, 2022

Spacecraft Testing Topics

- Need & philosophy of testing
- Types of testing & sequence
- Testing examples
  - Mechanical testing: SA deployment, spin, vibration (sine & random), shock, acoustic, thermal-vacuum, radio frequency
  - Software testing
- Transport SC & integrate for launch

August 23, 2022

Course Layout

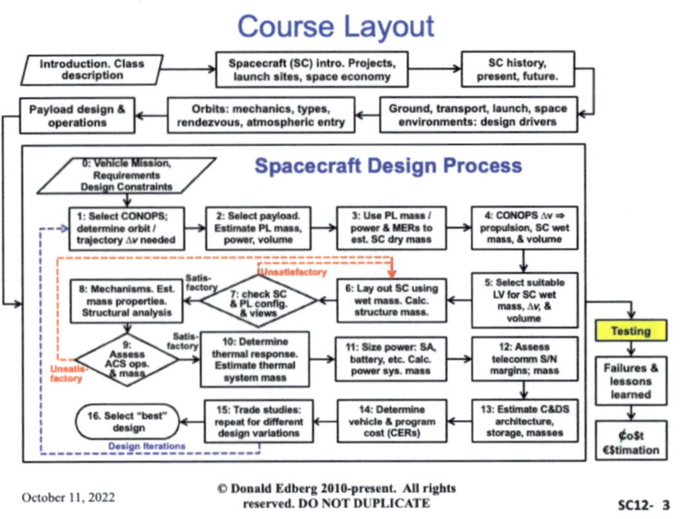

Introduction. Class description
Spacecraft (SC) intro. Projects, launch sites, space economy
SC history, present, future.
Ground, transport, launch, space environments: design drivers
Orbits: mechanics, types, rendezvous, atmospheric entry
Payload design & operations
Spacecraft Design Process
0: Vehicle Mission, Requirements Design Constraints
1: Select CONOPS; determine orbit / trajectory Δv needed
2: Select payload. Estimate PL mass, power, volume
3: Use PL mass / power & MERs to est. SC dry mass
4: CONOPS Δv ⇒ propulsion, SC wet mass, & volume
5: Select suitable LV for SC wet mass, Δv, & volume
6: Lay out SC using wet mass. Calc. structure mass.
Unsatisfactory
7: check SC & PL config. & views
Satis-factory
8: Mechanisms. Est. mass properties. Structural analysis
9: Assess ACS ops. & mass
Unsatis-factory
Satis-factory
10: Determine thermal response. Estimate thermal system mass
11: Size power: SA, battery, etc. Calc. power sys. mass
12: Assess telecomm S/N margins; mass
13: Estimate C&DS architecture, storage, masses
14: Determine vehicle & program cost (CERs)
15: Trade studies: repeat for different design variations
16. Select "best" design
Design Iterations
Testing
Failures & lessons learned
¢o$t €$timation

October 11, 2022

Spacecraft Testing Topics

- Need & philosophy of testing
- Types of testing & sequence
- Testing examples
  - Mechanical testing: SA deployment, spin, vibration (sine & random), shock, acoustic, thermal-vacuum, radio frequency
  - Software testing
- Transport SC & integrate for launch

August 23, 2022

Testing

- Use to locate errors in analytical models & problems in hardware
  - Mechanical (structural)
  - Vibration, shock, & acoustics (VS&A)
  - Thermal
  - RF, electrical, etc.
- Software
- Correct any/all as needed

August 23, 2022

Testing vs. Failure Mechanisms

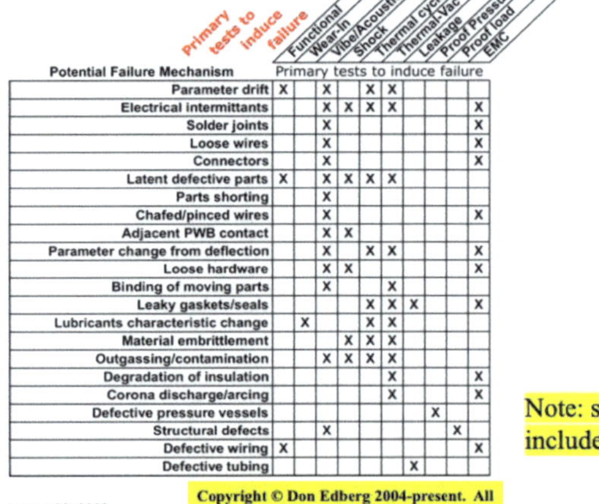

Primary tests to induce failure

Potential Failure Mechanism	Functional	Wear-In	Vibe/Acoustic	Shock	Thermal cycle	Thermal-Vac	Leakage	Proof Pressure	Proof load	EMC
Parameter drift	X		X		X	X				
Electrical intermittants			X	X	X	X				X
Solder joints			X							X
Loose wires			X							X
Connectors			X							X
Latent defective parts	X		X	X	X	X				
Parts shorting			X							
Chafed/pinced wires			X							X
Adjacent PWB contact			X	X						
Parameter change from deflection			X		X	X				X
Loose hardware			X	X						X
Binding of moving parts			X			X				
Leaky gaskets/seals					X	X	X			X
Lubricants characteristic change		X			X	X				
Material embrittlement				X	X	X				
Outgassing/contamination			X	X	X	X				
Degradation of insulation						X				X
Corona discharge/arcing						X				X
Defective pressure vessels								X		
Structural defects			X						X	
Defective wiring	X									X
Defective tubing							X			

Note: software not included in table

August 23, 2022

Strategic Testing Lessons

- Test enough to get out of "infant failure" regime, but not enough to wear things out
- Do independent sanity checks
- Examine test failures until resolved: NO WAIVERS

August 23, 2022

Failure Probability: "Bathtub" Curve

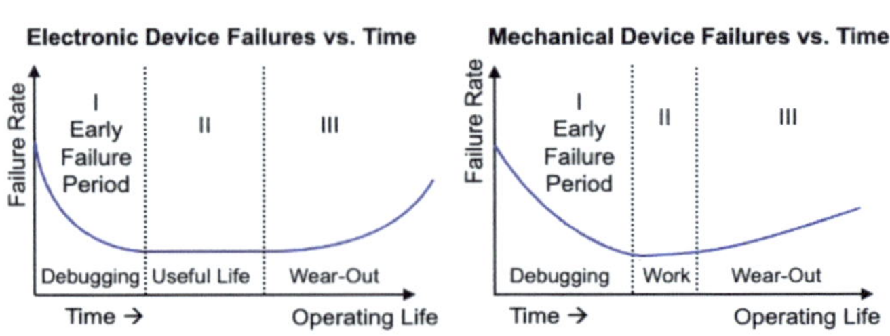

Electronic Device Failures vs. Time
Failure Rate
I
Early Failure Period
II
III
Debugging
Useful Life
Wear-Out
Time →
Operating Life
Mechanical Device Failures vs. Time
Failure Rate
I
Early Failure Period
II
III
Debugging
Work
Wear-Out
Time →
Operating Life

- Most failures tend to occur at the *beginning* or *end* of service
- Result is the "bathtub" curve of probability

courtesy: Fortescue 4[ed].

August 23, 2022

"Classical" Approach to Mechanical Testing

- Structural test article: representative of flight structure, dummy electronic boxes. Used for static, modal survey, and *qual*-level shake test
- Flight-like test article: shake, acoustic, shock testing at *flight* levels
- Engineering test article: for electrical software purposes, simulations, troubleshooting
  - May be from structural model (after mechanical tests completed)

August 23, 2022 SC12- 9

Protoflight Mechanical Testing Approach

- Structural test article
  - Used for static, sine-shake, and acoustic testing at *qualification*-levels. Also used for modal survey, and shock test
  - Refurbished after qual-testing to flight model
- Flight-level test article
  - Used for sine-shake and acoustic testing (qual-level acceptance duration) at *flight*-levels.

Fewer models = "lower cost"

August 23, 2022 SC12-10

Physical Testing & Test Levels

- Hardware is tested to levels determined by the customer
- Levels are chosen to be higher than expected flight levels but less than failure loads, i.e. qualification vs. workmanship testing
- *Testing is an art* and can cause great grief if done improperly

August 23, 2022 SC12-11

Spacecraft Testing Topics

- Need & philosophy of testing
- Types of testing & sequence
- Testing examples
  - Mechanical testing: SA deployment, spin, vibration (sine & random), shock, acoustic, thermal-vacuum, radio frequency
  - Software testing
- Transport SC & integrate for launch

August 23, 2022 SC12-12

Types of Testing

- Subcomponent level tested individually
  - Parts, subassemblies
  - Software
- Whole-vehicle
  - LV: scale model testing (WT, acoustic, etc.); integrated vehicle testing, on-pad tests
  - SC: integrated vehicle testing on test stands and in environmental chambers

August 23, 2022 SC12-13

Typical Test Sequence

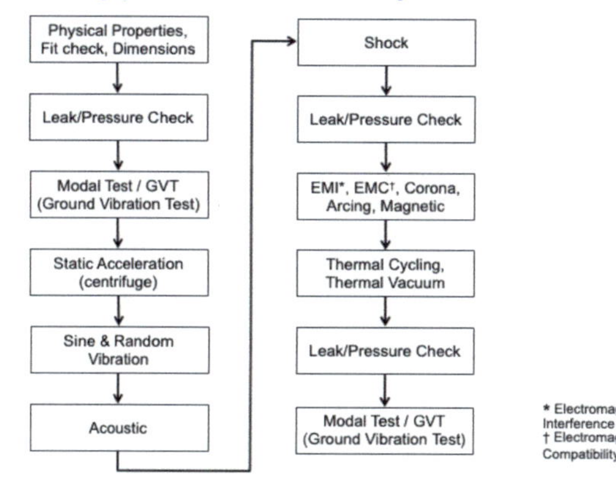

8/23/22 SC12-14

"Chamber of Horrors" Video
(courtesy GSFC)

NASA-Spacecraft-Chamber-of-Horrors.mp4

August 23, 2022 SC12-15

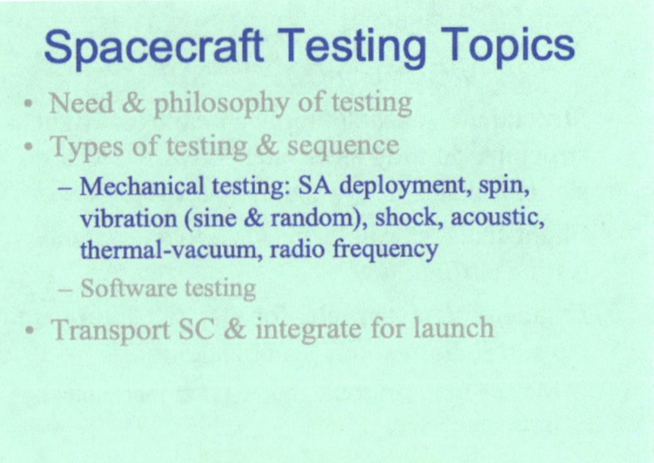

Spacecraft Testing Topics

- Need & philosophy of testing
- Types of testing & sequence
  - Mechanical testing: SA deployment, spin, vibration (sine & random), shock, acoustic, thermal-vacuum, radio frequency
  - Software testing
- Transport SC & integrate for launch

August 23, 2022 SC12-16

Subcomponent Testing (Solar Array Deployment)

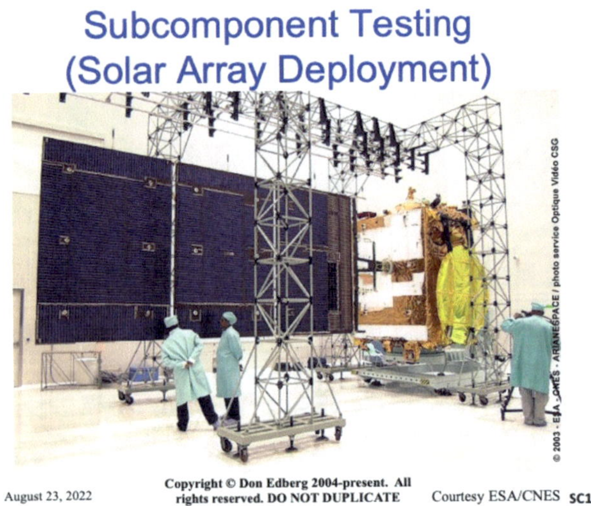

August 23, 2022 Courtesy ESA/CNES SC12-17

Solar Array (SA) Install & Test

AquariusUpdate-Solar-Panels-Attached.mp4 courtesy JPL

August 23, 2022 SC12-18

Full Spacecraft Mechanical & Shock Testing

- Spin testing & Centrifuge tests
- Static proof testing
- Low-level sine sweep or random (modes)
- High-level sine sweep or
- High-level random
- Shock testing (simulates separation shocks, etc.)
- Acoustic testing
- *Repeat* low-level sine after testing (why?)

August 23, 2022 SC12-19

Spin Testing

- Spin testing used to dynamically balance the SC
- Ensures proper moments of inertia and control system performance
- Same balancing as you pay for at the tire store!

MGS Spin Test May 8th, 1996

MGS spin balance.mpg

August 23, 2022 Courtesy Mullard Space Science Laboratory, Surrey SC12-20

Spacecraft Component Vibe Testing on Slip Table

Aerospace

Accelerometers mounted on test specimen measure dynamic responses

Control accelerometers mounted at base of the test fixture (location represents interface to the launch vehicle)

August 23, 2022 SC12-21

Why Vibration Test?

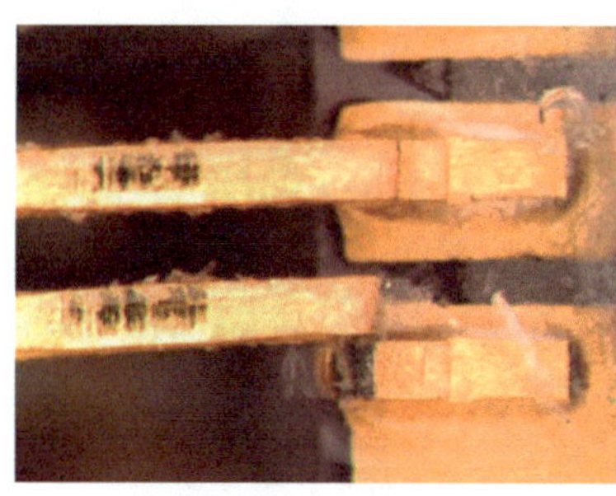
Courtesy Aerospace *Crosslink*

- Cracked field-programmable gate array (FPGA) lead from an electronic unit caused a functional failure
- Testing coupled with stress analyses showed fatigue failure caused by dynamic resonant coupling between circuit card & chassis

August 23, 2022 SC12-22

Vibration Power Spectral Densities (PSDs)

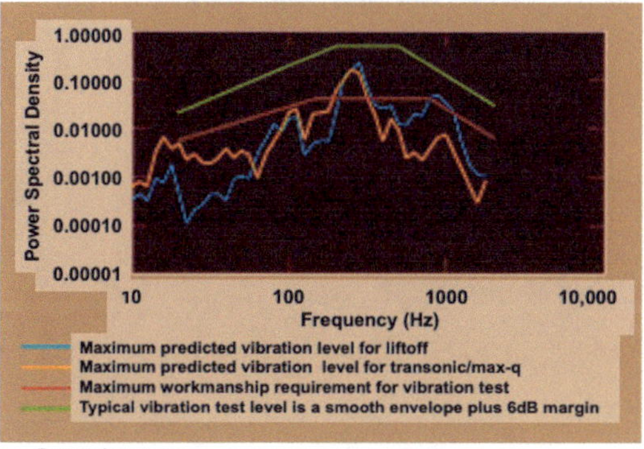

Source: Aerospace

Typical vibration test level used to simulate the LV environment. A 6-dB qualification margin is typically added to the maximum predicted environment to ensure that hardware is sufficiently robust

August 23, 2022 SC12-23

Typical Full-SC Vibration Testing

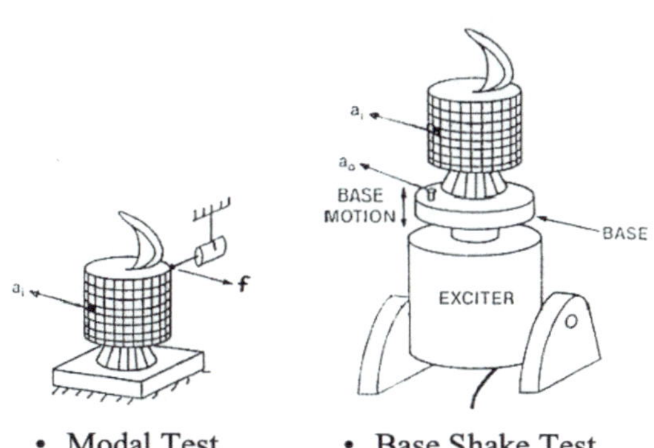

- Modal Test
- Base Shake Test

August 23, 2022 SC12-24

Sirius-1 Spacecraft & Lateral Vibration Shake Table

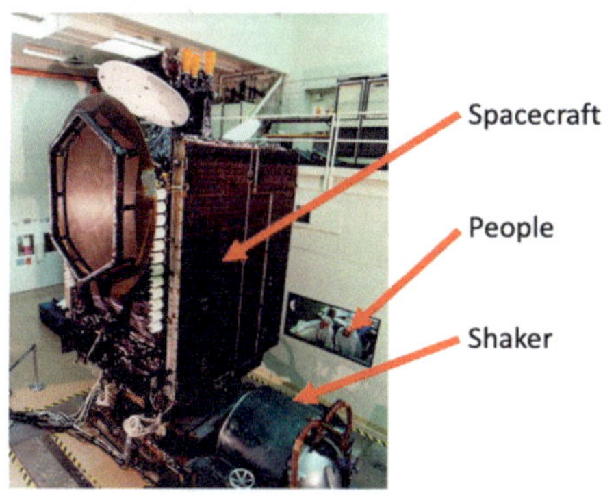

August 23, 2022 SC12-25

Random Base Shake Vibration Test (courtesy D-Orbit)

Base shake at D-Orbit

August 23, 2022 SC12-26

Herschel Axial Sine Sweep Vibration Test (courtesy ESA)

Herschel_Shakertest.mp4

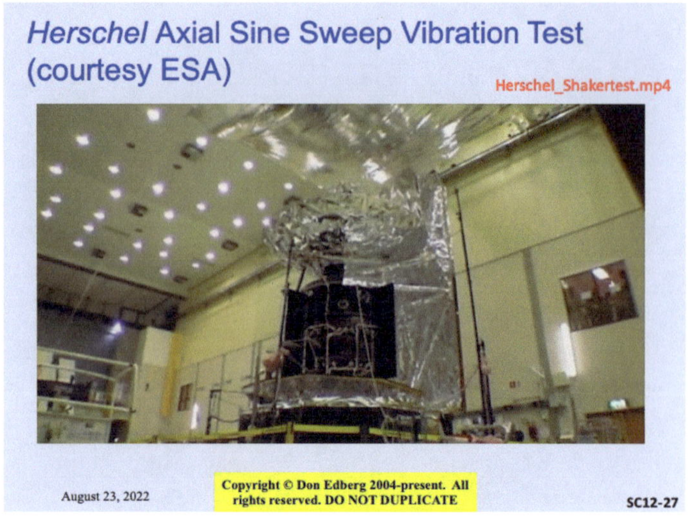

August 23, 2022

SC12-27

Curiosity Random Shake Test

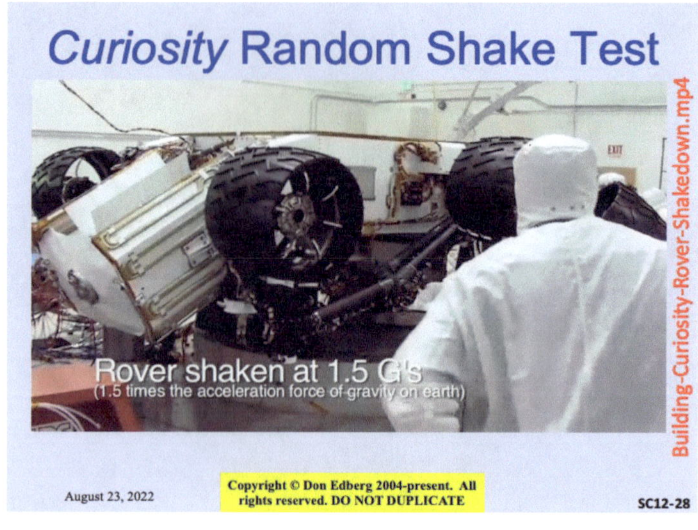

Building-Curiosity-Rover-Shakedown.mp4

August 23, 2022

SC12-28

Shock: Time History & SRS

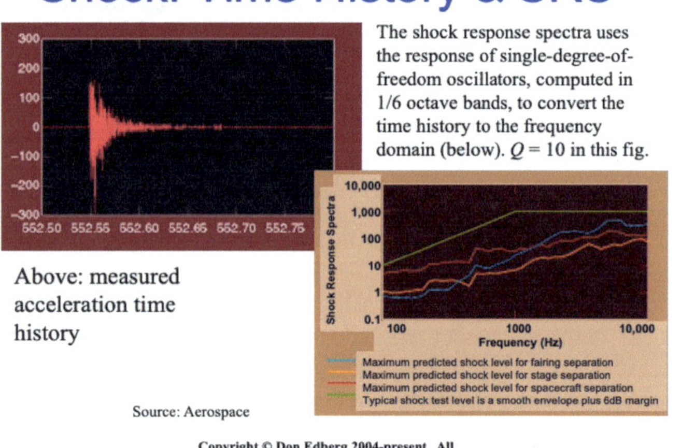

The shock response spectra uses the response of single-degree-of-freedom oscillators, computed in 1/6 octave bands, to convert the time history to the frequency domain (below). $Q = 10$ in this fig.

Above: measured acceleration time history

Source: Aerospace

August 23, 2022

SC12-29

BepiColumbo Shock Test (ESA)

BepiColombo&LVA-sep-test.mov

August 23, 2022

SC12-30

Acoustic Pressure Test Levels

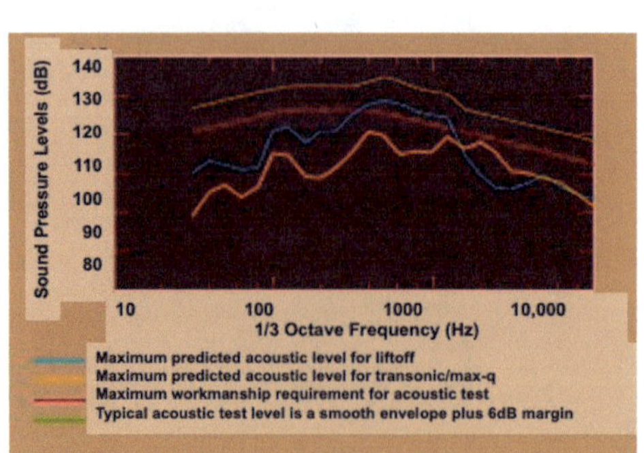

- Simulated launch vehicle environment
- Spectrum is divided into 1/3-octave bands
- Sound pressure level specified in dB, each band
- Frequency range typically from 30 – 10,000 Hz.

Source: Aerospace

August 23, 2022

SC12-31

Acoustic Test Criteria

- Listed in Payload Planner's guide
- Levels and durations depend on type of test
- Acceptance testing: add 0 dB to the maximum predicted environment
- Protoflight testing: +3 dB
- Qualification testing ("qual" testing): +6 dB
- Duration of test? Too long causes fatigue

August 23, 2022

SC12-32

4.2.4.2 Acoustic Testing

- The max flight level acoustic environments defined in § 4.2.3.3 are **increased by 3 dB** for acoustic **qualification** and **protoflight** testing. The acoustic test duration is 120 s for qualification testing and 60 s for protoflight testing. For spacecraft acoustic acceptance testing, acoustic test levels are equal to the max flight level acoustic environments defined in § 4.2.3.3. The acoustic acceptance test duration is 60 s. The acoustic qualification, acceptance, and protoflight test levels for the *Delta II* launch vehicle configurations are defined in Tables 4-9, 4-10, and 4-11.

Delta II payload planner's guide

Acoustic Testing

- Air horns in chamber generate a prescribed sound pressure into confined space and spacecraft
- Microphones located around spacecraft are used to monitor and control pressure levels

Vega-C PLF Acoustic Testing

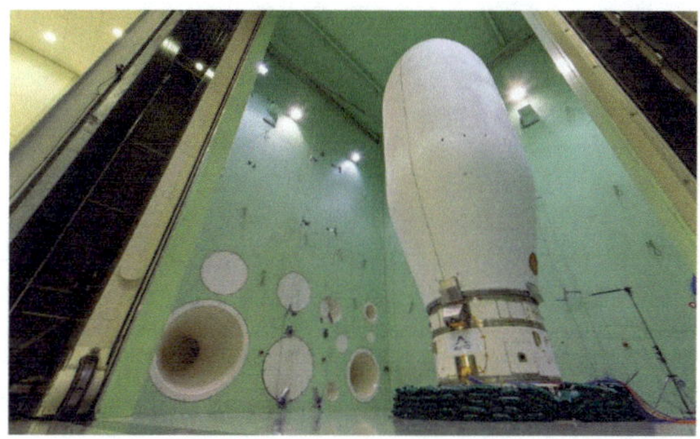

Courtesy ESA

Thermal Vacuum Testing

- Evacuate chamber containing spacecraft to simulate space vacuum
- LN_2 chills walls to simulate space temperature
- Shine quartz lamps to simulate sun

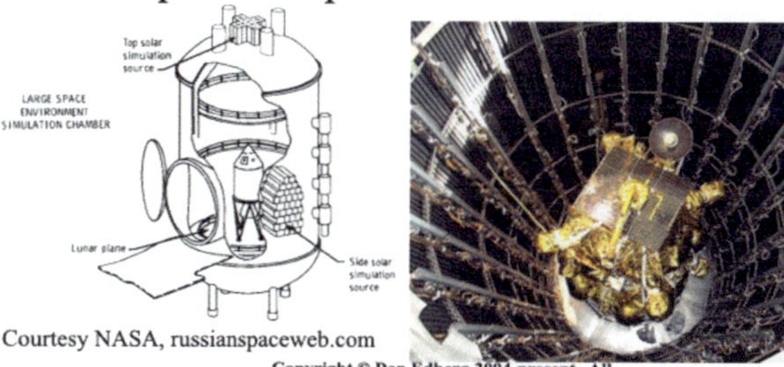

Courtesy NASA, russianspaceweb.com

Thermal Testing Goals & Types

- Screen out components with physical flaws
- Demonstrate that a device can activate & operate in extreme & changing temperatures

The four most common thermal tests are

- *Thermal cycling*
- *Thermal vacuum testing*
- *Thermal balance testing*
- *Burn-in testing*

To 'speed up' thermal testing, *thermal acceleration factors* are utilized

Thermal Testing Types

1. Thermal cycling

- Subjects test article to cycles at hot & cold temperatures in *ambient air* or *gaseous nitrogen* (*not* vacuum)
- Convection enables relatively rapid cycling between hot & cold levels

2. Thermal vacuum testing

- Same thing in a vacuum chamber; cycles are slower, but method provides the most realistic simulation of flight conditions

3. Thermal balance testing

- Conducted in vacuum
- Use dedicated test phases simulating flight conditions to obtain steady-state temperature data that are compared to predictions
- Provides thermal control subsystem verification & data gathering for correlation with thermal analytic models

4. Burn-in testing

- Typically part of thermal cycling tests
- Additional test time is allotted: item is made to operate while temperature is cycled or held at an elevated level

GLL in Thermal-Vac Test

- Simulate space environment
- Avionics powered on/off
- Solar & eclipse simulated

Source: JPL

August 23, 2022

SC12-39

Curiosity Thermal-Vacuum Test

Building-Curiosity-Go-From-Shake-to-Bake.mp4

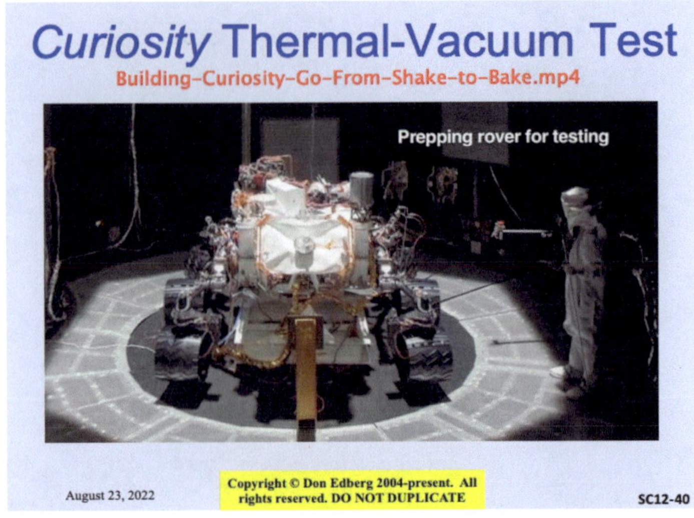

August 23, 2022

SC12-40

Thermal Test Profile Example

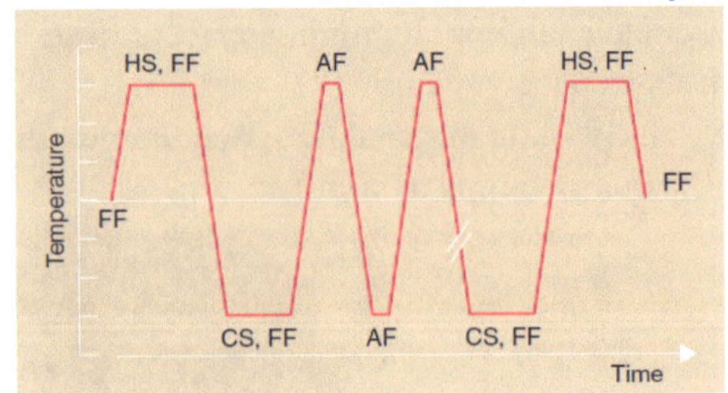

HS = Hot starts
CS = Cold starts
FF = Full functional tests
AF = Abbreviated functional tests
Typically performed at temperature plateaus

Source: Aerospace

August 23, 2022

SC12-41

Thermal Acceleration Factors (*AF*)

- Test: 1000 cycles, ΔT_{test} = +125° – (–55°) = 180 C
- Space application ΔT_{space} = +55° – (–30°) = 85 C
  - Relative humidity assumed equal
  - Difference of relatively short dwell times at upper temps ignored
- For bond wire fatigue failure, $q = 4$ is most appropriate*. So, using the Coffin-Manson relationship†:
  Acceleration Factor $AF = (180 \div 85)^4 = 20.1$
- A 1000-cycle temperature test simulates 20,100 cycles in space, about 3.4 – 4.2 years of LEO ops (in a period of 2 – 2.5 months)
- Mission time simulated is even greater for deep space missions with a minimum of planetary shadowing
- Note: thermal stress levels must not introduce failure modes not existing in field, i.e. temps above insulation outgassing temp or above PC boards' glass transition temps.

Courtesy: *Eric Hoffman; †Wayne Nelson

August 23, 2022

SC12-42

Radio Frequency Testing

- EMI Testing: be sure that ElectroMagnetic *Interference* does not cause problems
- EMC Testing: ensures that different systems are ElectroMagnetically *Compatible*
- Testing usually done in antenna test range or anechoic chamber

Galaxy 14 courtesy Northrop Grumman

August 23, 2022

SC12-43

Electromagnetic Compatibility Test at D-Orbit

August 23, 2022

SC12-44

Compact Range Test of Payload

- Test onboard equipment:
  - Range safety receivers
  - Antennas
  - Etc.

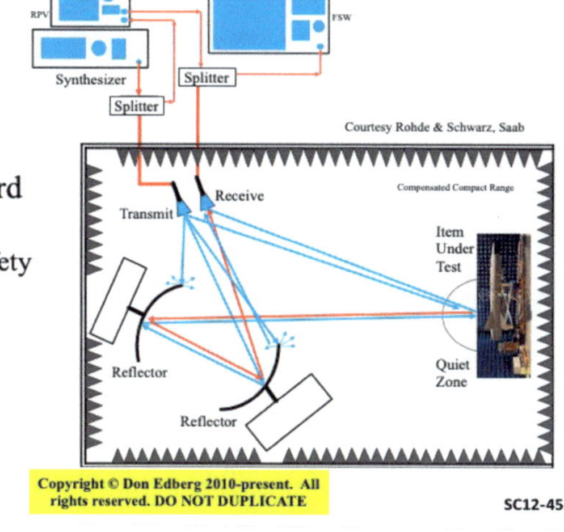

8/23/22 SC12-45

Spacecraft Testing Topics

- Need & philosophy of testing
- Types of testing & sequence
  - Mechanical testing: SA deployment, spin, vibration (sine & random), shock, acoustic, thermal-vacuum, radio frequency
  - Software testing
- Transport SC & integrate for launch

August 23, 2022 SC12-46

Software Design & Testing

Software design parallels hardware process

- Requirements used to develop SOW

After software is completed & successfully compiled:

- Testing
- Walkthrough/Inspection
- Verification and validation (V&V), preferably by independent organization (IV&V)
- Test on spacecraft hardware or simulator

August 23, 2022 SC12-47

Software Testing

- Testing typically include:
  - Some sort of *nominal* testing: test cases are designed to mimic normal operation, &
  - *negative* testing, in which test cases are selected to try to "break" the program
- For example, the software might be run using input values of correct type & within expected range to verify conformance with nominal requirements
- It might also be run using input values & data rates beyond expected ranges

August 23, 2022 SC12-48

"White Box" Software Testing

- Based on detailed knowledge of design, accounts for internal software structure in formulation of test cases & completion criteria
- Most common types
  - Branch testing, which runs through every instruction in each conditional statement in a program
  - Path testing, which runs through every set of conditional statements or branches
- Typically conducted at unit level (i.e., smallest testable component of software) & at unit integration level
- Example: programmer testing his/her own module

Courtesy Eric Hoffman; Aerospace

August 23, 2022 SC12-49

"Black Box" Software Testing

- Based on functional requirements (specification) only
- Disregards software's internal structure & implementation
- Test data, completion criteria, & procedures developed solely to test whether system meets requirements without consideration of how software is coded
- Used at all levels of testing
- Particularly applicable at higher levels of integration where underlying components are no longer visible.
- Example: a Red Team conducting a test

Courtesy Eric Hoffman; Aerospace

August 23, 2022 SC12-50

Software Defect Testing

- Design tests to cause the system to perform incorrectly and expose a defect.
- Interface tests — use knowledge of functional specification, structure, and implementation to design tests that will exercise each object & message type in the system.
- Never permit defect testing to replace static verification (e.g., code walkthroughs, formal methods).

Courtesy Eric Hoffman, ATI SC QA course

Limitations of Software Testing

- Only a partial solution to creating reliable software
- Testing is done to show a program has bugs, but cannot by itself provide an assurance of failure-free operations
- Testing must be pursued in conjunction with other practices in systems engineering, requirements definition, & software development such as inspection, use of automated development aids, static source code & design analysis, & peer review
- Testing cannot occur until after the code is written—about halfway or more through project development.
- Cost of fixing errors rises dramatically as project progresses because more deliverables are affected
  - Requirements errors cost 10 × more to fix in code phase than in requirements phase
- Methods of software verification other than testing (under the broad categories of inspection, analysis, or demonstration) must be used to catch errors in the earlier phases of design.

Software "Error Seeding"

- Process of intentionally adding known faults in a program
  - Use to monitor the rate of detection and removal
  - Yields an estimate the number of faults remaining in the program.
- **Don't forget** to *remove the test faults!* (***Red Tag*** items)

Code Walkthrough/Inspection

- Requires study of the requirements, design, & code *prior* to the actual review, after 1st good compilation
- Includes some or all of the following players:
  - Presenter (lead reader, usually the designer/programmer)
  - Moderator (coordinator, chairman)
  - Recorder (scribe, secretary)
  - 1-2 other technical reviewers (* = optional)
  - Maintenance oracle*
  - Standards bearer*
  - User representative*
  - System liaison (system engineer)*
- Perform module by module
- Can be highly effective

Ref: M Fagan, "Design and Code Inspection," *IEEE Trans. Software Engr.*, July 1986

Software Reviews Pay Off

Types of errors found in ~6,877,000 debugged SLOC* on 28 projects
(*software lines of code, including comments)

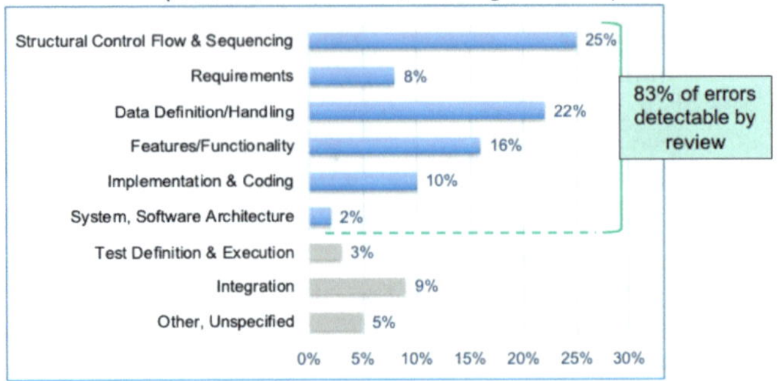

Ref: Blum, **Software Engineering: A Holistic View**

Curiosity Software Testing

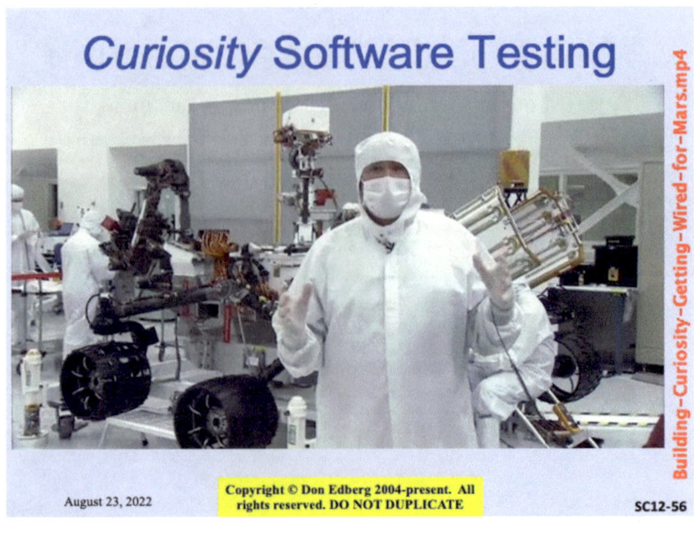

Building-Curiosity-Getting-Wired-for-Mars.mp4

Spacecraft Testing Topics

- Need & philosophy of testing
- Types of testing & sequence
- Testing examples
  - Mechanical testing: SA deployment, spin, vibration (sine & random), shock, acoustic, thermal-vacuum, radio frequency
  - Software testing
- **Transport SC & integrate for launch**

August 23, 2022

SC12-57

Curiosity: Prepare, Pack, & Ship

Building-Curiosity-The-Big-Move.mp4

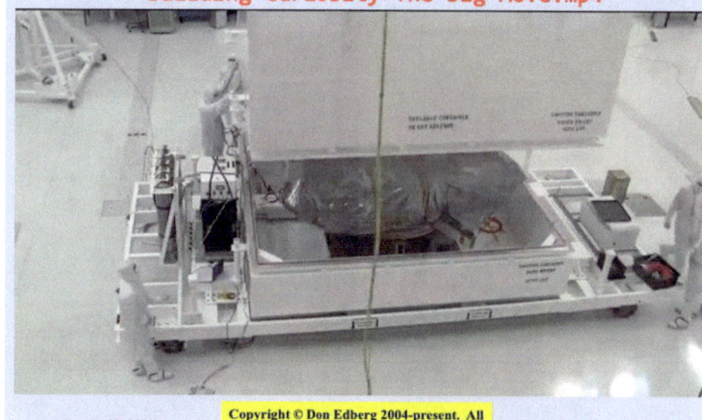

August 23, 2022

SC12-58

After Successful Test: Transport to Launch Site, Test Again, & Integrate with LV

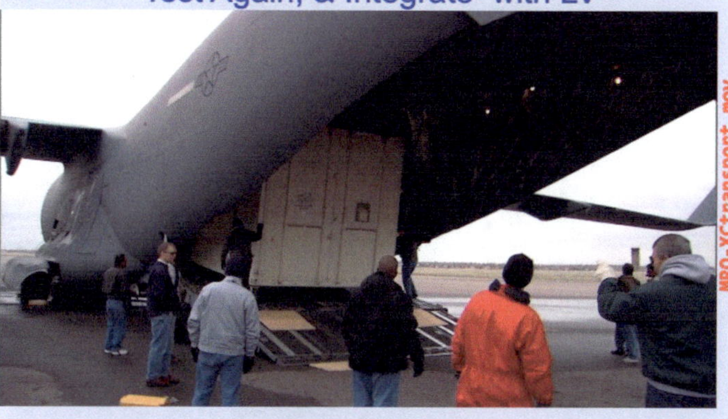

August 23, 2022

SC12-59

Gremlin Did His Best to Trip Up Mars rover

His purpose: To make sure the mission team was prepared to deal with space junk, lost signals and other challenges.

BY AMINA KHAN, **Los Angeles Times**, Aug. 4, 2012

The Curiosity rover is making its final descent to the Martian surface, and all eyes in the Jet Propulsion Laboratory's control room are glued to computer screens. Except for Rob Manning's. The chief engineer for the rover mission is glancing at his watch, eager for his window of opportunity.

He sneaks out of the control room where the simulation is underway and heads down the street. There, in a cavernous building, an SUV-sized rover sits on a bed of sand.

Manning reaches for a radio connected to the rover and flips a switch.

Back in the control room, the signal cuts out. The flummoxed engineers sit in silence.

ROB MANNING, chief engineer for the mission, spends part of his time thinking up challenges to hurl at the Mars Science Laboratory team.

Manning congratulates himself on a job well done.

Regardless of whether the Curiosity rover lands safely on Mars on Sunday night, Manning can say, with a clear conscience, that he did his best to ruin it.

Manning heads a team of hundreds of engineers who design, test and operate the Mars Science Laboratory, as the rover mission is officially known. But for nine days, he becomes his own team's worst enemy. The cheerful veteran of two previous rover missions devises horrible scenarios for the team to face. He throws solar flares at the rover and pokes holes in its fuel system.

You know Murphy's law, that anything that can go wrong will go wrong? "Well, I'm Murphy," Manning said.

August 23, 2022

SC12-60

This sabotage has a purpose: to make sure the rover team members can deal with any challenge that space throws at them.

Manning's high jinks "push the team and your capabilities to the limit," said deputy mission manager Brian Portock, one of Manning's victims. "That causes you some stress, and you try and prepare the best you can for that."

Botching a fake landing takes work. Months in advance, Manning pulled about a dozen engineers off their normal duties to help him plan his devious scenarios. To do the job right, he must find ways to outsmart himself. "Being a gremlin allows me to soul-search and look at all the things that I missed," he said.

With help from his temporary rogues, he hatches another plan: push the rover off course.

In this scenario, a micrometeorite the size of a large grain of sand gets past the spacecraft's protective aluminum-and-Mylar blankets and punches a tiny hole in a fuel line. It doesn't put much of a dent in the fuel levels, but in a matter of days the tiny bit of thrust from the leaking hydrazine nudges the spacecraft off its course.

To make the error look believable, team members adjust the pressure in the fake rover's fuel tank and modify the Doppler signal that tells the craft how far it is from Earth. As a bonus, there's a Catch-22: To push the spacecraft back on course, the engineers need to give the spacecraft some gas. But using a broken fuel system could cause more problems.

"That's pretty evil," said victim Jeff Simmonds, who manages the payload of instruments aboard the rover.

Manning's deviousness knows few bounds. He turns some of his victims into spies so he can maximize his mayhem.

August 23, 2022

SC12-61

"I'm really good at playing dumb," said double agent Tracy Neilson, a fault protection engineer who sent an instant-message to her purported adversaries when it was time to stick a new kink in the system.

Manning also has the power to pull key problem-solvers off the fix-it team and force the remaining engineers to devise their own solutions, said Beth Fabinsky, one of Manning's co-gremlins. Rather than pretending these employees are out sick — Mars team members tend to show up for work even when ill — the story line has them winning the lottery or taking off to Tahiti.

With fake data emanating from the test rover's many instruments, sometimes things don't match up and the simulation goes awry in ways Manning never intended. In those cases, he and his minions issue a card telling engineers to ignore the problem and offering a facetious explanation. One card explains that the lab's radio antennas came under attack by "evil possessed clowns."

The cards ensure the engineers pay attention to the right mistakes. The goal is not to see the team fail, but to teach them how to deal with unforeseen challenges.

It's also a valuable team-building exercise, Manning said: "You need to figure out who the core people are who can solve your problem, listen to them carefully — and then trust them."

Still, for all his good intentions, the chief engineer seems to take an unholy glee in his stint on the dark side.

Shortly after the radio signal cuts out in the control room, the team calls a meeting to figure out what happened. As they gather around a conference table, their eyes turn to Manning, knowing he's to blame.

He pulls out a pair of red sequined devil horns, sticks them onto his head and grins.

August 23, 2022

SC12-62

Spacecraft Integration: Attach to LV

Terra spacecraft being encapsulated into payload fairing. MOPITT instrument is outlined. (Photo courtesy of NASA).

NASA

Payload Encapsulation

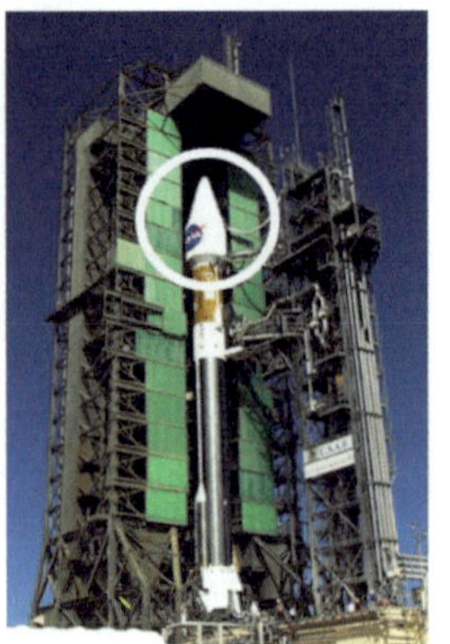

Courtesy Lockheed Martin

Scale Model Dynamic Testing

In general, geometric scale factors associated with physical parameters may be expressed as a length scale factor raised to an appropriate power (see ref. II-4) if similarity of materials is maintained between the model and prototype structures. The physical parameters of importance in the present paper may be expressed in terms of length l, mass m, and acceleration a, and the following table of scale factors for the 1/10-scale model is valid:

Parameter	m-l-a relationship	Ratio of model to full scale	Parameter	m-l-a relationship	Ratio of model to full scale
Length, displacement, etc.	l	10^{-1}	Velocity	$(la)^{1/2}$	1
Mass	m	10^{-3}	Frequency	$(a/l)^{1/2}$	10
Acceleration	a	10	Angular velocity	$(a/l)^{1/2}$	10
Force	ma	10^{-2}	Area moment of inertia	l^4	10^{-4}
Longitudinal stiffness	ma	10^{-2}	Mass moment of inertia	ml^2	10^{-5}
Stress	ma/l^2	1	Bending stiffness	mal^2	10^{-4}
Pressure	ma/l^2	1	Torsion stiffness	mal^2	10^{-4}
Modulus of elasticity	ma/l^2	1			

Table from NASA TN-D-5831 refers to Regier, Arthur: The Use of Scaled Dynamic Models in Several Aerospace Vehicle Studies. Paper presented at ASME Colloquium on Use of Models and Scaling in Simulation of Shock and Vibration (Philadelphia, Pa.), Nov. 1963.

Blank slide

Lessons To Be Learned From Some Spacecraft Failures

Genesis: Courtesy NASA

August 23, 2022

(50) SC13- 1

Failures & Lessons Learned: Topics

- Reliability and failures
- Five common mistakes to look out for
  - Could the sign be wrong?
  - How will last-minute configuration changes be verified?
  - Can vehicle survive computer crash?
  - Circuit overcurrent protection?
  - Pyros cause unexpected damage?
- Research history & more examples
- Redundancy
- Best practices to avoid failure

August 23, 2022

SC13- 2

Course Layout

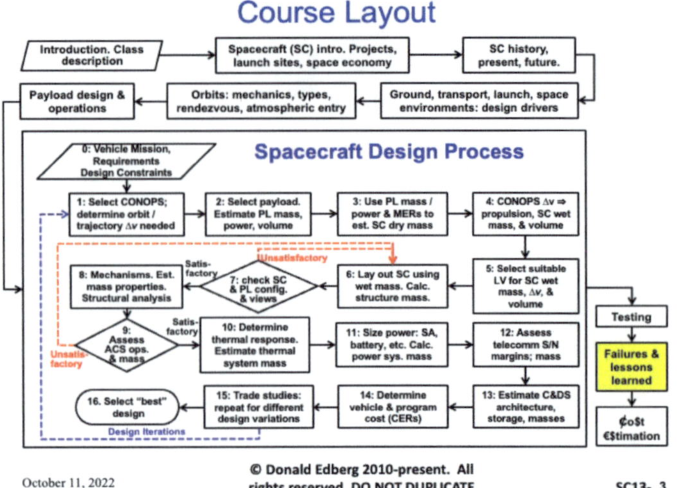
Introduction. Class description
Spacecraft (SC) intro. Projects, launch sites, space economy
SC history, present, future.
Payload design & operations
Orbits: mechanics, types, rendezvous, atmospheric entry
Ground, transport, launch, space environments: design drivers
Spacecraft Design Process
0: Vehicle Mission, Requirements Design Constraints
1: Select CONOPS; determine orbit / trajectory Δv needed
2: Select payload. Estimate PL mass, power, volume
3: Use PL mass / power & MERs to est. SC dry mass
4: CONOPS Δv ⇒ propulsion, SC wet mass, & volume
Unsatisfactory
Satis-factory
8: Mechanisms. Est. mass properties. Structural analysis
7: check SC & PL config. & views
6: Lay out SC using wet mass. Calc. structure mass.
5: Select suitable LV for SC wet mass, Δv, & volume
Testing
Satis-factory
9: Assess ACS ops. & mass
Unsatis-factory
10: Determine thermal response. Estimate thermal system mass
11: Size power: SA, battery, etc. Calc. power sys. mass
12: Assess telecomm S/N margins; mass
Failures & lessons learned
16. Select "best" design
15: Trade studies: repeat for different design variations
14: Determine vehicle & program cost (CERs)
13: Estimate C&DS architecture, storage, masses
Design Iterations
¢o$t €$timation

Failures & Lessons Learned: Topics

- Reliability and failures
- Five common mistakes to look out for
  - Could the sign be wrong?
  - How will last-minute configuration changes be verified?
  - Can vehicle survive computer crash?
  - Circuit overcurrent protection?
  - Pyros cause unexpected damage?
- Research history & more examples
- Redundancy
- Best practices to avoid failure

Over 40 Years, Reliability Record Remains Spotty

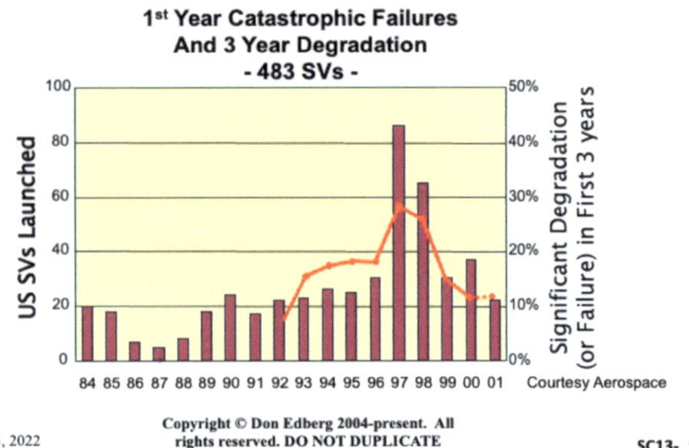
1st Year Catastrophic Failures
And 3 Year Degradation
- 483 SVs -
US SVs Launched
Significant Degradation (or Failure) in First 3 years
0
20
40
60
80
100
0%
10%
20%
30%
40%
50%
84 85 86 87 88 89 90 91 92 93 94 95 96 97 98 99 00 01
Courtesy Aerospace

Why Do Spacecraft Fail?

Studies/surveys show, in order of importance:

1. Poor design
2. Misjudged environments
3. Software
4. Human error, particularly mission ops
5. Interconnects
6. Mechanically-deployed systems
7. Piece-part failure

Refs: H. Hecht and R. Fleeter

Failures & Lessons Learned: Topics

- Reliability and failures
- Five common mistakes to look out for
  - Could the sign be wrong?
  - How will last-minute configuration changes be verified?
  - Can vehicle survive computer crash?
  - Circuit overcurrent protection?
  - Pyros cause unexpected damage?
- Research history & more examples
- Redundancy
- Best practices to avoid failure

Five Common Mistakes to Look Out For

- Most common mistakes gathered in an Aerospace report, "Five Common Mistakes Reviewers Should Look Out For."
- Available to U.S. government agencies and contractors
- Includes reviewer's checklist titled "100 Questions for Technical Review" & all lesson bulletins published through June 2007

The Five Questions:

1. Could the sign be wrong?
2. How will last-minute configuration changes be verified?
3. Can the vehicle survive a computer crash?
4. Is the circuit overcurrent protection adequate?
5. Can pyros cause unexpected damage?

August 23, 2022 SC13- 9

1. Could The Sign Be Wrong?

- *TERRIERS* spacecraft lost because one torque coil had to be installed with a phase opposite of that of the other two — change was not incorporated into flight software
- *Skipper* solar panels connected backward, draining battery, mission failed 1996. Electric current *magnitude* verified, but not direction! (Note: foreign unit utilized a positive ground.)

August 23, 2022 SC13-10

Genesis

- Mission nearly flawless until recovery attempted
- Parachutes did not deploy
- Cause: g-switches installed *backwards*!

August 23, 2022 *Genesis*: Courtesy NASA SC13-11

G-Switch Orientation

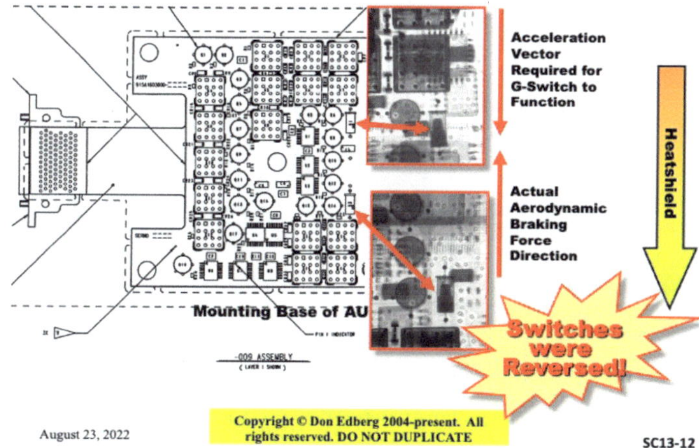

August 23, 2022 SC13-12

Genesis String of Events

- Schematic copied from **Stardust**
- Box CDR lacked technical content
- Verification requirements not clear
  - Centrifuge test expected (in CDR package), but not required. Verification matrix had test, but no detail
  - Systems Engineering did not have to sign off on Subsystem plans
- Designer verified **function** (open/close) of switches; Systems Engineering believed **orientation** of switches were verified
- Electrical designer incorrectly performed orientation verification via Mechanical drawing inspection
- Red Team review assumed design was correct because it was a "heritage" design
- Systems Engineering did not close the loop with the designer
  - Systems Engineering not required to review test result

August 23, 2022 SC13-13

2. How Will Last-Minute Configuration Changes Be Verified?

- SC frequently modified after factory testing
  - Placeholder blankets swapped out, flight connectors mated, databases updated, brackets installed to secure hardware at launch site
  - Non-flight items (test plugs & dust covers) removed
  - Last-minute changes made in the heat of a countdown have caused several failures: late installations and removals difficult to verify

August 23, 2022 SC13-14

Tethered Satellite System (*TSS*) Problem

- Late analysis of *TSS* uncovered a structural shortfall in the deployment mechanism
- Necessary to add a bolt to design
- Original design engineer, thousands of miles away, could not see how the modified hardware fit
- Bolt protruded against a traveling ball nut
- *TSS* could not deploy

Common Problems

- Mistakes are often repeated
  - Incomplete requirement implementation
  - Improper changes or reuse
  - Inadequate configuration management process
  - ……

"We do not invent new mistakes, we just repeat old mistakes." *— Bill Ballhaus (CEO, Aerospace)*

What Is Configuration Management?

- Design Specs
- Purchase Specs
- Interface Control Documents
- Design Reviews
- Drafting Standards
  - Content & format
  - Checking
  - Release
  - Changes
- Change Control. & Incorporation
- Change Control Board
- Software Problem Reports
- SW Unit Dev. Folders
- Drawing Nos., Serial Nos.
- Fabrication Controls
  - Processes
  - Fabrication control cards
  - Workmanship standards
- Parts & material traceability
- Non-conformances
- Deviations &Waivers
- Material Review Board
- Configuration Accounting
- Test plans, procedures, data
- sheets
- Configuration Audits
  - Functional
  - Physical
- As-Built Documentation

Courtesy Eric Hoffman

NOAA-N 'Drop Test' Sept. 2003: D'oh!

Be Careful! ($135M Repair Bill)

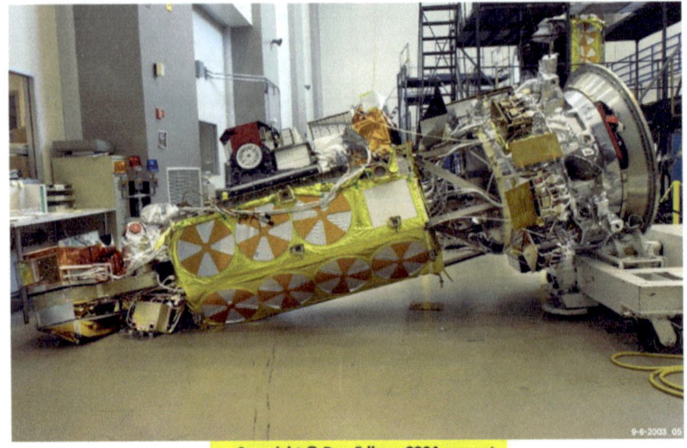

NOAA-N Sequence of Events

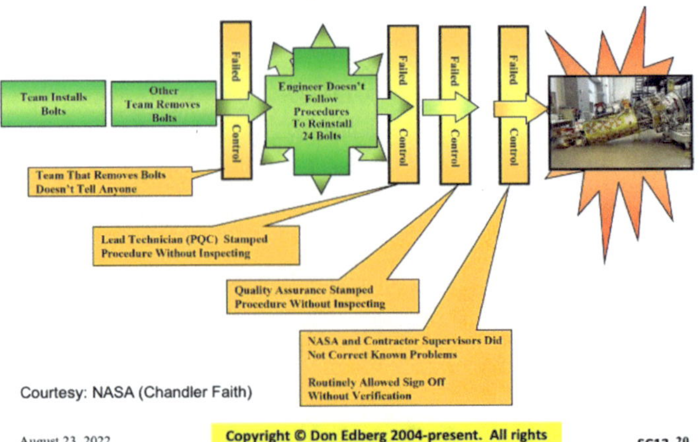

Courtesy: NASA (Chandler Faith)

NOAA-N Dropped at Lockheed 9/6/2003

- Spacecraft fell while being repositioned from vertical to horizontal
- 24 bolts missing from "turn over cart" due to two errors:
  - Technicians from another satellite program removed bolts from NOAA cart without proper documentation.
  - NOAA team failed to follow procedure to verify configuration of "turn over cart" since they had used it a few days earlier.
- Impact [*ha-ha*] on program and schedule:
  - Fall caused **$135M** damage. Significant rework & retest required.
- Immediate safety actions:
  - Prevent spacecraft from rolling
  - Discharge batteries
  - Depressurize the propulsion system
- Corrective actions
  - Lockheed Martin formed an Accident Review Team with GSFC
  - NOAA-N under guard, all records impounded, and personnel interviewed
  - After safety issues addressed, spacecraft damage assessed

August 23, 2022 SC13-21

3. Can Vehicle Survive Computer Crash?

- *Clementine* computer froze immediately after a thruster firing, causing all propellant to be used
- *Near Earth Asteroid Rendezvous* (*NEAR*) suffered a similar computer crash during which its thrusters fired thousands of times, but each firing instantly cut off by the still-operative watchdog timer: NEAR survived.

August 23, 2022 SC13-22

Computer Errors

- Computer errors often caused by subtle timing or memory glitches
- Every vehicle should be able to gracefully handle a computer fault:
  - Reverting to the "last known good state"
  - Reboot without being stuck in endless reset cycles
  - Remain in a safe mode
  - Recover from low bus voltage
  - Etc.

August 23, 2022 SC13-23

4. Circuit Overcurrent Protection?

- Shorts, plasma arcing cause many failures. Arcing most frequently results from pure tin plating that grows conductive filaments (commonly known as "whiskers")
- Numerous satellites, such as *Galaxy VII*, had been disabled because whiskers grew on the relays of their control processors, causing arcing that blew the fuses
- See *space environment effects* slides for whisker information

August 23, 2022 SC13-24

Fuses, Circuit Breakers, Etc.

- According to **Space Systems Failures**: "It may seem absurd that a satellite should have fuses at all, since it is generally impractical to replace them. However, fuses can protect a satellite during ground processing, when all manner of mishandling can occur."
- Careful design required to prevent the safety devices from misbehaving
- *Solar Max* fuses burned when residual gas leaked out, allowing filament to burn out

August 23, 2022 SC13-25

IMAGE Satellite

- Single-string transponder was protected by a circuit breaker
- Radiation tripped the breaker, turning the transponder off
- Unfortunately, a design flaw caused the breaker to misreport itself as "on"
- Onboard fault-management software prevented from resetting the circuit
- Malfunction disabled uplink, making ground rescue impossible

August 23, 2022 SC13-26

5. Pyros Cause Unexpected Damage?

- Premature firings caused the Wide-field Infrared Explorer (*WIRE*) spacecraft to open its telescope cover
- A logic controller chip unexpectedly asserted all outputs during power-on
- Both "arm" and "fire" inhibits were commanded by the same chip (a common design oversight), failed together
- All cryogens vented, mission ended 1st orbit

August 23, 2022 SC13-27

Unexpected Pyro Firing 2

1964 *Orbiting Solar Observatory* being mated to 3rd stage at Cape Canaveral

- Static electricity triggered igniter
- Searing gases burned 11 technicians and killed three

August 23, 2022 SC13-28

Unexpected Pyro Firing 3

Launch readiness work on Brazilian *VLS-1* V03 launch vehicle in Alcantara, Brazil, 2003:

- Electrostatic discharge introduced an arc in the unshielded pyro circuits, setting off the initiator for solid rocket motor
- Motor had no *safe-arm* devices to contain an accidental flash
- Vehicle caught fire & exploded on the pad, killing 21 workers, injuring 24

Courtesy: Wikipedia

8/23/22 SC13-29

Solid Propellant Accident in Brazil, 2003

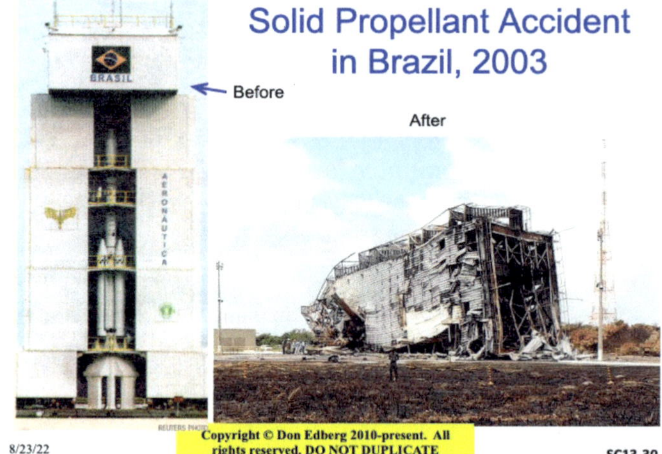

8/23/22 SC13-30

RSS "Safe-Arm" Device (*Saturn* S-IC)

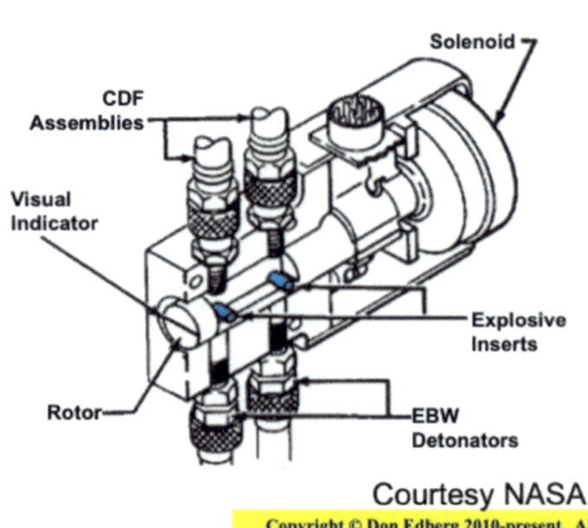

Courtesy NASA

- Remotely-controlled electromechanical ordnance device
- Used to make safe & arm PDSs
- Completed & interrupted explosive train by remote control
- Provides position indications to remote monitoring equipment

8/23/22 SC13-31

Failures & Lessons Learned: Topics

- Reliability and failures
- Five common mistakes to look out for
  - Could the sign be wrong?
  - How will last-minute configuration changes be verified?
  - Can vehicle survive computer crash?
  - Circuit overcurrent protection?
  - Pyros cause unexpected damage?
- **Research history & more examples**
- Redundancy
- Best practices to avoid failure

August 23, 2022 SC13-32

Researching History

- *WIRE* mission failed because of a start-up quirk in a logic controller previously described in NASA's "Application Notes"
- A *Maxus* rocket crashed because drained hydraulic fluid started a fire that burned through a guidance cable. Several programs redesigned fluid drains or added cable insulation. *Athena* program did not do so: maiden flight suffered exactly the same failure.

August 23, 2022 SC13-33

Galileo (GLL) Spacecraft (1989-2003)

HIGH-GAIN ANTENNA

LOW-GAIN ANTENNA
SUN SHIELDS
EXTREME ULTRAVIOLET SPECTROMETER
STAR SCANNER
THRUSTERS (2 places)
PROBE RELAY ANTENNA
JUPITER ATMOSPHERIC PROBE
RADIOSOTOPE THERMOELECTRIC GENERATORS (RTG) (2 places)
Engineering
Fields and Particles
Probe
Remote Sensing
PLASMA-WAVE ANTENNA
MAGNETOMETER SENSORS
ENERGETIC PARTICLES DETECTOR
PLASMA SCIENCE
HEAVY ION COUNTER (BACK)
DUST DETECTOR
RETROPROPULSION MODULE
ABOVE: SPUN SECTION
BELOW: DESPUN SECTION
RTG
SCAN PLATFORM, CONTAINING:
- ULTRAVIOLET SPECTROMETER
- SOLID-STATE IMAGING CAMERA
- NEAR-INFRARED MAPPING SPECTROMETER
- PHOTOPOLARIMETER RADIOMETER

8/23/22 34 urtesy JPL

Galileo at Jupiter Planned vs. Actual (Jammed Antenna)!

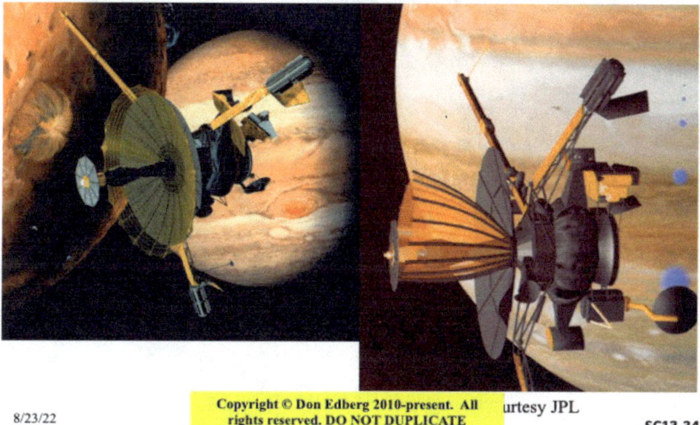

8/23/22 urtesy JPL SC13-34

Galileo Antenna Failure Blamed on Transportation Loads

- GLL made *three* trips from California to KSC before launch
- Transportation loads believed to have caused GLL HGA (high gain antenna) to jam from removal of lubrication due to vibration during transport
- Solution: no more deployable antennae on deep-space missions!

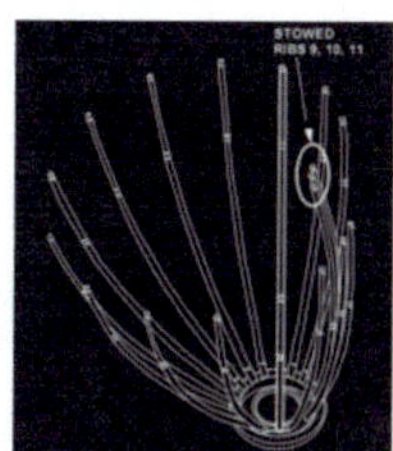

Galileo jammed antenna figure courtesy JPL

8/23/22 SC13-35

LISTEN To Your Hardware: *SEASAT*

- Slip ring design flown before, but:
- Previous project failed powered-on vibe
  - Debris across adjacent rings with opposite polarities shorted
  - Rewired to remove alternating polarities
- Notified of the redesign, SEASAT program felt the change did not apply *because its slip rings are unpowered during launch*
  - Deleted thermal vacuum test of slip rings

A massive arc destroyed the power system on-orbit

Conductive Springs
Ring Separators
Brush
Rings
Rotor
Solar Array
Bus Power Distribution Unit

August 23, 2022 Courtesy Aerospace SC13-37

Hubble Mirror Aberration

- A "null corrector" guided mirror polishing
  - Corrector set up with a metering rod
  - A non-reflective cap covered the rod's tip

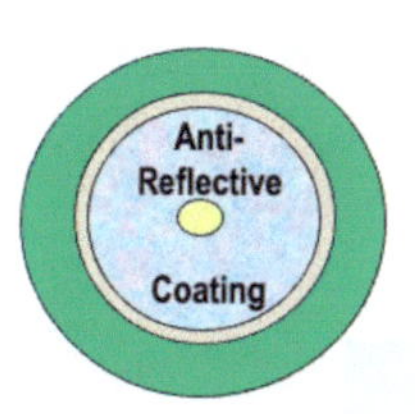

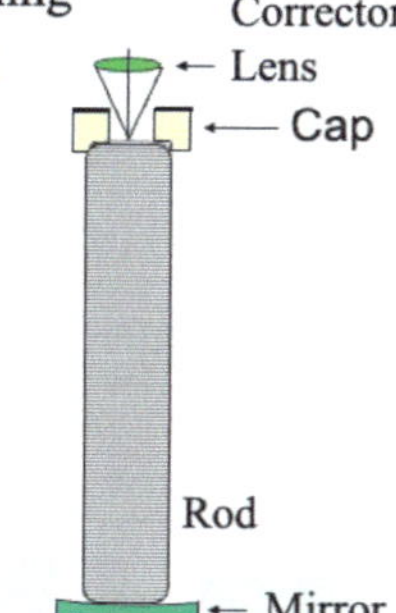

August 23, 2022 Courtesy Aerospace SC13-38

Independent Check (ii)

- Missing speck of paint induced error: laser beam focused on end cap instead of end of rod itself, producing incorrect length by 1.3 mm
- Fixture manufactured incorrectly, and several cheap washers were added to displace the cap to "proper" distance!
- Two independent tests indicated shape was *incorrect* but contradictory indicators were disregarded due to lack of confidence in *test fixture*!
- Caused spherical aberration to be ground into mirror

Chipped Coating

1.3 mm

As Designed — Actual

Courtesy Aerospace

August 23, 2022 SC13-39

Mars Climate Orbiter (MCO): RIP 1999

- Contractor-builder Lockheed-Martin; JPL-monitor
- Contractor used English units (in/s); JPL in SI (cm/s)
- Non-conversion of English units caused guidance error
- Too late to make sufficient course correction, spacecraft burned up in Mars' atmosphere

Courtesy NASA

August 23, 2022 SC13-40

Schematic MCO Encounter Diagram

Not to Scale

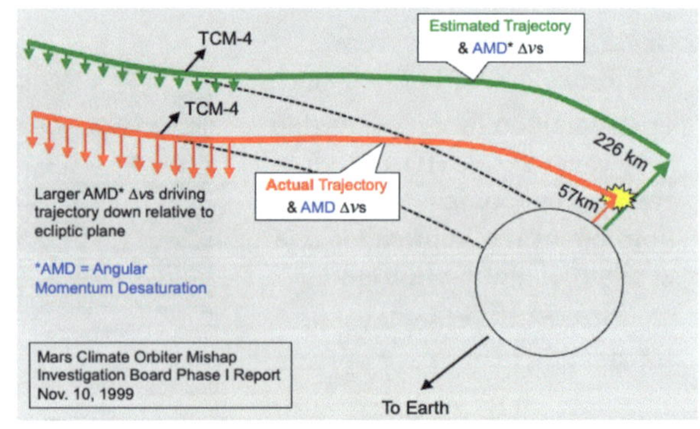

August 23, 2022 SC13-41

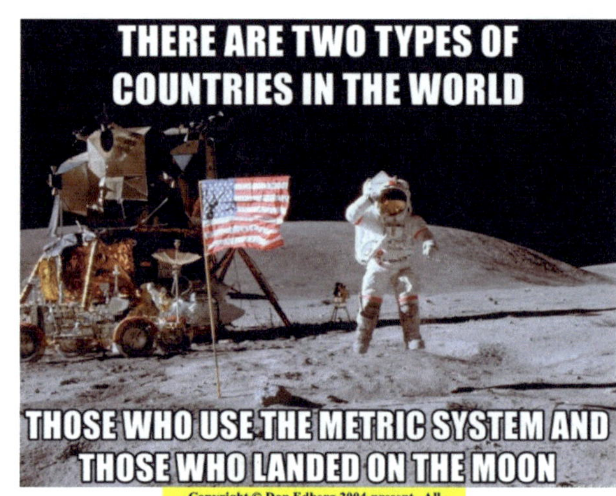

August 23, 2022 SC13-42

Mars Polar Lander (MPL) Mission (1999)

Lander crashed into Mars surface

- Touchdown Sensing System sensed leg deployment vibrations & interpreted them as 'touchdown,' causing premature engine shutdown
- *Mars Polar Lander* test program flaw
  - Touchdown Sensors wiring error in system test
  - System test not repeated with wiring correction
- Software design did not include detection/protection for spurious signals
- Lost: $165M

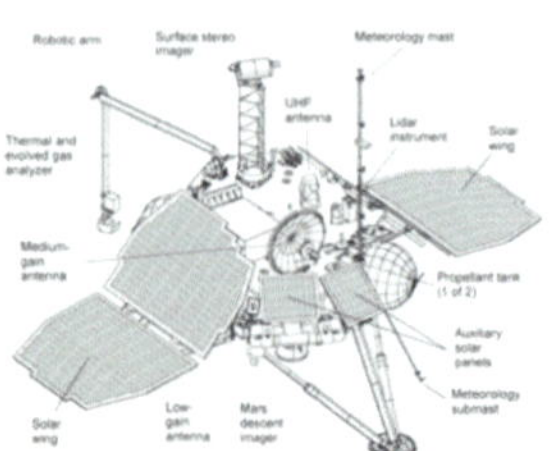

August 23, 2022 SC13-43

Phobos-Grunt (2011)

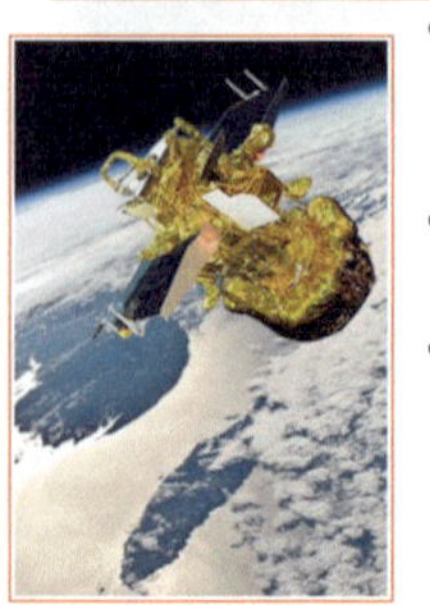

- Russian Mars' moon sample return
- 6 months before launch electrical & flight sequence testing not done
- L-8: main engine steering polarity found reversed in SW; another problem required update *not tested*
- Auto orbit-changing engine burns did not occur; failed to leave Earth
- Post-failure tests (apparently simulating in-flight conditions) revealed that the processor usage level in main flight control computer >90% capacity, perhaps lead to crashes and rebooting

Courtesy Anatoly Zak

August 23, 2022 SC13-44

Phobos-Grunt (cont.)

- Following the initial failure, ground controllers apparently succeeded in activating the onboard X-band transmitter
- Transmit power ~40 watts, power consumption ~200 watts
- Transmitter did not deactivate during eclipse periods of time
- Rechargeable batteries slowly drained, leading to a complete failure of all onboard systems on 11-28-2011

Huygens Relay at Titan/Cassini

- Huygens: transmit data during descent to Titan as Cassini approached at 6 km/s; Doppler shift a concern but not tested on ground
- Relay test to occur when SC flew by Earth 1999 resisted due to cost and "futility," but carried out. Test showed *no data recovery – receiver failed to sync*
- Failure predicted in documentation but not flowed to designers – prospect of large data loss
- Could not alter signal strength, other parameters, so trajectory change planned to reduce Doppler shift
- Early pre-heating of probe would cause ultra-precise oscillators to warm up & shift frequency to solve
- Fortunately: plenty of time between Earth & Saturn to implement changes; sufficient battery margin for pre-heating change. *Disaster averted!*

Schiaparelli Mars Lander Crash Sparked By Bad Inertial Data (2016)

LYON, France: European Space Agency (ESA) engineers believe they've found the root cause of the Oct. 19 crash of Schiaparelli Mars lander—erroneous information from IMU resulting in incorrect altitude reading.

Shortly after the lander's parachute deployed, the IMU "saturated" – it could not measure higher rotation rates – and stayed that way for about 1 second. When fed into the navigation system, the data generated an estimated altitude that was below ground level, according to ESA.

This triggered a premature release of the parachute and backshell. Braking thrusters then fired for the minimum preselected time of about 3 seconds, instead of the nominal 60 seconds. Finally, on-ground systems were activated as if Schiaparelli had already landed, whereas it was still 3.7 km (12,000 ft) from the surface.

The scenario has been "clearly reproduced in computer simulations of the control system's response to the erroneous information," ESA says.

Courtesy: AvWeek

Astronomy Satellite Break-Up – April 2016

- Attitude control problems on Japan's *Hitomi* X-ray observatory caused it to spin out of control, shed bits
- IMU detected 21.7 deg/hour rotation (but actually stable). RWAs attempted to counteract spin. Mag torquers unable to desaturate RWAs, SC went into "safe mode": tried to use thrusters to aim SA toward the sun; solar sensor was unable to locate sun.
- Rocket firings inadvertently caused faster rotation *due to a bad setting in the thruster system*
- New parameters uplinked were inadequate, did not properly account for the deployed structure
- SC shed 10 pieces. Efforts abandoned after determining "it was 'highly likely' that two solar arrays had broken off at their bases."
- Director of JAXA's Institute of Space and Astronautical Science said "There were human errors. But a bigger problem lies with our entire system as we were not able to detect those errors."

Software Increasingly Matters

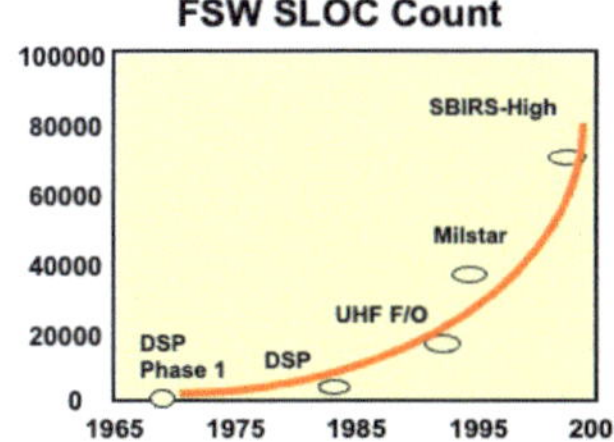

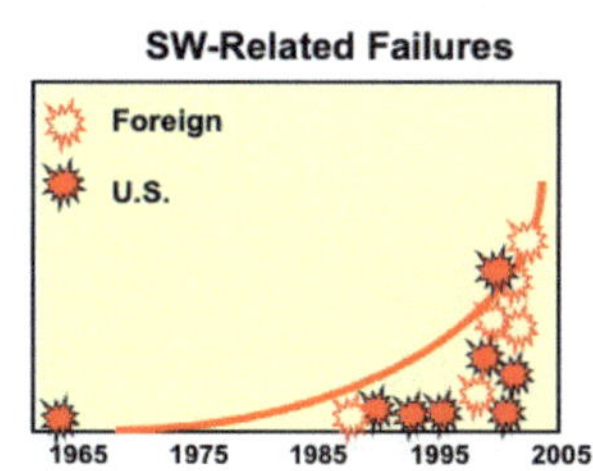

Over half of 2002-05 failures involved software

FSW SLOC = Flight Software Source Lines of Codes

Note: poorly defined SW reqts. greatly affect performance!

Courtesy Aerospace

Failures & Lessons Learned: Topics

- Reliability and failures
- Five common mistakes to look out for
  - Could the sign be wrong?
  - How will last-minute configuration changes be verified?
  - Can vehicle survive computer crash?
  - Circuit overcurrent protection?
  - Pyros cause unexpected damage?
- Research history & more examples
- **Redundancy**
- Best practices to avoid failure

Redundancy

- Extra components added so that if one fails, the mission can still proceed
  - Example: your car
  - Interesting fact: "13% of >1 million vehicles sold in US *did not offer* spare tire as standard equipment." LA Times, 2011 Jun 20, A13.
- Various types of redundancy
- Use for mechanisms, valves, switches, electronic components, etc.
- Consider in great detail when designing spacecraft's subsystems

August 23, 2022

SC13-51

Redundancy – Propulsion/Fluids

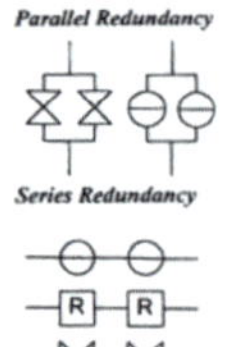

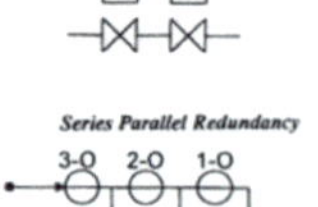

Series Parallel Redundancy

3-O 2-O 1-O

3-C 2-C 1-C

- Parallel redundancy: eliminates failure to open, but more possible leakage
- Series redundancy: reduced leakage, but failure to open increased
- Series-parallel allows protection from failure to open and close, and reduces leakage. Negatives: higher mass and cost

August 23, 2022

Ref.: Fortescue

SC13-52

Redundancy – Electronics

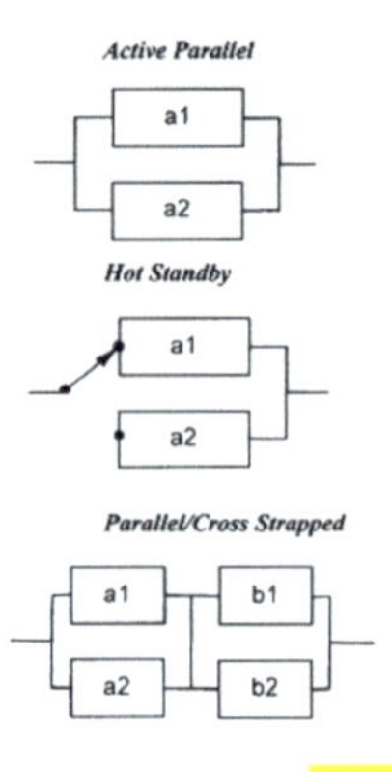

- Active Parallel: good if failure improbable or not common to both
- Hot standby: reduces failure probability if switch VERY reliable or if failure improbable or not common to both
- Parallel/Cross-strapped tolerates many more failures but adds complexity and requires additional fault analysis. Increased testing time needed.

August 23, 2022

Ref.: Fortescue

SC13-53

Example: Heater Strip Requires Redundancy

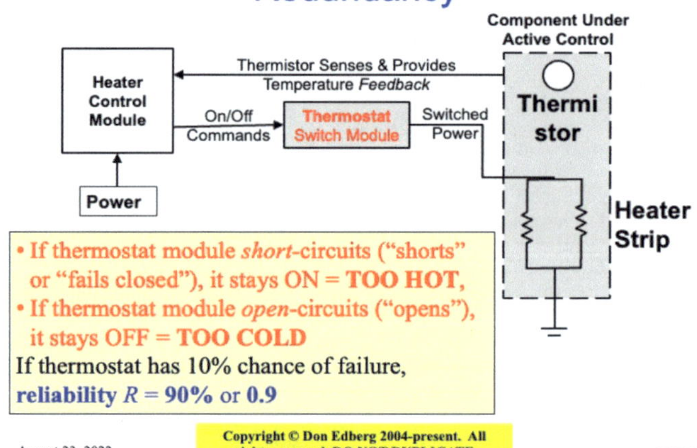

- If thermostat module *short*-circuits ("shorts" or "fails closed"), it stays ON = **TOO HOT**,
- If thermostat module *open*-circuits ("opens"), it stays OFF = **TOO COLD**

If thermostat has 10% chance of failure, **reliability R = 90%** or **0.9**

August 23, 2022

SC13-54

Series Redundant Heater Circuit

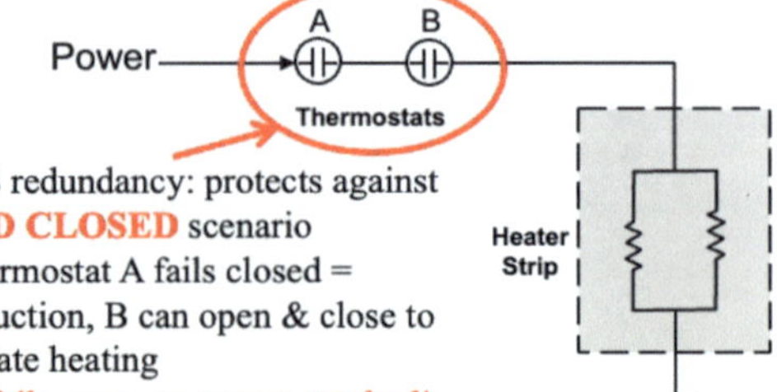

SERIES redundancy: protects against **FAILED CLOSED** scenario

- If thermostat A fails closed = conduction, B can open & close to regulate heating
- If A fails open, no power, too bad!

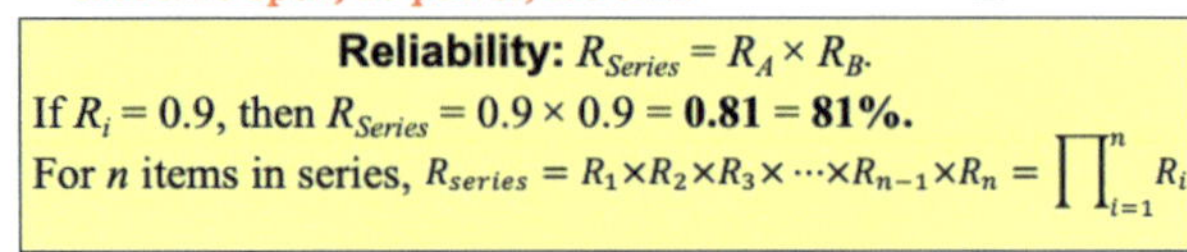

Reliability: $R_{Series} = R_A \times R_B$.

If $R_i = 0.9$, then $R_{Series} = 0.9 \times 0.9 = \mathbf{0.81} = \mathbf{81\%}$.

For n items in series, $R_{series} = R_1 \times R_2 \times R_3 \times \cdots \times R_{n-1} \times R_n = \prod_{i=1}^{n} R_i$

August 23, 2022

Ref: Conley

SC13-55

Parallel Redundant Heater Circuit

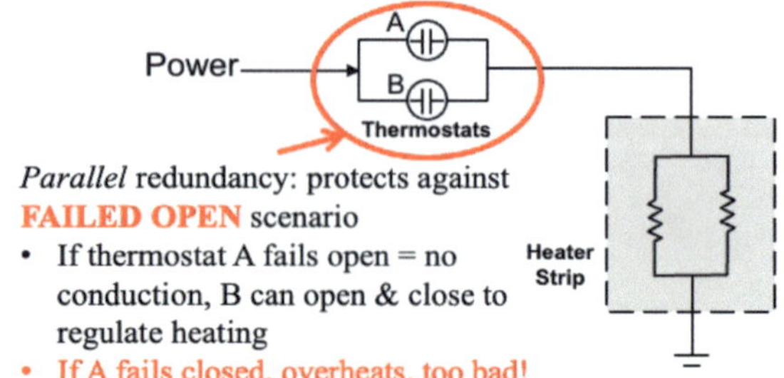

Parallel redundancy: protects against **FAILED OPEN** scenario

- If thermostat A fails open = no conduction, B can open & close to regulate heating
- If A fails closed, overheats, too bad!

Reliability: $R_{Parallel} = 1 - (1 - R_A)(1 - R_B)$.

If $R_i = 0.9$, then $R_{parallel} = 1\text{-}(1\text{-}0.9)(1\text{-}0.9) = 1\text{-}(0.1)^2 = \mathbf{0.99} = \mathbf{99\%}$

For n parallel items, $R_p = 1 - (1 - R_1)(1 - R_2)(1 - R_3)\ldots(1 - R_n) = 1 - \Pi_{i=1}^{n}(1 - R_i)$

August 23, 2022

Ref: Conley

SC13-56

Series-Parallel Redundant Heater Circuit

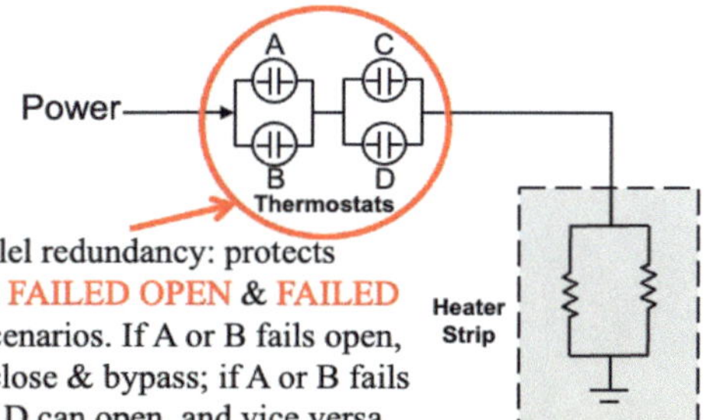

Series-Parallel redundancy: protects against both FAILED OPEN & FAILED CLOSED scenarios. If A or B fails open, B or A can close & bypass; if A or B fails closed, C & D can open, and vice versa.

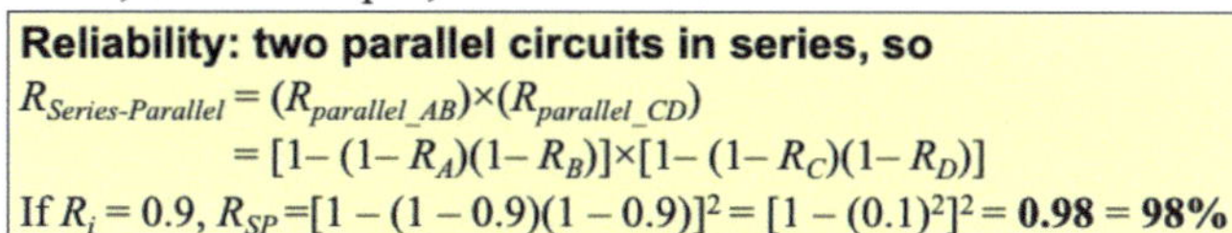

Reliability: two parallel circuits in series, so

$R_{Series\text{-}Parallel} = (R_{parallel\_AB}) \times (R_{parallel\_CD})$

$= [1-(1-R_A)(1-R_B)] \times [1-(1-R_C)(1-R_D)]$

If $R_i = 0.9$, $R_{SP} = [1-(1-0.9)(1-0.9)]^2 = [1-(0.1)^2]^2 = \mathbf{0.98 = 98\%}$

Ref: Conley

Parallel-Series Redundant Heater Circuit

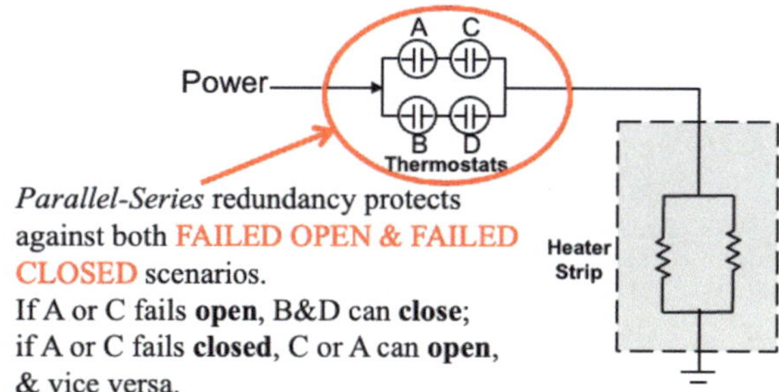

Parallel-Series redundancy protects against both FAILED OPEN & FAILED CLOSED scenarios.
If A or C fails **open**, B&D can **close**; if A or C fails **closed**, C or A can **open**, & vice versa.

Reliability: $R_{Parallel\text{-}Series} = 1-(1-R_{Series\_AB}) \times (1-R_{Series\_CD})$

$= 1-\{1-(R_A \times R_B)\} \times [1-(R_C \times R_D)]\}$

If $R_i = 0.9$, then $R_S = 1-[1-(0.9)(0.9)]^2 = 1-[0.19]^2 = \mathbf{0.9639} = \mathbf{96.4\%}$

Ref: Conley

Failures & Lessons Learned: Topics

- Reliability and failures
- Five common mistakes to look out for
  - Could the sign be wrong?
  - How will last-minute configuration changes be verified?
  - Can vehicle survive computer crash?
  - Circuit overcurrent protection?
  - Pyros cause unexpected damage?
- Research history & more examples
- Redundancy
- Best practices to avoid failure

Best Practices To Avoid Failure

- Remember history, use & add to lessons-learned database
- Use institutional expertise; utilize experienced successful project managers
- Use adequate design margins & redundancy when needed
- Have competent analysis, verification, independent reviews, oversight, & thorough testing
- Identify risks effectively & communicate frankly — don't be afraid to ask questions & offer opinions
- Contractor: notify customer of project risk, any deviations from acceptable practice
- Know validity of specifications, certifications, & limitations of your equipment
- Don't let schedule pressure outweigh mission safety
- Demand proof: DO NOT LAUNCH unless ready
- Provide telemetry coverage of critical events

NASA *Public Lessons Learned* System

- System provides access to official, reviewed lessons learned from NASA programs & projects
- Lessons made available to public by NASA Office of Chief Engineer & NASA Engineering Network
- Each lesson describes original driving event & provides recommendations
- `https://llis.nasa.gov/`

Get These Books (If you can)

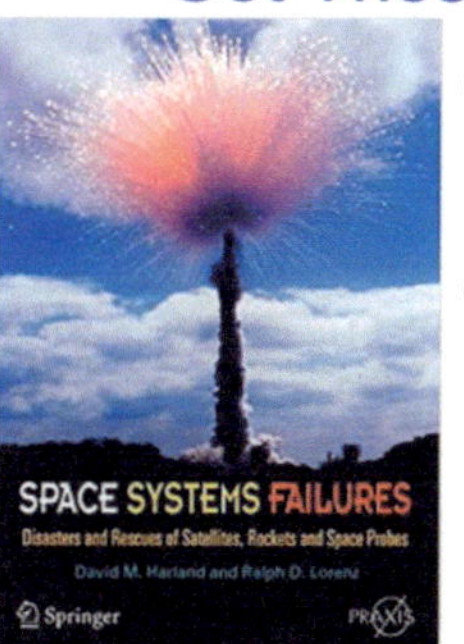

- Harland, D. M., and Lorenz, R. D., Space Systems Failures, Springer, Berlin, 2005 (left)
- Cheng, P. G., "Five Common Mistakes Reviewers Should Look Out For," Aerospace Report No. TOR-2007(8617)-1, 29 June 2007 [Available to U.S. government agencies and contractors only]

"Fools say that they learn by experience. I prefer to profit by others' experience." — Otto Bismarck

Blank

Spacecraft & Mission Cost Estimation

August 23, 2022

SC14- 1

Topics

- Cost estimation needs & proposal process
  - The importance of good decision-making
- The phases of design, timelines, reviews
- Cost estimation discussion & methods
- Cost estimating relationships (CERs)
- Cost models
- Inflation, labor costs, software development, propellant cost
- Launch costs
- Non-vehicle costs: mission ops, delays, extensions, insurance, DSN

August 23, 2022

SC14- 2

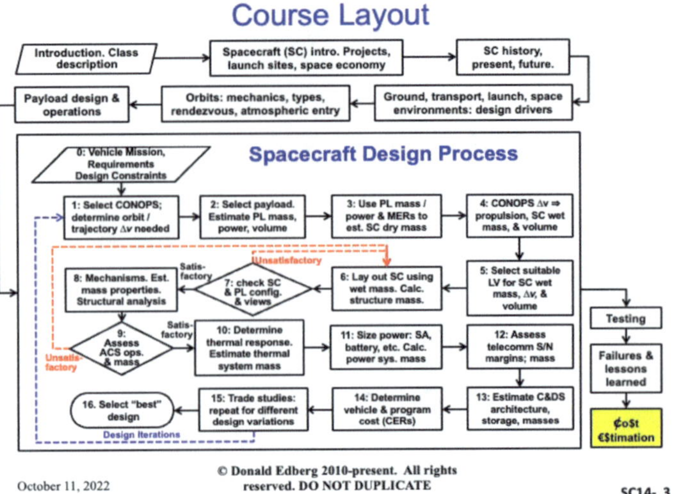
Course Layout
Introduction. Class description
Spacecraft (SC) intro. Projects, launch sites, space economy
SC history, present, future.
Ground, transport, launch, space environments: design drivers
Orbits: mechanics, types, rendezvous, atmospheric entry
Payload design & operations
Spacecraft Design Process
0: Vehicle Mission, Requirements Design Constraints
1: Select CONOPS; determine orbit / trajectory Δv needed
2: Select payload. Estimate PL mass, power, volume
3: Use PL mass / power & MERs to est. SC dry mass
4: CONOPS Δv ⇒ propulsion, SC wet mass, & volume
5: Select suitable LV for SC wet mass, Δv, & volume
6: Lay out SC using wet mass. Calc. structure mass.
Unsatisfactory
7: check SC & PL config. & views
Satis-factory
8: Mechanisms. Est. mass properties. Structural analysis
9: Assess ACS ops. & mass
Unsatis-factory
Satis-factory
10: Determine thermal response. Estimate thermal system mass
11: Size power: SA, battery, etc. Calc. power sys. mass
12: Assess telecomm S/N margins; mass
13: Estimate C&DS architecture, storage, masses
14: Determine vehicle & program cost (CERs)
15: Trade studies: repeat for different design variations
16. Select "best" design
Design Iterations
Testing
Failures & lessons learned
¢o$t €$timation

October 11, 2022

Notable Quotables

- "Crash programs fail because they are based on the theory that, with nine women pregnant, you can get a baby a month." **Wernher Von Braun**
- "There is just one thing I can promise you about the outer-space program — your tax-dollar will go further." **Wernher Von Braun**
- "It's a very sobering feeling to be up in space and realize that one's safety factor was determined by the lowest bidder on a government contract." **Alan Shepherd**

23-Aug-22

Topics

- Cost estimation needs & proposal process
  - The importance of good decision-making
- The phases of design, timelines, reviews
- Cost estimation discussion & methods
- Cost estimating relationships (CERs)
- Cost models
- Inflation, labor costs, software development, propellant cost
- Launch costs
- Non-vehicle costs: mission ops, delays, extensions, insurance, DSN

August 23, 2022

Proposal Process Timeline

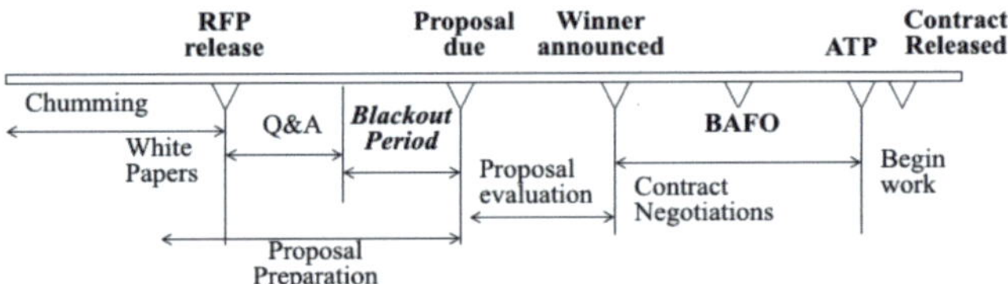
RFP release
Proposal due
Winner announced
ATP
Contract Released
Chumming
White Papers
Q&A
Blackout Period
Proposal evaluation
BAFO
Contract Negotiations
Begin work
Proposal Preparation

RFP = Request for Proposal
Q&A = Question and answer period
BAFO = Best And Final Offer
ATP = Authority to Proceed

23-Aug-22

Pre-RFP Preparation

- Knowing customer & that RFP is pending allows planning
- Meet company management
- Establish funding for proposal team effort ("B&P" = bid and proposal)
- Gather up critical individuals – knowledge & experience – for proposal effort
- Schedule a proposal "kickoff" meeting
- Choose proposal manager and "book boss."
- Begin storyboarding proposal

23-Aug-22

Proposal Physical Layout

- Book boss & proposal manager plan layout
- Storyboards created for each page of proposal; placed on wall
- Authors assigned
- Graphics planned ("Action captions")
- Style is first-person, "sales document," not a 3rd-person report. You are selling yourself!

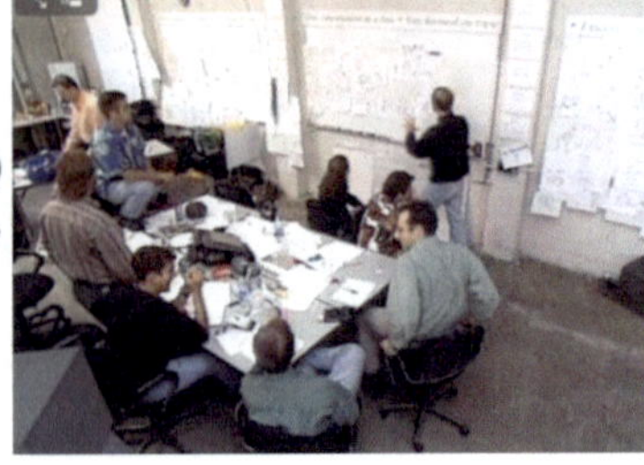

23-Aug-22

Software (& Hardware) Design Flow Must Be Planned

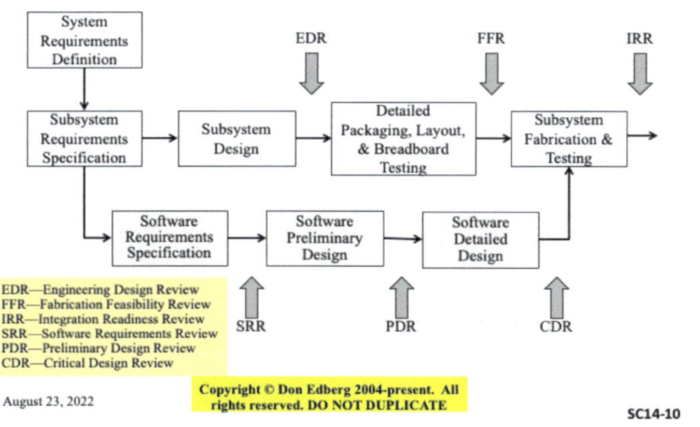

August 23, 2022

Topics

- Cost estimation needs & proposal process
  - The importance of good decision-making
- The phases of design, timelines, reviews
- Cost estimation discussion & methods
- Cost estimating relationships (CERs)
- Cost models
- Inflation, labor costs, software development, propellant cost
- Launch costs
- Non-vehicle costs: mission ops, delays, extensions, insurance, DSN

August 23, 2022

Designer Judgment is Important

- Little or no data available at the outset
- All design teams have essentially the same data and technology base to work with
- Early decisions have large downstream effects
- Choice of approach and success of the design are rooted in designers' knowledge and experience, and available methodology

23-Aug-22

Life Cycle Costs: Good Early Estimates Important

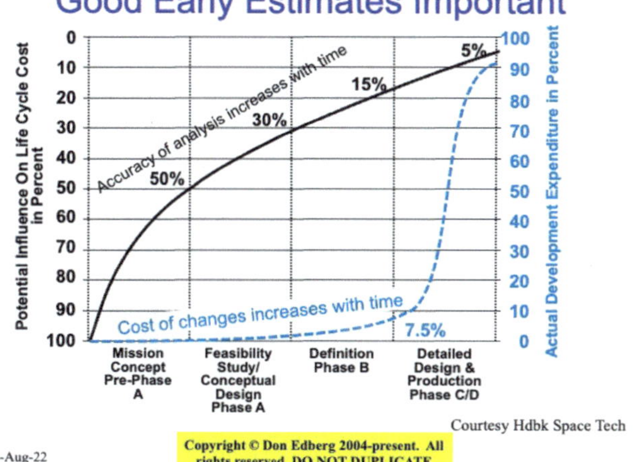

Courtesy Hdbk Space Tech

23-Aug-22

Design Decision-Making

- Mission & merit guide decisions: features should be evaluated in:
  - Mission environment & operations
  - Technology readiness levels (TRLs)
- Mission first, then configuration
  - No single magic configuration
- Don't fall in love with a concept too early:
  - Concept can end up driving whole design
- Lives can depend on making the right decisions (inhabited missions)
- Beware of "requirements creep"

23-Aug-22

Technology Readiness Levels (TRLs)

- JWST original cost estimate: ~$500M
- Final cost: ~$10B! 20 times higher
- Why the difference? TRL
  - Super lightweight segmented primary mirror for telescope then at TRL 4 but bid at TRL 6!
  - Not an honest appraisal, *we paid for it*!

9	Actual system "flight proven" through successful mission operations
8	Actual system completed & "flight qualified" through test & demo (ground or space)
7	System prototype demonstration in space envt.
6	System/subsystem model or prototype demo in a relevant environment (gnd. or space)
5	Component and/or breadboard validation in relevant environment
4	Component and/or breadboard validation in lab environment
3	Analytical & experimental critical function and/or characteristic proof-of-concept
2	Tech. concept and/or application formulated
1	Basic principles observed & reported

23-Aug-22

Basic Assumptions Memorandum (BAM)

- A document that defines design ground rules
- BAM elements:
  - Mission Statement
  - Mission Profile
  - Requirements
  - Constraints
  - Goals
  - Other assumptions that will affect design
- BAM is a "living document" that should reflect changes in assumptions as design proceeds

23-Aug-22 SC14-16

What is Good?

- Flight survival & structural integrity are ALWAYS REQUIRED
- More of something not needed is useless*
  *unless it enables something useful later
- Compromises & tradeoffs are inevitable
- Design Requirements:
  - Must meet constraints
  - Goals: Highly desirable
  - Figures of merit: more is better

23-Aug-22 SC14-17

Starship Evolution

23-Aug-22 Courtesy Quora SC14-18

Elon Musk's Five-Step Design Philosophy

1. Break down requirements into steps and simplify them (Musk used the term "make them less dumb")
   Attach responsibility for developing these requirements to a single individual, not a department
   - This means the individual cannot pass responsibility off for lack of results and must then question the validity of the requirements in order to act as a second level of design filtering (everybody makes mistakes at some point, and no specification can be considered unquestionable).
2. Delete any non-essential parts or processes, then reassess the system to see what changes might need to be made
3. Accelerate the cycle time, but not before applying the first two steps.
4. Automate the process

23-Aug-22 SC14-19

Successful Proposal Team: Boeing's *Saturn V* S-IC Team

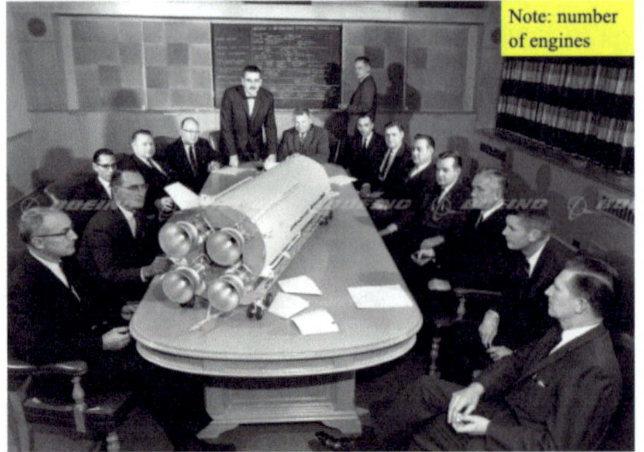

Note: number of engines

23-Aug-22 SC14-20

Early Space Shuttle Concepts

23-Aug-22 SC14-21

Proposal vs. Actual!

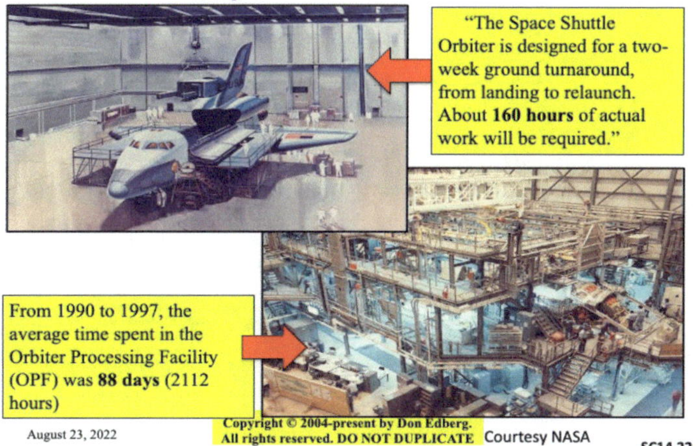

Courtesy NASA

Costs vs. Income vs. Breakeven

When non-recurring expenses are paid for, and profits are made.

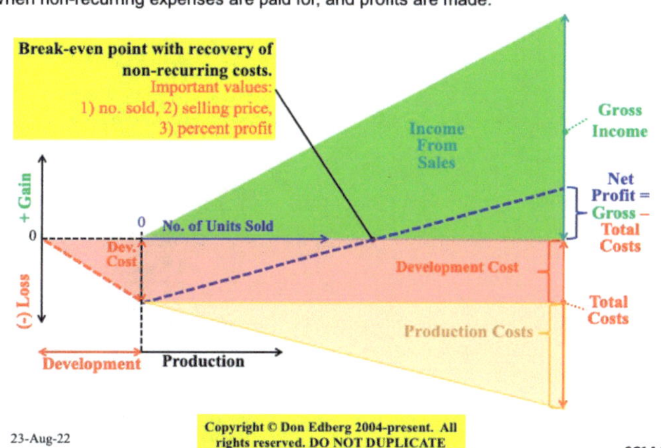

You Win The Proposal!

- Now what?
- Negotiations! Meet customer & hammer out terms
  - Customer attempts to reduce price
  - Proposer attempts to stand ground
- Result: "Best and final offer" (BAFO)
- Customer either takes it, or leaves it.
- If taken, get "Authority to Proceed" ATP
- Design work begins (or continues, now with authority!)

Topics

- Cost estimation needs & proposal process
  - The importance of good decision-making
- **The phases of design, timelines, reviews**
- Cost estimation discussion & methods
- Cost estimating relationships (CERs)
- Cost models
- Inflation, labor costs, software development, propellant cost
- Launch costs
- Non-vehicle costs: mission ops, delays, extensions, insurance, DSN

Phases of Design

- Conceptual
  - Requirements definition
  - Initial configuration concepts defined & evaluated
  - Basic design trades
  - Initial weight, cost, performance estimates
- Preliminary
  - Configuration downselect & freeze
  - Testing and analysis
  - Major component & software layout & design
  - "Real" cost
- Detail
  - Design parts to be built
  - Design tooling & software
  - Major component & systems testing
  - Refinement of financial & risk analyses

Stages of Mission Development

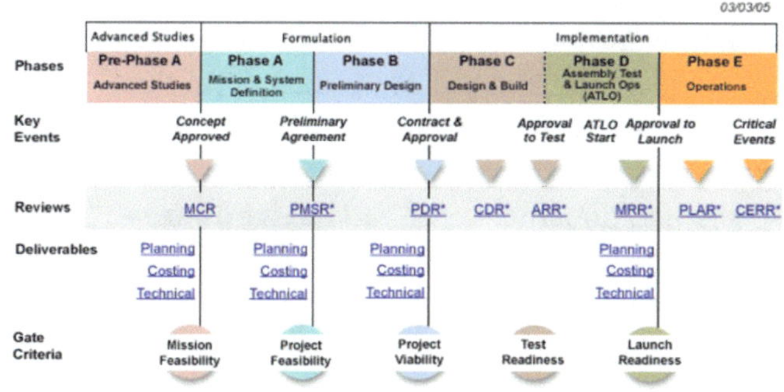

Details: NASA/SP-2007-6105 **Systems Engineering Handbook**

Program & Project Life Cycles (per NASA)

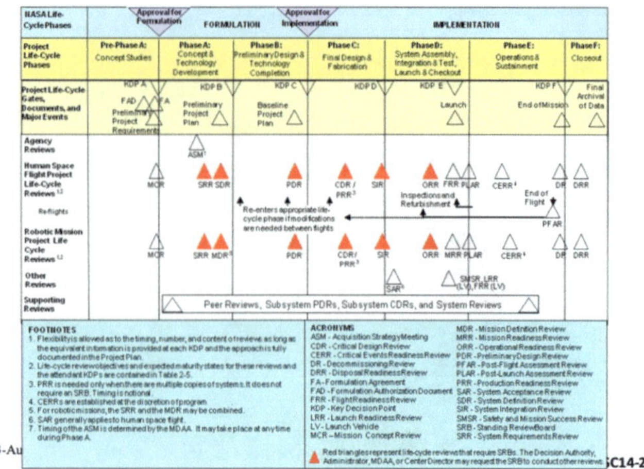

23-Aug-22

SC14-28

Purposes of NASA Reviews

Review	Title	Purpose
P/SRR	Program Requirement Review	The P/SRR is used to ensure that the program requirements are properly formulated and correlated with the Agency and mission directorate strategic objectives
P/SDR	Program Definition Review, or System Definition Review	The P/SDR ensures the readiness of the program for making a program commitment agreement to approve project formulation startups during program Implementation phase.
MCR	Mission Concept Review	The MCR affirms the mission need and examines the proposed mission's objectives and the concept for meeting those objectives
SRR	System Requirement Review	The SRR examines the functional and performance requirements defined for the system and the preliminary program or project plan and ensures that the requirements and the selected concept will satisfy the mission
MDR	Mission Definition Review	The MDR examines the proposed requirements, the mission architecture, and the flow down to all functional elements of the mission to ensure that the overall concept is complete, feasible, and consistent with available resources
SDR	System Definition Review	The SDR examines the proposed system architecture and design and the flow down to all functional elements of the system.
PDR	Preliminary Design Review	The PDR demonstrates that the preliminary design meets all system requirements with acceptable risk and within the cost and schedule constraints and establishes the basis for proceeding with detailed design. It will show that the correct design options have been selected, interfaces have been identified, and verification methods have been described
CDR	Critical Design review	The CDR demonstrates that the maturity of the design is appropriate to support proceeding with full-scale fabrication, assembly, integration, and test. CDR determines that the technical effort is on track to complete the flight and ground system development and mission operations, meeting mission performance requirements within the identified cost and schedule constraints.
PRR	Production Readiness Review	A PRR is held for FS&GS projects developing or acquiring multiple or similar systems greater than three or as determined by the project. The PRR determines the readiness of the system developers to efficiently produce the required number of systems. It ensures that the production plans; fabrication, assembly, and integration enabling products; and personnel are in place and ready to begin production.

23-Aug-22

SC14-29

Definitions of All the Reviews You'll Need!

ARR - Assembly, Test & Launch Ops Readiness Review
ASM – Acquisition Strategy Meeting
ASP – Acquisition Strategy Planning Meeting
CDR - Critical Design Review
CERR - Critical Events Readiness Review
DADR - Data Archive Delivery Review
DR - Decommissioning Review
EIRR - External Independent Readiness Review
ETRR - Environmental Test Readiness Review
FAD – Formulation Authorization Document
FIRDR - Final Interface Requirements & Design Review
FPCDR - Fault Protection Critical Design Review
FPPDR - Fault Protection Preliminary Design Review
FPR - Flight Parameters Review
FRR - Flight Readiness Review
FSCDR - Flight System Critical Design Review
FSPDR - Flight System Preliminary Design Review
HRCR - Hardware Review & Certification Record Review
IA - Independent Assessment
IAR - Independent Annual Review
IMAR - Independent Mission Assurance Review
KDP- Key Decision Point
LRR - Launch Readiness Review
MCR - Mission Concept Review
MDR - Mission Definition Review
MESB - Mission Executive Summary Briefing
MOS/GDS PDR /CDR - Mission Ops System/Ground Data System Preliminary/Critical Design Review
Msn/NavS PDR/CDR - Mission/Navigation System Preliminary/Critical Design Review
MRR - Mission Readiness Review
NAR – Non-Advocate Review
ORR - Operational Readiness Review
PDR - Preliminary Design Review
PETR - Post Environmental Test Review
PFAR – Post-Flight Assessment Review
PIRDR – Prelim. Interface Requirements & Design Review
PLAR – Post-Launch Assessment Review
PNAR – Preliminary Non-Advocate Review
PMSR - Project Mission System Review
PRR – Production Readiness Review
PSR - Pre Ship Review
SAR - System Acceptance Review
SDR - System Definition Review
SIR – System Integration Review
SMSR – Safety And Mission Success Review
SRCR - Software Review & Certification Record Review
SRR - System Requirements Reviews
TRR - Test Readiness Review

23-Aug-22

SC14-30

Topics

- Cost estimation needs & proposal process
  - The importance of good decision-making
- The phases of design, timelines, reviews
- **Cost estimation discussion & methods**
- Cost estimating relationships (CERs)
- Cost models
- Inflation, labor costs, software development, propellant cost
- Launch costs
- Non-vehicle costs: mission ops, delays, extensions, insurance, DSN

August 23, 2022

SC14-31

Cost Estimation

- Approximate at best unless based on existing hardware
- Subject to unknowns at mission definition
- Allows *relative* comparison of different options
- Labor-related costs (engineering, management, testing, documentation) add to hardware & SW

Provides:

- A historical perspective
- A basis for comparing or challenging bids
- A relative sense of mission component cost
- An estimate of cost uncertainty

August 23, 2022

SC14-32

Methods of Cost Modeling

1. Detailed, bottom-up estimate: use when a detailed design exists (*USELESS for advanced studies*)
2. Analogous systems estimate: used for conceptual phase, but not useful for trades
3. Parametric estimate: relies on "cost estimating relationships" (*CER*s) that express cost as a function of design, size, or performance parameters. MOST USEFUL FOR CONCEPTUAL DESIGN & TRADE STUDIES

August 23, 2022

SC14-33

Cost Estimation Must Include Entire Life Cycle

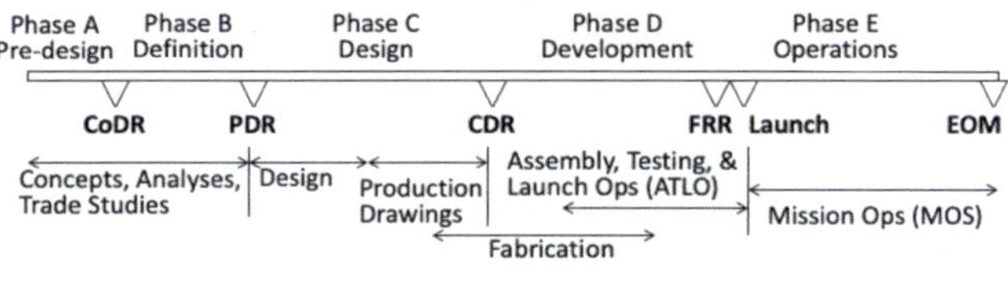

- Design/Draw/Build/Test Deliver Hardware & Software
- Launch Vehicle (+ delays)
- Operations (Telecomm, DSN, etc.)
- (Refurbishment/Disposal)

August 23, 2022

SC14-34

Topics

- Cost estimation needs & proposal process
  - The importance of good decision-making
- The phases of design, timelines, reviews
- Cost estimation discussion & methods
- **Cost estimating relationships (CERs)**
- Cost models
- Inflation, labor costs, software development, propellant cost
- Launch costs
- Non-vehicle costs: mission ops, delays, extensions, insurance, DSN

August 23, 2022

SC14-35

Cost Estimating Relationships (CERs) For Traditional vs. "FBC" Missions

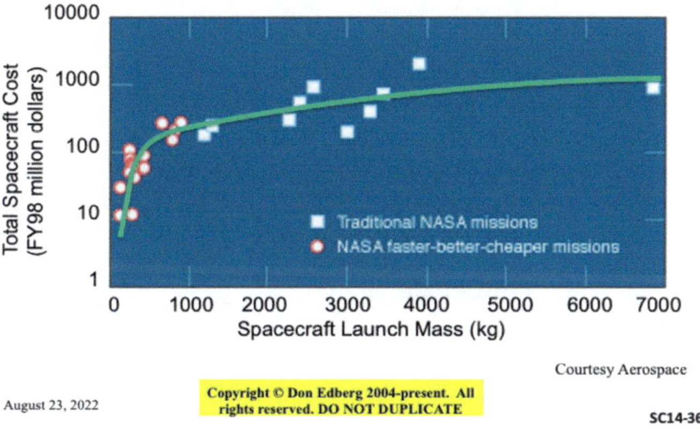

Courtesy Aerospace

August 23, 2022

SC14-36

Cost vs. Mass Study

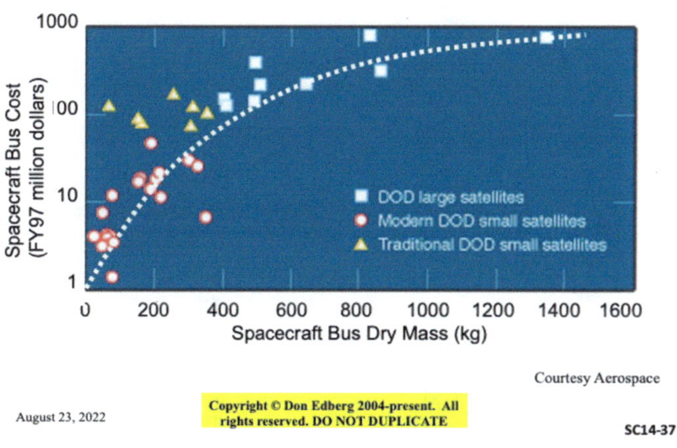

Courtesy Aerospace

August 23, 2022

SC14-37

Use Database to Produce CERs for Different Independent Variables

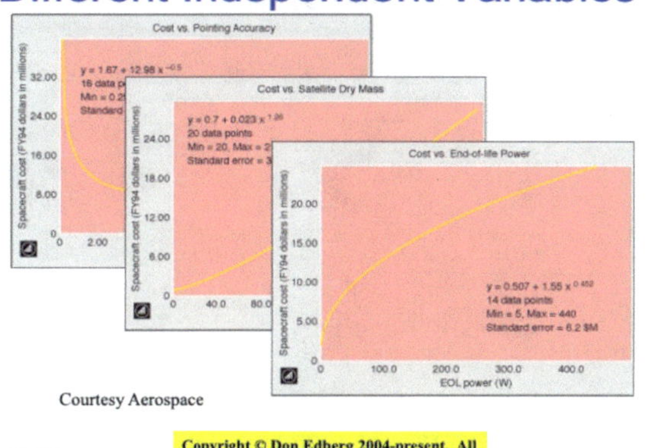

Courtesy Aerospace

August 23, 2022

SC14-38

Topics

- Cost estimation needs & proposal process
  - The importance of good decision-making
- The phases of design, timelines, reviews
- Cost estimation discussion & methods
- Cost estimating relationships (CERs)
- **Cost models**
- Inflation, labor costs, software development, propellant cost
- Launch costs
- Non-vehicle costs: mission ops, delays, extensions, insurance, DSN

August 23, 2022

SC14-39

Some Available Cost Models (2022)

- Project Cost Estimating Capability, or PCEC (replaces NAFCOM (NASA-Air Force Cost Model). At: http://software.nasa.gov/software/MFS-33187-2
- USCM (Unmanned Spacecraft Cost Model): https://www.uscmonline.com/
- SSCM (Aerospace Corp. Small Satellite Cost Model, New SMAD T 11-11), also available: https://aerospace.org/how-obtain-sscm
- NASA Cost Estimating Handbook: available from https://www.nasa.gov/offices/ooe/CAD/nasa-cost-estimating-handbook-ceh/
- SMAD table 20.17 (Traditional Cost Model)
- USAF Space Planners Guide 1965 (not computerized – manual calculations; PDF available from instructor or online)

Direct Operations Cost Elements

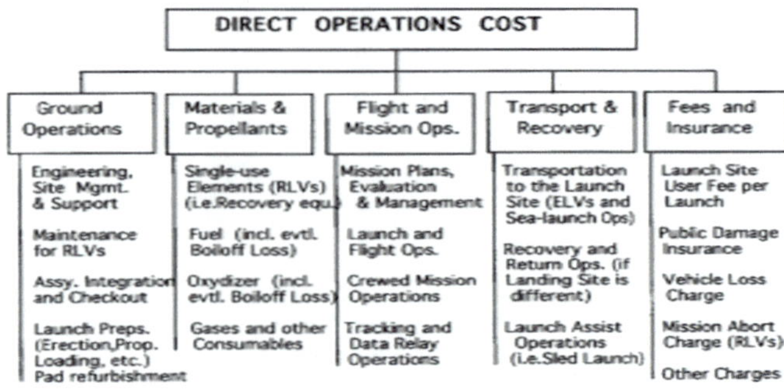

Courtesy: Koelle

Indirect Operations Cost Elements

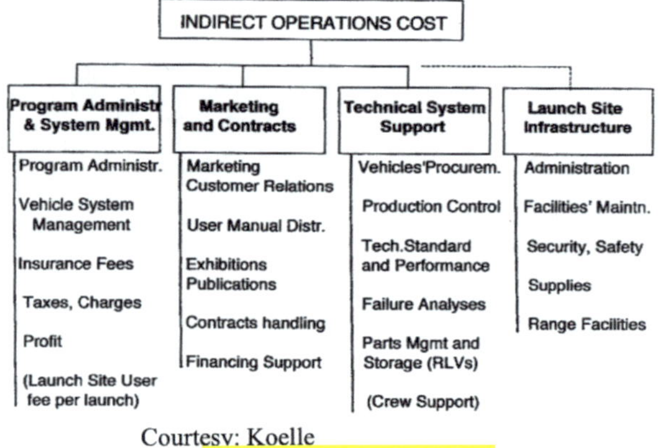

Courtesy: Koelle

QuickCost: New SMAD Table 11-12

Total development cost + protoflight flight unit, 2010$

$$= 2.2829 \times DryMass^{0.457} \times BOL\text{-}Power^{0.157} \times 2.718^{[(0.171 \times Data\%)+(0.00209 \times DesignLifeMonths) +(1.52 \times New)+(0.258 \times Planetary:1=\text{yes},0=\text{no}) +[0.0145(ATPyear - 1960)]+(0.467\ instr\text{-}comp\%) -(0.237 \times team)]}$$

41% standard error

QuickCost Definitions

Data%: use <0.5 for low, 0.5 median, >0.5 for high relative to state-of-the-art

New = Percent New: 0.2-0.3 simple mod; 0.3-0.7 extensive mod; 0.7-1.0 new, >1 for new technology

Instr-comp = instrument complexity: relative to "average" instruments: enter <0.5 for low, 0.5 median, >0.5 for high, relative to state-of-the-art

Team = experience: 1 = unfamiliar, 2 = mixed, 3 = Normal, 4 = Extensive

New SMAD Quick Cost Model: Table 11-12 (Live Calc)

Non-recurring cost for development plus one qualification unit			
SME-SMAD WBS Element	Estimated Cost in FY2010 [$K]	Absolute Standard Error of the Estimate (SEE) [%]	Estimated Cost in $K for Fiscal Year: 2045
SV for Unmanned Robotic Mission	$3,919	41.0%	$13,514

Cost Driver	Input value between Lower Limit &Upper Limit in Value Column		
	Lower Limit	Value	Upper Limit
Dry Mass (kg)	76	37,324	40,000
Power (W) @ BOL	90	10,000	10,000
Data Rate Percentile (%) relative to the state-of-the-art at Authority to proceed (ATP) >50% for higher data rates]	0%	70%	100%
Advertised Design Life (months)	220	240	300
Percentage of New (%) [20%-30% for Simple Mod, 30%-70% for Extensive Mod, 70%-100% for New Technology	28%	100%	130%
Planetary? [YES/NO]	Input 0 for NO	0	Input 1 for YES
ATP Years [YEAR minus 1960]	2015	2025	2025
Instrument Complexity Percentile (%) relative to "average" instrument complexity [>50% for higher complexity]	0%	60%	100%
Team Experience [3 for Normal]	1	3	4

New SMAD Hardware Cost Model Table 11-14 (Live Calc)

NICM Nonrecurring Hardware Development Plus One Protoflight Unit Cost for Remote Sensing Instrument Payload								
SME-SMAD WBS Element	Cost Driver(s)	Input a value between the Lower Limit and Upper Limit in the Value Column			Estimated Cost in FY2010 [$K]	Standard Error of the Estimate (SEE)	R^2	Estimated Cost in $K for Fiscal Year:
	Cost Driver	Lower Limit	Value*	Upper Limit				2045
Optical Earth-Orbiting (instruments on spacecraft in geocentric orbits) Payload (e.g., cameras, spectrometers, interferometers)	Instrument Total Mass (kg)	10	350	350	$174,126	35%	1	$506,900
	Maximum Instrument Power (W)	0.5	400	400				
	Total Data Rate (kbps)	0.1	30,000.0	30,000				

August 23, 2022

USCM8 Model: Non-Recurring (2010$, 1 of 2)

CER Category:	CER: Y = cost, FY 2010 K$	Cost Drivers	Input Range
SC Bus (alternate CER when no component info is available)	$Y = 110\ X_1$	X_1 = Spacecraft Mass (kg)	114-5127 kg
Structure & Thermal Control	$Y = 646\ X_1^{0.684}$	X_1 = Structure + Thermal mass (kg)	59-501 kg
Attitude Determination & Control System (ADCS)	$Y = 324\ X_1$	X_1 = ADCS Mass (kg)	35-524 kg
Electrical Power System (EPS)	$Y = 64.3\ X_1$	X_1 = EPS Mass (kg)	47-1065 kg
Propulsion (Reaction Control)	$Y = 20\ X_1^{0.485}$	X_1 = Total RCS tank volume (cc)	NA
Telemetry, Tracking, & Command (TT&C)	$Y = 26916$	Y = Average TT&C cost (there is no statistical CER for this element)	Based S-band TM
Communications payload	$Y = 618\ X_1$	X_1 = Communications Subsystem Mass	160-395 kg
Integration, Assembly, & Test (of bus & payload into SC) (IA&T)	$Y = 0.195\ X_1$	X_1 = SC Bus + Payload Nonrecurring cost ($K)	$3.6-545M
Program Level (for Comm sat)	$Y = 0.236\ X_1$	X_1 = SV + IA&T Nonrecurring cost ($K)	$7.9-354M
Program Level (other than Comm)	$Y = 0.414\ X_1$	X_1 = SV + IA&T Nonrecurring cost ($K)	$7.9-354M
Aerospace Ground Equipment	$Y = 0.421\ X_1^{0.907} * 2.44^{X2}$	X_1 = SC Bus Nonrecurring cost ($K); comm sats: $X_2 = 0$; non-comm: $X_2 = 1$	$7.9-354M

Courtesy New SMAD T11-8. SV = Space Vehicle.
Inflation factors follow.

August 23, 2022

USCM8 Model: Recurring (2010$) (2 of 2)

CER Category: *RECURRING*	CER: Y = cost, FY 2010 K$	Cost Drivers	Input range
SC Bus (alternate CER when no component info is available)	$Y = 289.5\ X_1^{0.716}$	X_1 = Spacecraft Mass (kg)	288-7398 kg
Structure & Thermal control	$Y = 22.6\ X_1$	X_1 = Structure + Thermal mass (kg)	59-501 kg
Attitude Determination & Control System (ADCS)	$Y = 795\ X_1^{0.593}$	X_1 = ADCS Mass (kg)	27-524 kg
Electrical Power System (EPS)	$Y = 32.4\ X_1$	X_1 = EPS Mass (kg)	111-1479 kg
Propulsion (Apogee Kick Motor)	$Y = 29\ X_1 + 0.024\ X_2$	X_1 = AKM Mass (kg) X_2 = Burn Time (s)	81-966 kg
Telemetry, Tracking, & Command (TT&C)	$Y = 883.7\ X_1^{0.491} 1.13^{X2}$	X_1 = TT&C Mass (kg) X_2 = Geosync Xfer orbit (0=no, 1=yes)	12-76 kg for S-band
Communications payload	$Y = 189\ X_1$	X_1 = Communications Payload Mass	38-928 kg
Integration, Assembly, & Test (of bus & payload into SC) (IA&T)	$Y = 0.124\ X_1$	X_1 = SC Bus + Payload Nonrecurring cost ($K)	$3.6-142M
Program Level (for Comm sat)	$Y = 0.234\ X_1$	X_1 = SV + IA&T Recurring cost ($K)	$13.3-268M
Program Level (for other than Comm sat)	$Y = 0.320\ X_1$	X_1 = SV + IA&T Recurring cost ($K)	$13.3-268M
Launch Operations & Orbital Support (LOOS)	$Y = 5850$	Y = Average LOOS cost ($K)	Not given

August 23, 2022

Courtesy Charles Brown

SSCM Model: Non-Recurring + 1 Protoflight Unit ('10$)

CER Category:	CER: Y = cost, FY 2010 K$	Cost Drivers	Input Range
SC Bus (alternate CER when no component info is available)	$Y = 1064 + 35.5\ X^{1.261}$	X = Spacecraft Bus Dry Mass (kg)	20-400 kg
Structure	$Y = 407 + 19.3\ X \ln X$	X = Structure mass (kg)	5-100 kg
Thermal control	$Y = 335 + 5.7\ X^2$	X = Thermal control mass (kg)	5-12 kg
Attitude Determination & Control System (ADCS)	$Y = 1850 + 11.7\ X^2$	X = ADCS Dry Mass (kg)	1-25 kg
Electrical Power Supply (EPS)	$Y = 1261 + 539\ X^{0.72}$	X = EPS Mass (kg)	7-70 kg
Propulsion (Reaction Control)	$Y = 89 + 3.0\ X^{1.261}$	X = SC Bus Dry Weight (kg)	20-400 kg
Telemetry, Tracking, & Command (TT&C)	$Y = 486 + 5.5\ X^{1.35}$	X = TT&C mass (kg)	3-30 kg
Command & Data Handling (C&DH)	$Y = 658 + 75\ X^{1.35}$	X = C&DH Mass	3-30 kg
Payload	$Y = 0.4\ X$	X = SC Bus Total cost ($K)	$2.6K-69 M
Integration, Assembly, & Test (IA&T)	$Y = 0.139\ X$	X = SC Bus Total cost ($K)	$2.6K-69 M
Program Level	$Y = 0.229\ X$	X = SC Bus Total cost ($K)	$2.6K-69 M
Ground Support Equipment	$Y = 0.66\ X$	X = SC Bus Total cost ($K)	$2.6K-69 M
Launch Operations & Orbital Support (LOOS)	$Y = 0.61\ X$	X = SC Bus Total cost ($K)	$2.6K-69 M

Useful for SC < 500 kg. Courtesy New SMAD T11-11. SC = Spacecraft.
Inflation factors follow.

NASA *Instrument* Cost Model (NICM)

WBS Element	CER MPE	Cost Drivers	Inputs
Optical – Planetary: Cameras, Spectrometers, Interferometers	$Y = 328\ M^{0.426}\ P^{0.414}\ DL^{0.375}$	M = Total Instrument Mass (kg) P = Max Instrument Power (W) DL = Design Life (months)	1-75 kg 1-75 W 10-150 mo.
Optical – Earth Orbit: Cameras, Spectrometers, Interferometers	$Y = 1163\ M^{0.328}\ P^{0.357}\ DR^{0.092}$	M = Total Instrument Mass (kg) P = Max Instrument Power (W) DR = Data Rate (kbps)	10-350 kg 5-400 W 0.1-30 kbps
Active & Passive Microwave Payload: radars, altimeters, sounders, GPS receivers	$Y = 23620\ M^{0.284}\ P^{0.335}\ DR^{0.090}\ T^{-1.296}$	M = Total Instrument Mass (kg) P = Max Instrument Power (W) DR = Data Rate (kbps) T = Inst. Tech Readiness Level	10-50 kg 10-600 W 0.1-1Mbps TRL 4-9
Particles Payload: plasma detectors, plasma wave det.	$Y = 980\ M^{0.327}\ P^{0.525}\ DL^{0.171}$	M = Total Instrument Mass (kg) P = Max Instrument Power (W) DL = Design Life (months)	1-75 kg 1-75 W 10-150 mo.
Fields Payload: electric field detectors, magnetic field detectors	$Y = 1130\ M^{0.184}\ P^{0.238}\ DL^{0.274}$	M = Total Instrument Mass (kg) P = Max Instrument Power (W) DL = Design Life (months)	0.1-35 kg 0.1-25 W 1-100 mo.

Courtesy New SMAD T11-14.
Inflation factors follow.

August 23, 2022

USAF Space Planners Guide

- 1965 vintage
- Contains detailed estimations for inhabited SV & LV costs
- Get numbers from plots
- Inflate to needed year

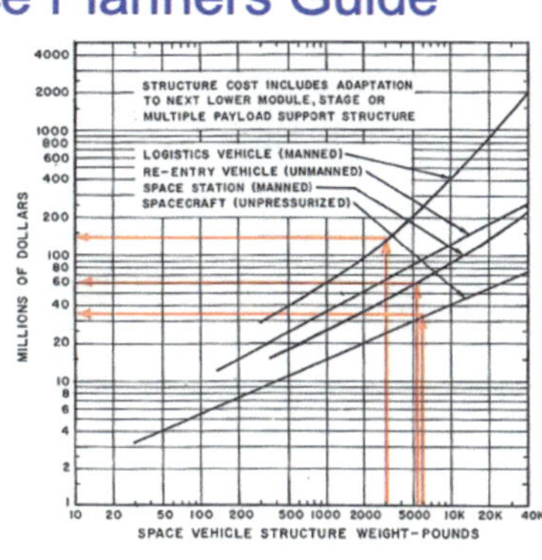

Figure VII. B-11 Space Vehicle Structure DT&E Cost

August 23, 2022

Topics

- Cost estimation needs & proposal process
  - The importance of good decision-making
- The phases of design, timelines, reviews
- Cost estimation discussion & methods
- Cost estimating relationships (CERs)
- Cost models
- **Inflation, labor costs, software development, propellant cost**
- Launch costs
- Non-vehicle costs: mission ops, delays, extensions, insurance, DSN

August 23, 2022

SC14-52

Inflation Factors, or "Other-Year" $

Either use CPI Inflation Calculator at `www.bls.gov/data/inflation_calculator.htm`

or Bureau of Labor Statistics data

or compute Inflation Factor:

- 3.1% = Average Inflation 1926 to 1992
- *Inflation Factor* = $(1.031)^n$ where n = no. years since reference year

Example: Costs figured in 2010$, need 2020$ (=10 years later):

➔2020 cost = 2010 cost × $(1.031)^{10}$ = 2010 cost × 1.357

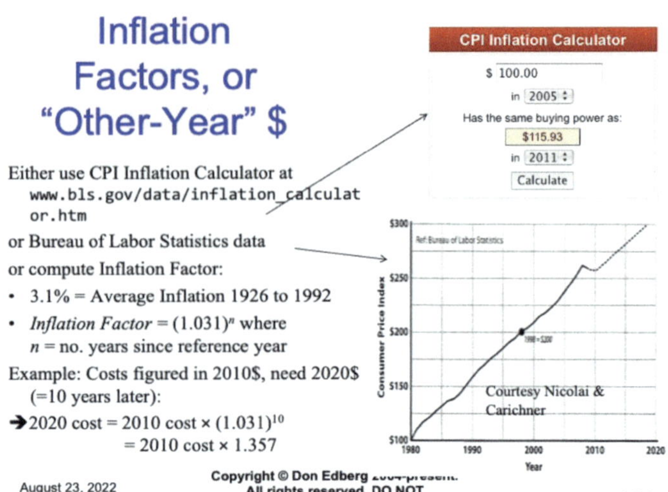

August 23, 2022

SC14-53

What is a "Work-Year" (WY)?

- Instead of $, some cost models use *Work-years* (WY) or *Man-Years* (MY – *yes, politically incorrect*!) of effort as cost value (TRANSCOST)
- Transformed by using a cost conversion valued to equivalent US$ for FY2000 (fiscal year 2000)

WY values depend on type of work & year done:

- Development: 1 MY = $205K
- Production: 1 MY = $200K
- Operation: 1 MY = $220K
- Other tasks: use 1 MY = $208K

Inflate values from *FY2000* to year of application

August 23, 2022

SC14-54

Aircraft Hourly Rates by Trade

Tooling: $R_T = 2.883\,y - 5666$

Engineering: $R_E = 2.576\,y - 5058$

Quality Control: $R_{QC} = 2.60\,y - 5112$

Manufacturing: $R_M = 2.316\,y - 4552$

Courtesy: Nicolai & Carichner

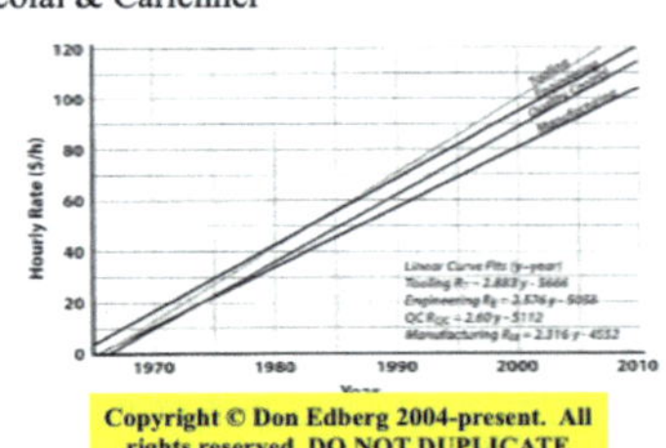

August 23, 2022

SC14-55

Selected Mission Costs (Real-Year US$)

Name	Launch Date	Wet Mass (kg)	Dev. Cost $M	Launch Support $M	Mission Ops $M	Tracking /Data $M	Total $M
Cassini	1997	5,712	1,422	422	755	54	3,313*
Mars Observer	1992	345	479	293	41		813
Mars Pathfinder	1996	798	175	50	14	1	265†
Mars Global Surveyor	1996	1,051	131	53	90	1	273
Mars Climate Orbiter & Polar Lander	1998-9	629 576	193	92	43	1	329
Near-Earth Asteroid Rendezvous (NEAR)	1996	798	125	44	42	1	212
Lunar Prospector	1998	295	31	26	37		64
Stardust	1999	385	127	45	37		210
Deep Space 1	1998	490	99	45	11		156
Deep Space 2	1999	3.6	25	2	1		27

*$660M foreign contribution
† $25M rover

Courtesy McCurdy: "Faster, Better, Cheaper – Low-Cost Innovation in the U.S. Space Program"

August 23, 2022

SC14-56

Selected Mission Masses

Name	Launch Date	Component Weights (kg)				Total (kg)
Cassini	1997	Orbiter: 2,125	Huygens Probe: 320	PAF: 135	Propellant: 3,132	5,712
Mars Pathfinder	1996	Lander: 269	—	—	@ Mars Entry: 459	798
Mars Global Surveyor	1996	Spacecraft: 671	—	—	Propellant: 381	1,051
Mars Climate Orbiter	1998-9	Spacecraft: 338	—	—	Propellant: 291	629
NEAR	1996	Spacecraft: 479	—	—	Propellant: 319	798
Lunar Prospector	1998	Spacecraft: 158	—	—	Propellant: 137	295
Stardust	1999	Spacecraft: 300*	—	—	Propellant: 85	385
Deep Space 1	1998	Spacecraft: 377	—	Xe 82	Hydrazine: 31	490
Deep Space 2	1999	Ground Station: 1.7	Penetrator: 0.7	Aeroshell: 1.2	—	3.6

*Includes 46 kg return capsule & parachute

Courtesy McCurdy

August 23, 2022

SC14-57

Selected SC Development Costs (1999 US$)

Name	Launch Date	Mass (lb)	Dev. Cost ($M)	Cost per lb (1999 $K)
Mariners 1, 2, 3, 4	1962, 1964	446, 574	955	936
Mariners 6 & 7	1969	838	711	848
Mariners 8 & 9	1971	2,196	667	304
Viking	1975	7,700	3,700	480
Cassini	1997	5,551	1,650	297
Mars Observer	1992	2,240	663	296
Mars Pathfinder	landed 1997	1,256	220	175
Mars Global Surveyor	1996	1,479	145	98
Chandra X-Ray Obs.	1999	10,560	1,818	172
Space Infrared Telescope Facility (Spitzer)	2002	1,650	473	287

Courtesy McCurdy: "Faster, Better, Cheaper – Low-Cost Innovation in the U.S. Space Program"

Software Development Costs

- Number of lines of code (LOC) used to estimate cost
- Average cost/LOC ~$10-$25 for fully-tested line of code in 2008, GNC section of JPL*
- "A professional programmer can produce *ten lines* of fully-tested, bug-free code per day." †
- On-line cost models for estimating software development costs:
  `http://cost.jsc.nasa.gov/COCOMO.html`
  `http://csse.usc.edu/research/COCOMOII/`
- REVIC = Revised Intermediate COCOMO

*Amy Attiyah, Supervisor, Navigation & Mission Design Sys. Engr., JPL
†Brian Muirhead, **High Velocity Leadership: The Mars Pathfinder Approach**, p. 115

Software Complexity

System	Lines of Code (LOC)[1]	Software Cost at $25/LOC
Human comprehension	~5,000	$125K
BIRD & TerraSAR satellites	90 K	$2.3M
Vega (ESA Launch Vehicle, 2013)[2]	100 K	$2.5M
A310 control system	400 K	$10M
Curiosity Mars Rover (2012)[5]	500 K	$12.5M
Cellular phone	1 M	$25M
Automobile	2 M	$50M
Space Shuttle	3 M	$75M
X-47B UCAV (NGC)	3.4 M[3]	$85M
B-2 Stealth Bomber	4 M (1.8 M[4])	$100M ($45M)
A340 control system	20 M	$500M
Linux	30 M	$750M
Windows XP (not necessarily bug-free!)	45 M	$1.1B

1 = Most values courtesy **Handbook of Space Flight**, p. 372, 2 = **ESA Bulletin**, 3 = **Aviation Week**. 4 = personal conversation, Thomas Colangelo. 5 = NASA Video

Propellant & Gas Costs – Almost Negligible

Propellant / Gas	$US/kg (2020)
Liquid Hydrogen (LH_2)	8.80*
Liquid Oxygen (LO_2)	0.26*
Liquid Methane (LCH4)	1.46**
Kerosene/RP-1	3.93*
Hydrogen Peroxide, 85% (H_2O_2)	5.67†
Dinitrogen Tetroxide (N_2O_4, a.k.a. "NTO")	108.15*
Hydrazine (N_2H_4; also MMH, UDMH, Aerozine 50)	417.88*
Solid propellant (ATK, Hercules large motors)	11.06*
Helium gas	18.56†
Nitrogen gas	0.11†
Hydrogen gas	10.93†

*Values inflated from 2009 values Transcost/Koelle. **S. Dinkin, The Space Review, Jan. 2016
†Values from Willie Costa
Some prices may be found at [illegible] and [illegible]

Topics

- Cost estimation needs & proposal process
  - The importance of good decision-making
- The phases of design, timelines, reviews
- Cost estimation discussion & methods
- Cost estimating relationships (CERs)
- Cost models
- Inflation, labor costs, software development, propellant cost
- **Launch costs**
- Non-vehicle costs: mission ops, delays, extensions, insurance, DSN

U.S. Launch Vehicles	Launch Mass (T)	LEO Payload	Payload to Polar	GTO Payload	Price/Flight US$M	Work Years
Atlas V 401	342	9600	7300	4750	130-140 (163*)	420-460
Atlas V 431	482	15400	11300	7500	140-160	460-530
Atlas V 551	587	18700	15000	8900	165-180	550-600
~~Delta II 7320~~	~~152~~	~~2690~~	~~1580~~	–	~~75-90~~	~~250-300~~
~~Delta II 7925~~	~~232~~	–	~~3200~~	~~1840~~	~~110-130~~	~~370-400~~
~~Delta IVM~~	~~250~~	~~8100~~		~~4200~~	~~150~~	~~500~~
~~Delta IVM+4,2~~	~~329~~	~~10400~~		~~5800~~	~~170~~	~~570~~
Delta IVH	733	23000		13130	360	1200
~~Falcon 1~~	~~28~~	~~420~~	~~420~~		~~8.9 (2009)~~	~~(57)~~
~~Falcon 1e~~	~~35~~	~~1010~~			~~10.9 (2010)~~	~~(62)~~
Falcon 9	333	10450	7000	4500	54-59.5 (2011)	(290)
Minotaur	36	600	335	–	21-24	70-80
Pegasus XL	23	450	200	–	20-25	80-100
Taurus XL	73	1590	860	557	45-50	150-160
Taurus II	240	5000			(45)	(150)
~~Saturn IB~~	~~583~~	~~15000~~	–	–	~~43 (1963)~~	~~1500~~
~~Saturn V~~	~~2850~~	~~127000~~	–	~~45000~~	~~220 (1974)~~	~~4280~~
~~Space Shuttle~~	~~2040~~	~~24000~~	–	–	~~1250-3000~~	~~2200~~

* Aviation Week May 21-June 3, 2018, p. 33

Courtesy: Koelle, **Handbook of Cost Engineering**

Non-U.S. Launch Vehicles	Launch Mass (T)	LEO Payload	Payload to Polar	GTO Payload	Price/Flight	Work Years
Ariane 5 ES	746	21000	10000	—	152-160 M €	620-650
Ariane 5 ECA	780	25000	14200	9500	165-175M €	675-720
CZ-2C Long March	213	2800	15000	8900	25-30 M US$	
CZ-3B	430			5100	70-80 M US$	
CZ-2E/HO	461	8800			60-76 M US$	
CZ-2F	464	8400				
Dnepr (Russia)	209	3700	300	—	12 M US$	
H-IIA-202 (Japan)	298	9900	3600	4100	16 B ¥	540
H-IIB	551	19000		8000	23 B ¥	730
ISRO GSLV (India)	418	4500	1800	2300	40 M US$	
ISRO PSLV	294	3500	1350	1050	32 M US$	
Kosmos-3M	109	1400	760	—	15 M US$	
M-5 (Japan)	140	1850	600	—	7 B ¥	280
Proton-K	694	20700			100 M US$	
Proton-M-Briz-M	705			6150	120 M US$	
Rockot	107	1850	1300	—	30 M €	
Soyuz/Molniya	305	6200	2700	1700	65 M US$	
Soyuz-2 Kourou	310	6200	3900	2800	75-80 M €	
Tsyklon-3	190	3600	2100	—	30-35 M US$	
Vega	139	2400	1500	—	50 M €	
Zenit 3SL SeaLaunch	462	—	—	6000	150 M US$	
Zenit 2	445	13700	5000	—	120 M US$	

Courtesy: Koelle, **Handbook of Cost Engineering**

Launch Vehicle Notes

- LVs cost from $6M to $600M each
  - $1K to $120K per pound to orbit
  - Rule of thumb is **$5K/pound**
  - But $/lb or $/kg is *deceptive*!
- Regardless of the size of your payload, you have to pay for ***INTEGER 1*** LV (unless you are dual-manifested or ridesharing)
- NASA on-line cost models for LV cost estimating: `https://www.nasa.gov/offices/ooe/CAD/nasa-cost-estimating-handbook-ceh/`

Better Yet: Actual Price Quote!

- ULA has online cost calculator at https://www.rocketbuilder.com (as of Apr. '21)

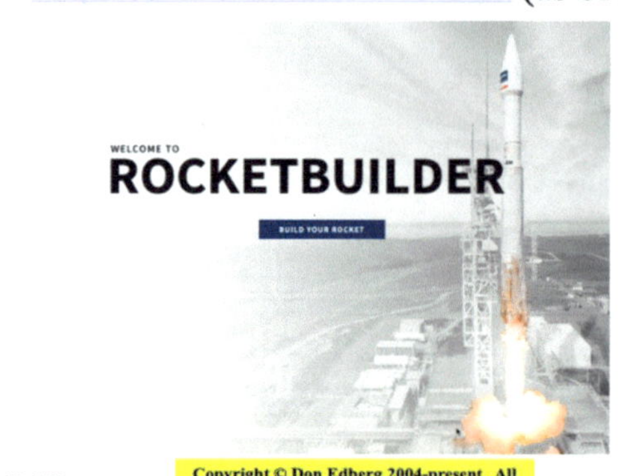

Topics

- Cost estimation needs & proposal process
  - The importance of good decision-making
- The phases of design, timelines, reviews
- Cost estimation discussion & methods
- Cost estimating relationships (CERs)
- Cost models
- Inflation, labor costs, software development, propellant cost
- Launch costs
- **Non-vehicle costs: mission ops, delays, extensions, insurance, DSN**

More Costs: Mission Operations

- *Mission Ops* can consume up to ***60%*** of the life-cycle cost!
  - JPL suggests **$50K/day** = $18.25M/year (uninhabited missions, '08$)
  - For manned missions: estimate no. of ops personnel × $1,000/day × mission days ('08$)

Cost of Extensions/Delays

- A data point: the *Deep Impact* spacecraft received a 4-year mission extension ~2008. The cost was US$32M or US$8M/year
- Another data point: NASA's Mars mission *InSight*'s launch was delayed from Mar. 2016 to May 5, 2018. "The delay added $150M to the project's cost ..." This is $69M/year.

(Aviation Week May 21-June 3, 2018, p. 33)

Other Cost: Insurance

- Most commercial payloads insure: LVs only 90-95% reliable
- Insurance cost ~10% of gross cost (LC+SC)

Three risks covered:

1. Relaunch SC if launch operation fails
2. Replace SC if destroyed or positioned in an improper orbit, or SC fails on-orbit
3. Liability for damage to 3rd parties caused by SC or LV

Other Costs (cont.)

- Launch delays ~ $1M/week
- Deep space network utilization costs for planetary missions
  - Telecommunications notes
  - "NASA's Mission Operations and Communications Services" paper (Oct. 2014, no document number),
  - Or `http://cost.jsc.nasa.gov/CECM.html`

To Estimate DSN Costs:

- $AF = RB[AW(0.9 + FC/10)]$, where:
  - AF = weighted Aperture Fee, per hour of use
  - RB = contact-dependent hourly rate, adjusted annually ($1057/hr for FY09)
  - AW = aperture weighting
    - = 0.80 for 34 m High-Speed Beam Waveguide (HSB) station
    - = 1.00 for all other 34 m stations (i.e. 34 BWG & 34 HEF)
    - = 2.00 for a two 34 m station array
    - = 3.00 for a three 34 m station array
    - = 4.00 for a four 34 m station array (70 m equivalent)
    - = 4.00 for 70 m stations
  - FC = number of station contacts per calendar week

(From NASA's Mission Operations & Communications Services)

Summary

- Cost modeling both difficult & critical
- Parametric modeling useful to identify cost drivers & relative costs
- Use engineering judgment to temper results
- Consider cost throughout the design process
- Launch vehicle & mission ops add considerable costs

Conclusions: Spacecraft Design, Development, & Operations Short Course

- Designing, developing, and operating spacecraft IS rocket science (or rocket *engineering*!)
- Systems engineering vitally important: everything must work together
- There are many experts in the field; consult with them and use them
- Use redundancy; learn from failures
- This class should help you understand how they work – hope it was helpful!

August 23, 2022

Questions? Comments?

Thank you:

- For attending
- For your insightful questions
- For filling out class evaluation forms with your comments

You may also email comments to <dedberg@cpp.edu>

23-Aug-22

Made in United States
Troutdale, OR
02/15/2025

28991173R00204